T0254162

MicroRNAs

FROM BASIC SCIENCE TO DISEASE BIOLOGY

Edited by

Krishnarao Appasani
GeneExpression Systems, Inc.

Forewords by

Victor R. Ambros
Dartmouth Medical School

and

Sidney Altman
Yale University
Winner of the Nobel Prize in Chemistry, 1989

CAMBRIDGE
UNIVERSITY PRESS

CAMBRIDGE UNIVERSITY PRESS
Cambridge, New York, Melbourne, Madrid, Cape Town, Singapore, São Paulo, Delhi

Cambridge University Press
The Edinburgh Building, Cambridge CB2 8RU, UK

Published in the United States of America by Cambridge University Press, New York

www.cambridge.org
Information on this title: www.cambridge.org/9780521118552

© Cambridge University Press 2008

This publication is in copyright. Subject to statutory exception
and to the provisions of relevant collective licensing agreements,
no reproduction of any part may take place without the written
permission of Cambridge University Press.

First published 2008
This digitally printed version 2009

A catalogue record for this publication is available from the British Library

ISBN 978-0-521-86598-2 hardback
ISBN 978-0-521-11855-2 paperback

Additional resources for this publication at www.cambridge.org/9780521118552

Cambridge University Press has no responsibility for the persistence or
accuracy of URLs for external or third-party Internet websites referred to in
this publication, and does not guarantee that any content on such websites is,
or will remain, accurate or appropriate.

MicroRNAs

MicroRNAs (miRNAs) are evolutionary conserved RNA molecules that regulate gene expression, and their recent discovery is currently revolutionizing both basic biomedical research and drug discovery. Expression levels of miRNAs have been found to vary between tissues and with developmental stages and hence evaluation of the global expression of miRNAs potentially provides opportunities to identify regulatory points for many different biological processes. This wide-ranging reference work, written by leading experts from both academia and industry, will be an invaluable resource for all those wishing to use miRNA techniques in their own research, from graduate students, post-docs and researchers in academia to those working in research and development in biotechnology and pharmaceutical companies who need to understand this emerging technology. From the discovery of miRNAs and their functions to their detection and role in disease biology, this volume uniquely integrates the basic science with industry application towards drug validation, diagnostic and therapeutic development.

KRISHNARAO APPASANI is the Founder and Chief Executive Officer of Gene Expression Systems, a gene discovery company, focusing on functional genomics in cancer research. He is the Editor of *RNA Interference: From Basic Science to Drug Development* (2005).

To my wife Shyamala for her 25 years companionship

Contents

Contributors

Hatim T. Allawi
Third Wave Technologies, Inc.
502 S. Rosa Rd
Madison, WI 53719-1256, USA

Sidney Altman
Department of Molecular, Cell and Developmental Biology
Yale University
New Haven, CT 06520, USA

Ines Alvarez-Garcia
The Wellcome Trust/Cancer Research UK
Gurdon Institute and Department of Biochemistry
The Henry Wellcome Building of Cancer and Developmental Biology
University of Cambridge
Tennis Court Road
Cambridge CB2 1QN, United Kingdom

Victor R. Ambros
Department of Genetics
Dartmouth Medical School
Maynard Street
Hanover, NH 03755, USA

Parker B. Antin
Department of Cell Biology and Anatomy
University of Arizona
PO Box 245044
Life Sciences North 462
1501 N. Campbell Avenue
Tucson, AZ 85724-5044, USA

Krishnarao Appasani
GeneExpression Systems, Inc.
PO Box 540170
Waltham, MA 02454, USA

Norio Asou
Department of Hematology
Kumamoto University School of Medicine
1-1-1 Honjo
Kumamoto 860-8556, Japan

John Barnes
Department of Human Genetics
Emory University
615 Michael Street
Atlanta, GA 30322, USA

Ergin Beyret
Department of Cell Biology and Yale Stem Cell Center
Yale University School of Medicine
333 Cedar Street, SHM I-213
New Haven, CT 06511, USA

Graham J. Brock
Ordway Research Institute
150 New Scotland Avenue
Albany, NY 12208, USA

George Adrian Calin
Center for Human Cancer Genetics
The Ohio State University Comprehensive Cancer Center
385L Comprehensive Cancer Center
410 West 10th Avenue
Columbus, OH 43210, USA

Jinhong Chang
Fox Chase Cancer Center
333 Cottman Avenue
Philadelphia, PA 19111, USA

Yue-Qin Chen
Biotechnology Research Center
Key Laboratory of Gene Engineering of the Ministry of Education
Zhongshan University
Guangzhou, 510275, People's Republic of China

Caifu Chen
Research & Development
Applied Biosystems
850 Lincoln Centre Drive
Foster City, CA 94404, USA

Silvia Anna Ciafrè
Experimental Medicine and Biochemical Sciences
University of Rome
Via Montpellier 1, Rome 00133, Italy

Amelia Cimmino
Center for Human Cancer Genetics
The Ohio State University Comprehensive Cancer Center
385L Comprehensive Cancer Center
410 West 10th Avenue
Columbus, OH 43210, USA

Carlo Maria Croce
Center for Human Cancer Genetics
The Ohio State University Comprehensive Cancer Center
385L Comprehensive Cancer Center
410 West 10th Avenue
Columbus, OH 43210, USA

Dean J. Danner (deceased)
Department of Human Genetics
Emory University
615 Michael Street, Room 305C
Atlanta, GA 30322, USA

Diana K. Darnell
Department of Cell Biology and Anatomy
University of Arizona
PO Box 245044
Life Sciences North 462
1501 N. Campbell Avenue
Tucson, AZ 85724-5044, USA

Sean Davey
Department of Cell Biology and Anatomy
University of Arizona
PO Box 245044
Life Sciences North 462
1501 N. Campbell Avenue
Tucson, AZ 85724-5044, USA

Matthias Eder
Hanover Medical School
Center for Internal Medicine
Department of Hematology and Oncology
Carl-Neuberg Strasse 1
D-30623 Hanover, Germany

Anton J. Enright
Computational and Functional Genomics
The Wellcome Trust Sanger Institute
Wellcome Trust Genome Campus
Hinxton, Cambridge CB10 1SA, United Kingdom

Muller Fabbri
Center for Human Cancer Genetics
The Ohio State University Comprehensive Cancer Center
385L Comprehensive Cancer Center
410 West 10th Avenue
Columbus, OH 43210, USA

Yoichi R. Fujii
Molecular Biology and Retroviral Genetics Group
Division of Nutritional Sciences
Graduate School of Pharmaceutical Sciences
Nagoya City University
Nagoya 467-8603, Japan

Michael Gallo
Epic Therapeutics, Inc.
220 Norwood Park
Norwood, MA 02062, USA

Joanne Garver
US Genomics, Inc.
12 Gill Street, Suite 4700
Woburn, MA 01801, USA

Ramiro Garzon
Center for Human Cancer Genetics
The Ohio State University Comprehensive Cancer Center
385L Comprehensive Cancer Center
410 West 10th Avenue
Columbus, OH 43210, USA

Michael Greenberg
Department of Neurobiology
Children's Hospital, 300 Longwood Avenue
Harvard Medical School
Boston, MA 02115, USA

Finn Grey
Vaccine & Gene Therapy Institute
Oregon Health & Sciences University
Portland, OR 97201, USA

Dominic Grün
Center for Functional Comparative Genomics
Department of Biology, NYU, 1009 Main Building
100 Washington Square East
New York, NY 10003-6688, USA

Sam Griffiths-Jones
Computational and Functional Genomics
The Wellcome Trust Sanger Institute
Wellcome Trust Genome Campus
Hinxton, Cambridge CB10 1SA, United Kingdom

Steven R. Gullans
RxGen, Inc.
New Haven, CT, USA

Maria Hackett
Novartis Institutes for Biomedical Research
250 Massachusetts Avenue
Cambridge, MA 02139, USA

Joshua W. Hagen
Memorial Sloan-Kettering Institute
Department of Developmental Biology
521 Rockefeller Research Labs
1275 York Avenue, Box 252
New York, NY 10021, USA

Annick Harel-Bellan
Laboratoire Oncogenese, Differenciation et Transduction du Signal
CNRS UPR 9079
Institut Andre Lwoff,
Batiment B, 1er Etage
7 rue Guy Moquet
94800 Villejuif, France

Brian D. Harfe
Department of Molecular Genetics & Microbiology
University of Florida College of Medicine
1600 SW Archer Road
Gainesville, FL 32610-0266, USA

John Harris
US Genomics, Inc.
12 Gill Street, Suite 4700
Woburn, MA 01801, USA

Artemis G. Hatzigeorgiou
University of Pennsylvania
Center for Bioinformatics
1407 Blockley Hall/6021
Philadelphia, PA 19104-6021, USA

Alec J. Hirsch
Vaccine & Gene Therapy Institute
Oregon Health & Sciences University
Portland, OR 97201, USA

Hristo B. Houbaviy
Department of Cell Biology
University of Medicine & Dentistry of New Jersey
Two Medical Center Drive
Stratford, NJ 08084-1489, USA

David T. Humphreys
Molecular Genetics Program
Victor Chang Cardiac Research Institute
384 Victoria Street, Darlinghurst
Sidney, NSW 2010 Australia

Gregory J. Hurteau
Ordway Research Institute
150 New Scotland Avenue
Albany, NY 12208, USA

Tyler Jacks
Massachusetts Institute of Technology
Department of Biology and Center for Cancer Research
40 Ames Street, E17-518
Cambridge, 02139 MA, USA

Peng Jin
Department of Human Genetics
Emory University School of Medicine
615 Michael Street, Room 325.1
Atlanta, GA 30322, USA

Sakari Kauppinen
Wilhelm Johannsen Centre for Functional Genome Research
Institute of Medical Biochemistry and Genetics
University of Copenhagen
Blegdamsvej 3
DK-2200 Copenhagen North, Denmark
and
Santaris Pharma
Boge Alle 3
Horshol DK-2970, Denmark

Simran Kaur
Department of Cell Biology and Anatomy
University of Arizona
PO Box 245044
Life Sciences North 462
1501 N. Campbell Avenue
Tucson, AZ 85724-5044, USA

Jay H. Konieczka
Department of Cell Biology and Anatomy
University of Arizona
PO Box 245044
Life Sciences North 462
1501 N. Campbell Avenue
Tucson, AZ 85724-5044, USA

Tomoko Kozu
Division of Cancer Treatment
Research Institute for Clinical Oncology
Saitama Cancer Center
818 Komuro, Ina, Saitama 362-0806, Japan

Madhu S. Kumar
Massachusetts Institute of Technology
Department of Biology and Center for Cancer Research
40 Ames Street, E17-518
Cambridge, 02139 MA, USA

Eric C. Lai
Memorial Sloan-Kettering Institute
Department of Developmental Biology
521 Rockefeller Research Labs
1275 York Avenue, Box 252
New York, NY 10021, USA

Kaiqin Lao
Advanced Research Technology
Applied Biosystems
850 Lincoln Center Drive
Foster City, CA 94404, USA

Yu Liang
Research & Development
Applied Biosystems
850 Lincoln Centre Drive
Foster City, CA 94404, USA

Bing Lim
Stem Cell and Developmental Biology
Genome Institute of Singapore
60 Biopolis Street, #02-01, Genome
Singapore 138672

Haifan Lin
Department of Cell Biology and Yale Stem Cell Center
Yale University School of Medicine
333 Cedar Street, SHM I-213
New Haven, CT 06511, USA

Shi-Lung Lin
Department of Cell and Neurobiology
Keck School of Medicine
University of Southern California
1333 San Pablo Street, BMT-403
Los Angeles, CA 90033-9112, USA

Yu-Chun Luo
Biotechnology Research Center
Key Laboratory of Gene Engineering of the Ministry of Education
Zhongshan University
Guangzhou, 510275, People's Republic of China

Victor I. Lyamichev
Third Wave Technologies, Inc.
502 S. Rosa Rd
Madison, WI 53719-1256, USA

Danielle M. Maatouk
Department of Molecular Genetics & Microbiology
University of Florida College of Medicine
1600 SW Archer Road
Gainesville, FL 32610-0266, USA

David I. K. Martin
Molecular Genetics Program
Victor Chang Cardiac Research Institute
384 Victoria Street, Darlinghurst
Sidney, NSW 2010, Australia

Stephen McLaughlin
US Genomics, Inc.
12 Gill Street, Suite 4700
Woburn, MA 01801, USA

Joshua T. Mendell
Program in Human Genetics and Molecular Biology
Institute of Genetic Medicine
Johns Hopkins University School of Medicine
Baltimore, MD 21205, USA

Molly Megraw
University of Pennsylvania
Center for Bioinformatics
1407 Blockley Hall/6021
Philadelphia, PA 19104-6021, USA

Benjamin Mersey
Department of Human Genetics
Emory University
615 Michael Street
Atlanta, GA 30322, USA

Joseph D. Miller
Department of Cell and Neurobiology
Keck School of Medicine
University of Southern California
1333 San Pablo Street, BMT-403
Los Angeles, CA 90033-9112, USA

Eric A. Miska
The Wellcome Trust/Cancer Research UK
Gurdon Institute and Department of Biochemistry
The Henry Wellcome Building of Cancer and Developmental Biology
University of Cambridge
Tennis Court Road
Cambridge CB2 1QN, United Kingdom

Mark Nadel
US Genomics, Inc.
12 Gill Street, Suite 4700
Woburn, MA 01801, USA

Irina Naguibneva
Laboratoire Oncogenese, Differenciation et Transduction du Signal
CNRS UPR 9079
Institut Andre Lwoff,
Batiment B, 1er Etage
7 rue Guy Moquet
94800 Villejuif, France

Lori A. Neely
Technology & Pre-Development
Millipore Corp.
80 Ashby Rd.
Bedford, MA 01730, USA

Jay A. Nelson
Vaccine & Gene Therapy Institute
Oregon Health & Sciences University
Portland, OR 97201, USA

Kathryn A. O'Donnell
Program in Human Genetics and Molecular Biology
Institute of Genetic Medicine
Johns Hopkins University School of Medicine
Baltimore, MD 21205, USA

Javier F. Palatnik
Molecular Biology Division
IBR – Institute of Molecular and Cellular Biology of Rosario
Suipacha 531
2000 Rosario, Argentina

Sonal Patel
US Genomics, Inc.
12 Gill Street, Suite 4700
Woburn, MA 01801, USA

Anna Polesskaya
Laboratoire Oncogenese, Differenciation et Transduction du Signal
CNRS UPR 9079
Institut Andre Lwoff,
Batiment B, 1er Etage
7 rue Guy Moquet
94800 Villejuif, France

Thomas Preiss
Molecular Genetics Program
Victor Chang Cardiac Research Institute
384 Victoria Street, Darlinghurst
Sidney, NSW 2010, Australia

Liang-Hu Qu
Biotechnology Research Center
Key Laboratory of Gene Engineering of the Ministry of Education
Zhongshan University
Guangzhou, 510275, People's Republic of China

Nikolaus Rajewsky
Assistant Prof. of Biology and Mathematics
New York University
Center for Functional Comparative Genomics
Department of Biology, NYU, 1009 Main Building
100 Washington Square East
New York, NY 10003-6688, USA
and
MDC/Charite Medical University
Max Delbruck Centrum for Molecular Medicine
Berlin-Buch
13092 Berlin, Germany

Marc Rehmsmeier
Universität Bielefeld
Center for Biotechnology (CeBiTec)
33594 Bielefeld, Germany

Jason R. Rock
Department of Molecular Genetics & Microbiology
University of Florida College of Medicine
1600 SW Archer Road
Gainesville, FL 32610-0266, USA

Ramiro E. Rodriguez
Molecular Biology Division
IBR – Institute of Molecular and Cellular Biology of Rosario
Suipacha 531
2000 Rosario, Argentina

Pål Sætrom
Interagon AS
Laboratoriesenteret
Medisinsk Teknisk Senter
NO-7489 Trondheim, Norway

Michaela Scherr
Hanover Medical School
Center for Internal Medicine
Department of Hematology and Oncology
Carl-Neuberg Strasse 1
D-30623, Hanover, Germany

Carla Schommer
Department of Molecular Biology
Max Plank Institute for Developmental Biology
Tubingen 72076, Germany

Gerhard Schratt
University of Heidelberg
Interdisciplinary Center for Neurosciences
Im Neuenheimer Feld 345, Room 160
D-69120 Heidelberg, Germany

Praveen Sethupathy
University of Pennsylvania
Center for Bioinformatics
1407 Blockley Hall/6021
Philadelphia, PA 19104-6021, USA

Neil R. Smalheiser
University of Illinois at Chicago
Psychiatric Institute
1601 W. Taylor Street
Chicago, IL 60612, USA

Ola Snøve Jr.
Interagon AS
Laboratoriesenteret
Medisinsk Teknisk Senter
NO-7489 Trondheim, Norway

Takashi Sonoki
Department of Hematology and Oncology
Wakayama Medical University
Kimi-idera 811-1
Wakayama 641-8510, Japan

Simon D. Spivack
Human Toxicology & Molecular Epidemiology
Wadsworth Center
NYS Department of Health
Albany, NY 12201, USA

Stacey Stanislaw
Department of Cell Biology and Anatomy
University of Arizona
PO Box 245044
Life Sciences North 462
1501 N. Campbell Avenue
Tucson, AZ 85724-5044, USA

M. Azim Surani
The Wellcome Trust/Cancer Research UK
Gurdon Institute of Cancer and Developmental Biology
University of Cambridge
Tennis Court Road
Cambridge CB2 1QR, United Kingdom

Keith Szulwach
Department of Human Genetics
Emory University School of Medicine
615 Michael Street, Room 325.1
Atlanta, GA 30322, USA

Ruoying Tan
Research & Development
Applied Biosystems
850 Lincoln Centre Drive
Foster City, CA 94404, USA

Fuchou Tang
The Wellcome Trust/Cancer Research UK
Gurdon Institute of Cancer and Developmental Biology
University of Cambridge
Tennis Court Road
Cambridge CB2 1QR, United Kingdom

Yvonne Tay
Stem Cell and Developmental Biology
Genome Institute of Singapore
60 Biopolis Street, #02-01, Genome
Singapore 138672

John M. Taylor
Fox Chase Cancer Center
333 Cottman Avenue
Philadelphia, PA 19111, USA

Andrew M. Thomson
Stem Cell and Developmental Biology
Genome Institute of Singapore
60 Biopolis Street, #02-01, Genome
Singapore 138672

Vetle I. Torvik
University of Illinois at Chicago
Psychiatric Institute
1601 W. Taylor Street
Chicago, IL 60612, USA

Andrea Ventura
Massachusetts Institute of Technology
Department of Biology and Center for Cancer Research
40 Ames Street, E17-518
Cambridge, 02139 MA, USA

Jia-Fu Wang
Biotechnology Research Center
Key Laboratory of Gene Engineering of the Ministry of Education
Zhongshan University
Guangzhou, 510275, People's Republic of China

Belinda J. Westman
Molecular Genetics Program
Victor Chang Cardiac Research Institute
384 Victoria Street, Darlinghurst
Sidney, NSW 2010 Australia

Duncan Whitney
US Genomics, Inc.
12 Gill Street, Suite 4700
Woburn, MA 01801, USA

Linda Wong
Research & Development
Applied Biosystems
850 Lincoln Centre Drive
Foster City, CA 94404, USA

Tatiana A. Yatskievych
Department of Cell Biology and Anatomy
University of Arizona
PO Box 245044
Life Sciences North 462
1501 N. Campbell Avenue
Tucson, AZ 85724-5044, USA

Shao-Yao Ying
Department of Cell and Neurobiology
Keck School of Medicine
University of Southern California
1333 San Pablo Street, BMT-403
Los Angeles, CA 90033-9112, USA

Hui Zhou
Biotechnology Research Center
Key Laboratory of Gene Engineering of the Ministry of Education
Zhongshan University
Guangzhou, 510275, People's Republic of China

Foreword

Sidney Altman

Twenty-five years ago it was possible to predict that many more RNAs that were known at that time would be found: Some of them would be catalytic and some of them would not be but would serve other functions. The prediction turned out to be spectacularly true and this book deals with one class of novel RNAs, the microRNAs (miRNAs). The task now is to elucidate the nature and function of all these new RNAs.

This book is a compendium of experimental methods, *in silico* and the traditional kind, of analyzing the miRNAs. There are many chapters on miRNAs on intra-cellular functions, in developmental events and in disease. A newcomer would do well to have this volume handy for purposes of reference and for its educational value *per se*. There are not many books on miRNAs that have such an extensive and complete view of the field as it now exists. My hope is that the problems presented here will be worked on assiduously to pave the way for a new synthesis of various RNAs as important regulatory genes in eukaryotes.

Yale University
New Haven, CT, USA

Foreword

Victor R. Ambros

MicroRNA research is enjoying an inflationary period of astonishingly rapid progress that began in 2001 after a relatively long gestation. The idea that small RNAs could regulate gene expression by base-pairing to messenger RNAs can be traced back more than 45 years. In a 1961 *Journal of Molecular Biology* paper, Jacob and Monod proposed an antisense RNA base-pairing model for the *lac* repressor/operator interaction that they had defined genetically. The *lac* repressor turned out to be a protein, but the field of antisense RNA mediated gene regulation grew steadily through studies of authentic antisense regulatory RNAs in bacteria. The identification in 1993 of the *lin-4* microRNA and its antisense regulation of *lin-14* did not trigger a surge of interest in antisense gene regulation in eukaryotes, primarily because *lin-4* did not exhibit any evident conservation outside nematodes (although we now know that *lin-4* is related to vertebrate mir-125, but is not close enough to be detected by the methods available in the era before the availability of genome sequences). Even the identification in 2000 of *let-7*, a second microRNA in *C. elegans*, did not immediately stimulate a sense among biologists that these exceptionally small RNAs could represent a general phenomenon. I must admit that I was among the skeptical majority.

The modern era of microRNA biology began with the finding from the Ruvkun laboratory, reported in a 2000 *Nature* paper, that the *let-7* microRNA has been conserved almost precisely in all its 21 nt for more than 400 million years: since the common ancestor of all bilaterally symmetric animals. Note that this report of the conservation of *let-7* closely followed the identification in 1999 by the Baulcombe laboratory of similar *c.*22 nt RNAs associated with gene silencing in plants. These convergent results from the Ruvkun and Baulcombe labs really rang the bells around our ears, announcing undeniably that the *lin-4* and *let-7* RNAs must be part of a widespread and ancient suite of genetic regulatory phenomena involving small RNAs. The realization that there must be more small RNAs like *lin-4* and *let-7*, spurred several labs to immediately try to identify other microRNAs encoded in eukaryotic genomes.

Within a year or two of the discovery of *C. elegans let-7*, numerous novel microRNAs were identified in flies, worms, vertebrates and plants, including many others that, like *let-7*, are deeply conserved. The subsequent years have brought remarkable progress towards understanding the diversity of microRNAs, their evolution, biogenesis, expression patterns, biological functions, and molecular mechanisms. MicroRNAs continue to surprise and delight us in unexpected ways. These little regulators seem to be engaged in practically every aspect of metazoan biology, regulating diverse processes including viral replication, cell fate, morphogenesis, differentiation, physiology, and disease. This versatility of microRNAs is reflected by the diverse array of topics included in this volume, which capture the status of understanding of microRNA biology in 2008. These chapters cover the identification and detection of microRNAs *in vivo*, the computational analysis of mRNA microRNA-target interactions, and the molecular mechanisms of gene regulation by microRNAs. Further and reflecting an area of particular excitement currently, a large portion of the volume describes genetic and molecular studies of the roles of microRNAs in development and human disease. The full scope of the involvement of microRNAs in human disease is difficult to envision at this early stage, but it is already clear from the work reported here that microRNAs will figure significantly in the molecular pathology of many diseases.

This volume marks the state of the field at this moment, circa 2008, but we can be assured that there are still major surprises and important principles to emerge from future research on microRNAs. This is apparent from the many questions that arise from the work reported here. Each chapter addresses important fundamental questions, but in so doing also leaves us with an enhanced appreciation of the unanswered questions and technical challenges before us. Much more work is required to fully understand the developmental and physiological roles of particular microRNAs. How well do genetic studies using model organisms inform about the functions in humans of the conserved microRNAs? Indeed, we have no idea how is it that the entire 21–22 nucleotides of a microRNA such as *let-7* can be precisely conserved over vast evolutionary time; what selective forces could have constrained every single nucleotide of an RNA for so long? MicroRNA genes can also evolve rapidly, either *de novo* or by gene duplication; so how do the evolution of microRNAs and their targets affect human population diversity and disease susceptibility? Finally, improvements in microRNA detection technology and computational methods will certainly lead to the discovery of important new microRNA genes and microRNA–target interactions, particularly for rare cell types and stem cell niches. Progress in microRNA research has been astonishingly rapid over the past six years or so, and we can be sure that the future will see an acceleration of this pace of discovery, yielding a wealth of exciting new biological insights and human health applications.

Dartmouth Medical School
Hanover, NH, USA

Introduction

Krishnarao Appasani

It is now clear an extensive miRNA world was flying almost unseen by genetic radar.
Gary Ruvkun, Professor, Harvard Medical School; *Cell*, **S116**, S95, 2004

RNomics is a newly emerging sub-discipline of genomics that categorically studies the structure, function and processes of non-coding ribo nucleic acids in a cell. MicroRNomics is a sub-field of RNomics that describes the biogenesis and mechanisms of tiny RNA regulators, and their involvement in the processes of development, differentiation, cell proliferation, cell death, chromosomal segregation and metabolism. The discovery of microRNAs in species ranging from *Caenorhabditis elegans* (*C. elegans*) to humans, and their regulatory functions, opened a new tier of understanding of gene expression. In addition, the study of RNA interference (RNAi) and its mechanisms propelled the elucidation of various functions of microRNAs. The development and brisk progress in microRNAs research over the past few years allowed us to comprehend the genomics at a newer scale. Therefore, it is now an ideal time to examine the available evidence in a systematic way in order to know where this impressive field is going.

MicroRNAs: From Basics to Disease Biology is mainly intended for readers in the fields of molecular cell biology, genomics, biotechnology and molecular medicine. This book may be useful in more advanced graduate level courses. This book, which focuses on the concepts of microRNA biology, key implications in various forms of diseases, and applications in diagnostic and drug development, consists of thirty-eight chapters, grouped into six sections. Most of the chapters are written by the original discoverers or their associated scientists from academia, biotech and pharma, exclusively from the microRNomics field.

Although several reviews have been published covering these tiny molecules, there is no single volume currently available in the marketplace. This is the first book that integrates the academic science with industrial applications from the diverse role of microRNAs development to disease biology, covering bioinformatics, quantitation, prognosis and diagnostics as well drug development and validation. This book will serve as a reference for graduate students, post-doctoral researchers, and professors from academic research institutions who wish to initiate microRNA research in their own laboratories. Additionally, this book will serve

MicroRNAs: From Basic Science to Disease Biology, ed. Krishnarao Appasani. Published by Cambridge University Press. © Cambridge University Press 2008.

as a descriptive and in-depth scientific explanatory analysis for executives and scientists in research and development from biotech and pharmaceutical companies. It will provide a valuable executive scientific summary for investors and those responsible for business development in the life sciences, who need to keep abreast with the microRevolution. Most importantly, my hope is that this volume will serve both as a prologue to the field for newcomers, as well as an update to those already active in the field.

Sydney Brenner from the MRC Molecular Biology Laboratory in Cambridge opened the field of C. *elegans* biology (mid 1970s), when the *lin-4* mutant was originally discovered. Later (his post-doctoral fellow) Howard Robert Horvitz from Massachusetts Institute of Technology isolated *lin-14*. Although both these genes were discovered in the study of developmental timing defects of the worm their nature and functions were not explored until Horovitz's post doctoral fellows, Ambros and Ruvkun, studied them. Fourteen years ago, Victor Ambros and his colleagues from Dartmouth Medical School identified for the first time that *lin-4* was the first short non-coding microRNA that regulates gene expression in C. *elegans*. At the same time Gary Ruvkun and his colleagues from the Harvard Medical School demonstrated *lin-14* to be the first microRNA target gene. Together, these two seminal discoveries identified a novel mechanism of post-transcriptional gene regulation. A successful initial collaboration and subsequent independent contributions by Ambros and Ruvkun paved the way to our current understanding of microRNA biology. The microRNAs world did not take off until the discovery of RNAi, and *let-7*, a second microRNA discovered by Ruvkun and his colleagues in 2000. The highly conserved nature of *let-7* attracted a great deal of attention to microRNAs research and since then many groups have identified microRNAs in various organisms, from protozoans to humans. The paradigm from the heterochromic *lin-4/lin-14* genes remains the model; miRNAs have now been shown to control mRNA abundance in plants, and are thought to regulate many more steps beyond translation. Despite considerable attention having been paid to microRNAs over the past five years, a lot of questions remain unanswered. The mechanism of action of microRNAs is still controversial and only a few in vivo functions of microRNAs have been demonstrated in any organism to date. In this book I intend to bring the various aspects of microRNA research together, including the biology, applications, their role in stem cell and disease biology. As the field of microRNomics is rapidly growing, any volume that describes the material might be little outdated. The role of microRNAs in immunology, cardiology, diabetes, and unicellular organisms has been recently reported, after I communicated all the original chapters to the press. Therefore I have added two new chapters reviewing these topics at the end of this book.

As Ambros mentioned in an earlier paper, the *lin-4* story is one of persistent curiosity, luck, timing, and the generosity of colleagues. Indeed, it is a key breakthrough in the modern molecular biology. In the coming years, microRNA research will necessitate the rewriting of our textbooks and alter the basic concept of the central dogma. In the near future we will be able to use microRNA-based molecular signatures or panels of multiplexed microRNA biomarkers to identify whether a tumor is malignant or benign, its site of origin, its prognostic subtype, and even predict its response to therapy. Clearly, high-throughput approaches and bioinformatics will have a primary role in the future of clinical oncology and pathology. In addition, microRNAs will be used in targeted cardiovascular therapy and they will be a central tenet in 'translational medicine'. To fully realize the potential of microRNAs as biomarkers to aid in drug development, industry must implement best practices for biomarker development, and promote translational research strategies. The biggest obstacles to turning discovery from bench to bedside are not only, in the advancement of technology, but also the necessary regulatory approvals. In a nutshell, microRNA array technology has the potential to accommodate all required assay formats on one testing platform, and to provide better reagents for pathological diagnosis in the future. It is hoped that by fully understanding the functions of microRNAs in various forms of human diseases and their subsequent development as theranostics, so we will be the new era of 'translational medicine'.

Many people have contributed to making my involvement in this project possible. I thank my teachers throughout my life for their excellent teaching, guidance and mentorship, which caused me to become a scientist. I am thankful to all of the contributors to this book. Without their commitment this book would not have emerged for the reader. Many people have had a hand in the launch of this book. Each chapter has been passed back and forth between the authors for criticism and revision, so that each chapter represents a joint composition. Thanks to you, the reader, who makes the hours I have spent putting together this volume worthwhile, if you find value in the hours you spend with this book. I am indebted to the staff of Cambridge University Press, and in particular Katrina Halliday for her generosity and efficiency throughout the editing of this book; she truly understands the urgency and need of this volume. I would specially like to thank Nobel Laureate Sidney Altman and Victor Ambros for their kindness and support in writing the Forewords to this book. Last, but not least, I thank my two wonderful sons, Raakish and Raghu, for their understanding and cooperation during the development of this book.

A portion of the royalties will be contributed to the Dr Appasani Foundation, a non-profit organization devoted to bring social change through education of youth in developing nations.

I

Discovery of microRNAs in various organisms

1 The microRNAs of *C. elegans*

Ines Alvarez-Garcia and Eric A. Miska*

Discovery

MicroRNA (miRNA) is a class of short non-coding RNA that regulates gene expression in many eukaryotes. MicroRNA was first discovered in *Caenorhabditis elegans* by Victor Ambros' laboratory in 1993 (Lee *et al.*, 1993). At the same time Gary Ruvkun's laboratory identified the first microRNA target gene (Wightman *et al.*, 1993). Together, these two seminal discoveries identified a novel mechanism of post-transcriptional gene regulation. The history of these events has been reported recently, as remembered by the main researchers involved (Lee *et al.*, 2004; Ruvkun *et al.*, 2004). The realization of the importance of the discovery of microRNA, however, took about seven years and was preceded by the identification of a second microRNA in *C. elegans* by the Ruvkun and Horvitz laboratories and also by the rise in interest in another class of short RNA, siRNA, involved in the process of RNAi and related phenomena in plants and animals (Hamilton and Baulcombe, 1999; Zamore *et al.*, 2000). Although the discovery of microRNA was unexpected, theoretical work on the mechanism of the *lac* repressor by Jacob and Monod had postulated a microRNA-like mechanism forty years earlier, see Figure 1.1 (Jacob and Monod, 1961).

Biogenesis and mechanism of action

From studies in *C. elegans*, *Drosophila melanogaster* and mammalian cell culture a model for microRNA biogenesis in animals is emerging (see Figure 1.2a and Figure 1.2b): microRNAs are transcribed by RNA polymerase II as long RNA precursors (pri-miRNAs) (Kim, 2005), which are processed in the nucleus by the RNase III enzyme Drosha and DGCR8/Pasha, to form the approximately 70 base pre-miRNA (Lee *et al.*, 2003; Denli *et al.*, 2004; Gregory *et al.*, 2004; Han *et al.*, 2004; Landthaler *et al.*, 2004). Pre-miRNAs are exported from the nucleus by Exportin-5 (Lund *et al.*, 2004) and processed by the RNase III enzyme, Dicer, and incorporated into an Argonaute containing silencing complex (RISC) (Du and Zamore, 2005). From this

---

* Author to whom correspondence should be addressed.

MicroRNAs: From Basic Science to Disease Biology, ed. Krishnarao Appasani. Published by Cambridge University Press. © Cambridge University Press 2008.

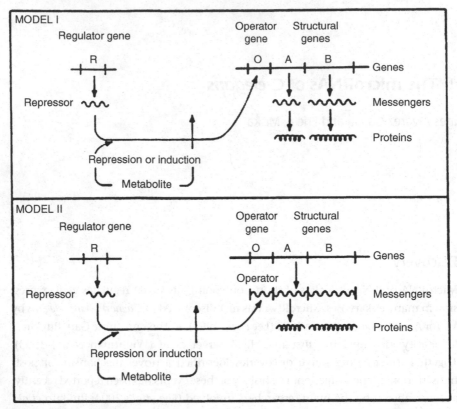

Figure 1.1. Models for the regulation of protein synthesis. Reproduced from Jacob and Monod (1961). Model II depicts a microRNA-like mechanism. Both models were proposed for the *lac* repressor and turned out to be false, as the *lac* repressor is a protein transcription factor. However, model I and II may underlie short RNA mediated gene regulation at the transcriptional and post-transcriptional level respectively.

general model a number of factors have been directly implicated in microRNA function in *C. elegans*. Drsh-1 (Drosha) and Pash-1 (DGCR8/Pasha) are required for pri-miRNA processing (Denli *et al.*, 2004). Dcr-1 (Dicer) is required for pre-miRNA processing (Grishok *et al.*, 2001; Ketting *et al.*, 2001; Knight and Bass, 2001). Interestingly, two Argonaute family members, Alg-1 and Alg-2 are also required for the generation of the mature microRNA from pre-miRNA, suggesting that at least in *C. elegans* Argonaute proteins not only play roles in microRNA mechanism of action but also microRNA biogenesis (Grishok *et al.*, 2001). In animals microRNAs are thought to function through the inhibition of effective mRNA translation of target genes through imperfect base-pairing with the 3′-untranslated region (3′UTR) of target mRNAs (see Figure 1.2c) (Bartel, 2004; Pillai *et al.*, 2005). However, alternative mechanisms of action include direct mRNA cleavage (Mansfield *et al.*, 2004; Yekta *et al.*, 2004) and mRNA stability (Bagga *et al.*, 2005; Jing *et al.*, 2005; Giraldez *et al.*, 2006). *In vivo* microRNA targets are largely unknown, but estimates range from one to hundreds of target genes for a given microRNA, based on microRNA target predictions using a variety of bioinformatics approaches (Lewis *et al.*, 2003; Stark *et al.*, 2003; John *et al.*, 2004; Rajewsky and Socci, 2004; Brennecke *et al.*, 2005; Farh *et al.*, 2005; Lewis *et al.*, 2005; Stark *et al.*, 2005; Xie *et al.*, 2005; Lall *et al.*, 2006).

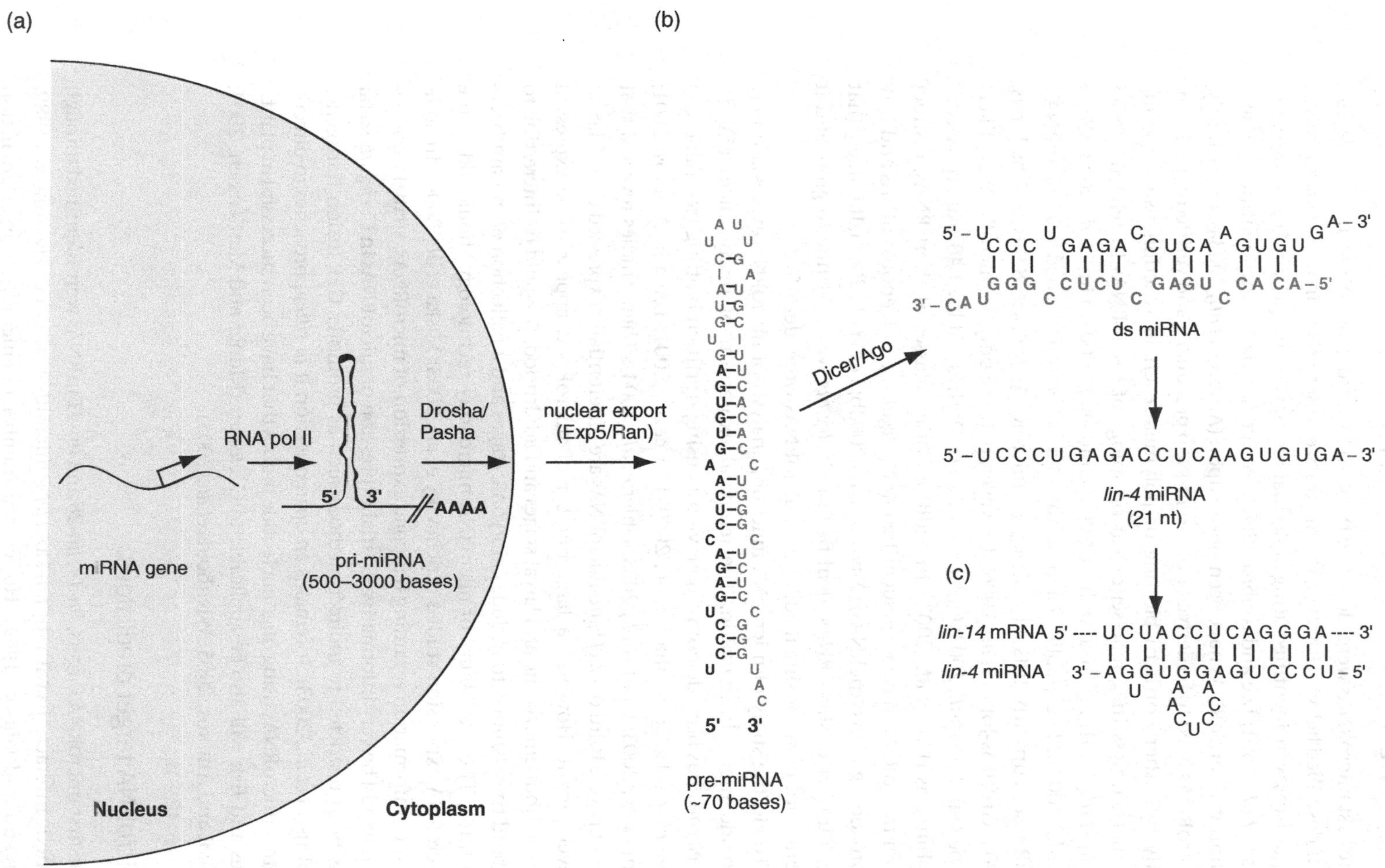

Figure 1.2. MicroRNA biogenesis and mechanism of action. A current model of the biogenesis of microRNAs in animals. This model is a generic model of the biogenesis of microRNAs in animals. The *lin-4* miRNA and the *lin-14* mRNA are shown as examples. The corresponding factors that have been implicated in the biogenesis of microRNAs in *C. elegans* are Drsh-1 (Drosha), Pash-1 (Pasha), Dcr-1 (Dicer) and Alg-1/2 (Ago).

(a)

Nucleus

Cytoplasm

mRNA gene

RNA pol II

5' 3' //AAAA

pri-miRNA
(500–3000 bases)

Drosha/
Pasha

nuclear export
(Exp5/Ran)

(b)

pre-miRNA
(~70 bases)

5' 3'

Dicer/Ago

5'- U CCC U GAGA C CUCA A GUGU G A-3'
         | | | | | | | | | | | | | | |
     3'- CAU GGG C CUCU GAGU CACA-5'

ds miRNA

5'- UCCCUGAGACCUCAAGUGUGA-3'

lin-4 miRNA
(21 nt)

(c)

lin-14 mRNA 5'---- UCUACCUCAGGGA ----3'
                    | | | | | | | | | | | |
lin-4 miRNA   3'- AGGUGGAGUCCCU-5'

MicroRNA gene identification and expression analysis

The first microRNAs were identified through forward genetic screens in *C. elegans*, as is described below. However, the majority of microRNAs in *C. elegans* and other organisms were identified using molecular biology techniques and bioinformatics (Lau *et al.*, 2001; Lee and Ambros, 2001). Strict criteria for the identification and naming of microRNAs have been agreed upon (Ambros *et al.*, 2003a). Importantly, microRNAs need to be derived from stem-loop precursors and verified experimentally by either northern blotting or polymerase chain reaction using specific primers, unless they represent clear orthologs of microRNAs in related species (Lau *et al.*, 2001; Lee and Ambros, 2001; Grad *et al.*, 2003; Lim *et al.*, 2003; Ohler *et al.*, 2004). The public database that has been established for microRNAs, miRBase, currently lists 115 *C. elegans* microRNAs (release 8.0)(Griffiths-Jones, 2004; Griffiths-Jones *et al.*, 2006). However, of these only 98 microRNAs are likely to be real (Lim *et al.*, 2003; Ohler *et al.*, 2004). Of these, at least 40% are conserved in humans (Lim *et al.*, 2003). In addition, other classes of short RNAs, distinct from microRNAs have been identified in *C. elegans*, endogenous siRNAs and tiny non-coding RNAs (tncRNAs) (Ambros *et al.*, 2003b; Lee *et al.*, 2006). It is likely that the number of short RNAs identified in *C. elegans* will continue to grow and it would not be surprising if additional microRNAs were identified.

To aid the study of microRNA function, analysis of microRNA expression may provide useful information. The temporal expression of many microRNAs in *C. elegans* has been determined previously using northern blotting (see Table 1.1) (Lee *et al.*, 1993; Reinhart *et al.*, 2000; Lau *et al.*, 2001; Lee and Ambros, 2001; Lim *et al.*, 2003; Grad *et al.*, 2003; Ohler *et al.*, 2004). These studies revealed that possibly one third of *C. elegans* microRNAs are differentially expressed during larval development. However, a large number of microRNAs appear to be expressed throughout embryonic and larval stages and adulthood. It would be interesting to extend these studies to include embryonic stages and a collection of mutant backgrounds. The availability of microRNA microarray technology should aid future microRNA expression studies (Krichevsky *et al.*, 2003; Miska *et al.*, 2004). In addition to information regarding temporal expression of microRNAs, spatial information would be very informative. Spatial expression of microRNAs in *C. elegans* so far has been restricted to indirect methods such as promoter GFP fusion transgenes (Johnson *et al.*, 2003). It remains an open question if *in situ* approaches to directly detect microRNAs using oligonucleotide probes that have been successful in plants, flies and fish will also be applicable to *C. elegans* (Kidner and Martienssen, 2004; Sokol and Ambros, 2005; Wienholds *et al.*, 2005).

MicroRNA target predictions

The first microRNA targets, *lin-14*, *lin-28* and *lin-41* mRNAs were identified through genetic interactions with the *lin-4* and *let-7* microRNAs respectively (Ambros, 1989; Ruvkun and Giusto, 1989). However, a number of microRNA target prediction algorithms have been described (Lewis *et al.*, 2003; Stark *et al.*, 2003; John *et al.*,

Table 1.1. MicroRNA expression during larval development in *C. elegans*

microRNA	E	L1	L2	L3	L4	A
let-7				**	**	**
lin-4			**	**	**	**
miR-1	*	*	*	*	*	*
miR-2	**	**	**	**	**	**
miR-34		*	*	*	**	**
miR-35	**					**
miR-36	**					**
miR-37	**					**
miR-38	**					**
miR-39	**					**
miR-40	**					**
miR-41						
miR-42	**	*	*	*	*	
miR-43	**	*	*	*	*	
miR-44/45	**	**	**	**	**	*
miR-46	**	**	**	**	**	**
miR-47	**	**	**	**	**	**
miR-48				**	**	**
miR-49	**	**	*	*	*	
miR-50	**	**	**	**	**	**
miR-51	**	**	**	**	**	**
miR-52/53	**	**	**	**	**	**
miR-54	**	**	**	**	**	**
miR-55	**	**	**	**	**	**
miR-56	**	**	**	**	**	**
miR-57	**	**	**	**	**	**
miR-58	**	**	**	**	**	**
miR-59	*				**	
miR-60	*	*	*	*	*	*
miR-61	**	**	**	**	**	**
miR-62	*	*	*	*	*	*
miR-63	**	**	**	**	**	**
miR-64/65	**	**	**	**	**	**
miR-66	**	**	**	**	**	**
miR-67	**	**	**	**	**	**
miR-70	*	*	**	**	**	**
miR-71		*	*	*	*	*
miR-72	**	**	**	**	**	**
miR-73	**	**	**	**	**	**
miR-74	**	**	**	**	**	**
miR-75	*	**	**	**	**	**
miR-76	*	*	*	*	*	*
miR-77			*	*	**	**
miR-78						
miR-79	**	**	**	**	**	**
miR-80	**	**	**	**	**	**
miR-81/82	**	**	**	**	**	**
miR-83	*	*	*	*	*	*
miR-84				**	**	**

Table 1.1. (cont.)

microRNA	E	L1	L2	L3	L4	A
miR-85					**	**
miR-86	**	**	**	**	**	**
miR-87	**	**	**	**	**	**
miR-90	**	**	**	**	**	**
miR-124	**	**	**	**	**	**
miR-228	**	**	**	**	**	**
miR-229	**	**	**	**	**	**
miR-230	*	*	*	*	*	*
miR-231	**	**	**	**	**	**
miR-232	**	**	**	**	**	**
miR-233	**	**	**	**	**	**
miR-234	*	**	*	*	*	*
miR-235	*	*	*	*	*	*
miR-236	**	**	**	**	**	**
miR-237				**	**	**
miR-238	*	**	**	**	**	**
miR-239a/b	*	**	**	**	**	**
miR-240					*	*
miR-241				**	**	**
miR-242	*	*	*	*	*	*
miR-243	**	**	**	**	**	**
miR-244	**	**	**	**	**	**
miR-245					*	*
miR-246					*	*
miR-247				*		
miR-248		*	*	*	*	*
miR-249						
miR-250	*	*	*	*	*	*
miR-251	*	*	*	*	*	*
miR-252	*	*	*	*	*	*
miR-253						
miR-254						
miR-255						

MicroRNA expression during *C. elegans* larval development has been studied systematically using only northern blotting (Lau *et al.*, 2001; Lee and Ambros, 2001; Lim *et al.*, 2003).
* indicates low expression, ** indicates high expression levels.

2004; Rajewsky and Socci, 2004; Brennecke *et al.*, 2005; Farh *et al.*, 2005; Lewis *et al.*, 2005; Stark *et al.*, 2005; Xie *et al.*, 2005; Lall *et al.*, 2006; Watanabe, 2006). Although it remains unclear how many of the predicted microRNA/microRNA target interactions are important *in vivo*, successes of microRNA target prediction include the identification of an *in vivo* target for the *D. melanogaster* microRNA encoded by *bantam* (Brennecke *et al.*, 2003). To date two microRNA target prediction studies have been published for *C. elegans* (Lall *et al.*, 2006; Watanabe, 2006). It will be of great interest to see the integration of these and new target predictions with microRNA and target mRNA expression data in *C. elegans*, as has recently been done for *D. melanogaster* and humans (Farh, 2005; Stark *et al.*, 2005).

MicroRNAs control the timing of development

Not the search for non-coding RNA, but an interest in the genetic program controlling developmental timing in *C. elegans* (Horvitz and Sulston, 1980; Chalfie *et al.*, 1981; Ambros and Horvitz, 1984) led to the cloning of the first microRNA, *lin-4* miRNA (Lee *et al.*, 1993), and the identification of the first microRNA target, *lin-14* mRNA (Wightman *et al.*, 1993)(see Figure 1.2). The developmental timing pathway regulates stage-specific processes during *C. elegans* larval development (see Figure 1.3a). This pathway is often referred to as the heterochronic pathway, a term borrowed from evolutionary biology (Gould, 2000). The heterochronic pathway has recently been reviewed in detail elsewhere (Rougvie, 2005). Study of the heterochronic pathway in *C. elegans* has been focussing on the developmental fate of several blast cells in the lateral hypodermis, collectively referred to as the seam cells (see Figure 1.3b). The seam cells undergo a cell division pattern that is synchronized with the four larval molts of the animal. Only at the adult stage will the seam cells exit mitosis and terminally differentiate. Terminal differentiation of the seam cells is followed by secretion of an adult-specific cuticle structure, the lateral alae (Figure 1.3c) (Cox *et al.*, 1981). In *lin-4* mutant animals the seam cells repeat the cell division pattern that exemplifies the first larval stage (L1) and fail to differentiate. This mutant phenotype has been interpreted as a heterochronic change; the developmental clock being stuck in the L1 stage, resulting in developmental "retardation." Gain-of-function mutations in the *lin-4* miRNA target *lin-14* lead to the identical phenotype, whereas loss-of-function mutations in *lin-14* result in an opposite "precocious" phenotype, where the seam cells skip the cell division of the first larval stage. The *lin-4/lin-14* gene products therefore act as a developmental switch controlling the L1 to L2 transition.

Three microRNAs of the *let-7* family, *mir-48*, *mir-84* and *mir-241* act redundantly to control the next developmental transition, from the L2 to the L3 stage (Reinhart *et al.*, 2000; Lau *et al.*, 2001; Abbott *et al.*, 2005; Li *et al.*, 2005). Loss-of-function mutations in these three microRNAs lead to the repetition of the cell division pattern of the second larval stage, whereas a gain-of-function mutation in *mir-48* results in a precocious phenotype. A likely target of *mir-48*, *mir-84* and *mir-241* during this transition is the *C. elegans hunchback* ortholog *hbl-1*. The microRNA *let-7*, the second microRNA to be identified (Reinhart *et al.*, 2000), controls the transition from the fourth larval to the adult stage and two of its targets in the heterochronic pathway are the *lin-41* and *hbl-1* genes, both of which are also heterochronic genes (Slack *et al.*, 2000; Abrahante *et al.*, 2003; Lin *et al.*, 2003). More recently, two additional *let-7* target genes, the transcription factors *daf-12* and *pha-4*, were identified using a combination of target prediction through bioinformatics and RNA interference (RNAi) analyses (Grosshans *et al.*, 2005). Of these, at least *daf-12* is also a regulator of the heterochronic pathway controlling seam cell fate (Antebi *et al.*, 1998). Finally, the two *let-7* family microRNAs *mir-48* and *mir-84* also control the cessation of the larval molting cycle at the adult stage, with *mir-48*; *mir-84* double mutant animals undergoing a supernumerary molt at the adult stage (Abbott *et al.*, 2005). It is striking that at

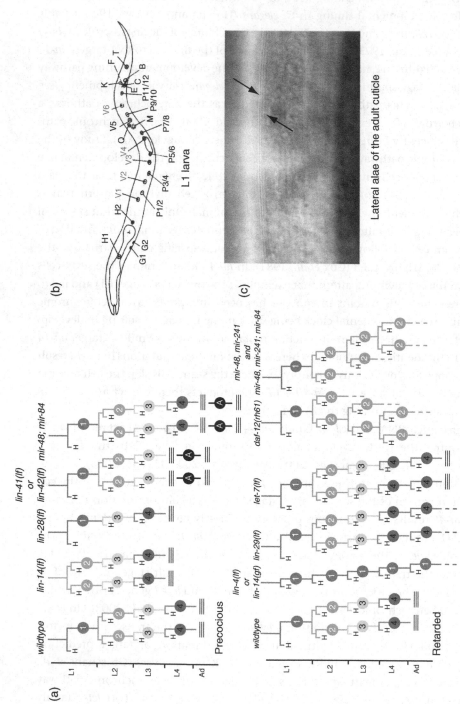

Figure 1.3. MicroRNAs control developmental timing in *C. elegans*. (a) *C. elegans* cell lineage diagram for the seam cells V1–V4 and V6. H, cells fused to hypodermis. Triple lines: differentiation of seam cells and alae formation. L1–L4, Ad, larval stages one to four and adult stage. (b) Position of the seam cells V1 to V6 in the lateral hypodermis of a L1 larva. (c) DIC image of an adult cuticle with arrows indicating the lateral alae.

least two microRNA families and at least four microRNAs are involved in the control of developmental timing in *C. elegans*. As the *lin-4* and *let-7* microRNA families are conserved, it is tempting to speculate that these two families may play similar roles in other organisms. This notion is supported by the temporal regulation of *let-7* expression in several species (Pasquinelli *et al.*, 2000).

MicroRNAs in the organogenesis of the vulva

Evidence for a role of microRNAs in organogenesis comes from studies of vulval development in *C. elegans*. The *C. elegans* vulva is a ring-like structure that forms the connection between the hermaphrodite gonadal arms and the exterior and is essential for egg-laying and sperm entry. The vulva is derived from a group of cells in the ventral hypodermis that are induced to undergo a series of cell divisions and differentiation by a signal from the gonadal anchor cell (Sulston and Horvitz, 1977). Vulval induction requires RAS/LET-60 signalling (Beitel *et al.*, 1990); *let-7* loss-of-function mutants die by bursting at the vulva (Reinhart *et al.*, 2000; Slack *et al.*, 2000) and this bursting is suppressed by RNAi against the *C. elegans RAS* ortholog *let-60* (Johnson *et al.*, 2005). Furthermore, over-expression of the *let-7* family microRNA miR-84 led to a multivulva phenotype that is reminiscent of *RAS/let-60* gain-of-function mutants. Recent 3'UTR reporter experiments suggest that RAS/LET-60 expression levels are regulated post-transcriptionally and may be directly regulated by the *let-7* family of microRNAs (Johnson *et al.*, 2005). In addition to the *let-7* family, a distinct microRNA, miR-61, has also been implicated in vulval development. It was shown that over-expression of miR-61 leads to the ectopic expression of reporter genes indicative of secondary cell fates of the Pn.p cells (Yoo and Greenwald, 2005). Although no actual cell fate changes were observed in this study, these observations raise the possibility that several microRNA families are involved in the coordination of vulval cell fates.

MicroRNAs control left/right asymmetry in the nervous system

As in the case of *lin-4* and *let-7*, forward genetic screens identified a third microRNA, *lsy-6*, involved in left/right asymmetry in the *C. elegans* nervous system (Johnston and Hobert, 2003). In *C. elegans* two bilateral taste receptor neurons, ASE left (ASEL) and ASE right (ASER), display a left/right asymmetrical expression pattern of *gcy-5*, *gcy-6* and *gcy-7*, three putative chemoreceptor genes (Hobert *et al.*, 2002; Chang *et al.*, 2003) (see Figure 1.4). In a genetic screen for mutants in which the normally ASEL-specific expression of *gcy-7* is disrupted, the microRNA gene *lsy-6* was isolated (Chang *et al.*, 2003). In *lsy-6* mutants ASEL do not express *gcy-7*, but instead express the ASER-specific *gcy-5* gene. Genetic interaction and GFP reporter studies showed that *lsy-6* is a negative regulator of the Nkx-type homeobox gene *cog-1*, which was identified in the same genetic screen. Interestingly, a second microRNA, miR-273, may act upstream in the same pathway as a regulator of *die-1*, which encodes a C2H2 zinc finger transcription factor (Chang *et al.*, 2004). The transcription factor *die-1* shows ASEL-specific expression and acts upstream of *lsy-6*.

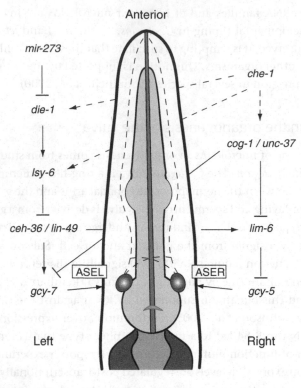

Figure 1.4. MicroRNAs control differentiation. Neuronal specification of ASEL and ASER cells by the microRNAs *lsy-6* and miR-273. *che-1*, *unc-37* and *ceh-36/lin-49* are required for the neuronal determination of the pair of ASE neurons. The ASE cells differ in the asymmetric expression of two candidate chemoreceptor genes, *gcy-7* (in ASEL) and *gcy-5* (in ASER), which also define different chemosensory functions. The microRNA *lsy-6* represses the translation of *cog-1* mRNA, which results in the expression of *lim-6* and *gcy-7*. In ASER miR-273 represses the translation of *die-1* mRNA, which leads to the expression of *cog-1* and *gcy-5*.

MicroRNAs in ageing

C. elegans life span is highly controlled by several genetic pathways including an insulin/IGF-1-like signalling cascade. Recently, heterochronic or developmental timing genes were implicated in the regulation of life span in *C. elegans* (Boehm and Slack, 2005). Loss-of-function mutants of *lin-4* have a shortened life span. Conversely, a partial loss-of-function mutant of *lin-14*, a direct target of *lin-4*, shows life span extension. In addition, RNAi of *lin-14* suppresses the shorter life span observed in *lin-14* mutant animals. The shortened life span of *lin-4* mutants is phenocopied by a *lin-14* gain-of-function mutation, which lacks the *lin-4* complementary sites in its 3′UTR. Together these experiments suggest that the *lin-4* microRNA suppresses senescence through repression of *lin-14* in adult *C. elegans* (Boehm and Slack, 2005). Interestingly, expression of the *lin-4* microRNA appears to be regulated by another factor involved in *C. elegans* life span regulation, DAF-16 (Baugh and Sternberg, 2006). DAF-16, a FOXO transcription factor, regulates *lin-4* expression in a food-dependent manner at the L1 stage, as assayed

Table 1.2. Function of *C. elegans* microRNAs

Process	microRNA	Targets	Function	Evidence	References
developmental timing	*lin-4* microRNA	*lin-14*, *lin-28*	stem cell differentiation	LOF	Lee *et al.*, 1993; Wightman *et al.*, 1993
developmental timing	*let-7* microRNA	*lin-41*, *hbl-1*, *daf-12*, *pha-4*	stem cell differentiation	LOF	Reinhart *et al.*, 2000; Abrahante *et al.*, 2003; Lin *et al.*, 2003; Slack *et al.*, 2000; Grosshans *et al.*, 2005
developmental timing	miR-48, miR-84, miR-241	*hbl-1*	stem cell differentiation	LOF	Abbott *et al.*, 2005; Lin *et al.*, 2003
developmental timing	miR-48, miR-84	unknown	cessation of molting	LOF	Abbott *et al.*, 2005
organogenesis	miR-84	*let-60*	differentiation/proliferation	GOF	Johnson *et al.*, 2005
differentiation	*lsy-6* microRNA	*cog-1*	left/right asymmetry	LOF	Johnston and Hobert, 2003
differentiation	miR-273	*die-1*	left/right asymmetry	GOF	Chang *et al.*, 2004
life span	*lin-4*	*lin-14*	longevity	LOF + GOF	Boehm and Slack, 2005

This table includes all microRNAs that have been analyzed *in vivo* using loss-of-function studies or microRNAs for which a likely function has been demonstrated using an indirect approach, e. g. mis-expression experiments. This table does not contain microRNAs for which a target mRNA has been predicted and validated using over-expression experiments, but for which no further functional characterization at the cellular or organismal level has been carried out.

by a *lin-4::yfp* reporter transgene. Although much is still unclear it is exciting that the developmental timing and life span pathways appear to be inter-connected.

Conclusions

Over thirteen years ago the genetic analysis of developmental timing in *C. elegans* led to the discovery of microRNAs. Despite considerable attention to microRNAs over the past five years, a lot of questions remain unanswered. The mechanism of action of microRNAs is still controversial and only few *in vivo* functions of microRNAs have been demonstrated in any organism to date. In *C. elegans* several developmental roles for microRNAs are emerging (see Table 1.2). Given that *C. elegans* offers powerful forward and reverse genetic approaches combined with a phenotypic analysis at the cellular level, one might hope that *C. elegans* will continue to make important contributions to our understanding of microRNA biology.

REFERENCES

Abbott, A. L., Alvarez-Saavedra, E., Miska, E. A. *et al.* (2005). The *let-7* MicroRNA family members *mir-48*, *mir-84*, and *mir-241* function together to regulate developmental timing in *Caenorhabditis elegans*. *Developmental Cell*, **9**, 403–414.

Abrahante, J. E., Daul, A. L., Li, M. *et al.* (2003). The *Caenorhabditis elegans hunchback-like gene lin-57/hbl-1* controls developmental time and is regulated by microRNAs. *Developmental Cell*, **4**, 625–637.

Ambros, V. (1989). A hierarchy of regulatory genes controls a larva-to-adult developmental switch in *C. elegans*. *Cell*, **57**, 49–57.

Ambros, V. and Horvitz, H. R. (1984). Heterochronic mutants of the nematode *Caenorhabditis elegans*. *Science*, **226**, 409–416.

Ambros, V., Bartel, B., Bartel, D. P. *et al.* (2003a). A uniform system for microRNA annotation. *RNA*, **9**, 277–279.

Ambros, V., Lee, R. C., Lavanway, A., Williams, P. T., and Jewell, D. (2003b). MicroRNAs and other tiny endogenous RNAs in *C. elegans*. *Current Biology*, **13**, 807–818.

Antebi, A., Culotti, J. G. and Hedgecock, E. M. (1998). daf-12 regulates developmental age and the dauer alternative in *Caenorhabditis elegans*. *Development*, **125**, 1191–1205.

Bagga, S., Bracht, J., Hunter, S. *et al.* (2005). Regulation by *let-7* and *lin-4* miRNAs results in target mRNA degradation. *Cell*, **122**, 553–563.

Bartel, D. P. (2004). MicroRNAs: genomics, biogenesis, mechanism, and function. *Cell*, **116**, 281–297.

Baugh, L. R. and Sternberg, P. W. (2006). DAF-16/FOXO regulates transcription of cki-1/Cip/Kip and repression of lin-4 during *C. elegans* L1 arrest. *Current Biology*, **16**, 780–785.

Beitel, G. J., Clark, S. G. and Horvitz, H. R. (1990). *Caenorhabditis elegans ras* gene *let-60* acts as a switch in the pathway of vulval induction. *Nature*, **348**, 503–509.

Boehm, M. and Slack, F. (2005). A developmental timing microRNA and its target regulate life span in *C. elegans*. *Science*, **310**, 1954–1957.

Brennecke, J., Hipfner, D. R., Stark, A., Russell, R. B. and Cohen, S. M. (2003). bantam encodes a developmentally regulated microRNA that controls cell proliferation and regulates the proapoptotic gene hid in *Drosophila*. *Cell*, **113**, 25–36.

Brennecke, J., Stark, A., Russell, R. B. and Cohen, S. M. (2005). Principles of microRNA-target recognition. *Public Library of Science Biology*, **3**, e85.

Chalfie, M., Horvitz, H. R. and Sulston, J. E. (1981). Mutations that lead to reiterations in the cell lineages of *C. elegans*. *Cell*, **24**, 59–69.

Chang, S., Johnston, R. J., Jr. and Hobert, O. (2003). A transcriptional regulatory cascade that controls left/right asymmetry in chemosensory neurons of *C. elegans*. *Genes & Development*, **17**, 2123–2137.

Chang, S., Johnston, R. J., Jr., Frokjaer-Jensen, C., Lockery, S. and Hobert, O. (2004). MicroRNAs act sequentially and asymmetrically to control chemosensory laterality in the nematode. *Nature*, **430**, 785–789.

Cox, G. N., Staprans, S. and Edgar, R. S. (1981). The cuticle of *Caenorhabditis elegans*. II. Stage-specific changes in ultrastructure and protein composition during postembryonic development. *Developmental Biology*, **86**, 456–470.

Denli, A. M., Tops, B. B., Plasterk, R. H., Ketting, R. F. and Hannon, G. J. (2004). Processing of primary microRNAs by the Microprocessor complex. *Nature*, **432**, 231–235.

Du, T. and Zamore, P. D. (2005). microPrimer: the biogenesis and function of microRNA. *Development*, **132**, 4645–4652.

Farh, K. K., Grimson, A., Jan, C. *et al.* (2005). The widespread impact of mammalian microRNAs on mRNA repression and evolution. *Science*, **310**, 1817–1821.

Giraldez, A. J., Mishima, Y., Rihel, J. *et al.* (2006). Zebrafish MiR-430 promotes deadenylation and clearance of maternal mRNAs. *Science*, **312**, 75–79.

Gould, S. J. (2000). Of coiled oysters and big brains: how to rescue the terminology of heterochrony, now gone astray. *Evolution & Development*, **2**, 241–248.

Grad, Y., Aach, J., Hayes, G. D. *et al.* (2003). Computational and experimental identification of *C. elegans* microRNAs. *Molecular Cell*, **11**, 1253–1263.

Gregory, R. I., Yan, K. P., Amuthan, G. *et al.* (2004). The Microprocessor complex mediates the genesis of microRNAs. *Nature*, **432**, 235–240.

Griffiths-Jones, S. (2004). The microRNA Registry. *Nucleic Acids Research*, **32**, D109–111.

Griffiths-Jones, S., Grocock, R. J., van Dongen, S., Bateman, A. and Enright, A. J. (2006). miRBase: microRNA sequences, targets and gene nomenclature. *Nucleic Acids Research*, **34**, D140–144.

Grishok, A., Pasquinelli, A. E., Conte, D. *et al.* (2001). Genes and mechanisms related to RNA interference regulate expression of the small temporal RNAs that control *C. elegans* developmental timing. *Cell*, **106**, 23–34.

Grosshans, H., Johnson, T., Reinert, K. L., Gerstein, M. and Slack, F. J. (2005). The temporal patterning microRNA *let-7* regulates several transcription factors at the larval to adult transition in *C. elegans*. *Developmental Cell*, **8**, 321–330.

Hamilton, A. J. and Baulcombe, D. C. (1999). A species of small antisense RNA in posttranscriptional gene silencing in plants. *Science*, **286**, 950–952.

Han, J., Lee, Y., Yeom, K. H. *et al.* (2004). The Drosha-DGCR8 complex in primary microRNA processing. *Genes & Development*, **18**, 3016–3027.

Hobert, O., Johnston, R. J., Jr. and Chang, S. (2002). Left-right asymmetry in the nervous system: the *Caenorhabditis elegans* model. *Nature Reviews Neuroscience*, **3**, 629–640.

Horvitz, H. R. and Sulston, J. E. (1980). Isolation and genetic characterization of cell-lineage mutants of the nematode *Caenorhabditis elegans*. *Genetics*, **96**, 435–454.

Jacob, F. and Monod, J. (1961). Genetic regulatory mechanisms in the synthesis of proteins. *Journal of Molecular Biology*, **3**, 318–356.

Jing, Q., Huang, S., Guth, S. *et al.* (2005). Involvement of microRNA in AU-rich element-mediated mRNA instability. *Cell*, **120**, 623–634.

John, B., Enright, A. J., Aravin, A. *et al.* (2004). Human microRNA targets. *Public Library of Science Biology*, **2**, e363.

Johnson, S. M., Lin, S. Y. and Slack, F. J. (2003). The time of appearance of the *C. elegans let-7* microRNA is transcriptionally controlled utilizing a temporal regulatory element in its promoter. *Developmental Biology*, **259**, 364–379.

Johnson, S. M., Grosshans, H., Shingara, J. *et al.* (2005). RAS is regulated by the *let-7* microRNA family. *Cell*, **120**, 635–647.

Johnston, R. J. and Hobert, O. (2003). A microRNA controlling left/right neuronal asymmetry in *Caenorhabditis elegans*. *Nature*, **426**, 845–849.

Ketting, R. F., Fischer, S. E., Bernstein, E. *et al.* (2001). Dicer functions in RNA interference and in synthesis of small RNA involved in developmental timing in *C. elegans*. *Genes & Development*, **15**, 2654–2659.

Kidner, C. A. and Martienssen, R. A. (2004). Spatially restricted microRNA directs leaf polarity through ARGONAUTE1. *Nature*, **428**, 81–84.

Kim, V. N. (2005). MicroRNA biogenesis: coordinated cropping and dicing. *Nature Reviews Molecular Cell Biology*, **6**, 376–385.

Knight, S. W. and Bass, B. L. (2001). A role for the RNase III enzyme DCR-1 in RNA interference and germ line development in *Caenorhabditis elegans*. *Science*, **293**, 2269–2271.

Krichevsky, A. M., King, K. S., Donahue, C. P., Khrapko, K. and Kosik, K. S. (2003). A microRNA array reveals extensive regulation of microRNAs during brain development. *RNA*, **9**, 1274–1281.

Lall, S., Grun, D., Krek, A. *et al.* (2006). A genome-wide map of conserved microRNA targets in *C. elegans*. *Current Biology*, **16**, 460–471.

Landthaler, M., Yalcin, A. and Tuschl, T. (2004). The human DiGeorge syndrome critical region gene 8 and Its *D. melanogaster* homolog are required for miRNA biogenesis. *Current Biology*, **14**, 2162–2167.

Lau, N. C., Lim, L. P., Weinstein, E. G. and Bartel, D. P. (2001). An abundant class of tiny RNAs with probable regulatory roles in *Caenorhabditis elegans*. *Science*, **294**, 858–862.

Lee, R. C. and Ambros, V. (2001). An extensive class of small RNAs in *Caenorhabditis elegans*. *Science*, **294**, 862–864.

Lee, R. C., Feinbaum, R. L. and Ambros, V. (1993). The *C. elegans* heterochronic gene *lin-4* encodes small RNAs with antisense complementarity to *lin-14*. *Cell*, **75**, 843–854.

Lee, R., Feinbaum, R. and Ambros, V. (2004). A short history of a short RNA. *Cell*, **116**, S89–92, 81 p following S96.

Lee, R. C., Hammell, C. M. and Ambros, V. (2006). Interacting endogenous and exogenous RNAi pathways in *Caenorhabditis elegans*. *RNA*, **4**, 589–597.

Lee, Y., Ahn, C., Han, J. *et al.* (2003). The nuclear RNase III Drosha initiates microRNA processing. *Nature*, **425**, 415–419.

Lewis, B. P., Shih, I. H., Jones-Rhoades, M. W., Bartel, D. P. and Burge, C. B. (2003). Prediction of mammalian microRNA targets. *Cell*, **115**, 787–798.

Lewis, B. P., Burge, C. B. and Bartel, D. P. (2005). Conserved seed pairing, often flanked by adenosines, indicates that thousands of human genes are microRNA targets. *Cell*, **120**, 15–20.

Li, M., Jones-Rhoades, M. W., Lau, N. C., Bartel, D. P. and Rougvie, A. E. (2005). Regulatory mutations upstream of *mir-48*, a *C. elegans let-7* family microRNA, cause developmental timing defects. *Developmental Cell*, **3**, 415–422.

Lim, L. P., Lau, N. C., Weinstein, E. G. *et al.* (2003). The microRNAs of *Caenorhabditis elegans*. *Genes & Development*, **17**, 991–1008.

Lin, S. Y., Johnson, S. M., Abraham, M. *et al.* (2003). The *C. elegans hunchback* homolog, *hbl-1*, controls temporal patterning and is a probable microRNA target. *Developmental Cell*, **4**, 639–650.

Lund, E., Guttinger, S., Calado, A., Dahlberg, J. E. and Kutay, U. (2004). Nuclear export of microRNA precursors. *Science*, **303**(5654), 95–98.

Mansfield, J. H., Harfe, B. D., Nissen, R. *et al.* (2004). MicroRNA-responsive 'sensor' transgenes uncover Hox-like and other developmentally regulated patterns of vertebrate microRNA expression. *Nature Genetics*, **36**, 1079–1083.

Miska, E. A., Alvarez-Saavedra, E., Townsend, M. *et al.* (2004). Microarray analysis of microRNA expression in the developing mammalian brain. *Genome Biology*, **5**, R68.

Ohler, U., Yekta, S., Lim, L. P., Bartel, D. P. and Burge, C. B. (2004). Patterns of flanking sequence conservation and a characteristic upstream motif for microRNA gene identification. *RNA*, **10**, 1309–1322.

Pasquinelli, A. E., Reinhart, B. J., Slack, F. *et al.* (2000). Conservation of the sequence and temporal expression of *let-7* heterochronic regulatory RNA. *Nature*, **408**, 86–89.

Pillai, R. S., Bhattacharyya, S. N., Artus, C. G. *et al.* (2005). Inhibition of translational initiation by *let-7* microRNA in human cells. *Science*, **309**, 1573–1576.

Rajewsky, N. and Socci, N. D. (2004). Computational identification of microRNA targets. *Developmental Biology*, **267**, 529–535.

Reinhart, B. J., Slack, F. J., Basson, M. *et al.* (2000). The 21-nucleotide *let-7* RNA regulates developmental timing in *Caenorhabditis elegans. Nature*, **403**, 901–906.

Rougvie, A. E. (2005). Intrinsic and extrinsic regulators of developmental timing: from miRNAs to nutritional cues. *Development*, **132**, 3787–3798.

Ruvkun, G. and Giusto, J. (1989). The Caenorhabditis elegans heterochronic gene lin-14 encodes a nuclear protein that forms a temporal developmental switch. *Nature*, **338**, 313–319.

Ruvkun, G., Wightman, B. and Ha, I. (2004). The 20 years it took to recognize the importance of tiny RNAs. *Cell*, **116**, S93–96, 92 p following S96.

Slack, F. J., Basson, M., Liu, Z. *et al.* (2000). The *lin-41* RBCC gene acts in the *C. elegans* heterochronic pathway between the *let-7* regulatory RNA and the LIN-29 transcription factor. *Molecular Cell*, **5**, 659–669.

Sokol, N. S. and Ambros, V. (2005). Mesodermally expressed *Drosophila* microRNA-1 is regulated by Twist and is required in muscles during larval growth. *Genes & Development*, **19**, 2343–2354.

Stark, A., Brennecke, J., Russell, R. B. and Cohen, S. M. (2003). Identification of *Drosophila* microRNA targets. *Public Library of Science Biology*, **3**, E60.

Stark, A., Brennecke, J., Bushati, N., Russell, R. B. and Cohen, S. M. (2005). Animal microRNAs confer robustness to gene expression and have a significant impact on 3'UTR evolution. *Cell*, **123**, 1133–1146.

Sulston, J. E. and Horvitz, H. R. (1977). Post-embryonic cell lineages of the nematode, *Caenorhabditis elegans. Developmental Biology*, **56**, 110–156.

Watanabe, Y., Yachie, N., Numata, K. *et al.* (2006). Computational analysis of microRNA targets in *Caenorhabditis elegans. Gene*, **365**, 2–10.

Wienholds, E., Kloosterman, W. P., Miska, E. *et al.* (2005). MicroRNA expression in zebrafish embryonic development. *Science*, **309**, 310–311.

Wightman, B., Ha, I. and Ruvkun, G. (1993). Posttranscriptional regulation of the heterochronic gene *lin-14* by *lin-4* mediates temporal pattern formation in *C. elegans. Cell*, **75**, 855–862.

Xie, X., Lu, J., Kulbokas, E. J. *et al.* (2005). Systematic discovery of regulatory motifs in human promoters and 3' UTRs by comparison of several mammals. *Nature*, **434**, 338–345.

Yekta, S., Shih, I. H. and Bartel, D. P. (2004). MicroRNA-directed cleavage of HOXB8 mRNA. *Science*, **304**, 594–596.

Yoo, A. S. and Greenwald, I. (2005). LIN-12/Notch activation leads to microRNA-mediated down-regulation of Vav in *C. elegans. Science*, **310**, 1330–1333.

Zamore, P. D., Tuschl, T., Sharp, P. A. and Bartel, D. P. (2000). RNAi: double-stranded RNA directs the ATP-dependent cleavage of mRNA at 21 to 23 nucleotide intervals. *Cell*, **101**, 25–33.

2 Non-coding RNAs – development of man-made vector-based intronic microRNAs (miRNAs)

Shao-Yao Ying*, Joseph D. Miller and Shi-Lung Lin

Introduction

The central dogma of molecular biology is that genomic DNA is transcribed into messenger RNA (mRNA) which is translated into proteins. This dogma has recently been challenged because some segments of the DNA transcribed into the mRNA precursor (pre-mRNA) are not necessarily translated into proteins. Instead, these RNAs regulate the expression of other genes. The segments of the DNA which function directly as regulatory RNAs rather than coding for protein products are called non-coding RNAs. In the human genome, the vast majority (nearly 95%) of DNA is of the non-coding variety. Frequently diseases are associated with malfunction of the non-coding RNA. We present here the historical background and significance of non-coding RNA research, with a particular eye to the current status of work on the microRNAs and future prospects for development of artificial intronic microRNAs (miRNAs). These miRNAs can play critical roles in development, protein secretion, and gene regulation. Some of them are naturally occurring antisense and hairpin RNAs whereas others have more complex structures. To understand the diseases caused by dysregulation of these miRNAs, a tissue-specific expression system is needed to recreate the function and mechanism of individual miRNA *in vitro* and *in vivo*.

Non-coding RNAs

The non-coding RNA (ncRNA) can be defined as segments of a RNA molecule that are not translated into a protein, but function in modulating the synthesis of proteins. Compared with the large segments of RNA molecules that are translated into proteins, the non-coding RNAs usually are small, less than 300 nucleotides (nt) in length; therefore, they are also called small RNA (sRNA). Less frequently, they are called non-messenger RNA (nmRNA), small non-messenger RNA

---

* Author to whom correspondence should be addressed.

MicroRNAs: From Basic Science to Disease Biology, ed. Krishnarao Appasani. Published by Cambridge University Press. © Cambridge University Press 2008.

(snmRNA), tiny non-coding RNA (tncRNA), small modulatory RNA (smRNA), or small regulatory RNA. Even though the non-coding RNAs translate no proteins, they modulate the function of other genes. Hence, the DNA sequence from which a non-coding RNA is transcribed is often called an RNA gene or non-coding regulatory gene. In other words, the non-coding RNA is "a RNA gene controlling other protein-coding genes."

Here we will confine our discussion to small RNAs; that is, transcripts of less than 300 nucleotides (nt) that participate directly in RNA processing and degradation, but indirectly in protein synthesis and gene regulation. Because RNA polymerases type II (Pol-II) are inefficient in generating small RNAs of this size, the small RNAs are either directly transcribed by RNA polymerases type III (Pol-III) or indirectly processed from a large transcript of Pol-II.

The diversity of small non-coding RNAs

Transfer RNA

The most prominent example of non-coding RNA is transfer RNA (tRNA) which is involved in the process of translation and is the first type of small RNA identified and characterized (Holley, 1965). Transfer RNA is RNA that transfers a specific amino acid to a growing polypeptide chain at the ribosomal site of protein synthesis during translation. The tRNA is a small RNA, 74–93 nucleotides long, consisting of amino-acid attachment and codon recognition sites, allowing translation of specific amino acids into a polypeptide. The secondary and tertiary structure of tRNAs are cloverleaves with 4–5 domains and an L-shaped three-dimensional (3D) structure, respectively.

Nucleolar RNA

Another example of non-coding RNA is ribosomal RNA. Ribosomal RNA (rRNA) is the primary constituent of ribosomes. The rRNA is transcribed from DNA and in eukaryotes it is processed in the nucleolus before being transported through the nuclear membrane. This type of RNA may produce small nucleolar RNA (snoRNAs), the second type of small RNA. Many of the newly discovered snoRNAs are synthesized in an intron-processing pathway. Several snoRNAs and snoRN proteins (RNPs) are needed for processing of ribosomal RNA, but precise functions remain to be defined. In principle, snoRNAs could have several roles in ribosome synthesis including: folding of pre-rRNA, formation of rRNP substrates, catalyzing RNA cleavages, base modification, assembly of pre-ribosomal subunits, and export of product rRNP particles from the nucleus to the cytoplasm.

The snoRNA acts as a guide to direct pseudouridylation and 2′-O-ribose methylation of ribosomal RNA (rRNA) in the nucleolus. Consequently, the snoRNA guides the snoRNP complex to the modification site of the target rRNA via sequence hybridization. The proteins then catalyze the modification of bases in the rRNA. Therefore, this type of RNA is also called guided RNA.

The snoRNA is also associated with proteins forming part of the mammalian telomerase as well as with proteins involved in imprinting on the paternal chromosomes. It is encoded in introns of genes transcribed by eukaryotic Pol-II, even when some of the host genes do not code for proteins. As a result, the intron, but not the exon, of these genes is evolutionarily conserved in vertebrates. For this reason, some of the introns of the genes employed in plants or invertebrates are still functioning in vertebrates.

The structure of snoRNAs consists of conserved sequences base-paired to their target RNAs. Nearly all vertebrate guide snoRNAs originate from introns of either protein-coding or noncoding RNAs transcribed by Pol-II whereas only a few yeast guide snoRNAs derive from introns, suggesting that introns accumulated during evolution reflect the conservation of transgenes incorporated into the introns, as mentioned above (Filipowicz, 2000; Maxwell and Fournier, 1995; Tycowski *et al.*, 1996). These introns are processed through pathways involving endonucleolytic cleavage by RNase III-related enzymes, exonucleolytic trimming and possibly RNA-mediated cleavage, which occur in large complexes called exosomes (Allmang *et al.*, 1999; van Hoof and Parker, 1999).

Nuclear RNA

Nuclear RNA (snRNA) is a class of small RNA molecules that are found within the nuclei of eukaryotic cells. They are involved in a variety of important processes such as RNA splicing (removal of introns from mRNA precursor) and maintaining the telomeres. They are always associated with specific proteins, and the complexes are referred to as small nuclear ribonucleoproteins (snRNP). Some examples of small nuclear RNA are U2 snRNAs, pre-5S rRNAs, and U6 snRNAs. U2 snRNAs in embryonic stem (ES) cells and pre-5S rRNAs in *Xenopus* oocytes facilitate cell survival after UV irradiation by binding to conserved protein R_0. Eukaryotic U6 snRNAs are the 5 types of spliceosomal RNA involving in mRNA splicing (U1–U6). These small nuclear RNAs have a secondary structure consisting of a stem loop, internal loop, a stem-closing internal loop, and the conserved protein binding site (Frank *et al.*, 1994).

Phage and viral RNA

Another form of small RNA is 30 ribonucleotides in length and functions as a priming initiator for bacteriophage F1 DNA replication (Stavianopoulos *et al.*, 1971; 1972). This function is solely to initiate a given site on the phage DNA, suggesting a primitive defense against foreign pathogen invasion. The phage T4-derived intron is involved in a RNA-RNA interaction in the inhibition of protein synthesis (Wank and Schroeder, 1996).

Small interfering RNA (siRNA)

The small interfering RNA (siRNA) is a small double-stranded RNA molecule, 20–25 nt in length, that suppresses the expression of genes via RNA interference (RNAi) involving the enzyme Dicer. The story of siRNAs began with the observation of pigment expression in the *Petunia* plant. A few years ago, van der Krol *et al.* (1990) tried to intensify flower pigmentation by introducing additional genes, but unexpectedly observed reduced floral pigmentation in some plants, suggesting that gene silencing may be involved in naturally occurring regulation of gene function. This introduction of multiple transgenic copies of a gene into the *Petunia* plant resulted in gene silencing of not only the transgenic, but also the endogenous gene copy, as has been observed by others (Napoli *et al.*, 1990). This suggests co-suppression of homologous genes (the transfer gene and the endogenous gene) in part through methylation (Matzke *et al.*, 1989; Napoli *et al.*, 1990). This phenomenon is termed RNA interference (RNAi). Note that the transgene introduced to the *Petunia* plant is a double-stranded RNA (dsRNA) which is perfectly complementary to the target gene.

When dsRNA was injected into *C. elegans*, Fire and his co-workers noticed gene silencing and RNAi (Fire *et al.*, 1998). It is thought that the dsRNA, once it enters the cells, is cut up by an RNAse III-familial endonuclease, known as Dicer. Dicer consists of an amino terminal helicase domain, a PAZ domain, two RNAse III motifs, and a dsRNA binding motif. Therefore, Dicer binds to the dsRNA and excises the dsRNA into small interfering RNA (siRNAs). These siRNAs locate other single-stranded RNA molecules that are completely complementary to either strand of the siRNA duplex. Then, the RNA-degrading enzymes (RNAses) destroy the RNAs complementary to the siRNAs. This phenomenon is also named posttranscriptional gene silencing (PTGS) or transgene quelling. In other words, gene silencing can be activated by introducing transgenes, RNA viruses or dsRNA sequences that are completely complementary to the targeted gene transcripts.

In mammals, dsRNAs longer than 30 nt will activate an antiviral response, which will lead to the nonspecific degradation of RNA transcripts, the production of interferon, and the overall shutdown of host cell protein synthesis (Shi, 2003). As a result, long dsRNA will not produce gene-specific RNAi activity in mammalian cells (Sui *et al.*, 2002).

Several terms have been used to describe the same or similar phenomenon in different biological systems of different species, including short interfering (si) RNAs (Elbashir *et al.*, 2001), small temporal (st) RNAs (Pasquinelli *et al.*, 2000), heterochromatic siRNAs (Reinhart and Bartel, 2002), and small modulatory dsRNAs (Kuwabara *et al.*, 2004).

MicroRNA (miRNA)

MicroRNAs (miRNAs) are small single-stranded RNA genes possessing the reverse complement of another protein-coding gene's mRNA transcript. These miRNAs can inhibit the expression of the target protein-coding gene. MiRNA was first observed in *C.elegans* as RNA molecules of 18–23 nt that are complementary to the 3′ untranslated regions of the target transcripts, including *lin-4* (Lee *et al.*, 1993) and *let-7* (Lau *et al.*, 2001) genes (see also Figure 2.1). The development of the worm was regulated by these RNA genes. Lee and his co-workers first observed that *lin-4*, a miRNA, is essential for the normal temporal control of diverse postembryonic developmental events by negatively regulating the level of LIN-14 protein (Hong *et al.*, 2000). Another example of miRNA is *let-7* in *C. elegans*, which is required for the transition from late larva to adult by preventing the expression of the *lin-41* gene. The *let-7* RNA was found to be highly conserved in late larval

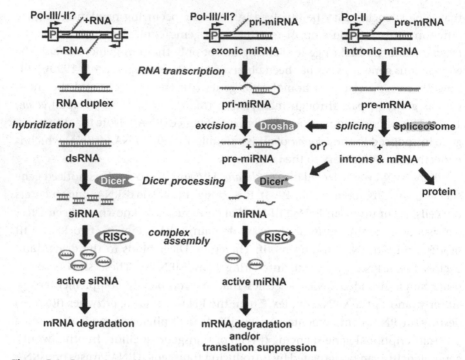

Figure 2.1. Comparison of biogenesis and RNAi mechanisms among siRNA, intergenic (exonic) miRNA and intronic miRNA. SiRNA is likely formed by two perfectly complementary RNAs transcribed from two different promoters (remains to be determined) and further processing into 19–22 base pair duplexes by the RNase III-familial endonuclease, Dicer. The biogenesis of intergenic miRNAs, e.g. *lin-4* and *let-7*, involves a long transcript precursor (pri-miRNA), which is probably generated by Pol-II or Pol-III RNA promoters, while intronic miRNAs are transcribed by the Pol-II promoters of its encoded genes and co-expressed in the intron regions of the gene transcripts (pre-mRNA). After RNA splicing and further processing, the spliced intron may function as a pri-miRNA for intronic miRNA generation. In the nucleus, the pri-miRNA is excised by Drosha RNase to form a hairpin-like pre-miRNA template and then exported to cytoplasm for further processing by Dicer* to form mature miRNAs. The Dicers for siRNA and miRNA pathways are different. All three small regulatory RNAs are finally incorporated into a RNA-induced silencing complex (RISC), which contains either strand of siRNA or the single-strand of miRNA. The effect of miRNA is considered to be more specific and less adverse than that of siRNA because only one strand is involved. On the other hand, siRNAs primarily trigger mRNA degradation, whereas miRNAs can induce either mRNA degradation or suppression of protein synthesis depending on the sequence complementarity to the target gene transcripts.

stages in several species, implicating it as a common regulatory gene which silences the expression of specific genes (Hutvagner *et al.*, 2001). Subsequently, miRNAs were found to occur in diverse organisms ranging from worms, to flies, to humans (Lagos-Quintana *et al.*, 2003), suggesting these molecules represent a gene family that has evolved from an ancient ancestral small RNA gene.

The miRNA is thought to be transcribed from DNA that is not translated but regulates the expression of other genes. Primary transcripts of the miRNA genes, pri-miRNAs, are long RNA transcripts consisting of at least a hairpin-like miRNA precursor. Pri-miRNAs are processed in the nucleus to pre-miRNAs by the ribonuclease Drosha with the help of Microprocessor (Lee *et al.*, 2003) and exported from the nucleus by Exportin-5 (Lund *et al.*, 2003). The 60–90 nt miRNA precursors form the stem and loop structures, and the cytoplasmic RNAse III enzyme Dicer excises the miRNA from the pre-miRNA hairpin stem region. MiRNAs and siRNAs seem to be closely related, especially considering the dsRNA and hairpin structures. The siRNA can be considered as a duplex form of miRNA in which the RNA molecule contains both miRNA and its reverse complement. Therefore, one may consider siRNAs as a kind of miRNA precursor.

MiRNAs suppress gene expression based on their complementarity to a part of one or more messenger RNAs (mRNAs), usually at a site in the 3'-UTR. The annealing of the miRNA to the target mRNA inhibits protein translation. In some cases, the formation of dsRNA through the binding of miRNA triggers the degradation of the mRNA transcript through a process similar to RNAi, though in other cases it is believed that the miRNA complex blocks the protein translation machinery or otherwise prevents protein translation without causing the mRNA to be degraded.

Because most of the miRNA suppresses gene function based on partial complementarity, conceivably one miRNA may target more than one mRNA, and many miRNAs may act on one mRNA, coordinately modulating the intensity of gene expression in various tissues and cells. Therefore, miRNAs may have a broad function in fine-tuning the protein-coding genes. Indeed, the discovery of miRNAs has revolutionized our understanding of gene regulation in the post-genome era.

Intronic miRNA

Introns account for the largest proportion of non-coding sequences in the protein-coding DNA of the genome. The transcription of the genomic protein-coding DNA generates precursor messenger RNA (pre-mRNA), which contains four major parts including the 5'-untranslated region (UTR), the protein-coding exon, the non-coding intron and the 3'-UTR. In broad terms, both 5'- and 3'-UTR can be seen as a kind of intron extension; however, their processing during mRNA maturation is different from that of the intron located between two protein-coding exons, termed the in-frame intron. The in-frame intron was originally thought to be a huge genetic wasteland in gene transcripts, but this stereotypic misconception was abandoned because of the finding of intronic miRNAs. To this day, the biogenesis of intronic miRNAs remains to be fully elucidated.

Some small regulatory RNAs are produced from intronic RNA fragments. For example, snoRNAs are produced from intronic segments from genes encoding ribosomal proteins and nucleolar proteins. In addition, some small RNAs are produced from genes in which exons no longer have the capacity to encode proteins. This type of intron-processing involves RNase III-related enzymes, exonucleolytical trimming, and possibly RNA-mediated cleavage. Thus, intronic miRNA is a new class of miRNAs derived from the processing of the introns of a protein-coding gene.

The major difference between the intronic miRNAs and previously described intergenic miRNAs is the requirement of type II RNA polymerases (Pol-II) and spliceosomal components for the biogenesis of intronic miRNAs (Ying and Lin, 2005a, b). Both intronic and intergenic miRNAs may share the same assembly process, namely the RNA-induced silencing complex (RISC), the effector of RNAi-related gene silencing. Although siRNA-associated RISC assembly has been used to predict miRISC assembly, the link between final miRNA maturation and RISC assembly remains to be determined. The characteristics of Dicer and RISC in siRNA versus miRNA mechanisms are distinctly different (Lee *et al.*, 2004a, b; Tang, 2005).

The intronic miRNAs need to fulfill the following requirements. First, they must share the same promoter with their encoding gene transcripts. Second, they are located in the non-protein-coding region of a primary gene transcript (the pre-mRNA). Third, they are co-expressed with the gene transcripts. Lastly, they are removed from the transcript of their coding genes by nuclear RNA splicing and excision processes to form mature miRNAs.

Certain of the currently known miRNAs are encoded in the genomic intron region of a gene, but they are of an orientation opposite to that of the protein-coding gene transcript. Therefore, these miRNAs are not considered to be intronic miRNAs because they do not share the same promoter with the gene and they are not released from the protein-coding gene transcript by RNA splicing. The promoters of these miRNAs are located in the antisense direction to the gene, probably using the gene transcript as a potential target for antisense miRNAs. A good example of this type of miRNA is *let-7c*, which is an intergenic miRNA located in the antisense region of the intron of a gene.

MiRNAs vs. siRNAs

Differing from the double-stranded siRNA, miRNA is a form of single-stranded RNA which is about 18–25 nt long, but derived from a long RNA transcript of non-coding DNA, termed pri-miRNA. Another difference between siRNAs and miRNAs is that a miRNA is partially complementary to a part of one or more mRNAs, usually at a site in the 3′-UTR whereas siRNA is completely complementary to a protein-coding part of an mRNA. In addition, the formation of the double-stranded RNA through the binding of the miRNA sometimes triggers the degradation of the mRNA transcript through a process similar to RNA interference (RNAi), though in other cases it is believed that the miRNA complex blocks the protein translation machinery or otherwise prevents protein translation without causing the mRNA to be degraded.

The third difference between siRNAs and miRNAs is in their biogenesis (see below). The biogenesis of miRNAs and siRNAs shares some features but also differs in certain respects. Both are cleaved from highly structured or double-stranded RNA precursors by an RNase III (Dicer). However, recent evidence suggests that a different member of the Dicer family may be involved in specific cleavage of pre-miRNAs in both the nucleus and cytoplasm (Khvorova *et al.*, 2003). Furthermore, members of the Agonaunte (AGO) gene family have been implicated in the formation and/or function of both siRNAs and miRNAs. AGO I is a key protein that is required for both the siRNA and miRNA pathways and is likely the endonuclease that cleaves the mRNA targeted by the RISC (Okamura *et al.*, 2004; Kidner and Martienssen, 2005). Elucidating the roles of the full complement of the AGO protein family will reveal further modulation of the RISC, and small RNA regulation more generally. Additional proteins involved in the RISC at the convergence of the PTGS and miRNA pathways have been reported (Meister *et al.*, 2004; Liu *et al.*, 2005).

The siRNA and miRNA biogenesis pathways are also distinct in that the viral suppressor of silencing, Helper component Proteinase (Hc-Pro), eliminates siRNAs but enhances miRNA accumulation (Mallory *et al.*, 2002). Furthermore, there are different genetic requirements for the biogenesis of siRNAs and miRNAs. siRNAs are ~20–22 base pairs in length and mediate RNA silencing. Double-stranded RNA generated during viral replication or derived from aberrant transgenes is digested into siRNAs that direct sequence specific degradation of the corresponding viral or transgene RNA. RNA silencing functions, therefore, to protect the host from viral infection and aberrant gene transfection. In contrast, miRNA over evolutionary time selects segments of transposons for incorporation into the genome for fine-tuning of gene regulation in vertebrates, including human beings. Both siRNAs and miRNAs are also implicated in transcriptional gene silencing and heterochromatin inactivation through a slightly different mechanism (Tang, 2005). This hypothesis for the differences between siRNA and miRNA gene silencing may provide a clue toward explaining the prevalence of native siRNAs in invertebrates but relative scarcity in mammals.

In plants, both siRNAs and miRNAs trigger DNA methylation as well as RNAi (Jones *et al.*, 1999; Béclin *et al.*, 2002; Vaistij *et al.*, 2002). Another function of miRNAs is a specialized non-coding RNA molecule which is X chromosome-encoded. This non-coding RNA is named Xist. Xist is preferentially expressed from only one of the two female X chromosomes and builds up in cis along the chromosome from which it was transcribed. That X chromosome is tightly packaged in transcriptionally inactive heterochromatin; therefore, only one female X-chromosome is active. This phenomenon is associated with DNA methylation. By the same token, the viruses, transgenes and transposons, which have been incorporated into the introns of the mammalian genome during evolution, may take advantage of these characteristics by splicing the pri-miRNAs and incorporating them into Dicer-like proteins for gene silencing and mRNA degradation.

MicroRNAs (miRNAs) are encoded in the genomes of plants and animals and are processed from highly structured precursors. Interestingly, these miRNAs mediate the expression of important developmental regulators, as discussed above. Most miRNAs regulate gene expression at the level of translation, and some appear to alter mRNA stability. Considering the large number of miRNA genes in diverse species, it is likely that some miRNAs regulate gene expression at other levels, such as transcription, mRNA translocation, RNA processing, or even genome accessibility.

The miRNAs have advantages over siRNAs as potential therapeutic reagents (Ying and Lin, 2005a, b). First, double-stranded RNAs (dsRNA) longer than 25 base pairs or high doses of siRNAs produce interferon-associated non-specific

RNA degradation in mammalian cells. Second, short siRNAs are unstable *in vivo* and require pharmacological doses for therapeutic responses in mice. Third, highly stringent complementarity between siRNA and its RNA target is required for RNAi induction, making it impossible to overcome the problem of frequent viral mutation. Fourth, vector-based Pol-III-directed siRNAs occasionally synthesize RNA products longer than the desired siRNA, producing unexpected interferon cytotoxicity. The competition between the Pol-III promoter and another viral promoter of the vector, such as LTR or CMV, can also result in unexpected interferon cytotoxicity. For these reasons, we have developed an intronic miRNA system using Pol-II RNA transcription (see below).

Transposons and intronic miRNA

The intronic and other non-coding RNAs may have evolved to provide a second level of gene expression in eukaryotes, enabling a fine-tuning of the complex network of gene activity. In bacterial and organellar genomes, group II introns contain both catalytic RNAs and retrotransposable elements. The retrotransposable elements make this type of intron mobile. Therefore, these introns are reversely spliced directly into a DNA target site and subsequently reverse transcribed by the intron-encoded gene. After insertion into the DNA, the introns are spliced out of the gene transcript to minimize damage to the host.

There is a potential evolutionary relationship between group II introns and both eukaryotic spliceosomal introns and non-LTR-retrotransposons. Taking advantage of this feature, it is feasible to design mobile group II introns to be incorporated into gene-targeting vectors as "targetrons," to specifically target various genes (Lambowitz and Zimmerly, 2004). There is evidence that introns in *Caenorhabditis* genes are recently gained and some of them are actually derived from "donor" introns present in the same genome. Furthermore, a few of these new introns apparently derive from other introns in the same gene (Coghlan and Wolfe, 2004). Perhaps the splicing machinery determines where introns are added to genes. On the other hand, some newly discovered brain-specific small nucleolar RNAs of unknown function are encoded in introns of tandem repeats and the expression of these introns is paternally imprinted.

From an evolutionary vantage, transposons are probably very old and may exist in the common ancestor genome. They may enter the host multiple times as parasites. This feature of transposons is similar to that of retroviruses. Too much transposon activity can destroy a genome. To counterattack the activity of transposons and viruses, some organisms developed a mechanism to remove and/or silence the activity of transposons and viruses. For example, bacteria frequently delete their genes so that transposons and retroviruses incorporated in the genome are removed. In eukaryotes, miRNA is a way of reducing transposon activity. Conceivably, miRNA may be involved in resistance against viruses, similar to the diversity of antibody production in an immune system.

Identical twins derived from the same zygote have the same genetic information in their nuclear DNA. Any differences between monozygotic twins later in life are mostly the result of environmental influences rather than genetic inheritance. But monozygotic twins may not share all of their DNA sequences. Female monozygotic twins can differ because of differences in X-chromosome inactivation. Consequently, one female twin can have an X-linked condition such as muscular dystrophy and the other twin can be free of it. Monozygotic twins frequently demonstrate slightly, but definitely distinguishing, disease susceptibility and physiology more generally. For example, myotonic dystryophy (DM) is a dominantly inherited, multisystemic disease with a consistent constellation of seemingly unrelated and rare clinical features including myotonia, muscular dystrophy, cardiac conduction defects, posterior iridescent cataracts, and endocrine disorders (Harper, 1989). DM2 (Type 2) is caused by a CCTG expansion (mean ∼5000 repeats) located in intron 1 of the zinc finger protein 9 (*ZNF9*) gene (Liquori *et al.*, 2001). It is possible that monozygotic twins with this disorder display symptom heterogeneity because of varying intronic miRNAs or different levels of insertion of intronic genes.

Class II transposons can cut and paste. The enzyme transposase binds to the ends of the transposon, which are repeats, and the target site on the genome, which is cut to leave sticky ends. These two components are joined together by ligases. In this way, transposons increase the size of the genome because they leave multiple copies of themselves in the genome. It is very possible that transposons are selectively advantageous for the genome in modulating gene regulation via miRNAs. It is not too far-fetched to suggest that when transposons are inserted in the introns of the protein-coding gene, under appropriate conditions, they, a part of them, or their secondary structures, may become intronic miRNAs.

Biogenesis and mechanism of miRNAs

Although the biogenesis of microRNAs is going to be detailed in the later chapters of this book, the five steps that are involved in vertebrates are discussed briefly here. First, miRNA is generated as a long primary precursor miRNA (pri-miRNA), most likely mediated by RNA polymerases type II (Pol-II) (Lin *et al.*, 2003; Lee *et al.*, 2004a, b). The pri-miRNA is transcribed from the genome. Second, the long pri-miRNA is excised by Drosha-like RNase III endonucleases and/or spliceosomal components to form the ∼60–70 nt precursor miRNA (pre-miRNA). The pre-miRNA exhibits considerable secondary structure including regions of imperfectly paired dsRNA, which are sequentially cleaved to one or more miRNAs. This step depends on the origin of the pri-miRNA, whether located in an exon or an intron respectively (Lee *et al.*, 2003; Lin *et al.*, 2003). Third, the pre-miRNA is exported out of the nucleus by Ran-GTP and a receptor, Exportin-5 (Lund *et al.*, 2003; Yi *et al.*, 2003). In the cytoplasm, Dicer-like endonucleases cleave the pre-miRNA to form mature 18–25 nt miRNA. Lastly, the mature miRNA is incorporated into a ribonuclear particle (RNP) to form the RNA-induced gene silencing complex (RISC) which executes RNAi-related gene silencing (Khvorova *et al.*, 2003; Schwarz *et al.*, 2003). Only one of the two strands is the miRNA; the other counterpart is named miRNA*. The mature miRNA can block mRNA translation based on partial complementarity between the miRNA and the targeted mRNA, particularly via base pairing with the 3'-untranslated region of the mRNA. If there is a perfect complementarity between the miRNA and the targeted mRNA, mRNA degradation occurs similarly to that mediated by siRNA. Auto-regulatory negative feedback via miRNAs regulates some genes, including those involved in the RNA silencing mechanism itself.

The assembly of RISC for siRNA has been reported in an *in vitro* system, and a similar assembly probably also occurs for miRNA. But the link between final miRNA maturation and RISC assembly remains unknown. In recent studies using zebrafish, it was demonstrated that the stem loop structure of the pre-miRNAs is involved in strand selection for mature miRNA during RISC assembly. These findings further suggest that the duplex structure of siRNA may not be strictly required for the assembly of miRNA-associated RISC *in vivo*. Conceivably, the stem loop structure of miRNAs may facilitate the selection and maturation processes for the correct miRNA sequence. Proposed pathways for the biogenesis of miRNA are based on the *in vitro* model developed for siRNA. For these reasons, future work needs to focus on distinguishing the individual properties and differences in action of Dicer and RISC in siRNA and miRNA processing.

Identification

Systematic identification of microRNAs is detailed by Hagen and Lai in the following chapter of this book. Currently, there are four major ways to identify miRNAs. They are (a) direct cloning, (b) computer search of the genome, (c) miRNA microarray search in different species, and (d) artificial preparation of man-made miRNA for targeting known gene sequences.

Direct cloning

The conventional direct cloning of short RNA molecules, as in the cloning of *let-7* and *lin-4*, is still the method of choice to identify new miRNAs. Conceivably, one can isolate the small RNAs and sequence them individually. So far, results have been dominated by a few highly expressed miRNAs. But, once the miRNA is identified, its role in other organisms, including human beings, can be explored. For example, *let-7* was originally identified in *C. elegans*. Subsequently, reduced expression of the *let-7* microRNA (Takamizawa *et al.*, 2004) and Dicer (Karube *et al.*, 2005) in human lung cancers suggested that the alteration of *let-7* expression is associated with clinical and biological effects.

Computer search

There are numerous new computational methods that provide ways to estimate the total number of miRNA genes in different animals (McManus, 2003; Gesellchen and Boutros, 2004; Jin *et al.*, 2004; Xu *et al.*, 2004; and several chapters in Part III of this book). Fundamentally, each program identifies highly conserved genomic non-coding regions that possess stem loop structures with specific "seed" sequences, and complementarity of the first 8–10 nt. Then the secondary structure is examined in terms of both the forward and reverse complements of the sequence. In addition, the following criteria help to identify miRNAs: the longest helical arm, free energy of the arm, short internal loops, and asymmetric and bulged loops. The identified miRNAs usually are more heterogeneous than those discovered experimentally, suggesting traditional cloning has a high false negative or miss rate. However, computational techniques may suffer from a high false alarm rate. Therefore, validation of the identified miRNAs by Northern blot analysis and functional study is critical. These methods are still evolving and there is a possibility of one-to-many and many-to-one relationships between the miRNAs and their targets. Potentially thousands of mammalian targets may be identified with this approach.

Microarray search

To facilitate such investigations, an oligonucleotide microchip for genome-wide microRNA profiling in diverse tissues of various species was developed (Liu *et al.*, 2005; Miska *et al.*, 2004; Valoczi *et al.*, 2004; and several chapters in Part V of this book). Some of these chips use locked nucleic acid (LNA)-modified oligonucleotides so as to allow both miRNA *in situ* hybridization and miRNA expression profiling (Valoczi *et al.*, 2004; see also Chapter 17 by Kauppinen in this book). Again, this approach can identify regulation via a large class of microRNAs. A good example is the study of miRNAs regulating brain morphogenesis in zebrafish (Giraldez *et al.*, 2005).

Generation of man-made miRNA

In human and rat cells, we discovered that the biogenesis of miRNA-like precursors from the 5'-proximal intron regions of gene transcripts (i.e. pre-mRNAs) is mediated through the mammalian RNA polymerase type II (Pol-II). We have taken advantage of this observation by developing a unique method to silence gene expression using miRNA-inducing vector cassettes (Lin *et al.*, 2003). Using artificial introns containing hairpin-like miRNA precursors (pre-miRNA), we were able to generate mature miRNA molecules when the intron was inserted into a host gene. These miRNA molecules could trigger gene silencing in human prostate cancer cells, human cervical cancer cells, and rat neuronal stem cells, as well as in zebrafish, chickens, and mice *in vivo*.

Figure 2.2 shows the synthetic intron (SpRNAi) consisting of (from the 5'-end to the 3'-end) a 5'-splice donor site (DS), an inserted pre-miRNA oligonucleotide (insert), a branch-point domain (BrP), a poly-pyrimidine tract (PPT), and a 3'-splice acceptor site (AS). The exons (exon 1 and exon 2) of the red fluorescent membrane protein (rGFP) gene were disrupted because the artificial intron was inserted in between them to form a recombined SpRNAi-rGFP gene cassette. Therefore, the rGFP was not expressed owing to disruption of the rGFP by the insertion of the pre-miRNA. When the miRNA is formed and spliced out, the disrupted rGFP exons are ligated and rGFP is translated. By showing red fluorescence on the cell membranes, we were able to ascertain intron splicing and miRNA maturation. This red fluorescent emission serves as a visual indicator for the generation of intronic miRNAs. Both the intron-derived miRNA system and

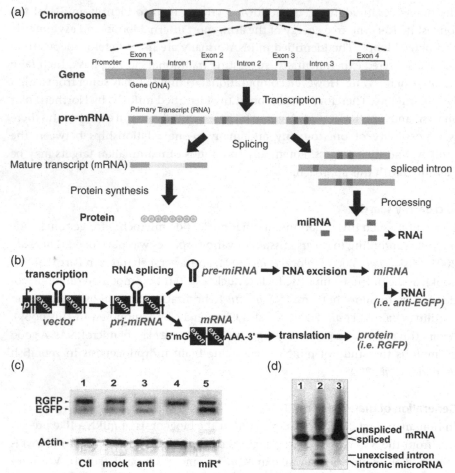

Figure 2.2. Biogenesis and function of intronic miRNAs. (a) The native intronic miRNA is co-transcribed with a precursor messenger RNA (pre-mRNA) by Pol-II and cleaved out of the pre-mRNA by an RNA splicing machinery, spliceosome. The spliced intron with hairpin-like secondary structures is further processed into mature miRNAs capable of triggering RNAi effects, while the ligated exons become a mature messenger RNA (mRNA) for protein synthesis. (b) We designed an artificial intron containing pre-miRNA, namely SpRNAi, mimicking the biogenesis processes of the native intronic miRNAs. (c) When a designed miR-EGFP (280–302)–stemloop RNA construct was tested in the EGFP-expressing Tg(UAS:gfp) zebrafishes, we detected a strong RNAi effect only on the target EGFP (lane 4). No detectable gene silencing effect was observed in other lanes from left to right: 1, blank vector control (Ctl); 2, miRNA–stemloop targeting HIV-p24 (mock); 3, miRNA without stemloop (anti); and 5, stemloop–miRNA* complementary to the miR-EGFP(280–302) sequence (miR*). The off-target genes such as vector RGFP and fish actin were not affected, indicating the high target specificity of miRNA-mediated gene silencing. (d) Three different miR-EGFP(280–302) expression systems were tested for miRNA biogenesis from left to right: 1, vector expressing intron-free RGFP, no pre-miRNA insert; 2, vector expressing RGFP with an intronic 5'-miRNA-stemloop-miRNA*-3' insert; and 3, vector similar to the 2 construct but with a defected 5'-splice site in the intron. In Northern blot analysis probing the miR-EGFP(280–302) sequence, the mature miRNA was released only from the spliced intron resulted from the vector 2 construct in the cell cytoplasm.

Pre-mRNA construct with SpRNAi:

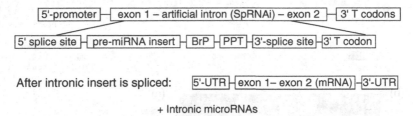

After intronic insert is spliced: 5'-UTR — exon 1 – exon 2 (mRNA) — 3'-UTR

+ Intronic microRNAs

Figure 2.3. Schematic construct of the artificial SpRNAi intron in a recombinant gene SpRNAi–RGFP for intracellular expression and processing. The components of the Pol-II-mediated SpRNAi system include several consensus nucleotide elements consisting of a 5'-splice site, a branch-point domain (BrP), a poly-pyrimidine tract (PPT), a 3'-splice site and a pre-miRNA insert located between the 5'-splice site and the Brp domain. The expression of the recombinant gene is under the regulation of either a mammalian Pol-II RNA promoter or a compatible viral promoter for cell-type-specific effectiveness. Mature miRNAs are released from the intron by RNA splicing and further Dicer processing.

the rGFP are activated, using specific Pol-II promoters. In this manner, we were able to select the gene targeted for silencing.

By the same token, we can use a cell-specific promoter so that the intronic miRNA is co-expressed with its encoding gene in a specific cell population. By activating the specific promoter and expressing the gene, we can generate cell-specific intronic miRNAs. The prerequisites for this type of miRNA generation are (a) the coupled interaction of nascent Pol-II-mediated pre-mRNA transcription and intron excision, occurring within certain nuclear regions proximal to genomic perichromatin fibrils, (b) formation of mature miRNAs after Pol-II RNA splicing and excision processing, and (c) the exons of the encoding gene transcript are ligated together to form a mature mRNA for reporter protein synthesis (e.g. rGFP); see Figures 2.3 and 2.4.

Potential application of man-made miRNAs

Man-made intronic miRNAs have potential applications in (a) analysis of gene function, (b) evaluation of the function and effectiveness of miRNA, (c) design and development of novel gene therapy, (d) design and development of anti-viral vaccines, and (e) development of loss-of-function transgenic animals.

Analysis of gene function

The human genome contains more than 22 billion bases. Analyzing and understanding these sequences is a challenge to all of the natural sciences. With promoter-, cell-specific, and intron-mediated cassettes in the form of plasmids or vectors, together with computer-identified gene sequences and potential miRNA sequences, one can explore the functional significance of these sequences both *in vitro* and *in vivo* with targeted candidates. In addition, the significant role of a particular molecule in the signal pathway or stage of development of a disease can be elucidated with this approach, particularly the downstream mechanism of action.

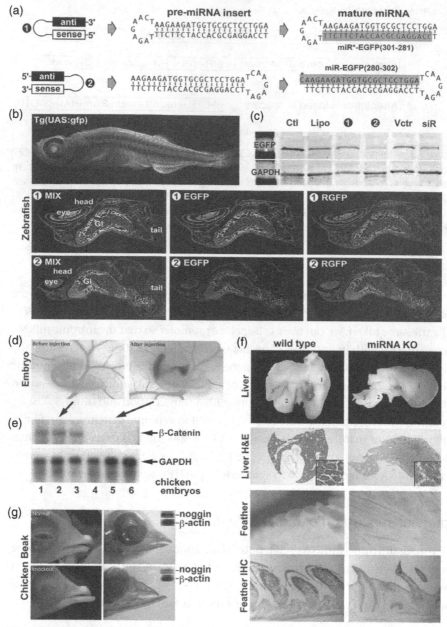

Figure 2.4. Intronic miRNA-mediated gene silencing effects *in vivo*. (a)–(c) Different preferences of RISC assembly were observed by transfection of 5'-miRNA*-stemloop-miRNA-3' (❶) and 5'-miRNA–stemloop–miRNA*–3' (❷) pre-miRNA structures in zebrafish, respectively. (a) One mature miRNA, namely miR-EGFP(280/302), was detected in the ❷-transfected zebrafishes, whereas the ❶-transfection produced another kind of miRNA, miR*–EGFP(301-281), which was partially complementary to the miR-EGFP(280/302). (b) The RNAi effect was only observed in the transfection of the ❷ pre-miRNA, showing less EGFP (green) in ❷ than ❶, while the miRNA indicator RGFP (red) was evenly present in all vector transfections. (c) Western blot analysis of the EGFP protein levels confirmed the specific silencing result of (b). No detectable gene silencing was observed in fishes without (Ctl) and with liposome only (Lipo) treatments. The transfection of either a U6-driven siRNA

Evaluation of miRNA function and effectiveness

Prediction of miRNA candidates using computer programming has identified thousands of genomic hairpin sequences. However, the function of these sequences remains to be determined. Because direct transfection of hairpin-like miRNA precursors (pre-miRNAs) in mammalian cells is not always sufficient to trigger effective RISC assembly, a key step for RNAi-related gene silencing, the utilization of the intronic miRNA-expressing system can solve this problem and successfully increased the efficiency and effectiveness of miRNA-associated RNA interference (RNAi) induction *in vitro* and *in vivo*.

Based on the strand complementarity between the designed miRNA and its target gene sequence, we have also developed a miRNA isolation protocol to purify and identify the mature miRNAs generated by the intronic miRNA-expressing system. Several intronic miRNA identities and structures have now been confirmed active *in vitro* and *in vivo*. According to this proof-of-principle method, we now have the knowledge needed to design more efficient and effective pre-miRNA inserts for the intronic miRNA-expressing system.

Design and development of novel gene therapy

The intronic miRNAs that trigger RNAi may be usable as drugs. For example, miRNA could repress essential genes in eukaryotic human pathogens or viruses that are dissimilar from any human genes. Intronic miRNA is inserted in the non-coding region of genomic DNA and must be activated by RNA transcription and splicing machineries of the host cell; this process is strictly controlled and compatible to the gene regulation mechanism of the cell. Proponents of therapies based on intronic RNA suggest that this approach may alleviate some patients' concerns about alteration of their normal genome activity. Therefore, this method of treatment would likely be safe because no protein-coding gene is affected. For this reason, intronic miRNAs and therapies based on this approach

---

Figure 2.4. (cont.)
vector (siR) or an empty vector (Vctr) without the designed pre-miRNA insert resulted in no gene silencing significance. (d)–(g) Silencing of endogenous β-catenin and noggin genes in chicken embryos. (d) The pre-miRNA construct and fast green dye mixtures were injected into the ventral side of chicken embryos near the liver primordia below the heart. (e) Northern blot analysis of extracted RNAs from chicken embryonic livers with anti-β-catenin miRNA transfections (lanes 4–6) in comparison with wild types (lanes 1–3) showed a more than 98% silencing effect on β-catenin mRNA expression, while the house-keeping gene, GAPDH, was not affected. (f) Liver formation of the β-catenin knockouts was significantly hindered (upper right two panels). Microscopic examination revealed a loose structure of hepatocytes, indicating the loss of cell-cell adhesion due to breaks in adherins junctions fromed between β-catenin and cell membrane E-cadherin in early liver development. In severely affected regions, feather growth in the skin close to the injection area was also inhibited (lower right two panels). Immunohistochemistry staining of β-catenin protein expression (brown) showed a significant decrease in the feather follicle sheaths. (g) The lower beak development was increased by the mandible injection of the anti-noggin pre-miRNA construct (down panel) in comparison to the wild type (up panel). Right panels showed bone (alizarin red) and cartilage (alcian blue) staining to demonstrate the outgrowth of bone tissues in the lower beak of the noggin knockout. Northern blot analysis (small windows) confirmed a ∼60% decrease of noggin mRNA expression in the lower beak area. (See color plate 1)

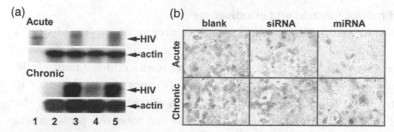

Figure 2.5. Silencing of HIV-1 genome replication using anti-*gag/pro/pol* miRNA transfections into CD4$^+$ T lymphocytes isolated from the acute and chronic phases of AIDS infections. (a) Northern blot analysis showed about 98% and 70% decreases of HIV genome in the acute and chronic infections after miRNA treatments (lanes 4), respectively. No effect was detected in the T cells transfected by miRNA* targeting the same *gag/pro/pol* region of the viral genome (lane 5). The size of pure HIV-1 provirus was measured as about 9700 nucleotide bases (lanes 1). RNA extracts from normal non-infected CD4$^+$ Th lymphocytes were used as a negative control (lanes 2), whereas those from HIV-infected T cells were used as a positive control (lanes 3). (b) Immunostaining of HIV p24 marker confirmed the results of (a). Since the *ex vivo* HIV-silenced T lymphocytes were resistant to any further infection by the same strains of HIV, they may be transfused back to the donor patient for eliminating HIV-infected cells.

may be useful or the design and development of non-invasive gene therapy against inheritable diseases such as myotonic dystrophy, fragile X mental retardation syndrome, Alzheimer's disease, Parkinson's disease and cancers.

Design and development of anti-viral vaccine

The problems in designing an effective vaccine to prevent viral infection such as HIV are as follows. First, the global prevalence of this epidemic is mainly caused by the high mutation rate of the HIV genome that gradually generates more resistant strains to highly active antiretroviral therapy. Second, the HIV provirus is capable of integrating into a host cellular genome to escape from the inhibitory effects of the treatment, which inactivates the viral replication cycle but does not destroy the latent viral genome. The result is an increased number of HIV carriers. Such an increase in drug-resistant HIV strains and their carriers has posed great challenges and financial burdens on AIDS prevention programs. In contrast with traditional anti-HIV drugs, miRNAs can trigger either translation repression or RNA degradation depending on their degree of complementarity with the target genes. For these reasons, our approach using vector-based miRNA to overcome the complications of HIV mutation has proven feasible (Lin *et al.*, 2005). We believe the knowledge obtained from this type of approach will facilitate the development of anti-viral drugs and vaccines against HIV infections (see Figure 2.5).

Development of loss-of-function transgenic animals

Transgenic animal models are valuable tools for testing gene functions and drug mechanisms *in vivo*. Using the intronic miRNA described above, one can establish loss-of-function models in zebrafish and mice. Loss-of-function in transgenic zebrafish cannot be achieved with siRNA, but has been demonstrated with intronic miRNAs (Lin *et al.*, 2005, Ying and Lin, 2005b). The zebrafish, possessing numerous features similar to human biological systems, is most suitable for genetic and pathological studies of human diseases, particularly study of the molecular mechanisms by which the loss of a specific gene causes a disease or disorder. In addition, all pharmaceutically developed drugs can be screened with this approach in a loss-of-function transgenic zebrafish model.

Concluding remarks

The unique features of the synthetic intronic miRNA cassette are as follows. First, it is compatible with Pol-II RNA transcription for tissue-specific and gene-specific targeting and gene silencing. Second, this approach is compatible with drug-inducible promoters such as the Tet-On or Tet-Off system. Third, the intronic miRNA described possesses interchangeable promoters and vectors. Fourth, the intronic pre-miRNA inserts are also interchangeable. Fifth, the cassettes can be made with multiple pre-miRNA (cluster) insertion so that a cluster of genes can be tested at the same time. Finally, the fluorescent indication of GFP proteins yields an instant visual aid representing miRNA levels.

ACKNOWLEDGMENTS

This study was supported by NIH/NCI grant CA-85722.

REFERENCES

Allmang, C., Kufel, J., Chanfreau, G. *et al.* (1999). Functions of the exosome in rRNA, snoRNA and snRNA synthesis. *European Molecular Biology Organization Journal*, **18**, 5399–5410.

Béclin, C., Boutet, S., Waterhouse, P. and Vaucheret, H. (2002). A branched pathway for transgene-induced RNA silencing in plants. *Current Biology*, **12**, 684–688.

Coghlan, A. and Wolfe, K. H. (2004). Origins of recently gained introns in Caenorhabditis. *Proceedings of the National Academy of Sciences of the USA*, **101**, 11 362–11 367.

Elbashir, S. M., Lendeckel, W. and Tuschl, T. (2001). RNA interference is mediated by 21- and 22-nucleotide RNAs. *Genes & Development*, **15**, 188–200.

Filipowicz, W. (2000). Imprinted expression of small nucleolar RNAs in brain: time for RNomics. *Proceedings of the National Academy of Sciences of the USA*, **97**, 14 035–14 037.

Fire, A., Xu, S., Montgomery, M. K. *et al.* (1998). Potent and specific genetic interference by double-stranded RNA in *Caenorhabditis elegans*. *Nature*, **391**, 806–811.

Frank, D. N., Roiha, H. and Guthrie, C. (1994). Architecture of the U5 small nuclear RNA. *Molecular and Cellular Biology*, **14**, 2180–2190.

Gesellchen, V. and Boutros, M. (2004). Managing the genome: microRNAs in *Drosophila*. *Differentiation*, **72**, 74–80.

Giraldez, A. J., Cinalli, R. M., Glasner, M. E. *et al.* (2005). MicroRNAs regulate brain morphogenesis in zebrafish. *Science*, **308**, 833–838.

Harper, P. S. (1989). *Myotonic Dystrophy*, 2nd edn. London: Saunders.

Holley, R. W. (1965). Structure of an alanine transfer ribonucleic acid. *The Journal of the American Medical Association*, **194**, 868–871.

Hong, Y., Lee, R. C. and Ambros, V. (2000). Structure and function analysis of *LIN-14*, a temporal regulator of postembryonic developmental events in *Caenorhabditis elegans*. *Molecular and Cellular Biology*, **20**, 2285–2295.

Hutvágner, G., McLachlan, J., Pasquinelli, A. E. *et al.* (2001). A cellular function for the RNA-interference enzyme Dicer in the maturation of the let-7 small temporal RNA. *Science*, **293**, 834–838.

Jin, P., Alisch, R. S. and Warren, S. T. (2004). RNA and microRNAs in fragile X mental retardation. *Nature Cell Biology*, **6**, 1048–1053.

Jones, L., Hamilton, A. J., Voinnet, O. *et al.* (1999). RNA–DNA interactions and DNA methylation in post-transcriptional gene silencing. *Plant Cell*, **11**, 2291–2301.

Karube, Y., Tanaka, H., Osada, H. *et al.* (2005). Reduced expression of Dicer associated with poor prognosis in lung cancer patients. *Cancer Science*, **96**, 111–115.

Khvorova, A., Reynolds, A. and Jayasena, S. D. (2003). Functional siRNAs and miRNAs exhibit strand bias. *Cell*, **115**, 209–216.

Kidner, C. A. and Martienssen, R. A. (2005). The role of ARGONAUTE1 (AGO1) in meristem formation and identity. *Developmental Biolology*, **280**, 504–517.

Kuwabara, T., Hsieh, J., Nakashima, K., Taira, K. and Gage, F. H. (2004). A small modulatory dsRNA specifies the fate of adult neural stem cells. *Cell*, **116**, 779–793.

Lagos-Quintana, M., Rauhut, R., Meyer, J., Borkhardt, A. and Tuschl, T. (2003). New microRNAs from mouse and human. *RNA*, **9**, 175–179.

Lambowitz, A. M. and Zimmerly, S. (2004). Mobile group II introns. *Annual Review of Genetics*, **38**, 1–35.

Lau, N. C., Lim, L. P., Weinstein, E. G. and Bartel, D. P. (2001). An abundant class of tiny RNAs with probable regulatory roles in *Caenorhabditis elegans*. *Science*, **294**, 858–862.

Lee, R. C., Feinbaum, R. L. and Ambros, V. (1993). The C. elegans heterochronic gene *lin-4* encodes small RNAs with antisense complementarity to *lin-14*. *Cell*, **75**, 843–854.

Lee, Y., Ahn, C., Han, J. *et al.* (2003). The nuclear RNase III Drosha initiates microRNA processing. *Nature*, **425**, 415–419.

Lee, Y., Kim, M., Han, J. *et al.* (2004a). MicroRNA genes are transcribed by RNA polymerase II. *European Molecular Biology Organization Journal*, **23**, 4051–4060.

Lee, Y. S., Nakahara, K., Pham, J. W. *et al.* (2004b). Distinct roles for *Drosophila* Dicer-1 and Dicer-2 in the siRNA/miRNA silencing pathways. *Cell*, **117**, 69–81.

Lin, S. L., Chang, D., Wu, D. Y. and Ying, S. Y. (2003). A novel RNA splicing-mediated gene silencing mechanism potential for genome evolution. *Biochemical and Biophysical Research Communications*, **310**, 754–760.

Lin, S. L., Huang, F. T., Chang, D., Ji, H. H. and Ying, S. Y. (2004). Combinational therapy for HIV-1 eradication and vaccination. *International Journal of Oncology*, **24**, 81–88.

Lin, S. L., Chang, D. and Ying, S. Y. (2005). Asymmetry of intronic pre-miRNA structures in functional RISC assembly. *Gene*, **356**, 32–38.

Liquori, C. L., Ricker, K., Moseley, M. L. *et al.* (2001). Myotonic dystrophy type 2 caused by a CCTG expansion in intron 1 of ZNF9. *Science*, **293**, 864–867.

Liu, J., Valencia-Sanchez, M. A., Hannon, G. J. and Parker, R. (2005). MicroRNA-dependent localization of targeted mRNAs to mammalian P-bodies. *Nature Cell Biology*, **7**, 719–723.

Lund, E., Guttinger, S., Calado, A., Dahlberg, J. E. and Kutay, U. (2003). Nuclear export of microRNA precursors. *Science*, **303**, 95–98.

McManus, M. T. (2003). MicroRNAs and cancer. *Seminars in Cancer Biology*, **13**, 253–258.

Mallory, A. C., Reinhart, B. J., Bartel, D., Vance, V. B. and Bowman, L. H. (2002). A viral suppressor of RNA silencing differentially regulates the accumulation of short interfering RNAs and micro-RNAs in tobacco. *Proceedings of the National Academy of Sciences of the USA*, **99**, 15 228–15 233.

Matzke, M. A., Primig, M. J., Trnovsky, J. and Matzke, A. J. M. (1989). Reversible methylation and inactivation of marker genes in sequentially transformed tobacco plants. *European Molecular Biology Organization Journal*, **8**, 643–649.

Maxwell, E. S. and Fournier, M. J. (1995). The small nucleolar RNAs. *Annual Review of Biochemistry*, **64**, 897–934.

Meister, G., Landthaler, M., Patkaniowska, A. *et al.* (2004). Human Argonaute2 mediates RNA cleavage targeted by miRNAs and siRNAs. *Molecular Cell*, **15**, 185–197.

Miska, E. A., Alvarez-Saavedra, E., Townsend, M. *et al.* (2004). Microarray analysis of microRNA expression in the developing mammalian brain. *Genome Biology*, **5**, R68.

Napoli, C., Lemieux, C. and Jorgensen, R. A. (1990). Introduction of a chimeric chalcone synthase gene into *Petunia* results in reversible co-suppression of homologous genes in trans. *Plant Cell*, **2**, 279–289.

Okamura, K., Ishizuka, A., Siomi, H. and Siomi, M. C. (2004). Distinct roles for Argonaute proteins in small RNA-directed RNA cleavage pathways. *Genes & Development*, **18**, 1655–1666.

Pasquinelli, A. E., Reinhart, B. J., Slack, F. *et al.* (2000). Conservation of the sequence and temporal expression of *let-7* heterochronic regulatory RNA. *Nature*, **408**, 86–89.

Reinhart, B. J. and Bartel, D. P. (2002). Small RNAs correspond to centromere heterochromatic repeats. *Science*, **297**, 1831.

Schwarz, D. S., Hutvagner, G., Du, T. *et al.* (2003). Asymmetry in the assembly of the RNAi enzyme complex. *Cell*, **115**, 199–208.

Shi, Y. (2003). Mammalian RNAi for the masses. *Trends in Genetics*, **19**, 9–12.

Stavianopoulos, J. G., Karkus J. D. and Charguff, E. (1971). Nucleic acid polymerase of the developing chicken embryos: a DNA polymerase preferring a hybrid template. *Proceedings of the National Academy of Sciences of the USA*, **68**, 2207–2211.

Stavianopoulos, J. G., Karkus J. D. and Charguff, E. (1972). Mechanism of DNA replication by highly purified DNA polymerase of chicken embryos. *Proceedings of the National Academy of Sciences of the USA*, **69**, 2609–2613.

Sui, G., Soohoo, C., Affar, el B. *et al.* (2002). A DNA vector-based RNAi technology to suppress gene expression in mammalian cells. *Proceedings of the National Academy of Sciences of the USA*, **99**, 5515–5520.

Takamizawa, J., Konishi, H., Yanagisawa, K. *et al.* (2004). Reduced expression of the let-7 microRNAs in human lung cancers in association with shortened postoperative survival. *Cancer Research*, **64**, 3753–2756.

Tang, G. (2005). siRNA and miRNA: an insight into RISCs. *Trends in Biochemical Sciences*, **30**, 106–114.

Tycowski, K. T., Shu, M. D. and Steitz, J. A. (1996). A mammalian gene with introns instead of exons generating stable RNA products. *Nature*, **379**, 464–466.

Vaistij, F. E., Jones, L. and Baulcombe, D. C. (2002). Spreading of RNA targeting and DNA methylation in RNA silencing requires transcription of the target gene and a putative RNA-dependent RNA polymerase. *Plant Cell*, **14**, 857–867.

Valoczi, A., Hornyik, C., Varga, N. *et al.* (2004). Sensitive and specific detection of microRNAs by northern blot analysis using LNA-modified oligonucleotide probes. *Nucleic Acids Research*, **32**, e175.

van der Krol, A. R., Mur, L. A., Beld, M., Mol, J. N. and Stuitje, A. R. (1990). Flavonoid genes in petunia: addition of a limited number of gene copies may lead to a suppression of gene expression. *Plant Cell*, **2**, 291–299.

van Hoof, A. and Parker, R. (1999). The exosome: a proteasome for RNA? *Cell*, **99**, 347–350.

Wank, H. and Schroeder, R. (1996). Antibiotic-induced oligomerisation of group I intron RNA. *Journal of Molecular Biology*, **258**, 53–61.

Xu, P., Guo, M. and Hay, B. A. (2004). MicroRNAs and the regulation of cell death. *Trends in Genetics*, **20**, 617–624.

Yi, R., Qin, Y., Macara, I. G. and Cullen, B. R. (2003). Exportin-5 mediates the nuclear export of pre-microRNAs and short hairpin RNAs. *Genes & Development*, **17**, 3011–3016.

Ying, S. Y. and Lin, S. L. (2005a). Intronic microRNAs (miRNAs). *Biochemical and Biophysical Research Communications*, **326**, 515–520.

Ying, S. Y. and Lin, S. L. (2005b). MicroRNA: fine-tunes the function of genes in zebrafish. *Biochemical and Biophysical Research Communications*, **335**, 1–4.

3 Seeing is believing: strategies for studying microRNA expression

Joshua W. Hagen and Eric C. Lai*

Introduction

Studies during the early 1990s uncovered a novel mechanism by which *lin-4* inhibits the nuclear factor encoded by *lin-14* to promote the transition between the first and second larval stages of *C. elegans* development. In particular, *lin-4* encodes a small RNA that binds to multiple sites in the 3′ untranslated region (3′-UTR) of the *lin-14* transcript, thereby negatively regulating *lin-14* at a post-transcriptional level (Lee *et al.*, 1993; Wightman *et al.*, 1993). Nearly a decade would pass before it became fully evident that *lin-4* was actually the prototype of a novel and extensive class of regulatory RNA, now collectively referred to as the microRNA (miRNA) family (Lagos-Quintana *et al.*, 2001; Lau *et al.*, 2001; Lee and Ambros, 2001; Reinhart *et al.*, 2000). These miRNAs are ~21–24 nucleotide RNAs that are processed from precursor transcripts containing a characteristic hairpin structure, and have been identified in diverse animals, plants and even viruses (Bartel, 2004; Griffiths-Jones *et al.*, 2006; Lai, 2003). MiRNAs now constitute one of the largest gene families known, with hundreds to perhaps a thousand or more genes in individual species.

Knowledge of temporal and spatial elements of gene expression is essential for a comprehensive understanding of gene function, whether in the context of normal physiology or pathology. With whole genome sequences and extensive databases of expressed sequences in hand, the systematic analysis of mRNA expression patterns using microarrays, *in situ* hybridization, and even promoter fusions is well underway. Only in recent years, however, have such techniques been successfully transferred to the study of miRNA expression. The goal of this chapter is to review these techniques, particularly as they have been applied to animal systems.

Biogenesis of miRNAs

Although this chapter will not focus on the mechanism of miRNA biogenesis, it is necessary to review this pathway briefly in order to place the different strategies

---

* Author to whom correspondence should be addressed.

MicroRNAs: From Basic Science to Disease Biology, ed. Krishnarao Appasani. Published by Cambridge University Press. © Cambridge University Press 2008.

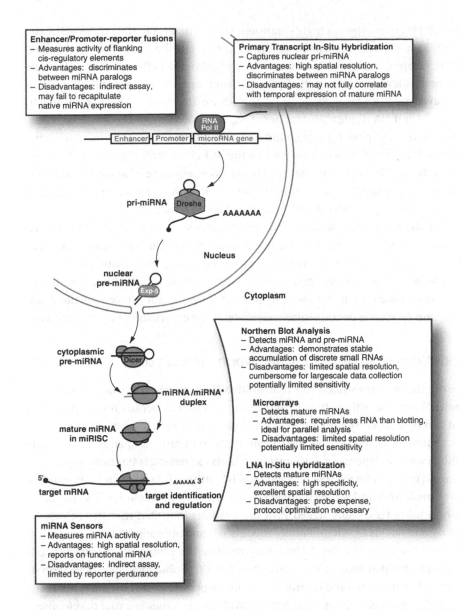

Figure 3.1. The microRNA biogenesis pathway. Primary miRNA transcripts are produced by RNA polymerase II and include a hairpin structure that is a substrate for the nuclear RNAse III Drosha. Cleaved pre-miRNA hairpins are exported from the nucleus by Exportin-5 (Exp-5). In the cytoplasm, pre-miRNA hairpins are cleaved again by Dicer and assemble into the miRISC, which mediates negative post-transcriptional gene regulation. The boxed texts summarize various miRNA detection techniques that are described in greater detail during this review.

for miRNA expression analysis in context (Figure 3.1). For further details on this pathway, see also Chapter 34 by Scherr and Eder in this book. A significant fraction of miRNA hairpin sequences reside on the sense strand of introns and may be processed from primary transcripts of their host mRNAs. However, independently transcribed miRNAs are also transcribed by RNA polymerase II and possess 5′ cap and 3′ polyadenylated tails (Lee *et al.*, 2004). The primary miRNA (pri-miRNA) transcript is a highly transient species that is rapidly processed by the class II RNase III enzyme Drosha (Lee *et al.*, 2003). Drosha cleaves within the base of the hairpin, thereby releasing the pre-miRNA and defining one end of the eventual miRNA. Following its export to the cytoplasm, the pre-miRNA hairpin is cleaved near its terminal loop by the RNase III enzyme Dicer, resulting in a ~21–24 miRNA duplex (Hutvagner *et al.*, 2001; Ketting *et al.*, 2001; Lee *et al.*, 2002). Generally speaking, one strand of the pre-miRNA duplex, thought to be the active miRNA species, accumulates to a much higher level than does the other strand, termed the miRNA* species. This is a consequence of asymmetric incorporation of one specific strand into the miRISC (Schwarz *et al.*, 2003), a silencing complex that is guided by the miRNA to its target transcripts.

Many steps along the pathway of miRNA biogenesis are amenable for functional analyses of miRNA expression, from detection of promoter activity, hybridization to the nuclear primary miRNA transcript or to the cytoplasmic mature miRNA, to visualization of the negative regulatory activity of the mature miRNA (Figure 3.1). As we shall discuss, each of these entry points has certain advantages and disadvantages, and each gives distinct information about the expression of a given miRNA.

miRNA detection by cloning and Northern analysis

The original forward strategy for miRNA identification was to clone and sequence small RNAs and map them back to the genome, in order to identify those small RNAs that can be associated with a deduced hairpin precursor transcript. As there is no strict cutoff on how extensive the hairpin must be in order to qualify as a genuine miRNA, some independent demonstration of its expression is typically required. A stringent criterion for expression validation is Northern blot analysis to show that a given small RNA indeed accumulates as a discrete and stable species. Northern analysis using temporally staged RNA preparations indicated that the expression of *lin-4* and *let-7* is developmentally modulated, and appropriate temporal activation of these miRNAs is key to how they regulate the timing of stage-specific cell lineages (Lee and Ambros, 2001; Reinhart *et al.*, 2000). Subsequent Northern analysis presented in the initial trio of miRNA cloning papers (Lagos-Quintana *et al.*, 2001; Lau *et al.*, 2001; Lee and Ambros, 2001) provided a clear message that developmentally modulated expression of miRNAs is frequently observed.

With invertebrate model organisms such as flies and worms, it is technically challenging to isolate large amounts of RNA from micro-dissected tissues. With vertebrates, however, it is straightforward to prepare sufficient RNA from specific tissues and organs for use in cloning and Northern analysis. Efforts of this nature in 2002 provided general evidence that the expression of individual miRNAs is often restricted to individual or limited organ subsets (Lagos-Quintana *et al.*, 2002). In this study, small RNA libraries were prepared from a variety of murine organs and tissues, including heart, liver, small intestine, colon, spleen, midbrain, brain cortex and cerebellum. The cloning frequency of certain miRNAs was strongly biased in certain tissues, which was taken as a reflection of tissue-specificity. For example, miR-122 represented 72% of all miRNAs cloned from liver,

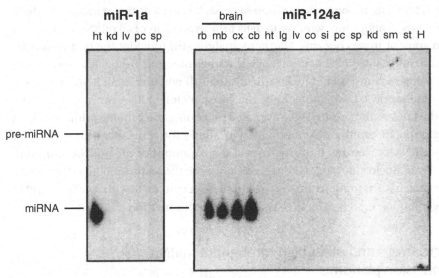

Figure 3.2. Northern analysis of organ-specific RNA preparations reveals tissue-specific miRNA expression. In this example, murine miR-1a exhibits heart-specific expression (left) while miR-124a is largely restricted to the brain. Note that while expression of the mature ~21 nt miRNA predominates, a small amount of ~65 nt pre-miRNA is also detected. Abbreviations: heart (ht), kidney (kd), liver (lv), pancreas (pc), spleen (sp), colon (co), small intestine (si), smooth muscle (sm), HeLa cells (H), lung (lg), midbrain (mb), cortex (cx), cerebellum (cb), and rest of brain (rb). Figure reproduced, with permission, from Lagos-Quintana *et al.* (2002).

suggesting that it is a major liver small RNA. Tissue-specificity was verified by Northern blots of individual miRNAs across the different tissues analyzed. For example, miR-122a was expressed solely in liver, miR-1 was restricted to heart, while miR-124 was found to be relatively specific to brain (Lagos-Quintana *et al.*, 2002; Lee and Ambros, 2001) (see Figure 3.2). miRNAs can also show regional expression within an individual organ. For example, mir-375, originally cloned from a pancreatic islet cell line, proved to be expressed specifically in pancreatic islet cells but not within exocrine pancreas (Poy *et al.*, 2004). Analysis of miRNAs by Northern blots reached a peak in 2004, when hundreds of blots were completed for a single study to generate detailed information on the tissue specificity of mammalian miRNAs (Sempere *et al.*, 2004).

Although a method that is tried-and-true, Northern blot analysis has a number of limitations. First is that it provides a relatively crude picture of spatially modulated gene expression. Many genes are expressed only in a subset of cells in a given tissue, and such spatially restricted information is lost when preparing RNA from a dissected organ. Moreover, the preparation of organ-specific RNA itself is often impractical, especially from smaller model organisms. Second, the restricted hybridization conditions dictated by the small size of miRNA probes prevents this method from distinguishing amongst closely related miRNA sequences. A third consideration is the somewhat labor-intensive nature of Northern analysis, which makes it less desirable for tackling genome-wide profiling efforts. Finally, not all miRNAs that can be detected by sequencing can be detected by Northern analysis. Whether this reflects that a given miRNA is generally expressed at very low levels, is expressed in only a few cells, or some combination of these, it is evident that standard Northern analysis is insufficiently sensitive to detect the expression of all miRNAs.

In spite of these concerns, Northern analysis still provides the best means for visualizing stable RNAs of discrete sizes. Other methods (to be described in the following sections) do not necessarily distinguish whether small RNAs of varying sizes are produced, which might be the case if a cloned RNA was not in fact the product of specific Drosha/Dicer-mediated processing of a genuine primary miRNA transcript. In addition, Northern analysis allows for convenient monitoring of pre-miRNA processing. In many cases, a small amount of pre-miRNA hairpin in the 65–75 nucleotide range is visualized along with the mature product (Figure 3.2). The effects of mutations or tissue/temporal-specific factors that affect miRNA biogenesis are therefore ideally visualized with this method.

Microarrays and other high-throughput miRNA expression platforms

A potential solution to the low-throughput and high RNA-input nature of Northern analysis is to adapt microarray technology. The major technical issue in using microarrays lies in efficiently labeling small RNAs. Unlike mRNAs, miRNAs cannot directly be reverse transcribed using dT-linked primers since they lack poly-adenylated tails, while their short length precludes the use of random priming. The first report described a miRNA microarray containing oligonucleotides spotted onto membranes, which were probed with small RNAs directly end-labeled with γ^{33}P dATP using polynucleotide kinase (Krichevsky *et al.*, 2003). Most subsequent microarray platforms, however, make use of oligo-spotted glass slides probed with fluorescently labeled small RNAs. This format allows microarrays to be analyzed with scanners designed for conventional microarrays, and has the advantage of allowing a reference probe set labeled with a different fluor to be hybridized simultaneously as a normalization control.

The tremendous interest in adapting microarray technology for miRNA expression analysis has resulted in many independent solutions for labeling and detecting small RNAs (see also Chapter 1 by Alvarez-Garcia and Miska). In some methods, small RNAs are labeled directly by ligating a fluorescently modified dinucleotide to their 3′ ends using T4 RNA ligase (Thomson *et al.*, 2004), or by labeling internal guanine residues using fluorescently labeled cisplatin derivatives (Babak *et al.*, 2004). Added sensitivity may be provided by incorporating multiple fluors using a 3′ tailing approach (Shingara *et al.*, 2005). In other methods, small RNAs are cloned and subjected to limited PCR amplification using fluorescently labeled primers (Barad *et al.*, 2004; Baskerville and Bartel, 2005; Miska *et al.*, 2004); this may potentially reduce the amount of input RNA required. Bead-based methods for solution hybridization of small RNA pools have also been devised (Barad *et al.*, 2004; Lu *et al.*, 2005). In this platform, capture beads are impregnated with dye mixtures that can be distinguished by flow cytometry; more than 100 distinct bead colors can be generated, one for each miRNA to be analyzed. Each bead is linked to oligonucleotides complementary to a different miRNA, and the amount of hybridization to each bead can be detected using a tag incorporated into the cloned small RNAs.

A concern of these high-throughput miRNA expression analyses is whether they accurately discriminate amongst related miRNAs, a known concern of Northern analysis. Members of a miRNA family typically differ the most at their 3′ ends, and the RAKE (RNA-primed, array-based Klenow enzyme) assay takes advantage of this property to differentiate small RNAs (Nelson *et al.*, 2004). In this method, small RNAs are hybridized to spotted DNA probes and extended on-slide with labeled nucleotides using Klenow enzyme. Because full complementarity at the miRNA 3′ end is required for Klenow extension to generate a labeled product, this method minimizes cross-detection of miRNAs with distinct 3′ ends. A modification of the Taqman assay that strongly discriminates amongst related miRNAs has also been described (Chen *et al.*, 2005). As reviewed in greater detail by Chen and colleagues in Chapter 20, a designed stem-loop primer with complementarity to the 3′ end of the miRNA is used for miRNA reverse transcription, followed by real-time PCR for quantitation.

Finally, the latest strategy for miRNA quantitation involves hybridizing two spectrally distinct probes to the 5′ and 3′ halves of a given miRNA (Neely *et al.*, 2006). These short probes include locked nucleic acid (LNA) substitutions that increase their affinity for complementary sequences; unbound probes are subsequently hybridized with complementary probes bearing fluorescent quenchers. A custom fluidics platform is then used to analyze single molecules of the hybridized solution in order to detect coincident events of the two fluors, which should correspond to an individual miRNA molecule hybridized to the 5′ and 3′ fluors. Because there is no amplification involved, and each miRNA is detected individually, this platform allows for higher sensitivity (in the 500 femtomolar range) and greater linear dynamic range (over three log orders) compared to most other miRNA profiling methods (see also Chapter 19 by Neely and colleagues).

Usage of these profiling methods has rapidly increased our knowledge of developmental and tissue-specific expression patterns of miRNAs. In addition, there is great current interest to use them to extend gene expression signatures associated with human diseases. The miRNA expression profiling of tumor cell lines and primary tumors has already established that unique miRNA expression profiles can effectively classify tumor types (Calin *et al.*, 2005; Hayashita *et al.*, 2005; He *et al.*, 2005a; Lu *et al.*, 2005; Murakami *et al.*, 2005); these profiles may even prove superior to mRNA profiling for differentiating tumor types (Lu *et al.*, 2005). Early classification and prognostic value of miRNA profiling may aid the rational treatment of cancers. Further discussion of this topic is provided in other chapters in this book (see, for example, Chapter 23 by Croce and colleagues).

One general conclusion of tumor profiling is that miRNA expression is broadly dampened in many tumors, relative to normal cells or tissues (Lu *et al.*, 2005). This is perhaps consistent with a viewpoint in which miRNAs promote or help enforce a differentiated state. On the other hand, there are select examples of miRNAs whose elevated expression appears causally linked to tumorigenesis. Microarray expression screening revealed that the polycistronic miRNA cluster miR-17-92 is amplified in certain tumors, including lung cancers and B-cell lymphomas (Hayashita *et al.*, 2005; He *et al.*, 2005b). Moreover, directed misexpression of this miRNA cluster enhances cell proliferation and can accelerate tumor formation in conjunction with c-myc (He *et al.*, 2005b). Therefore, this miRNA cluster displays properties of an oncogene.

miRNA sensors

Although miRNA cloning, Northern blot and microarray analyses all provide important information on miRNA expression, they provide only limited information on spatial patterns of expression. Therefore, methods to detect miRNAs or miRNA activity in wholemount preparations or sectioned tissues are highly desirable. Originally, the small size of the miRNA precluded conventional methods for detection by *in situ* hybridization. One strategy for bypassing this limitation is a genetically encoded miRNA "sensor." The backbone for the miRNA sensor is a ubiquitously expressed reporter transgene (such as *GFP* or *lacZ*) linked to a neutral 3′-UTR (such as the viral SV40 3′-UTR); this control transgene should exhibit spatially ubiquitous activity (Figure 3.3a). To modify this into a miRNA sensor, one introduces target sequences into its 3′-UTR that are perfectly complementary to a miRNA of interest. Although animal miRNAs do not typically have perfectly matched endogenous targets, miRNAs efficiently direct such artificial targets for cleavage and destruction by the RNAi pathway (Hutvagner and Zamore, 2002; Zeng *et al.*, 2002). Sensor expression therefore provides a "photographic negative" of the miRNA expression pattern, in which highest levels of the miRNA are found in cells that express lowest levels of the sensor (Figure 3.3a).

The sensor strategy was first employed to demonstrate spatially patterned expression of the *Drosophila* miRNA bantam (Brennecke *et al.*, 2003). In normal wing imaginal discs, bantam sensor levels are highest in regions of the disc known to be under mitotic arrest, suggesting that bantam miRNA levels are lowest in such cells. Conversely, double labeling of brains with bromodeoxyuridine (BrdU) and the bantam sensor revealed that actively proliferating cells (as marked by BrdU incorporation) downregulate the bantam sensor, indicating that these cells express elevated levels of bantam miRNA (Figure 3.3b–d) (Brennecke *et al.*, 2003). The spatial pattern of bantam expression inferred by its sensor thus provided important clues about the function of bantam in promoting cell proliferation.

The sensor strategy subsequently proved successful in a vertebrate system, and was used to demonstrate spatially modulated patterns of several mouse miRNAs (Mansfield *et al.*, 2004). For example, expression of two *let-7* isoforms was found to be modulated in limb buds, while the heart-specific expression miR-1 determined from Northern analysis could now be verified in wholemount embryos. Interestingly, the Hox miRNAs miR-10a and miR-196a were found to be expressed in Hox-like patterns that were modulated along the anterior–posterior axis, strongly suggesting that these miRNAs might also influence segmental identity.

The sensor strategy as outlined above provides a substantial improvement over the spatial resolution provided by Northern and microarray analysis. Another potential advantage of the miRNA sensor is that it reports on miRNA activity,

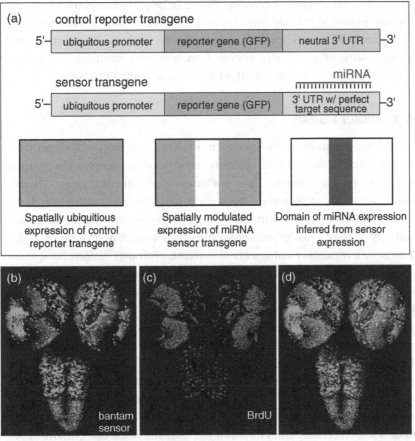

Figure 3.3. miRNA sensors reveal spatially modulated miRNA activity. (a) Sensor design and experimental interpretation. The control transgene consists of a ubiquitously expressed reporter transgene (depicted in green) lacking miRNA binding sites. The miRNA sensor includes perfect miRNA binding sites in its 3'-UTR, which cause it to be downregulated in miRNA-expressing cells in vivo (depicted as white stripe). Expression of the miRNA is therefore inferred to be in cells with low sensor expression (depicted as red stripe). (b)–(d) A *Drosophila* larval brain doubly labeled for bantam miRNA sensor activity (green, b) and bromodeoxyuridine (BrdU) to label proliferating cells (purple, c); merge (d). Note that bantam sensor levels are lowest in BrdU-positive cells, indicating that cells that are actively proliferating exhibit highest levels of the bantam miRNA. Images courtesy of Julius Brennecke and Stephen Cohen. (See color plate 2)

rather than the mere presence of the miRNA per se. This might be relevant should activity of the miRISC ever prove to be regulated. However, owing to reporter ubiquity, the sensor approach is unlikely to permit visualization of miRNAs with highly specific expression domains. Therefore, it would be preferable to design sensors that provide a positive, rather than negative, map of expression. While not yet implemented, a two-component strategy can be envisioned involving a ubiquitously expressed reporter transgene and a ubiquitously expressed reporter inhibitor that is linked to miRNA binding sites. In theory, miRNA-induced loss of the reporter inhibitor would permit reporter activation to positively mark miRNA-expressing cells. The sensor approach is also limited by reporter perdurance, as newly or transiently expressed miRNAs are ineffective at reducing previously accumulated reporter protein. This may be partially circumvented by placing reporter expression under the control of a conditionally active promoter, or by engineering the reporter protein to be unstable. Such modifications might improve the utility of genetically encoded sensors for examining the spatial expression of low abundance, transient, or spatially restricted miRNAs.

miRNA promoter fusions

As discussed earlier, miRNAs are products of RNA polymerase II-dependent transcription and often display specific and dynamic patterns of expression comparable to that of many developmentally important protein-encoding genes. The cis-regulatory control of complex gene expression patterns has traditionally been deconstructed using enhancer/promoter fusions to reporter constructs, an endeavor commonly referred to as "promoter-bashing." As it is not necessarily straightforward to isolate regulatory enhancers of interest, promoter fusions have most frequently been employed to understand the transcriptional regulation of genes with previously described expression patterns. In the case of miRNAs, however, for which spatial expression was lacking for some time, investigators have used promoter fusions as a forward approach to explore potential miRNA expression domains whilst simultaneously defining relevant cis-regulatory elements and transcription factor binding sites.

   miRNA promoter fusions were particularly insightful to the functional analysis of two *C. elegans* miRNAs that dictate cell fate asymmetry between a left/right pair of taste receptor neurons, ASE left (ASEL) and ASE right (ASER). In this system, *lsy-6* imposes the ASEL fate while miR-273 imposes the ASER fate (Chang *et al.*, 2004; Johnston and Hobert, 2003). Indeed, the activities of these miRNAs correlate with specific expression patterns as deduced from GFP fusions to the genomic regions upstream from these miRNAs: a *lsy-6*::GFP fusion is expressed in ASEL (Figure 3.4b–d), while the miR-273::GFP fusion is predominantly active in ASER. Asymmetric expression of these miRNAs in the left and right ASE neurons underlies their cell type-specific repression of transcription factors that control left–right ASE identity. Other studies have now defined promoter fragments that direct spatial and temporal aspects of several other miRNAs in *C. elegans*, *Drosophila* and mouse (Biemar *et al.*, 2005; Johnson *et al.*, 2003; Sokol and Ambros, 2005; Yoo and Greenwald, 2005; Zhao *et al.*, 2005).

   One potential advantage of promoter fusions is their capacity to differentiate between the expression of related miRNAs encoded by distinct genomic loci. For example, fragments from the unique miR-1-1 and miR-1-2 regulatory sequences show distinct expression patterns for these paralogous miRNAs (Zhao *et al.*, 2005). On the other hand, a general concern with enhancer analysis is the uncertainty of selecting sufficient genomic sequence to replicate native gene expression. For instance, a minigene including a 1 kilobase (kb) fragment from the miR-48 promoter region is temporally misexpressed in nematodes and induces a mutant phenotype (Li *et al.*, 2005), suggesting that this fragment lacks important repressive regulatory sequences. In cases where a miRNA mutant phenotype has been

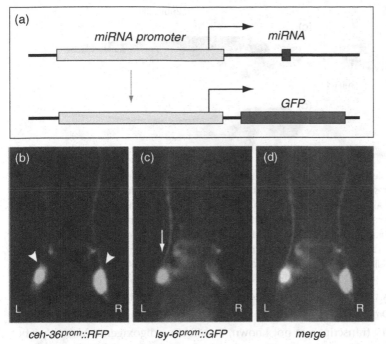

Figure 3.4. Enhancer/promoter-reporter fusions can reveal celltype-specific miRNA expression. (a) General strategy for constructing the reporter fusion. (b)–(d) A reporter fusion to regulatory sequences of the *lsy-6* miRNA reveals asymmetric expression of this miRNA in the ASE left/right chemosensory neuron pair. A *ceh-36* promoter-RFP fusion marks both left and right ASE neurons (b), while the *lsy-6* miRNA promoter-GFP transgene is active only in the left ASE neuron (c); merge is shown in (d). Images used with the permission of Oliver Hobert. (See color plate 3)

characterized, it is possible to use a transgene rescue approach to demonstrate that the functionally relevant cis-regulatory sequences have been included. However, few cases of mutant miRNA phenotypes have yet been described. Obtaining complete regulatory sequences may therefore present a challenge, especially for those miRNAs that fall within large intergenic regions.

Direct detection of miRNA primary transcripts by *in situ* hybridization

The aforementioned approaches generate valuable temporal and spatial miRNA expression data, but fall short of the ultimate goal of directly visualizing miRNAs *in situ*. Although standard methodologies exist to detect mRNA transcripts *in situ*, short DNA and RNA probes are inefficient at specifically recognizing mature miRNAs or pre-miRNA hairpins *in situ*. There are a number of possibilities for this: the specificity of 21 nucleotide probes may be insufficient, the capacity for self-hybridization might render pre-miRNA probes and/or targets ineffective, or the small number of labeled nucleotides in such short probes may be insufficient for effective detection. However, the realization that miRNA transcripts are processed from much longer primary transcripts (pri-miRNAs) suggested that these might make for more efficient hybridization targets.

fruitfly   zebrafish

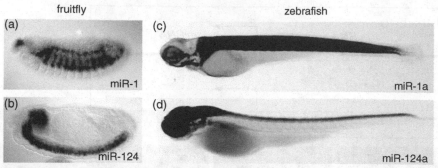

Figure 3.5. Wholemount *in situ* hybridization against primary miRNA transcripts and mature miRNAs. (a, b) *Drosophila* embryos hybridized with ~1 kilobase probes directed against primary miRNA (pri-miRNA) transcripts reveal miR-1 to be expressed in muscles (a) and miR-124 to be expressed in the central nervous system (b). Images used with permission of Aziz Aboobaker and Eric Lai. (c, d) Zebrafish hybridized with locked nucleic acid (LNA) probes directed against mature miR-1a and miR-124 reveal these miRNAs to be similarly expressed in muscles (c) and the central nervous system (d) in a vertebrate. Images used with permission of R. H. A. Plasterk.

This strategy was first reported in 2004 to successfully detect the expression pattern of miR-10 in *Drosophila* embryos (Kosman *et al.*, 2004). Although the structure of the primary miR-10 transcript was not known, an effective digoxigenin-labeled probe was prepared from a ~1 kb genomic region surrounding the miR-10 hairpin sequence. Detection of hybridized probes using fluorophore-conjugated anti-DIG antibody revealed strong hybridization to pairs of nuclear dots. A similar pattern is commonly observed when hybridizing intronic probes to tissue, and is inferred to represent nascent transcripts emanating from the two genomic loci. This strategy revealed that miR-10 is expressed in a Hox-like domain covering the thoracic and abdominal primordia in the early *Drosophila* embryo (Kosman *et al.*, 2004).

After this initial proof of principle, additional studies demonstrated that *in situ* hybridization to nascent miRNA transcripts can effectively delineate spatial patterns of miRNA expression (Biemar *et al.*, 2005; Ronshaugen *et al.*, 2005; Sokol and Ambros, 2005; Stark *et al.*, 2005). In particular, Aboobaker and colleagues used this method to systematically describe the spatial expression patterns of most miRNAs during *Drosophila* embryogenesis (Aboobaker *et al.*, 2005). These studies collectively reveal incredibly diverse and dynamic patterns of miRNA expression in blastoderm embryos, later in all three germ layers, and finally in a variety of differentiating organs. In several cases, the spatial expression pattern of miRNAs that have been conserved between invertebrates and vertebrates has been conserved. For example, a *Drosophila* ortholog of mammalian miR-1, known to be expressed in heart and skeletal muscle, is similarly expressed throughout the developing mesoderm and then in all differentiating muscles (Figure 3.5a). In addition, *Drosophila* miR-124 parallels that of its mammalian counterparts in that it is restricted to the central nervous system and brain (Figure 3.5b). In these studies, it is notable that knowledge of the nascent transcript sequence was not required; in most cases, effective probes were made using genomic probes 0.5–1 kb in length.

Detection of primary transcripts has the potential advantage of distinguishing amongst miRNAs of similar sequence. A salient example of this regards the four miR-2/miR-13 loci, which collectively encode eight nearly identical miRNAs. Spatially distinct expression was reported for three of these loci: the stand-alone miRNA cluster miR-13b-1/13a/2c is restricted to the central nervous system, the intronic miRNA miR-13b-2 is present in differentiating somatic muscles and the gut, and the intronic miR-2a-2/2a-1/2b-2 cluster is expressed in the epidermis and hindgut (Aboobaker *et al.*, 2005). Therefore, in contrast to the general presumption that similar miRNAs may have redundant functions, these *in situ* data indicate that these nearly identical miRNAs are actually deployed in largely non-overlapping territories in the embryo.

Detecting mature miRNAs with locked nucleic acid probes

Nascent miRNA transcript *in situ*s supply essential information on the spatial patterns of miRNA transcription. However, the apparent stability of mature miRNAs suggests they may persist in cell progeny long after transcription of the miRNA locus has ceased. Therefore, detection of the mature, ~21–24 nucleotide species is required to understand the full scope of their biological activity. As mentioned, mature miRNAs had for many years eluded direct detection by *in situ* hybridization. A breakthrough in achieving efficient hybridization with very short probes came with usage of locked nucleic acids (LNAs), which can strongly increase probe affinity. The chemistry and applications of LNAs with respect to miRNAs are separately detailed by Kauppinen in Chapter 17.

LNA-modified DNA probes were first successfully used for *in situ* hybridization to mature miRNAs in zebrafish (Wienholds *et al.*, 2005), yielding well-defined spatial expression information on miRNA expression (see Figure 3.5c, d). In contrast to nascent probes, which typically detect miRNA expression in nuclear dots, LNA probes detect expression predominantly in the cytoplasm, suggesting that they detect mature miRNAs and that these accumulate to much higher levels than do nascent transcripts. Systematic surveys of 115 miRNAs using these probes revealed that the majority (68%) of miRNAs in fish are expressed in tissue-specific patterns. Collectively, these include all organ systems, with many miRNAs displaying restricted expression within a given organ. For example, miR-217 and miR-7 were expressed by exocrine and endocrine pancreas, respectively. Some miRNAs displayed extremely celltype-specific expression, with expression in individual cells evident for some miRNA expressed in lateral line sensory organs. By and large, the most notable diversification of miRNA expression patterns involves the brain, with 43 miRNAs expressed throughout or within specific domains of the brain (Wienholds *et al.*, 2005). In theory, the complexity of brain miRNA expression may be related to the functional complexity and diversity of neurons. It will be interesting to understand whether all of these distinct brain miRNAs regulate development, differentiation, or neural activity.

LNA probes initially proved most successful for *in situ* analysis of zebrafish miRNAs. However, successful adaptation of protocols for LNA *in situ* hybridization have now been reported for an analysis of miR-1 during *Drosophila* mesoderm and muscle development, a study of miR-7 during *Drosophila* photoreceptor cell differentiation, and an examination of several miRNAs during murine embryonic development

(Kloosterman *et al.*, 2006; Li and Carthew, 2005; Sokol and Ambros, 2005). Successful *in situ*s from archival human brain tissue further suggest that LNA probes will be useful for assessing miRNA expression at the cellular level in human pathology specimens (Nelson *et al.*, 2006). Obernosterer and colleagues (2007) recently described detailed protocols for the usage of LNA probes for sectioned material. Therefore, this technology shows great promise for revealing detailed spatial expression domains of mature miRNAs *in situ*, information that has been so highly sought since their discovery.

Temporal comparisons between the patterns obtained using pri-miRNA and mature miRNA LNA probes for *Drosophila* miR-1 are potentially instructive; miR-1 is transcribed in the early blastoderm stage throughout the presumptive mesoderm, but mature miR-1 is not detected until later, during gastrulation (Sokol and Ambros, 2005). It remains to be seen if the delayed appearance of mature miR-1 reflects a genuine feature of miR-1 biogenesis or a technical detection issue. Nevertheless, is it clear that the ability to discriminate, *in situ*, between primary and mature transcripts will be essential to assess potential regulation of miRNA maturation. miRNA biogenesis has largely been assumed to be constitutive, but there is no particular reason to suspect that individual steps along the pathway are not regulated in certain settings. For instance, adenosine deaminase activity was recently suggested to interfere with pri-miR-142 processing by Drosha (Yang *et al.*, 2006). Double label analysis for pri-miRNA and mature miRNA transcripts may illuminate the kinetics of miRNA transcript processing and reveal novel regulatory steps of miRNA biogenesis.

The future of miRNA expression profiling

At this stage in the microRNA "revolution," we finally have at our fingertips a variety of powerful tools that allow us to examine miRNA expression at many levels. It is now abundantly clear that miRNAs exhibit diverse patterns of temporal and spatial expression, and that this knowledge has greatly informed the functional analysis of those miRNAs that have been well-characterized. We anticipate that high-throughput expression profiling and directed *in situ* expression analyses will continue to yield fresh insight as to the functions of miRNAs during normal development and physiology. In addition, with much information on the expression patterns of miRNAs in hand, we expect rapid progress to be made on understanding the cis-regulation of miRNAs. By integrating data on the upstream transcription factors that regulate miRNA expression with downstream miRNA target lists, we may eventually be able place miRNAs within larger genetic circuits and networks involving cell specification, differentiation, or metabolism. Ultimately, we look forward to understanding how perturbation of miRNA-related networks may apply to disease conditions, with the hope that they may lend diagnostic insight or lead to novel entrypoints for therapeutic intervention.

REFERENCES

Aboobaker, A. A., Tomancak, P., Patel, N., Rubin, G. M. and Lai, E. C. (2005). *Drosophila* microRNAs exhibit diverse spatial expression patterns during embryonic development. *Proceedings of the National Academy of Sciences USA*, **102**, 18 017–18 022.

Babak, T., Zhang, W., Morris, Q., Blencowe, B.J. and Hughes, T.R. (2004). Probing microRNAs with microarrays: tissue specificity and functional inference. *RNA*, **10**, 1813–1819.

Barad, O., Meiri, E., Avniel, A. *et al.* (2004). MicroRNA expression detected by oligonucleotide microarrays: system establishment and expression profiling in human tissues. *Genome Research*, **14**, 2486–2494.

Bartel, D. P. (2004). MicroRNAs. Genomics, biogenesis, mechanism, and function. *Cell*, **116**, 281–297.

Baskerville, S. and Bartel, D. P. (2005). Microarray profiling of microRNAs reveals frequent coexpression with neighboring miRNAs and host genes. *RNA*, **11**, 241–247.

Biemar, F., Zinzen, R., Ronshaugen, M. *et al.* (2005). Spatial regulation of microRNA gene expression in the *Drosophila* embryo. *Proceedings of the National Academy of Sciences USA*, **102**, 15 907–15 911.

Brennecke, J., Hipfner, D. R., Stark, A., Russell, R. B. and Cohen, S. M. (2003). *bantam* encodes a developmentally regulated microRNA that controls cell proliferation and regulates the proapoptotic gene *hid* in *Drosophila*. *Cell*, **113**, 25–36.

Calin, G. A., Ferracin, M., Cimmino, A. *et al.* (2005). A microRNA signature associated with prognosis and progression in chronic lymphocytic leukemia. *New England Journal of Medicine*, **353**, 1793–1801.

Chang, S., Johnston, R. J., Jr., Frokjaer-Jensen, C., Lockery, S. and Hobert, O. (2004). MicroRNAs act sequentially and asymmetrically to control chemosensory laterality in the nematode. *Nature*, **430**, 785–789.

Chen, C., Ridzon, D. A., Broomer, A. J. *et al.* (2005). Real-time quantification of microRNAs by stem-loop RT-PCR. *Nucleic Acids Research*, **33**, e179.

Griffiths-Jones, S., Grocock, R. J., van Dongen, S., Bateman, A. and Enright, A. J. (2006). miRBase: microRNA sequences, targets and gene nomenclature. *Nucleic Acids Research*, **34**, D140–144.

Hayashita, Y., Osada, H., Tatematsu, Y. *et al.* (2005). A polycistronic microRNA cluster, miR-17-92, is overexpressed in human lung cancers and enhances cell proliferation. *Cancer Research*, **65**, 9628–9632.

He, H., Jazdzewski, K., Li, W. *et al.* (2005a). The role of microRNA genes in papillary thyroid carcinoma. *Proceedings of the National Academy of Sciences USA*, **102**, 19 075–19 080.

He, L., Thomson, J. M., Hemann, M. T. *et al.* (2005b). A microRNA polycistron as a potential human oncogene. *Nature*, **435**, 828–833.

Hutvagner, G. and Zamore, P. D. (2002). A microRNA in a multiple-turnover RNAi enzyme complex. *Science*, **297**, 2056–2060.

Hutvagner, G., McLachlan, J., Pasquinelli, A. *et al.* (2001). A cellular function for the RNA-interference enzyme Dicer in the maturation of the *let-7* small temporal RNA. *Science*, **293**, 834–838.

Johnson, S. M., Lin, S. Y. and Slack, F. J. (2003). The time of appearance of the *C. elegans* let-7 microRNA is transcriptionally controlled utilizing a temporal regulatory element in its promoter. *Developmental Biology*, **259**, 364–379.

Johnston, R. J. and Hobert, O. (2003). A microRNA controlling left/right neuronal asymmetry in *Caenorhabditis elegans*. *Nature*, **426**, 845–849.

Ketting, R., Fischer, S., Bernstein, E. *et al.* (2001). Dicer functions in RNA interference and in synthesis of small RNAs involved in developmental timing in *C. elegans*. *Genes & Development*, **15**, 2654–2659.

Kloosterman, W. P., Wienholds, E., de Bruijn, E., Kauppinen, S. and Plasterk, R. H. (2006). In situ detection of miRNAs in animal embryos using LNA-modified oligonucleotide probes. *Nature Methods*, **3**, 27–29.

Kosman, D., Mizutani, C. M., Lemons, D. *et al.* (2004). Multiplex detection of RNA expression in *Drosophila* embryos. *Science*, **305**, 846.

Krichevsky, A. M., King, K. S., Donahue, C. P., Khrapko, K. and Kosik, K. S. (2003). A microRNA array reveals extensive regulation of microRNAs during brain development. *RNA*, **9**, 1274–1281.

Lagos-Quintana, M., Rauhut, R., Lendeckel, W. and Tuschl, T. (2001). Identification of novel genes coding for small expressed RNAs. *Science*, **294**, 853–858.

Lagos-Quintana, M., Rauhut, R., Yalcin, A. *et al.* (2002). Identification of tissue-specific microRNAs from mouse. *Current Biology*, **12**, 735–739.

Lai, E. C. (2003). microRNAs: runts of the genome assert themselves. *Current Biology*, **13**, R925–936.

Lau, N., Lim, L., Weinstein, E. and Bartel, D. P. (2001). An abundant class of tiny RNAs with probable regulatory roles in *Caenorhabditis elegans*. *Science*, **294**, 858–862.

Lee, R. C. and Ambros, V. (2001). An extensive class of small RNAs in *Caenorhabditis elegans*. *Science*, **294**, 862–864.

Lee, R. C., Feinbaum, R. L. and Ambros, V. (1993). The *C. elegans* heterochronic gene *lin-4* encodes small RNAs with antisense complementarity to *lin-14*. *Cell*, **75**, 843–854.

Lee, Y., Jeon, K., Lee, J. T., Kim, S. and Kim, V. N. (2002). MicroRNA maturation: stepwise processing and subcellular localization. *European Molecular Biology Organization Journal*, **21**, 4663–4670.

Lee, Y., Ahn, C., Han, J. *et al.* (2003). The nuclear RNase III Drosha initiates microRNA processing. *Nature*, **425**, 415–419.

Lee, Y., Kim, M., Han, J. *et al.* (2004). MicroRNA genes are transcribed by RNA polymerase II. *European Molecular Biology Organization Journal*, **23**, 4051–4060.

Li, M., Jones-Rhoades, M. W., Lau, N. C., Bartel, D. P. and Rougvie, A. E. (2005). Regulatory mutations of mir-48, a *C. elegans* let-7 family microRNA, cause developmental timing defects. *Developmental Cell*, **9**, 415–422.

Li, X. and Carthew, R. W. (2005). A microRNA mediates EGF receptor signaling and promotes photoreceptor differentiation in the *Drosophila* Eye. *Cell*, **123**, 1267–1277.

Lu, J., Getz, G., Miska, E. A. *et al.* (2005). MicroRNA expression profiles classify human cancers. *Nature*, **435**, 834–838.

Mansfield, J. H., Harfe, B. D., Nissen, R. *et al.* (2004). MicroRNA-responsive 'sensor' transgenes uncover Hox-like and other developmentally regulated patterns of vertebrate microRNA expression. *Nature Genetics*, **36**, 1079–1083.

Miska, E. A., Alvarez-Saavedra, E., Townsend, M. *et al.* (2004). Microarray analysis of microRNA expression in the developing mammalian brain. *Genome Biology*, **5**, R68.

Murakami, Y., Yasuda, T., Saigo, K. *et al.* (2005). Comprehensive analysis of microRNA expression patterns in hepatocellular carcinoma and non-tumorous tissues. *Oncogene, Dec5 e publication*.

Neely, L. A., Patel, S., Garver, J. *et al.* (2006). A single-molecule method for the quantitation of microRNA gene expression. *Nature Methods*, **3**, 41–46.

Nelson, P. T., Baldwin, D. A., Scearce, L. M. *et al.* (2004). Microarray-based, high-throughput gene expression profiling of microRNAs. *Nature Methods*, **1**, 155–161.

Nelson, P. T., Baldwin, D. A., Kloosterman, W. P. *et al.* (2006). RAKE and LNA-ISH reveal microRNA expression and localization in archival human brain. *RNA*, **12**, 187–191.

Obernosterer G., Martinez, J. and Alenius, M. (2007). Locked nucleic acid-based *in situ* detection of microRNAs in mouse tissue sections. *Nature Protocols*, **2**, 1508–1514.

Poy, M. N., Eliasson, L., Krutzfeldt, J. *et al.* (2004). A pancreatic islet-specific microRNA regulates insulin secretion. *Nature*, **432**, 226–230.

Reinhart, B. J., Slack, F., Basson, M. *et al.* (2000). The 21-nucleotide *let-7* RNA regulates developmental timing in *Caenorhabditis elegans*. *Nature*, **403**, 901–906.

Ronshaugen, M., Biemar, F., Piel, J., Levine, M. and Lai, E. C. (2005). The *Drosophila* microRNA iab-4 causes a dominant homeotic transformation of halteres to wings. *Genes and Development*, **19**, 2947–2952.

Schwarz, D. S., Hutvagner, G., Du, T. *et al.* (2003). Asymmetry in the assembly of the RNAi enzyme complex. *Cell*, **115**, 199–208.

Sempere, L. F., Freemantle, S., Pitha-Rowe, I. *et al.* (2004). Expression profiling of mammalian microRNAs uncovers a subset of brain-expressed microRNAs with possible roles in murine and human neuronal differentiation. *Genome Biology*, **5**, R13.

Shingara, J., Keiger, K., Shelton, J. *et al.* (2005). An optimized isolation and labeling platform for accurate microRNA expression profiling. *RNA*, **11**, 1461–1470.

Sokol, N. S. and Ambros, V. (2005). Mesodermally expressed *Drosophila* microRNA-1 is regulated by Twist and is required in muscles during larval growth. *Genes & Development*, **19**, 2343–2354.

Stark, A., Brennecke, J., Bushati, N., Russell, R. B. and Cohen, S. M. (2005). Animal microRNAs confer robustness to gene expression and have a significant impact on 3′UTR evolution. *Cell*, **123**, 1133–1146.

Thomson, J. M., Parker, J., Perou, C. M. and Hammond, S. M. (2004). A custom microarray platform for analysis of microRNA gene expression. *Nature Methods*, **1**, 47–53.

Wienholds, E., Kloosterman, W. P., Miska, E. *et al.* (2005). MicroRNA expression in zebrafish embryonic development. *Science*, **309**, 310–311.

Wightman, B., Ha, I. and Ruvkun, G. (1993). Posttranscriptional regulation of the heterochronic gene *lin-14* by *lin-4* mediates temporal pattern formation in *C. elegans*. *Cell*, **75**, 855–862.

Yang, W., Chendrimada, T. P., Wang, Q. *et al.* (2006). Modulation of microRNA processing and expression through RNA editing by ADAR deaminases. *Nature Structural Molecular Biology*, **13**, 13–21.

Yoo, A. S. and Greenwald, I. (2005). LIN-12/Notch activation leads to microRNA-mediated down-regulation of Vav in *C. elegans*. *Science*, **310**, 1330–1333.

Zeng, Y., Wagner, E. J. and Cullen, B. R. (2002). Both natural and designed micro RNAs can inhibit the expression of cognate mRNAs when expressed in human cells. *Molecular Cell*, **9**, 1327–1333.

Zhao, Y., Samal, E. and Srivastava, D. (2005). Serum response factor regulates a muscle-specific microRNA that targets Hand2 during cardiogenesis. *Nature*, **436**, 214–220.

4 MicroRNAs in limb development

Danielle M. Maatouk, Jason R. Rock and Brian D. Harfe*

Introduction

The vertebrate limb is a highly organized structure that must be patterned along three axes during development: anteroposterior, dorsoventral, and proximodistal (Tickle, 2003). For decades, the limb has served as a choice model system for developmental biologists because of the ease with which it can be manipulated and an organism's ability to survive with abnormal or absent limbs. Despite years of intense investigation, many of the molecules responsible for limb pattern formation are still not known.

Recently, a class of non-coding RNAs, the microRNAs (miRNAs), have been implicated in limb development. These molecules are ~22 nt in their mature form and can bind to mRNAs, leading to their degradation or inhibition of protein production (McManus and Sharp, 2002). The first miRNA to be discovered, *lin-4*, was identified in a forward genetic screen aimed at identifying developmental timing defects in *C. elegans* (Lee *et al.*, 1993). Nearly a decade after the discovery of *lin-4* in nematodes, a second miRNA, *let-7*, was identified in organisms ranging from *C. elegans* to humans (Pasquinelli *et al.*, 2000; Reinhart *et al.*, 2000). In the years since, at least 326 miRNAs have been validated in humans and 249 in mouse (Griffiths-Jones, 2004). Only a few of these miRNAs have known functions (reviewed in Harfe (2005)). Our lab is interested in the role miRNAs play in patterning the vertebrate limb.

MicroRNA processing

Mature miRNAs are produced through two cleavage events by members of the RNaseIII family of nucleases (Bernstein *et al.*, 2001; Hutvagner *et al.*, 2001; Ketting *et al.*, 2001; Lee *et al.* 2003). These enzymes show specificity for dsRNA substrates (Bernstein *et al.*, 2001; Grishok *et al.*, 2001; Ketting *et al.*, 2001).

* Author to whom correspondence should be addressed.

MicroRNAs: From Basic Science to Disease Biology, ed. Krishnarao Appasani. Published by Cambridge University Press. © Cambridge University Press 2008.

The primary RNA transcript (pri-miRNA) is cleaved by Drosha into a 70 nt RNA hairpin referred to as the premature RNA transcript (pre-miRNA) (Lee *et al.*, 2003). The pre-miRNA is exported to the cytoplasm by exportin-5 where it comes into contact with a second RNaseIII enzyme, Dicer, that further cleaves the hairpin to produce a ~22 nt mature miRNA (Bernstein *et al.*, 2001; Yi *et al.*, 2003; Lund *et al.*, 2004). The enzyme is a modular protein with two RNaseIII domains, a DExH/DEAH box RNA helicase domain, a PAZ domain, a dsRNA binding motif, and a conserved domain of unknown function (Bernstein *et al.*, 2001). The RNaseIII domains cleave dsRNA (Murchison and Hannon, 2004). It has been proposed that the RNA helicase domain is responsible for remodeling ribonucleoprotein complexes during miRNA processing (Schwer, 2001). The PAZ domain has ssRNA binding specificity (Lingel *et al.*, 2003; Song *et al.*, 2003; Yan *et al.*, 2003) and might direct the enzyme toward pre-miRNAs previously cleaved by Drosha with two-nucleotide 3' overhangs (Song *et al.*, 2003). The microRNAs biogenesis is also detailed in this text in a different chapter by Scherr and Eder.

Creating Dicer-deficient mouse models

The high level of sequence homology between miRNAs suggests that reverse genetic approaches will not be widely useful for identifying targets and functions of miRNAs. Since only one *Dicer* gene is found in the mouse genome (Nicholson and Nicholson, 2002), an alternative approach to removing individual, possibly redundant miRNAs is to delete the *Dicer* gene and, in effect, all mature miRNAs. *Dicer*-null mutations have been made and mice lacking *Dicer* die at embryonic day (E) 7.5, which is before most organ systems have started forming (Bernstein *et al.*, 2003). *Dicer*-deficient mice also show a reduction in the expression of *Oct4*, a marker of stem cells (Bernstein *et al.*, 2003). Taken together, these data suggest that *Dicer* (and miRNAs) plays an important role in vertebrate development and stem cell maintenance. However, *Dicer*-null mice are not useful for studying the possible functions of miRNAs in limb development because limbs do not begin outgrowth from the body until E9.5 (Martin, 1990).

To avoid this early embryonic lethality, we have taken advantage of conditional gene inactivation using Cre-mediated, site-specific recombination. A *Dicer* conditional allele was created by flanking an exon that encodes most of the second RNaseIII domain with loxP sites (Figure 4.1) (Harfe *et al.*, 2005). Cre-mediated recombination at the loxP sites excises the RNaseIII domain and results in a Dicer protein lacking one of its two cleavage domains. Irreversible recombination of a floxed allele will only occur in cells that express, or are exposed to, Cre protein. Many transgenic mouse lines have been published with different Cre expression patterns (http://www.mshri.on.ca/nagy/Cre-pub.html). Therefore, a floxed allele can be made null in a temporal, tissue, or lineage-specific manner (Kos, 2004).

The next sections will discuss our current understanding of the roles of miRNAs during limb development; more specifically, we discuss the role of Hox

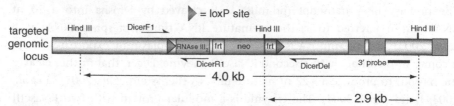

Figure 4.1. Construction of a Dicer conditional mouse allele. To target the Dicer locus, loxp sites were inserted around an exon that encodes most of the second RNase III domain allowing for Cre-mediated removal of this sequence. The 3′ probe was used to distinguish between wild type and targeted alleles in ES cells upon HindIII digestion. Dark gray boxes indicate exons, light gray boxes mark introns. The frt sites flanking the neo cassette can be used to remove the neo cassette using Flip recombinase. DicerF1, DicerR1 and DicerDel denote primers that are used to genotype the floxed Dicer allele (reproduced from Harfe *et al.* (2005); copyright of National Academy of Sciences USA).

cluster-embedded miRNAs in limb development and how the generation of a conditional *Dicer* allele has been a valuable tool for understanding the roles of miRNAs in the limb and other tissues.

Hox genes

Homeobox-containing genes, or Hox genes, are highly conserved genes that regulate the spatial organization and patterning of different body parts during development (Bateson, 1894; McGinnis *et al.*, 1984; Scott and Weiner, 1984). Mutations in these genes lead to homeotic transformations, where a region of an organism will take on the fate of a different region (Bateson, 1894). Hox genes are arranged in a cluster, and in the mouse, this cluster has been duplicated two times resulting in four Hox clusters (A, B, C and D) located on separate chromosomes (McGinnis and Krumlauf, 1992; Scott, 1992). Therefore, each gene in the Hox cluster exhibits high sequence conservation to genes in similar positions in other clusters. For example, the ninth gene in the Hox cluster is conserved in all four clusters in the mouse (*Hoxa9, b9, c9* and *d9*). Such genes are referred to as paralogs, and often exhibit overlapping expression patterns (Krumlauf, 1994). Intriguingly, genes within the Hox cluster exhibit both spatial and temporal colinearity. Genes located at the 3′ end of the cluster are expressed at earlier stages than 5′ genes, and are expressed in anterior portions of the embryo. In contrast, genes at the 5′ end of the cluster are expressed later, and are more posteriorly restricted (Lewis, 1978; Dolle and Duboule, 1989; Duboule and Dolle, 1989; Graham *et al.*, 1989; Izpisua-Belmonte *et al.*, 1991). Position in the Hox cluster also corresponds to the genes' responsiveness to retinoic acid (RA), as 3′ genes are more sensitive to RA than those located at the 5′ end of the cluster (Boncinelli *et al.*, 1991). The spatiotemporal specificity of each Hox gene sets up a gradient whereby different combinations of Hox gene products specify different regional identities within the developing embryo.

MicroRNAs in the Hox cluster

Two groups of miRNAs have been identified within the Hox cluster, *mir-10* and *mir-196* (Figure 4.2) (Lagos-Quintana *et al.*, 2003; Lim *et al.*, 2003). *mir-10a* is located upstream of *Hoxb4* while *mir-10b* is located upstream of *Hoxd4*. Similarly, *mir-196a-1*, *mir-196a-2* and *mir-196b* are located in similar regions in the Hox cluster, upstream of *Hoxb9*, *Hoxc9* and *Hoxa9* respectively. While genes of the Hox cluster are known to be expressed in a colinear manner, it was not known whether miRNAs within the cluster would follow this trend. To explore the expression patterns of most genes, it is customary to use RNA *in situ* hybridization. However, the small size of miRNAs makes it difficult to utilize such methods. Therefore, a technique originally developed in *Drosophila* was adapted for the mouse, where miRNA sensors use an RNAi-based mechanism to determine a miRNA's expression pattern (Brennecke *et al.*, 2003; Mansfield *et al.*, 2004). The miRNA sensor construct contains a ubiquitously-expressed reporter gene (*lacZ*) with two copies of a miRNA-complementary site introduced into the 3'UTR. When the *lacZ* transgene is expressed, the cells will produce β-galactosidase (β-gal); however, wherever the miRNA is expressed, the *lacZ* transcript will become inactivated by RNAi. Therefore, subsequent staining of transgenic embryos for β-gal will reveal blue cells where the miRNA is absent; cells expressing the miRNA will remain white.

Using this miRNA sensor method, the expression profiles for *mir-10a* and *mir-196b* were examined in E10.0 embryos (Mansfield *et al.*, 2004). The *mir-10a* sensor revealed a Hox-like expression pattern with an anterior limit similar to that of *Hoxb4*. This suggests that *mir-10a* and *Hoxb4* may be regulated by similar factors. Interestingly, a region in the posterior trunk that lacks *Hoxb4* expression coincides with a region of high *mir-10a* expression; however, miRNA binding sites for *mir-10a* have yet to be detected in the *Hoxb4* 3'UTR.

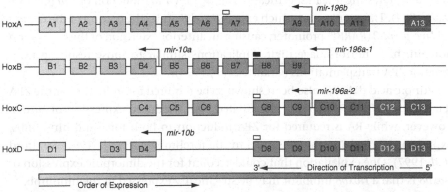

Figure 4.2. Schematic representation of the murine Hox cluster. The location and direction of gene transcription of miRNAs is marked by arrows. Target sites for miR-196 are indicated by rectangles. White rectangles represent sites of imperfect complementarity, black rectangles represent near-perfect complementarity to miRNAs. Genes represented by black boxes are transcribed later than those represented by white boxes. The order of expression and direction of transcription are indicated at the bottom of the figure. The figure is not drawn to scale.

A miRNA sensor was also constructed for *mir-196a* (Mansfield *et al.*, 2004). Unlike *mir-10a*, the expression pattern for *mir-196a* differed from that of its neighboring gene, *Hoxa9*, which has a more anterior expression limit. However, the anterior limit of *mir-196a* expression did coincide with the posterior limit of *Hoxb8* expression. The *Hoxb8* 3′UTR contains a near-perfect *mir-196b* complementary site and reporter constructs containing this 3′UTR are down-regulated in cells cultured in the presence of *mir-196a* (Mansfield *et al.*, 2004; Yekta *et al.*, 2004). RNA isolated from embryos between E11.5 and E17.0 contained *Hoxb8* cleavage products coinciding with the *mir-196a* binding site, indicating that silencing occurred by the RNAi mechanism. This suggests that *Hoxb8* is indeed a direct target of miR-196a (Mansfield *et al.*, 2004; Yekta *et al.*, 2004).

Hox genes and limb development

Development and patterning of the vertebrate limb requires the activity of at least two signaling centers. One signaling center, the apical ectodermal ridge (AER), forms as a thickening of cells along the most distal portion of the limb bud. The AER is responsible for limb outgrowth; embryonic manipulations generating an ectopic AER lead to ectopic limb formation while loss of the AER results in the absence of limb outgrowth (Tickle, 2003). A second signaling center, the zone of polarizing activity (ZPA), is located in the posterior limb bud and patterns the limb along the anteroposterior axis. Sonic hedgehog (Shh) is secreted from cells of the ZPA and ectopic expression of Shh in the anterior portion of the limb leads to a mirror-image duplication of the digits. This phenocopies ZPA transplantations to the anterior limb suggesting that Shh is the polarizing signal of the ZPA (Tickle, 2003).

Hox genes are known to be intricately involved in development of the limb. Genes within the HoxA and HoxD clusters are involved in both specifying the site of initial limb bud outgrowth and in controlling the later stages of limb patterning and growth (Dolle *et al.*, 1989; Nelson *et al.*, 1996). In the HoxB cluster, *Hoxb8* plays an interesting role in specification of the ZPA and induction of *Shh* (Charite *et al.*, 1994). Transgenic mice which express *Hoxb8* under the control of the *retinoic acid receptor B2* (*RARB2*) promoter, causing an anterior extension of the expression domain, have mirror-image digit duplications similar to those resulting from anterior ZPA transplantations (Charite *et al.*, 1994).

Retinoic acid (RA) has also been shown to be required for induction of the ZPA (Lu *et al.*, 1997). This response is mediated by RA-induced expression of *Hoxb8*; however, while RA is required for ZPA induction in both fore- and hindlimbs, expression of *Hoxb8* is only observed in the forelimbs (Lu *et al.*, 1997; Stratford *et al.* 1997). One explanation that could account for the dimorphic expression of *Hoxb8* is that a *Hoxb8* inhibitor may be specifically expressed in the hindlimb.

To determine if a miRNA could be the inhibitory factor, mice lacking *Dicer* specifically in the limb mesoderm were generated (Hornstein *et al.*, 2005). Limbs from control embryos (*Dicer^{floxed}/+*) and mutant embryos (*Dicer^{floxed}/Dicer^{floxed}; prxcre*) were cultured with or without RA and assayed for *Hoxb8* expression. As expected, control forelimbs expressed *Hoxb8* when exposed to RA, while RA

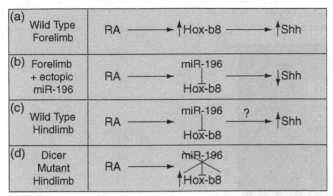

Figure 4.3. Different mechanisms regulate *Shh* induction in the limb. (a) Retinoic acid (RA) induces *Hoxb8*, and possibly additional genes, leading to expression of *Shh*. (b) Ectopic exposure of the forelimb to miR-196 leads to decreased expression of *Shh*. (c) *Shh* induction occurs independently of *Hoxb8* in the hindlimb. miR-196 expression in the hindlimb represses *Hoxb8* expression. RA induces *Shh* directly or through an unknown mechanism. (d) In the Dicer mutant hindlimb, miRNAs are not processed, leading to ectopic expression of *hoxb8*. Expression of Shh was not examined.

exposure to hindlimbs did not lead to *Hoxb8* induction. However, *Dicer* mutant hindlimbs did express *Hoxb8* after RA exposure, suggesting that a miRNA could be the hindlimb-specific inhibitory factor. *Hoxb8* is a known target of miR-196a and a microarray comparison of miRNAs in the forelimb and hindlimb found *mir-196* to be expressed 20-fold higher in the hindlimb compared to the forelimb (Mansfield *et al.*, 2004; Yekta *et al.*, 2004; Hornstein *et al.*, 2005). Furthermore, ectopic expression of miR-196 in the forelimb following RA exposure caused decreased accumulation of *Hoxb8* and *Shh* (Hornstein *et al.*, 2005). Therefore, two different mechanisms are responsible for the induction of *Shh* and the ZPA (summarized in Figure 4.3). In the forelimb, RA induces the expression of *Hoxb8*, allowing for Shh expression. However in the hindlimb, *Hoxb8* is targeted for degradation by miR-196, indicating that RA induces *Shh* directly or through an unknown factor. These findings indicate that miRNAs have important roles in the patterning of the vertebrate limb.

Although the interactions between RA, *Hoxb8*, *mir-196* and *Shh* seem to be important for proper formation of the ZPA (Charite *et al.*, 1994; Hornstein *et al.*, 2005), deletion of all three Hox-8 paralogs does not cause defects in limb development (van den Akker *et al.*, 2001). This may be because of functional redundancy with other Hox genes. Like *Hoxb8*, *Hoxa7* is preferentially expressed in the forelimb (Min *et al.*, 1998; Hornstein *et al.*, 2005) and contains a miR-196 complementary site in its 3′UTR (Lewis *et al.*, 2003). The generation of compound mutants that lack multiple paralogs of Hox genes, while complex, may be necessary to determine the significance of these interactions.

MicroRNAs in limb development

To further understand the roles of miRNAs during the development and patterning of the limb, we have used a conditional *Dicer* allele to remove mature miRNAs

WT 7.5   Dcr $^{flox\Delta/flox\Delta}$ E7.5
embryo   embryo

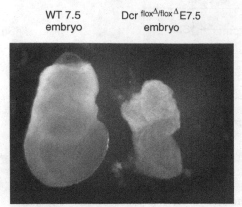

Figure 4.4. Deletion of *Dicer* results in embryonic lethality. Upon germline Cre-mediated recombination in mice homozygous for the floxed *Dicer* allele (*Dcr$^{flox\Delta/flox\Delta}$*), embryos resemble previously reported *Dicer* null embryos (Bernstein *et al.*, 2003; Harfe *et al.*, 2005). This confirms that the recombination of the conditional *Dicer* allele results in a null mutation (reproduced from Harfe *et al.* (2005); copyright of National Academy of Sciences USA).

Ctrl   Dcr $^{flox\Delta/flox\Delta}$

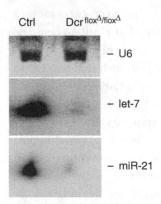

– U6

– let-7

– miR-21

Figure 4.5. Cells lacking functional *Dicer* are deficient in miRNA processing. RNA was extracted from primary fibroblasts from wild type mice and *Dcr$^{flox\Delta/flox\Delta}$* mice. Fibroblasts from the *Dcr$^{flox\Delta/flox\Delta}$* animals were infected with adenovirus expressing Cre protein to induce recombination of the floxed allele. The presence of small amounts of mature *let-7* and miR-21 in dicer null cells suggests that recombination did not occur in all fibroblasts or that a small amount of miRNA processing is *Dicer*-independent. The small nuclear RNA U6 was included as a loading control (reproduced from Harfe *et al.* (2005); copyright of National Academy of Sciences USA).

from specific regions of the vertebrate limb (Harfe *et al.*, 2005). Embryos homozygous for the floxed *Dicer* allele (*Dicer^{floxed}*/*Dicer^{floxed}*) were crossed to a germ-line Cre (β-*actin* Cre) to generate a null. These embryos arrested development at E7, consistent with the previously reported *Dicer* null phenotype (Figure 4.4) (Bernstein *et al.*, 2003; Harfe *et al.*, 2005). Northern blots performed with RNA extracted from wild type primary fibroblasts showed that *let-7* and miR-21 existed in their mature, processed form in wild type cells. However, in *Dicer* null fibroblasts which have undergone recombination following infection with a Cre-expressing adenovirus, mature miRNAs were only detected at very low levels (Figure 4.5) (Harfe *et al.*, 2005). This confirms that recombination of the

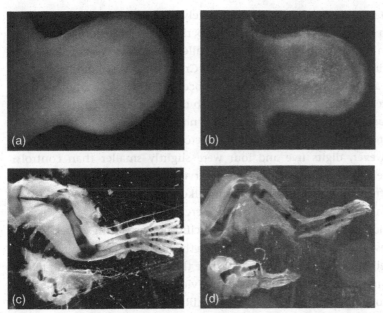

Figure 4.6. Limbs are morphologically abnormal in mice lacking functional *Dicer*. (a) and (b) Staining for cells undergoing cell death in E11.5 limbs using acridine orange shows a significant increase in cell death in mutant (b) compared to wild type (a) forelimbs. (c) and (d) Skeletal preparations of stage E20 wild type (top) and mutant (bottom) forelimbs (c) and hindlimbs (d). While *Dicer* does not appear to be required for patterning, there is a significant growth deficiency in mutant limbs (reproduced from Harfe *et al.* (2005); copyright of National Academy of Sciences USA). (See color plate 4)

Dicer^{floxed} allele resulted in the production of a null allele and significantly decreased the abundance of mature miRNAs. To examine the roles of miRNAs during limb development, mice homozygous for the conditional *Dicer* allele (*Dicer*^{floxed}/*Dicer*^{floxed}) were crossed to mice transgenic for the *prx1cre* allele. This transgene is expressed in the forelimb mesoderm at E9.5, and in the hindlimb mesoderm around E10.5 (Logan *et al.*, 2002). Therefore in the hindlimb, *Dicer* deletion occurs after early stages of limb development have been initiated.

Dicer removal from the limb mesoderm results in a loss of miRNA processing (Harfe *et al.*, 2005). Starting at E11.0, *Dicer* mutant limbs were smaller than those of wild type littermates. The size difference was more severe in the forelimb, possibly owing to the delayed expression of the *prx1cre* transgene in the hindlimb. To determine whether this decrease in size could be caused by loss of cells by apoptosis, embryos were stained with a marker of apoptotic cells. While no difference in staining was observed between wild type and mutant forelimbs at E10.0, by E10.5 *Dicer* mutants showed significant staining indicating increased cell death. In the hindlimb, cell death was not observed until E12.5 (Figure 4.6a). Therefore, the reduction in limb size can, in part, be attributed to an increase in cell death.

Remarkably, *Dicer*-deficient limbs form all the proper proximal limb components, including bones of the forelimb (humerus, radius, ulna) and hindlimb (femur, tibia, fibula) (Figure 4.6a) (Harfe *et al.*, 2005). The distal portion of the forelimb is most severely affected, with a reduction in digit number. This could be because of a defect in cell survival, or alternatively this could represent a patterning defect. To test whether the limb is improperly patterned when miRNAs are removed from the ZPA, the *Dicer* conditional allele was crossed to the *Shhgfpcre* allele. In the limb, the *Shhgfpcre* allele expresses Cre in the ZPA (Harfe *et al.*, 2004). Loss of *Dicer*, and as a consequence miRNA processing in the ZPA, did not have apparent patterning defects, as all five forelimb digits were formed (Harfe *et al.*, 2005). Additionally, Shh-expressing cells which normally contribute to digits five, four and part of three, were not found in ectopic locations in *Dicer* mutant embryos; however, digits five and four were slightly smaller than controls. Therefore, the severe reduction of digits observed in the forelimb when *Dicer* was removed from the entire limb mesoderm is likely not the result of a patterning defect, but can partially be explained by the increased cell death.

The conditional removal of *Dicer* from specific tissues during development allows us to investigate the roles miRNAs play in the initiation and differentiation of different cell lineages and tissue types. While these experiments do not tell us what specific miRNAs are important in an individual tissue, they do give us the opportunity to uncover possible downstream miRNA targets. While it was originally thought that the majority of miRNA target genes were down-regulated by translational inhibition, recent work by several groups has shown that miRNA targets do undergo mRNA degradation (Bagga *et al.*, 2005; Jakymiw *et al.*, 2005; Liu *et al.*, 2005). This degradation occurs by a different mechanism than transcripts that undergo RISC-mediated cleavage. However, the decrease in mRNA levels of miRNA targets can be exploited using methods that compare RNA levels.

Using the *prx1cre* transgene, we are currently undertaking experiments to identify genes that are misregulated when *Dicer* is removed from the limb mesoderm. By comparing the expression profiles between wild-type and mutant limbs using microarrays, we have uncovered over 250 genes that are misregulated when miRNAs are no longer processed (Maatouk and Harfe, unpublished data). miRNAs are negative regulators of their target genes, therefore direct targets of these genes should be up-regulated when *Dicer* is removed. However, once these genes are ectopically up-regulated they have the potential to affect the expression of other genes. The list of misregulated genes will therefore include two sets of genes, the first being up-regulated as a result of *Dicer* removal and the second being up or down-regulated as a secondary effect. Of the >250 genes we found to be misregulated in the *Dicer*-deficient limb mesoderm, over half were up-regulated. While down-regulated genes can be eliminated as direct miRNA targets, determining which of the up-regulated genes are direct miRNA targets is more challenging. It is possible to identify potential miRNA binding sites in target genes using recently developed computer programs such as TargetScan (Lewis *et al.*, 2003; Lewis *et al.*, 2005). These binding sites can then be tested *in vitro* for their ability to silence a reporter gene in the presence of the predicted miRNA. Additionally, knowledge of the specific miRNAs expressed in different tissue types will aid in the discovery and validation of target genes. Understanding which genes are misregulated as a result of *Dicer* removal will further our understanding of how miRNAs control developmental processes.

In addition to our data concerning miRNAs in limb morphogenesis, we have initiated collaborations to investigate the role of miRNAs in the development of a variety of other tissues. In one study, the roles of miRNAs during mammalian lung morphogenesis have been investigated using the *shhgfpcre* allele. The lungs of embryos lacking functional Dicer in Shh-expressing cells show a decrease in the characteristic branch morphology but no initial defect in epithelial cell proliferation (Harris *et al.*, 2006). This leads to the formation of several large epithelial pouches instead of numerous fine branches. Another defect in these embryos is the aberrant expression of *Fgf10*, a molecule known to be important during lung morphogenesis; however, the mechanism by which *Fgf10* expression is misregulated remains unclear as *Fgf10* does not appear to be a direct miRNA target (Harris *et al.*, 2006).

The conditional *Dicer* allele will be a powerful tool for identifying the general importance of miRNAs in many developmental paradigms. Researchers will be able to remove *Dicer* (and miRNAs) from any tissue in which Cre protein can be expressed. These experiments will greatly advance our knowledge of the functions of miRNAs during mammalian development.

REFERENCES

Bagga, S., Bracht, J., Hunter, S. *et al.* (2005). Regulation by let-7 and lin-4 miRNAs results in target mRNA degradation. *Cell*, **122**, 553–563.

Bateson, W. (1894). *Materials for the Study of Variation.* New York: Macmillan.

Bernstein, E., Caudy, A. A., Hammond, S. M. and Hannon, G. J. (2001). Role for a bidentate ribonuclease in the initiation step of RNA interference. *Nature*, **409**, 363–366.

Bernstein, E., Kim, S. Y., Carmell, M. A. *et al.* (2003). Dicer is essential for mouse development. *Nature Genetics*, **35**, 215–217.

Boncinelli, E., Simeone, A., Acampora, D. and Mavilio, F. (1991). HOX gene activation by retinoic acid. *Trends in Genetics*, **7**, 329–334.

Brennecke, J., Hipfner, D. R., Stark, A., Russell, R. B. and Cohen, S. M. (2003). Bantam encodes a developmentally regulated microRNA that controls cell proliferation and regulates the proapoptotic gene hid in Drosophila. *Cell*, **113**, 25–36.

Charite, J., de Graaff, W., Shen, S. and Deschamps, J. (1994). Ectopic expression of Hoxb-8 causes duplication of the ZPA in the forelimb and homeotic transformation of axial structures. *Cell*, **78**, 589–601.

Dolle, P. and Duboule, D. (1989). Two gene members of the murine HOX-5 complex show regional and cell-type specific expression in developing limbs and gonads. *European Molecular Biology Organization Journal*, **8**, 1507–1515.

Dolle, P., Izpisua-Belmonte, J. C., Falkenstein, H., Renucci, A. and Duboule, D. (1989). Coordinate expression of the murine Hox-5 complex homoeobox-containing genes during limb pattern formation. *Nature*, **342**, 767–772.

Duboule, D. and Dolle, P. (1989). The structural and functional organization of the murine HOX gene family resembles that of Drosophila homeotic genes. *European Molecular Biology Organization Journal*, **8**, 1497–1505.

Graham, A., Papalopulu, N. and Krumlauf, R. (1989). The murine and *Drosophila* homeobox gene complexes have common features of organization and expression. *Cell*, **57**, 367–378.

Griffiths-Jones, S. (2004). The microRNA registry. *Nucleic Acids Research*, **32**, D109–D111.

Grishok, A., Pasquinelli, A. E., Conte, D. *et al.* (2001). Genes and mechanisms related to RNA interference regulate expression of the small temporal RNAs that control *C. elegans* developmental timing. *Cell*, **106**, 23–34.

Harfe, B. D. (2005). MicroRNAs in vertebrate development. *Current Opinion in Genetics & Development*, **15**, 410–415.

Harfe, B. D., Scherz, P. J., Nissim, S. *et al*. (2004). Evidence for an expansion-based temporal Shh gradient in specifying vertebrate digit identities. *Cell*, **118**, 517–528.

Harfe, B. D., McManus, M. T., Mansfield, J. H., Hornstein, E. and Tabin, C. J. (2005). The RNaseIII enzyme Dicer is required for morphogenesis but not patterning of the vertebrate limb. *Proceedings of the National Academy of Sciences USA*, **102**, 10898–10903.

Harris, K. S., Zhang, Z., McManus, M. T., Harfe, B. D. and Sun, X. (2006). Dicer function is essential for lung epithelium morphogenesis. *Proceedings of the National Academy of Sciences USA* (in press).

Hornstein, E., Mansfield, J. H., Yekta, S. *et al*. (2005). The microRNA miR-196 acts upstream of Hoxb8 and Shh in limb development. *Nature*, **438**, 671–674.

Hutvagner, G., McLachlan, J., Pasquinelli, A. E. *et al*. (2001). A cellular function for the RNA-interference enzyme Dicer in the maturation of the let-7 small temporal RNA. *Science*, **293**, 834–838.

Izpisua-Belmonte, J. C., Falkenstein, H., Dolle, P., Renucci, A. and Duboule, D. (1991). Murine genes related to the Drosophila AbdB homeotic genes are sequentially expressed during development of the posterior part of the body. *European Molecular Biology Organization Journal*, **10**, 2279–2289.

Jakymiw, A., Lian, S., Eystathioy, T. *et al*. (2005). Disruption of GW bodies impairs mammalian RNA interference. *Nature Cell Biology*, **7**, 1167–1174.

Ketting, R. F., Fischer, S. E., Bernstein, E. *et al*. (2001). Dicer functions in RNA interference and in synthesis of small RNA involved in developmental timing in *C. elegans. Genes & Development*, **15**, 2654–2659.

Kos, C. H. (2004). Cre/loxP system for generating tissue-specific knockout mouse models. *Nutrition Reviews*, **62**, 243–246.

Krumlauf, R. (1994). Hox genes in vertebrate development. *Cell*, **78**, 191–201.

Lagos-Quintana, M., Rauhut, R., Meyer, J., Borkhardt, A. and Tuschl, T. (2003). New microRNAs from mouse and human. *RNA*, **9**, 175–179.

Lee, R. C., Feinbaum, R. L. and Ambros, V. (1993). The *C. elegans* heterochronic gene lin-4 encodes small RNAs with antisense complementarity to lin-14. *Cell*, **75**, 843–854.

Lee, Y., Ahn, C., Han, J. *et al*. (2003). The nuclear RNase III Drosha initiates microRNA processing. *Nature*, **425**, 415–419.

Lewis, B. P., Shih, I. H., Jones-Rhoades, M. W., Bartel, D. P. and Burge, C. B. (2003). Prediction of mammalian microRNA targets. *Cell*, **115**, 787–98.

Lewis, B. P., Burge, C. B. and Bartel, D. P. (2005). Conserved seed pairing, often flanked by adenosines, indicates that thousands of human genes are microRNA targets. *Cell*, **120**, 15–20.

Lewis, E. B. (1978). A gene complex controlling segmentation in *Drosophila. Nature*, **276**, 565–570.

Lim, L. P., Glasner, M. E., Yekta, S., Burge, C. B. and Bartel, D. P. (2003). Vertebrate microRNA genes. *Science*, **299**, 1540.

Lingel, A., Simon, B., Izaurralde, E. and Sattler, M. (2003). Structure and nucleic-acid binding of the *Drosophila* Argonaute 2 PAZ domain. *Nature*, **426**, 465–469.

Liu, J., Valencia-Sanchez, M. A., Hannon, G. J. and Parker, R. (2005). MicroRNA-dependent localization of targeted mRNAs to mammalian P-bodies. *Nature Cell Biology*, **7**, 719–723.

Logan, M., Martin, J. F., Nagy, A. *et al*. (2002). Expression of Cre Recombinase in the developing mouse limb bud driven by a Prxl enhancer. *Genesis*, **33**, 77–80.

Lu, H. C., Revelli, J. P., Goering, L., Thaller, C. and Eichele, G. (1997). Retinoid signaling is required for the establishment of a ZPA and for the expression of Hoxb-8, a mediator of ZPA formation. *Development*, **124**, 1643–1651.

Lund, E., Guttinger, S., Calado, A., Dahlberg, J. E. and Kutay, U. (2004). Nuclear export of microRNA precursors. *Science*, **303**, 95–98.

Mansfield, J. H., Harfe, B. D., Nissen, R. *et al.* (2004). MicroRNA-responsive 'sensor' transgenes uncover Hox-like and other developmentally regulated patterns of vertebrate microRNA expression. *Nature Genetics*, **36**, 1079–1083.

Martin, P. (1990). Tissue patterning in the developing mouse limb. *International Journal of Developmental Biology*, **34**, 323–336.

McGinnis, W. and Krumlauf, R. (1992). Homeobox genes and axial patterning. *Cell*, **68**, 283–302.

McGinnis, W., Levine, M. S., Hafen, E., Kuroiwa, A. and Gehring, W. J. (1984). A conserved DNA sequence in homoeotic genes of the *Drosophila* Antennapedia and bithorax complexes. *Nature*, **308**, 428–433.

McManus, M. T. and Sharp, P. A. (2002). Gene silencing in mammals by small interfering RNAs. *Nature Reviews Genetics*, **3**, 737–747.

Min, W., Woo, H. J., Lee, C. S. *et al.* (1998). 307-bp fragment in HOXA7 upstream sequence is sufficient for anterior boundary formation. *DNA and Cell Biology*, **17**, 293–299.

Murchison, E. P. and Hannon, G. J. (2004). miRNAs on the move: miRNA biogenesis and the RNAi machinery. *Current Opinion in Cell Biology*, **16**, 223–229.

Nelson, C. E., Morgan, B. A., Burke, A. C. *et al.* (1996). Analysis of Hox gene expression in the chick limb bud. *Development*, **122**, 1449–1466.

Nicholson, R. H. and Nicholson, A. W. (2002). Molecular characterization of a mouse cDNA encoding Dicer, a ribonuclease III ortholog involved in RNA interference. *Mammalian Genome*, **13**, 67–73.

Pasquinelli, A. E., Reinhart, B. J., Slack, F. *et al.* (2000). Conservation of the sequence and temporal expression of let-7 heterochronic regulatory RNA. *Nature*, **408**, 86–89.

Reinhart, B. J., Slack, F. J., Basson, M. *et al.* (2000). The 21-nucleotide let-7 RNA regulates developmental timing in *Caenorhabditis elegans*. *Nature*, **403**, 901–906.

Schwer, B. (2001). A new twist on RNA helicases: DExH/D box proteins as RNPases. *Nature Structural Biology*, **8**, 113–116.

Scott, M. P. (1992). Vertebrate homeobox gene nomenclature. *Cell*, **71**, 551–553.

Scott, M. P. and Weiner, A. J. (1984). Structural relationships among genes that control development: sequence homology between the Antennapedia, Ultrabithorax, and fushi tarazu loci of Drosophila. *Proceedings of the National Academy of Sciences USA*, **81**, 4115–4119.

Song, J. J., Liu, J., Tolia, N. H. *et al.* (2003). The crystal structure of the Argonaute2 PAZ domain reveals an RNA binding motif in RNAi effector complexes. *Nature Structural Biology*, **10**, 1026–1032.

Stratford, T. H., Kostakopoulou, K. and Maden, M. (1997). Hoxb-8 has a role in establishing early anterior–posterior polarity in chick forelimb but not hindlimb. *Development*, **124**, 4225–4234.

Tickle, C. (2003). Patterning systems – from one end of the limb to the other. *Developmental Cell*, **4**, 449–458.

van den Akker, E., Fromental-Ramain, C., de Graaff, W. *et al.* (2001). Axial skeletal patterning in mice lacking all paralogous group 8 Hox genes. *Development*, **128**, 1911–1921.

Yan, K. S., Yan, S., Farooq, A. *et al.* (2003). Structure and conserved RNA binding of the PAZ domain. *Nature*, **426**, 468–474.

Yekta, S., Shih, I. H. and Bartel, D. P. (2004). MicroRNA-directed cleavage of HOXB8 mRNA. *Science*, **304**, 594–596.

Yi, R., Qin, Y., Macara, I. G. and Cullen, B. R. (2003). Exportin-5 mediates the nuclear export of pre-microRNAs and short hairpin RNAs. *Genes & Development*, **17**, 3011–3016.

5 Identification of miRNAs in the plant *Oryza sativa*

Hui Zhou, Yue-Qin Chen, Yu-Chun Luo, Jia-Fu Wang and Liang-Hu Qu*

Introduction

MicroRNAs (miRNAs) are single-stranded small RNAs of *c.*22 nt in length that function as post-transcriptional negative regulators in plants and animals (Bartel, 2004; Kim, 2005). Among all the categories of endogenous small RNAs, these tiny RNAs have received the most notice (Sontheimer and Carthew, 2005). They act as small guides and direct negative regulations through sequence complementarity to the 3′-untranslated regions (UTRs) in animals or coding sequences in plants of an even larger number of target mRNAs (Grishok *et al.*, 2001; Lai, 2002; Bartel and Bartel, 2003). Now miRNA genes are recognized as a pervasive and widespread feature of animal and plant genomes. A large number of miRNAs have been characterized from different animals such as the worm *Caenorhabditis elegans*, the fly *Drosophila*, and mammals including the human, the mouse and the rat. The total number of miRNAs in multicellular organisms was estimated to represent about 1% of all genes (Lim *et al.*, 2003b; Grad *et al.*, 2003; Bartel, 2004). However, recent studies show the number of miRNAs in the primate is larger than initially believed (Berezikov *et al.*, 2005; Bentwich *et al.*, 2005). Bioinformatic analysis implies that 25% of vertebrate genes are conserved targets of the miRNAs (Lewis *et al.*, 2005). In total, multiple experimental and computational strategies all indicate that the true extent of miRNA regulation in complex organisms has not been fully recognized.

Although a large number of miRNAs have been identified from various animals, plant miRNA studies focused mainly on *Arabidopsis* before our study, and only a handful of miRNAs have been identified (Reinhart *et al.*, 2002; Park *et al.*, 2002; Llave *et al.*, 2002a). Rice is an economically important food crop. Recent progress in the Rice Genome Project has revealed 50 000–60 000 genes in the rice genome (Yu *et al.*, 2002), nearly twice the number of genes predicted for *Arabidopsis thaliana* (The *Arabidopsis* Genome Initiative, 2002), which made rice a favorite plant for functional genomic research. It is therefore interesting to

---

* Author to whom correspondence should be addressed.

MicroRNAs: From Basic Science to Disease Biology, ed. Krishnarao Appasani. Published by Cambridge University Press. © Cambridge University Press 2008.

investigate miRNAs in rice, which is regarded as a model of the monocotyledon. In this study, we developed a novel strategy to construct a rice cDNA library of size-fractionated RNAs (Wang *et al.*, 2004). Our results showed a very abundant and diverse population of small RNAs in rice, from which 20 miRNAs were further characterized.

Methodology

Nucleic acid isolation

Total RNAs from different tissues of *Oryza sativa* L. ssp. *indica*, *A. thaliana* and maize were isolated by guanidine thiocyanate/phenol–chloroform extraction as described by Chomczynski and Sacchi (1987). A fraction enriched with small-sized RNAs was obtained according to a protocol described previously (Park *et al.*, 2002). Briefly, 400 µl (1–2 mg) of total cellular RNAs were combined with 50 µl each of 50% PEG8000 and 5 M NaCl, incubated on ice for 2 h, and centrifuged at 15 000 g for 10 min. After adding 1/10 vol of 3 M sodium acetate and 2 vols of 95% ethanol to the supernatant, small-sized RNA was spun down at 15 000 g following incubation at − 20 °C for 2 h, washed with 75% ethanol, dried briefly, and re-suspended in RNase-free water.

Cloning of miRNAs from rice

Synthesis, cloning and sequencing of cDNAs were carried out as described previously (Chen *et al.*, 2003) with some modifications. Briefly, small RNAs with a size from 16 to 28 nt were recovered from denaturing 15% polyacrylamide gel fractionation (8 M urea, 13 TBE buffer), and then polyadenylated by using poly(A) polymerase (Takara). Synthesis of the first strand of cDNA was performed with 1 µg of poly(A)-tailed RNA in a 20 µl reaction mix containing 0.1 µg of primer HindIII (dT)16 5'-CCCCAAAGCT16-3' and 200 U of MMLV reverse transcriptase (Promega) for 45 min at 42 °C. The cDNAs were poly(G) tailed at the 3' end by using terminal deoxynucleotidyl transferase (Takara), and then amplified by PCR with primers HindIII (dT)$_{16}$ and BamHI (dC)$_{16}$, 5'-GGAATTCGGATC16-3' and cloned into plasmid pTZ18 as described previously (Zhou *et al.*, 2002). The recombinant plasmid-carrying fragments were sequenced with the BigDye terminator cycle sequencing kit (PE Applied Biosystems) and were analyzed on an ABI377 DNA sequencer.

Prediction of stem-loop structures

The cloned small RNA sequences were used to search the rice and *Arabidopsis* genome database using BLAST (http://www.ncbi.nlm.nih.gov/blast, http://btn.genomics.org.cn/rice). Only those sequences that comprised matches to one or more rice genome sequences were further analyzed. Secondary structures of RNA precursor were predicted by m-fold program (http://www.bioinfo.rpi.edu/applications/mfold/old/rna/form1.cgi) (Zuker, 2003). The RNA sequences were folded with flanking sequences: 150–400 bp upstream and 20 bp downstream or vice versa, 150 bp upstream and 150 bp downstream.

Prediction of miRNA targets

Potential targets for rice miRNAs are identified by BLAST analyses according to the search algorithm in which only three or fewer mismatches are allowed to be present in the complementarity between miRNAs and their targets, and gaps are not allowed (Rhoades *et al.*, 2002).

Northern blot analysis

A 30 µg aliquot of small-sized RNAs from different tissues of rice and other plants was separated on a denaturing 15% polyacrylamide–8 mol/l urea gel at 300 V for 3 h, then electrophoretically transferred to Zeta-probe GT membranes (Bio-Rad) by using a Trans-Blot Electrophoretic Transfer (Pharmacia LKB). After electroblotting, the RNAs were fixed to the membrane by UV cross-linking (1200 µJ, Stratalinker; Stratagene) and by baking in a vacuum oven at 80 °C for 1 h. DNA oligonucleotides complementary to miRNA sequences were synthesized (Sangon, Shanghai). The 5′ ends of the DNA probes were labeled with [γ-^{32}P]ATP (Yahui Co.) using T_4 poly-nucleotide kinase (Promega) and subjected to purification according to standard laboratory protocols as described previously (Sambrook *et al.*, 1989). The membrane was pre-hybridized in 7% SDS, 0.3 M NaCl, and 50 mM phosphate buffer (pH 7.2) at 42 °C for at least 1 h. Membranes were hybridized with ^{32}P-end-labeled oligonucleo-tide probes at a temperature of 10–15 °C below the calculated dissociation tempera-ture (Td) for 16 h. The blots were washed twice with $2 \times$ SSPE/0.5% SDS and once with $0.5 \times$ SSPE/0.1% SDS at 40 °C. The Northern blots were quantified by using a phosphorimager apparatus (Typhon 8600, Amersham Bioscience).

Results

A highly efficient method for miRNA cloning with 3′ end poly(A) strategy

To investigate the population of small RNAs in rice, we initiated a novel strategy to generate a cDNA library from rice that encodes candidate miRNAs (Figure 5.1). A cDNA library representing small RNAs was constructed from rice tissues of different developmental stages. A poly(A) tail was added to the ends of the size-fractionated RNA, followed by RT PCR with a primer pair of BamHI(dC) and HindIII(dT), then cloning and sequencing (see details in methodology). A total of 236 insert-contain-ing clones randomly selected from the library were analyzed, and 203 unique sequences were obtained. Although the sequences ranged in length from 15 to 30 nt, they had a much tighter length distribution, centering on 20–25 nt (Figure 5.2), which was coincident with the known specificity of Dicer processing. This profile was very similar to that of miRNA populations isolated from *A. thaliana* and *C. elegans*, and small interference RNA (siRNA) generated in a green fluorescent protein (GFP) and double-stranded GFP silencing system (Lau *et al.*, 2001; Llave *et al.*, 2002b).

Identification of 20 miRNAs from rice

The 203 cloned RNA sequences were subjected to BLAST analysis against the rice genome. Only two sequences originated from fragments of tRNA or rRNA. A total of 122 sequences had at least one nucleotide mismatch with the rice genome and

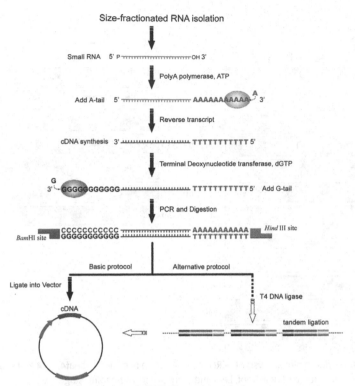

Figure 5.1. Schematic representation of cDNA library construction.

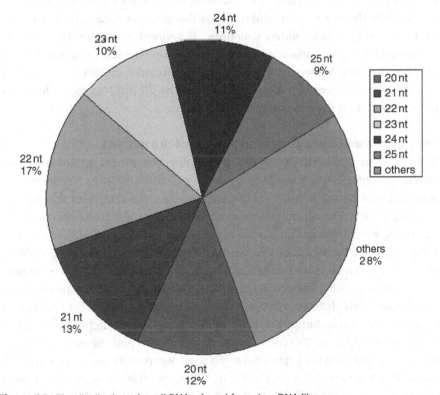

Figure 5.2. The distribution of small RNAs cloned from rice cDNA library.

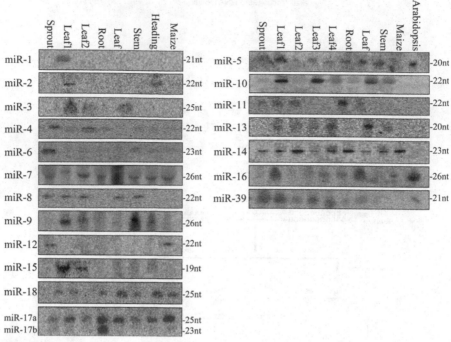

Figure 5.3. Northern blot analysis of miRNAs cloned from rice. Leaf1–Leaf4, 1–4 leaf stage of rice; Heading, heading stage of rice; Root, Leaf and Stem are from heading stage.

they were not further analyzed. The remaining 79 small RNA sequences matched perfectly with the rice genome, and their possible genomic locations were identified. In the analysis of secondary structures, 32 genomic sequences that contain putative miRNAs were capable of forming stem-loop structures characteristic of miRNA precursors. Subsequently, 20 of the 32 putative miRNAs were further confirmed by Northern blot (Figure 5.3). Thus the 20 miRNAs were identified from rice, and their sequences are shown in Table 5.1.

Tissue-specific and developmental expression of rice miRNAs

The expression of the rice miRNAs was examined by Northern blot with low molecular weight RNAs isolated from different tissues of rice. Northern blot confirmed that 20 miRNAs were stably expressed in rice cells from at least one tissue (Figure 5.3). The accumulations of some miRNAs appear to be developmental or tissue specific. For example, miR-1, miR-6 and miR-10 were somewhat more strongly expressed as seedlings compared with the adult plant, or vice versa. Note that miR-11 and miR-16 accumulate at a higher level in roots and leaves, respectively. A probe to miR-17 detected both miR-17a and miR-17b with a size of 25 and 23 nt, respectively. Interestingly, miR-17a accumulated in different tissues and developmental stages, but miR-17b with 23 nt is only most highly expressed in the rice roots. Other miRNAs were not uniformly expressed through all tissues, and large variations in the expression levels were observed. In some cases, miRNA precursors were detected as described in *C. elegans* (Lau *et al.*, 2001). Probes complementary to rice miRNAs were used to test total cellular RNAs from other

Table 5.1. Rice microRNAs identified by cloning

miRNA	Sequence (5' → 3')	Northern blotting	Length (nt)	Fold-back arm	Fold-back length	Chr.	Location in chromosome
miR-1	UGAAGAGUUUACCUUUUUACU	+	21	5'	102	1	intergenic
miR-2	GAAGGUUUGGUGUGCUUUUGUC	+	22	3'	82	1	intergenic
miR-3	UUAUUAAGAACAACAUUUUGUGUC	+	25	5'	119	1	intergenic
miR-4	UAAGCUAUUAAAUUGGCUAUAG	+	22	5'	362	1	intergenic
				5'	365	3	intergenic
				5'	362	4	intergenic
miR-5	UUCAUUAACAAGUUACACCA	+	20	5'	160	7	intergenic
miR-6	UCUAUGGAAAAUGGGUCAUUGCC	+	23	3'	68	5	exon
miR-7	GUGUGCGUGCCGCCAUGGGGAUGUC	+	26	3'	62	1	intron
				3'	64	4	intergenic
				3'	62	7	intergenic
				3'	64	10	intergenic
miR-8	UCUGUACAGUUUACAAUUCCCC	+	22	3'	180	1	Intergenic
miR-9	UCAGUAAAGAUUAUAUACUUCCUAC	+	24	5'	170	1	intergenic
miR-10	UAUCUGAAGUUCUUGCCAAGAAC	+	22	3'	60	2	intergenic
miR-11	UAUAGCCCUUGGUAGAGUUACU	+	22	5'	150	1	Intergenic
				5'	150	3	Intergenic
				5'	154	4	Intergenic
				5'	156	10	Intergenic
miR-12	UGAAUAAACGGCAGUUCCGUUG	+	22	3'	74	12	intergenic
miR-13	AACUCUUUGAUGAAAAGAUG	+	20	5'	114	1	Intergenic
				5'	114	3	Intergenic
				5'	114	10	Intergenic
				5'	114	12	Intergenic
miR-14	UGCCACGUUAGCAAAUAUUUUGC	+	23	5'	164	12	intergenic

Table 5.1. (cont.)

miRNA	Sequence (5′ → 3′)	Northern blotting	Length (nt)	Fold-back arm	Fold-back length	Chr.	Location in chromosome
miR-15	AUUUGAUUCGUCGAUCUUC	+	19	3′	63	2	Intergenic
				3′	63	4	Intergenic
				3′	63	6	Intergenic
miR-16	AAAAUGUAGUACAUCAUGUUUACUG	+	26	5′	71	7	intergenic
miR-17a	AAUGCCGAAUCGACGACCUAUGUAU	+	25	5′	124	4	Intergenic
				5′	124	10	intron
miR-17b	AAUGCCGAAUCGACGACCUAUGU	+	23	5′	124	4	Intergenic
				5′	124	10	intron
miR-18	UUAAUCCAUUUAAAUCAAUUAAAUC	+	25	3′	87	1	Intergenic
				3′	87	1	exon
				3′	87	3	Intergenic
				3′	87	4	Intergenic
				3′	76	7	Intergenic
				3′	87	8	Intergenic
				3′	87	8	Intergenic
				3′	76	10	intron
				3′	76	10	Intergenic
				3′	76	10	Intergenic
				3′	87	12	Intergenic
				3′	87	12	Intergenic
miR-39	**UGAUUGAGCCGUGCCAAUAUC**	+	21	3′	117	10	Intergenic
				3′	110	3	intergenic

The miRNA location in the predicted fold-back structure is specified (5′ or 3′ arm). The multiple copies of miRNAs whose precursors can form fold-back structures are also included in the table. The miRNA and its sequence conserved in *Arabidopsis* are shown in bold. The " + " represents that small RNA was validated by northern blotting.

plants in the same assays. The homologs of miR-39, miR-5 and miR-16 were found to be present in *A. thaliana*, and miR-7, miR-12, miR-14, miR-16, miR-17 and miR-18 were detected in maize (Figure 5.3), suggesting the conservation of these miRNAs in flowering plants.

Genomic organization of rice miRNAs

Among the 20 miRNAs, 11 were each encoded by a single copy in the rice genome, whereas 10 miRNAs had multiple loci in the genome (Table 5.1), probably because of duplications that were still active in the rice genome (Yu *et al.*, 2002). Comparative analyses showed that the multiple genomic sequences containing the miRNAs are often completely or partially conserved in the rice genome. For example, miR-18 corresponds to 12 genomic sequences in long terminus repeat sequences or in the Osr1 retrotransposon. Twenty-eight genomic loci representing 15 rice miRNAs were mapped to intergenic regions. Three copies of miRNAs were found in the sense orientation within introns of coding transcripts. Two copies of miRNAs were located in the exons of potential protein-coding genes. Another 13 copies from nine miRNAs were found in genomic regions that have no annotation in the genome. Therefore, miRNAs are either transcribed from their own promoters or derived from a pre-mRNA that frequently codes for an additional gene product.

Rice microRNA targets prediction

To identify potential targets, we searched for rice mRNAs with complementarity to the newly identified miRNAs by an algorithm employed upon the discovery of *Arabidopsis* miRNA targets, in which most antisense hits with three or fewer mismatches appeared to be potential miRNA targets (Rhoades *et al.*, 2002). Twenty-three unique rice genes were identified to be feasible targets for seven rice miRNAs, because there are high stringent complementarities with three or fewer mismatches between the miRNAs and their targets (Table 5.2). Among the seven miRNAs, three miRNAs are perfect antisenses matched to the mRNAs of protein-coding genes; miR-39 targets the coding region of mRNAs from four members of Scarecrow-like transcription factor that controls a wide range of developmental processes. This miRNA appears homologous to *Arabidopsis* miR-39 (or miR-171) that had been predicted to be present in rice by RNA hybridization (Reinhart *et al.*, 2002; Llave *et al.*, 2002b). In fact, miR-39 varies slightly in sequence (one nucleotide change) between the two plants, but a U/G pair still maintains a perfect duplex of miR-39 – target mRNA in rice. Interestingly, two rice miR-39 precursors display secondary structures very similar to that of *Arabidopsis* despite considerably divergent sequences outside the miRNAs, lending support to the evolutionary conservation of this miRNA in flowering plants. A novel target predicted for miR-5 is the mRNA of a protein-coding gene whose function remains annotated. Being different from the two precedents, miR-15 has a perfect antisense match to the fourth intron of PDR-like ABC transporter, implying a possible role for miRNA in the stage of RNA splicing. Lacking longer complementarity, the targets of 13 miRNAs remain to be further investigated.

Table 5.2. List of the potential targets for part of rice miRNAs

miRNA	Target gene	Number of mismatches	Target protein family
miR-1	Os.19366	3	Unknown protein
miR-4	Os.27038	3	MAP3K α protein kinase
	Os.23008	3	Mucin-like protein
	AJ535074	3	MRP-like ABC transporter
	Os.1798	3	Unknown protein
	Os.35458	3	Unknown protein
	Os.38137	3	Unknown protein
	AK062268	3	Unknown protein
	AK062967	3	Unknown protein
	AK068031	3	Unknown protein
	AK100013	3	Unknown protein
	AK100599	3	Unknown protein
miR-5	Os.25244	0	Trehalose-6-phosphate phosphatase
miR-10	Os.24878	3	Cf2/Cf5 disease resistance protein
	Os.11923	3	Unknown protein
miR-13	Os.9392	3	Aluminum-induced-like protein
	AK064680	3	Unknown protein
miR-15	AJ535053	0	PDR-like ABC transporter
miR-39	Os.2406	1	SCL transcription factor
	Os.20826	1	SCL transcription factor
	Os.21578	1	SCL transcription factor
	Os.31716	1	SCL transcription factor
	Os.46878	2	SCL transcription factor

Rice genes are labeled by their UniGene numbers or accession numbers in GenBank. Prediction was performed as described by Rhoades *et al.* (2002).

Discussion

Strategies for cloning microRNA

A pioneering strategy for microRNA identification was first established in the study of *C. elegans* and *Drosophila* in 2001 (Lagos-Quintana *et al.*, 2001; Lau *et al.*, 2001; Lee and Ambros, 2001). The brief cloning approaches are as follows: 5′ and 3′ adapter molecules were ligated to the ends of a size-fractionated RNA population, followed by RT-PCR amplification, digestion, concatamerization, cloning, and then sequencing (Lagos-Quintana *et al.*, 2001; Lau *et al.*, 2001). The protocol has matured and has been used by many labs. This "adapter protocol," however, has its limitations. It includes too many steps, including gel-purification, dephosphorylation, phosphorylation and ligation, which may lead to decrease of RNA, and as a result, may miss low abundant miRNAs. This protocol is also relatively complicated, and unskilled manipulations may cause excessive fragments of rRNA, tRNA, snoRNA and self-ligated adapter.

In order to simplify miRNA cloning manipulation and promote its efficiency, we developed a novel microRNA cloning strategy named "oligo protocol." In this protocol, 5′ and 3′ adapters were replaced by poly(A) and poly(C) (Figure 5.1). A poly(A) tailed, not a poly(G) tailed, strategy was used because poly(A) polymerase, low abundant miRNAs could be cloned efficiently with this protocol. Other advantages of this protocol are fewer steps and easy manipulation. Although poly(A)-tailed RNA resembles mRNA, which might result in some fragments derived from degraded mRNA in the cDNA library, use of the size-fractionated miRNA population can efficiently avoid contaminations of mRNA.

Our method can also be combined with 5' ligation of "adapter protocol," which can further optimize the procedure of the miRNA cloning. This has been adopted in the identification of miRNAs from human fetal liver (Fu *et al.*, 2005). Another alternative method for cloning is, following RT-PCR and digestion with restriction endonucleases, PCR products that were not directly ligated into vector, and instead are concatamerized with T_4 DNA ligase (Figure 5.1). Then, concatamers bigger than 300–500 bp are gel-purified and cloned. In this way, each clone would include the insertion of over 10 small RNAs, leading to the subsequent procedure of sequencing being more economical and labor saving.

Characterization of rice microRNA

An important criterion required for novel miRNAs is the secondary structures of precursors, which are stem-loop structures recognized and processed subsequently by the endonuclease Dicer. In animals the precursors are usually about 60–80 nt, but the sizes of plant precursors appear more variable (Park *et al.*, 2002; Reinhart *et al.*, 2002), which makes plant miRNA identifcation more complicated than that of an animal. Similar to *Arabidopsis* miRNAs (Reinhart *et al.*, 2002), hairpin structures of rice miRNAs we identified are variable in size. The predicted shortest hairpin precursor (miR10) is only 60 nt in length, whereas the length of miR4 precursor hits 365 nt (Table 5.1). We found that the mature miRNAs could be processed from either the 5' or 3' arm of the fold-back precursor. However, if miRNA matches to multiple sites of the genome, the miRNA always lies in the same arm of its potential precursors in all members of the gene family (Table 5.1), suggesting that these loci share a common origin through local or distant duplications in the rice genome. We noted that most of the cloned sequences had a composition preference that began with a uridine at the 5' terminus (Table 5.1), as had been observed in miRNAs from different organisms (Lee and Ambros, 2001; Lagos-Quintana *et al.*, 2002; Reinhart *et al.*, 2002). This preference for a U at the 5' terminus may help to unwind the miRNA:miRNA* duplex.

In *C. elegans* and mammals, many miRNA clusters have been found (Lagos-Quintana *et al.*, 2001; Lau *et al.*, 2001; Lim *et al.*, 2003a, b). Besides, a large number of animal miRNAs are located in the introns of pre-mRNAs and 25% of human miRNA genes are encoded within introns (Bartel, 2004). However, most rice miRNAs identified in this study are encoded by their own primary transcript. Just three of the total 47 rice miRNA loci were found to be located in introns of coding transcripts, which was consistent with the previous reports of *Arabidopsis* miRNAs (Llave *et al.*, 2002a; Park *et al.*, 2002).

MiRNA and siRNA in plants

A recent study reports the use of massively parallel signature sequencing (MPSS) technology to identify over 1.5 million small RNAs from *Arabidopsis thaliana*, which represent at least 75 000 unique sequences (C. Lu *et al.*, 2005). The results show a surprisingly abundant and complicated world of endogenous small RNAs in plants. However, it is notable that miRNAs belong to only a small fraction of small RNA species in plants and others are presumable endogenous siRNAs (Llave *et al.*, 2002a; Park *et al.*, 2002; Reinhart *et al.*, 2002; Bartel and Bartel, 2003). This point was also confirmed by our results. We analyzed 203 unique sequences of which only 20 were eventually identified to be miRNAs. Thus, biochemical approaches for identifying miRNAs in plants have to face a problem of how to distinguish miRNAs from numerous endogenous siRNAs. According to previous opinion (Bartel, 2004), stem-loop precursor, independent orgin and evolutionary conservation are three key features of miRNAs. However, a new kind of siRNAs named *trans*-acting siRNAs (ta-siRNAs), which derive from independent loci, was recently reported (Peragine *et al.*, 2004; Vazquez *et al.*, 2004; Allen *et al.*, 2005). Moreover, recent study in rice, *Arabidopsis* and *Populus trichocarpa* identified a large number of unconserved miRNAs (Wang *et al*, 2004; Lindow and Krogh, 2005; S. Lu *et al.*, 2005; Sunkar *et al.*, 2005). These show that stem-loop precursor is, so far, the feature remained to distinguish miRNAs from siRNAs (Chen, 2005). In addition, since the expression of miRNA is far more abundant than that of siRNA, positive signal in Northern blotting can serve as direct experimental proof of miRNA.

Regulatory role of rice microRNA

Plant miRNAs and their targets share a high degree of complementarity, which allows the target prediction to use algorithms that search the genome for mRNA–miRNA complementarity. Intriguingly, of the seven miRNAs whose targets had been predicted in this study, five miRNA targets (miR-39, miR-4, miR-10, miR-13, miR-15) fell into transcription factor. As in a previous study in *Arabidopsis* (Rhoades *et al.*, 2002), a large proportion of miRNA target genes consists of transcription factor members which have been regarded as an important regulatory level of gene expression. Now, regulation of microRNA reveals a novel RNA regulatory level that seems to be emergent earlier in the organic evolution. In *Arabidopsis*, miRNAs have been shown to regulate diverse developmental processes, including meristem function, organ polarity and vascular development, floral patterning and hormone response (Kidner and Martienssen, 2005). Compared with *Arabidopsis*, functional research of rice miRNAs falls behind, which may be because of a lack of mutants and difficulty of transgene in rice. Nevertheless, with the progress of rice functional genomics, rice miRNAs should be more fully elucidated in the coming years.

REFERENCES

Allen, E., Xie, Z., Gustafson, A. M. and Carrington, J. C. (2005). MicroRNA-directed phasing during trans-acting siRNA biogenesis in plants. *Cell*, **121**, 207–221.

Bartel, B. and Bartel, D. P. (2003). MiRNAs: at the root of plant development. *Plant Physiology*, **132**, 709–717.

Bartel, D. P. (2004). MiRNAs:genomics, biogenesis, mechanism, and function. *Cell*, **116**, 281–297.

Bentwich, I., Avniel, A., Karov, Y. *et al.* (2005). Identification of hundreds of conserved and nonconserved human miRNAs. *Nature Genetics*, **37**, 766–770.

Berezikov, E., Guryev, V., Belt, J. *et al.* (2005). Phylogenetic shadowing and computational identification of human microRNA genes. *Cell*, **120**, 21–24.

Chen, C. L., Liang, D., Zhou, H. *et al.* (2003). The high diversity of snoRNAs in plants: identification and comparative study of 120 snoRNA genes from *Oryza sativa*. *Nucleic Acids Research*, **31**, 2601–2613.

Chen, X. (2005). MicroRNA biogenesis and function in plants. *Federation of European Biochemical Society Letters*, **579**, 5923–5931.

Chomczynski, P. and Sacchi, N. (1987). Single-step method of RNA isolation by acid guanidinium thiocyanate–phenol–chloroform extraction. *Analytical Biochemistry*, **162**, 732–735.

Fu, H., Tie, Y., Xu, C. *et al.* (2005). Identification of human fetal liver miRNAs by a novel method. *Federation of European Biochemical Society Letters*, **579**, 3849–3854.

Grad, Y., Aach, J., Hayes, G. D. *et al.* (2003). Computational and experimental identification of *C. elegans* miRNAs. *Molecular Cell*, **11**, 1253–1263.

Grishok, A., Pasquinelli, A. E., Conte, D. *et al.* (2001). Genes and mechanisms related to RNA interference regulate expression of the small temporal RNAs that control *C. elegans* developmental timing. *Cell*, **106**, 23–34.

Kidner, C. A. and Martienssen, R. A. (2005). The developmental role of microRNA in plants. *Current Opinion in Plant Biology*, **8**, 38–44.

Kim, N. V. (2005). MicroRNA biogenesis: coordinated cropping and dicing. *Nature*, **6**, 376–385.

Lagos-Quintana, M., Rauhut, R., Lendeckel, W. and Tuschl, T. (2001). Identification of novel genes coding for small expressed RNAs. *Science*, **294**, 853–858.

Lagos-Quintana, M., Rauhut, R., Yalcin, A. *et al.* (2002). Identification of tissue-specific miRNAs from mouse. *Current Biology*, **12**, 735–739.

Lai, E. C. (2002). MiRNAs are complementary to 3′ UTR sequence motifs that mediate negative post-transcriptional regulation. *Nature Genetics*, **30**, 363–364.

Lau, N. C., Lim, L. P., Weinstein, E. G. and Bartel, D. P. (2001). An abundant class of tiny RNAs with probable regulatory roles in *Caenorhabditis elegans*. *Science*, **294**, 858–862.

Lee, R. C. and Ambros, V. (2001). An extensive class of small RNAs in *Caenorhabditis elegans*. *Science*, **294**, 863–864.

Lewis, B. P., Burge, C. B. and Bartel, D. P. (2005). Conserved seed pairing, often flanked by adenosins, indicates that thousands of human genes are microRNA targets. *Cell*, **120**, 15–20.

Lim, L. P., Lau, N. C., Weinstein, E. G. *et al.* (2003a). The miRNAs of *Caenorhabditis elegans*. *Genes & Development*, **17**, 991–1008.

Lim, L. P., Glasner, M. E., Yekta, S., Burge, C. B. and Bartel, D. P. (2003b). Vertebrate microRNA genes. *Science*, **299**, 1540.

Lindow, M. and Krogh, A. (2005). Computational evidence for hundreds of non-conserved plant miRNAs. *BMC Genomics*, **119**.

Llave, C., Kasschau, K. D., Rector, M. A. and Carrington, J. (2002a). Endogenous and silencing- associated small RNAs in plants. *Plant Cell*, **14**, 1605–1619.

Llave, C., Xie, Z., Kasschau, K. D. and Carrington, J. C. (2002b). Cleavage of scarecrow-like mRNA targets directed by a class of *Arabidopsis* miRNA. *Science*, **297**, 2053–2056.

Lu, C., Tej, S. S., Luo, S. *et al.* (2005). Elucidation of the small RNA component of the transcriptome. *Science*, **309**, 1567–1569.

Lu, S., Sun, Y., Shi, R. *et al.* (2005). Novel and mechanical stress-responsive miRNAs in *Populus trichocarpa* that are absent from *Arabidopsis*. *Plant Cell*, **17**, 2186–2203.

Park, W., Li, J., Song, R., Messing, J. and Chen, X. (2002). CARPEL FACTORY, a Dicer homolog and HEN1, a novel protein, act in microRNA metabolism in *Arabidopsis thaliana*. *Current Biology*, **12**, 1484–1495.

Peragine, A., Yoshikawa, M., Wu, G., Albrecht, H. L. and Poethig, R. S. (2004). SGS3 and SGS2/SDE1/RDR6 are required for juvenile development and the production of trans-acting siRNAs in *Arabidopsis*. *Genes & Development*, **18**, 2368–2379.

Reinhart, B. J., Weinstein, E. G., Rhoades, M. W., Bartel, B. and Bartel, D. P. (2002). MiRNAs in plants. *Genes & Development*, **16**, 1616–1626.

Rhoades, M. W., Reinhart, B. J., Lim, L. P. *et al.* (2002). Prediction of plant microRNA targets. *Cell*, **110**, 513–520.

Sambrook, J., Fritsch, E. F. and Maniatis, T. (1989). *Molecular Cloning: A Laboratory Manual.* Cold Spring Harbor, NY: Cold Spring Harbor Laboratory Press.

Sontheimer, E. J. and Carthew, R. W. (2005). Silence from within: endogenous siRNAs and miRNAs. *Cell*, **122**, 9–12.

Sunkar, R., Grike, T., Jain, P. K. and Zhu, J. K. (2005). Cloning and characterization of miRNAs from rice. *Plant Cell*, **17**, 1397–1411.

The *Arabidopsis* Genome Initiative. (2000). Analysis of the genome sequence of the flowering plant *Arabidopsis thaliana*. *Nature*, **408**, 796–815.

Vazquez, F., Vaucheret, H., Rajagopalan, R. *et al.* (2004). Endogenous trans-acting siRNAs regulate the accumulation of *Arabidopsis* mRNAs. *Molecular Cell*, **16**, 69–79.

Wang, J. F., Zhou, H., Chen, Y. Q., Luo, Q. J. and Qu, L. H. (2004). Identification of 20 microRNAs from *Oryza Sativa*. *Nucleic Acids Research*, **32**, 1688–1695.

Yu, J., Hu, S., Wang, J. *et al.* (2002). A draft sequence of the rice genome (*Oryza sativa* L. ssp. *indica*). *Science*, **296**, 79–92.

Zhou, H., Chen, Y. Q., Du, Y. P. and Qu, L. H. (2002). The *Schizosaccharomyces pombe* mgU6–47 snoRNA is required for the methylation of U6 snRNA at 41. *Nucleic Acids Research*, **30**, 894–902.

Zuker, M. (2003). Mfold web server for nucleic acid folding and hybridization prediction. *Nucleic Acids Research*, **31**, 3406–3415.

II

MicroRNA functions and RNAi-mediated pathways

6 Inhibition of translation initiation by a microRNA

David T. Humphreys*, Belinda J. Westman, David I. K. Martin
and Thomas Preiss

Introduction

MicroRNAs (miRNAs) are small (~22 nt) regulatory RNAs that are processed from
stem-loop-forming precursor transcripts (Bartel, 2004; He and Hannon, 2004;
Meister and Tuschl, 2004). In recent years, they have become the subject of intensive research, which quickly amassed a wealth of information on their biogenesis,
function and significance for gene regulation. It also became apparent that miRNAs
are an abundant class of gene regulators. The miRBase database (Release 7.1) lists
3424 miRNA sequence entries from various metazoa, plants and some viruses.
Entries for intensely studied organisms number in the hundreds of distinct
miRNA sequences, many of which exhibit phylogenetic conservation (Griffiths-
Jones, 2004). Bioinformatic analyses predict that each miRNA will target multiple
mRNAs (Lewis *et al.*, 2003; Bartel, 2004; Lai, 2004; Brennecke *et al.*, 2005; Krek *et al.*,
2005), suggesting that, collectively, these novel gene regulators affect the expression of large portions of the cellular transcriptome. The biological significance of
miRNAs is further corroborated by studies of individual examples, reporting their
roles in diverse cellular and developmental pathways (see Brennecke *et al.*, 2003;
Johnston and Hobert, 2003; Xu *et al.*, 2003; Chen *et al.*, 2004; Zhao *et al.*, 2005).

The miRNAs assemble into RNA–protein complexes, usually termed miRNP or
RISC (for RNA-induced silencing complex). Different purification schemes have
identified a number of resident proteins of miRNP/RISC complexes, with a member
of the Argonaute (Ago) protein family consistently found in each preparation
(Mourelatos *et al.*, 2002; Dostie *et al.*, 2003; Jin *et al.*, 2004; Chendrimada *et al.*,
2005; Meister *et al.*, 2005). The miRNA guides the miRNP/RISC complex to cytoplasmic mRNA targets, leading to an attenuation of their expression (Ambros, 2004;
Bartel, 2004; Cullen, 2004; He and Hannon, 2004). Target recognition takes place
by virtue of miRNA–mRNA base pairing and the degree of complementarity within
the duplex is important for both the efficiency of target recognition and the mode

* Author to whom correspondence should be addressed.

MicroRNAs: From Basic Science to Disease Biology, ed. Krishnarao Appasani. Published by
Cambridge University Press. © Cambridge University Press 2008.

of post-transcriptional control. Near-perfect complementarity between miRNA and target region favors cleavage of the bound mRNA by the associated Ago protein. Gaps and mismatches within the microRNA–mRNA duplex, particularly between bases 10 and 11 in the central region, preclude endonucleolytic cleavage and instead lead to miRNP-mediated repression of mRNA translation (see also below). Additional complexity is introduced by the fact that most organisms express several Ago protein isoforms, which carry out specialized functions. For instance, of four human Ago proteins, only Ago 2 appears to be able to cleave target mRNA (Meister *et al.*, 2004), while both Ago 2 and Ago 4 can repress translation (Pillai *et al.*, 2004).

Interestingly, plant miRNAs generally form near-perfect duplexes to coding region segments of their target mRNAs and thus trigger predominantly endonucleolytic cleavage, while animal miRNAs tend to partially match to sequences in the 3′ untranslated regions of their targets and thus are expected to primarily regulate mRNA translation (Bartel, 2004). The paradigm of animal miRNA function was set by the two miRNAs, *lin-4* and *let-7*. They were discovered in *Caenorhabditis elegans*, where they regulate developmental timing (Pasquinelli and Ruvkun, 2002; Ambros, 2004). Both *lin-4* and *let-7* interact with partially mismatched sequences in the 3′ UTR of their target mRNAs. Interaction between the miRNA *lin-4* and its target mRNAs *lin-14* and *lin-28* resulted in a clear reduction of protein expression with minor decreases in mRNA levels, leading to the model that the miRNA affected some stage of mRNA translation (Olsen and Ambros, 1999; Seggerson *et al.*, 2002). A recent report (Bagga *et al.*, 2005) indicates that *lin-4* or *let-7* can greatly decrease the level of their target mRNAs, although it is not obvious how these studies differ from previous work in *C. elegans*. Notwithstanding these unresolved issues, several studies with engineered miRNA/mRNA pairings have since further supported and characterized the relationship among base complementarity, mRNA cleavage, and translational repression (Doench *et al.*, 2003; Zeng and Cullen, 2003; Doench and Sharp, 2004; Humphreys *et al.*, 2005; Pillai *et al.*, 2005).

The complex process of mRNA translation is usually divided into an initiation, elongation and termination phase. Eukaryotic cells have evolved an intricate machinery to conduct initiation, a process that depends on multiple eukaryotic initiation factors (eIF), which co-operate to attract both subunits of the ribosome to the start of the mRNA coding region. Consistent with this evolutionary elaboration, most known mechanisms to control eukaryotic translation affect the initiation step, although regulation of later stages such as elongation, termination, or the release of the stable polypeptide has also been reported (Preiss and Hentze, 2003; Sonenberg and Dever, 2003; Gebauer and Hentze, 2004). Of note here, it was found that native complexes involving *C. elegans* *lin-14* and *lin-28* mRNAs sediment into the polysomal region of density gradients, even when isolated from larval stages where protein expression is repressed by the miRNAs (Olsen and Ambros, 1999; Seggerson *et al.*, 2002). EDTA sensitivity of these complexes and the observation that they disassembled when added to a heterologous translation reaction led to the expectation that miRNAs act at a step after translation initiation (Ambros, 2004; Bartel, 2004; Gebauer and Hentze, 2004;

He and Hannon, 2004). An important recent advance was the demonstration that mRNAs, when translationally repressed by miRNAs, collect within or adjacent to processing (P-) bodies (Liu *et al.*, 2005; Pillai *et al.*, 2005; Sen and Blau, 2005). P-bodies are large cytoplasmic aggregates that serve as sites of mRNA degradation. They are also known to contain translationally masked mRNA. How miRNAs trigger sequestration of the mRNA targets in the P-bodies is not known.

Thus, despite the early recognition of miRNAs as translational regulators, until about a year ago very little was known about the underlying molecular mechanism(s). We developed suitable conditions for direct transfection of an mRNA/miRNA cocktail into HeLa cells to recreate miRNA-mediated translational repression. In this system we show that a miRNA can affect translation initiation by inhibiting the roles of the mRNA cap structure and poly(A) tail. Within the initiation mechanism we define the function of the cap-binding protein eukaryotic initiation factor (eIF) 4E as a molecular target.

Methodology

Plasmid vectors and synthetic miRNA duplexes

The Renilla luciferase (R-luc) plasmid pRL-TK-4 sites was a gift from J. Doench and P. A. Sharp, MIT Cambridge, Massachusetts (Doench *et al.*, 2003; Doench and Sharp, 2004). For the construction of pEMCV-RL-TK-4 sites, we amplified by PCR the EMCV IRES sequence and ligated it into pRL-TK-4 sites, upstream of the R-luc coding region. To generate an unrelated Firefly luciferase (F-luc) transfection standard, we used either pGL3 (Promega; for plasmid co-transfection) or pT3luc, to produce an F-luc mRNA by *in vitro* transcription (Iizuka *et al.*, 1994, Preiss and Hentze, 1998). The CXCR4 miRNA duplex was as described (Doench *et al.*, 2003); the siCXCR4 duplex varied in the central bulge region to create perfect base-pairing matches with the targets.

Synthesis of mRNA transcripts

All plasmids were linearized with BamHI and purified by agarose gel electrophoresis and the Qiaex II kit (Qiagen). Linearized plasmids served as templates for *in vitro* transcription reactions using either the T7 or T3 MEGAscript® kit (Ambion). All transcripts were capped, either with $m^7G(5')ppp(5')G$ ("cap") or $A(5')ppp(5')G$ ("A-cap") (New England Biolabs). If required, the Poly(A) Tailing Kit® (Ambion) was used to add a > 150 nucleotide poly(A) tail to transcripts ("tail"). mRNAs were purified with the MEGAclear™ kit (Ambion), and concentrations estimated by A_{260} readings. mRNA quality was inspected by either denaturing agarose gel electrophoresis or by microfluidics using the RNA 6000 Nano LabChip® Kit on an Agilent 2100 Bioanalyzer.

Cell culture, mRNA transfections and dual luciferase assays

HeLa cells were maintained in DMEM with 5% fetal calf serum, supplemented with glutamine and penicillin/streptomycin. Cells were seeded in 24 well plates and transfected with 20 ng R-luc and 80 ng of F-luc mRNA (unless otherwise stated), with or without a miRNA, using Lipofectamine™ 2000 reagent (Invitrogen). Luciferase accumulation was detected using the Dual-Luciferase® Reporter Assay system (Promega) and measured in a FLUOstar Optima plate reader (BMG). The levels of R-luc activity varied systematically with different experimental mRNAs (Figure 6.3) but were well above background levels measured with extracts from cells that had been mock-transfected with a transfection mix containing only the F-luc-encoding control mRNA. For plasmid transfections HeLa cells were plated as described above and transfected with 1 µg pTK-R-luc-4-sites and 40 ng pGL3, with or without a miRNA, using lipofectamine 2000 reagent (Invitrogen).

Calculations of reporter expression and translational repression

After subtraction of background, R-luc activities were normalized against the corresponding F-luc transfection efficiency control. To compare baseline R-luc expression in the absence of miRNA, normalized R-luc activity from each of the different R-luc mRNA constructs was divided by the value for the "cap&tail" R-luc-4-sites mRNA. To calculate a measure of repression by a miRNA, the normalized R-luc activity from cells co-transfected with a miRNA was divided by the value from a control transfection without miRNA. Graphed values are averaged data from multiple experiments as indicated in figure legends. Full details of experimental methods can be found in the recent publication from our laboratory (Humphreys *et al.*, 2005).

Results and discussion

Using mRNA transfection to study miRNA function

In trying to unravel how a translational control mechanism under study works, a powerful approach is to alter the mode by which the test mRNA gets translated, for instance by insertion of internal ribosome entry sites (IRES; see below) (Ostareck *et al.*, 2001; Poyry *et al.*, 2004). The idea is that if one alters the aspect of mRNA translation that is usually targeted by the control mechanism, then regulation should be lost. To implement this approach in the context of miRNA function, we decided to employ a well-defined synthetic miRNA/mRNA target pairing that had been shown by Sharp and colleagues to exhibit all known aspects of miRNA-mediated translational control (Doench *et al.*, 2003; Doench and Sharp, 2004). In the published format, this system uses co-transfection of a synthetic RNA duplex (CXCR4) together with a plasmid (pRL-TK-4 sites) that expresses a *Renilla*-luciferase (R-luc) mRNA containing four imperfectly matching binding sites for the CXCR4 miRNA in its 3′ UTR. CXCR4 induces robust and specific repression (Figure 6.2d) of R-luc protein expression without loss of the plasmid-derived target mRNA control (Doench *et al.*, 2003; Doench and Sharp, 2004).

To facilitate the engineering of precise and predictable changes in the mode of translation occurring on the reporter mRNA, we decided to convert the CXCR4/pRL-TK-4 sites system from a plasmid transfection to a direct mRNA transfection approach. This requires the preparation of reporter mRNAs by *in*

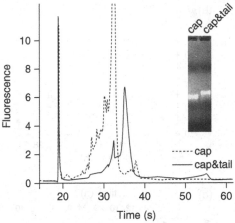

Figure 6.1. Quality control of *in vitro* transcribed R-luc mRNAs. Aliquots of "cap" and "cap&tail" R-luc-4 sites mRNA were analyzed on a Nano LabChip® using the Agilent 2100 Bioanalyzer or by conventional agarose gel electrophoresis and ethidium bromide staining (insert). The graph shows the electropherograms of both R-luc-4 sites mRNAs. Both analysis methods indicate good quality mRNA preparations (one major band/peak; size shift after polyadenylation).

vitro transcription of linearized plasmids (Figure 6.1). We made an R-luc mRNA using the linearized pRL-TK-4 sites plasmid as a template, prepared an unrelated firefly luciferase (F-luc) mRNA (Iizuka *et al.*, 1994) as a transfection control, and then co-transfected them into HeLa cells together with CXCR4 miRNA. To closely mimic the situation in plasmid-transfected cells, both transcripts were initially made with a physiological 5′ cap structure (m^{7}G(5′)ppp(5′)G) and a 3′ poly(A) tail.

We systematically varied several parameters of the mRNA transfection approach to optimize for detection sensitivity and extent of miRNA-mediated repression. First, serial ten-fold dilutions of the R-luc-4 sites mRNA were transfected into HeLa cells (Figure 6.2a). We found that R-luc activity was readily detectable even with the smallest amount of mRNA and that it increased approximately linearly with mRNA amounts up to 500 ng. To stay well below saturation, we decided to use 20 ng of R-luc mRNA for all other transfection studies. Next, we analyzed the influence of cell incubation time after transfection, by harvesting cells at different time points up to 32 h. Figure 6.2b shows that R-luc expression reaches a plateau between approximately 10–16 h and then gradually declines to low but still detectable levels at the latest time point tested. In parallel, transfections were also carried out with a reporter mRNA cocktail containing 2 nM CXCR4 miRNA, allowing a measurement of fold-repression by the miRNA at each time point. We note a time lag in the development of repression of R-luc mRNA expression, with repression reaching a plateau of approximately 5.5-fold around 16 h (Figure 6.2d). We further varied the concentration of CXCR4 miRNA in the transfection cocktail (Figure 6.2c) and registered a dose–response curve that indicated 2 nM as the lowest concentration for maximal repression. Finally, we measured CXCR4-mediated repression in cells transfected with the original plasmid pRL-TK-4

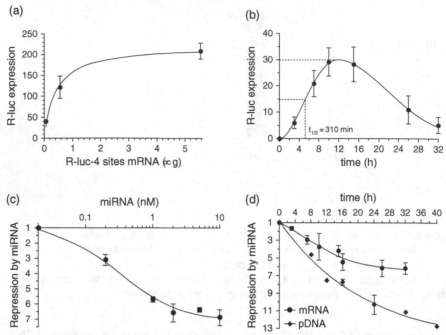

Figure 6.2. Optimization of the miRNA/mRNA co-transfection method. (a) Titration of "cap&tail" R-luc-4 sites mRNA amount. Multiple cultures of HeLa cells were transfected with a cocktail consisting of variable amounts of "cap&tail" R-luc-4 sites mRNA and 80 ng of F-luc mRNA. Cells were harvested and lysed 5 h after transfection. Normalized R-luc values are shown. (b) Time course of R-luc expression. Cultures were transfected with 20 ng of "cap&tail" R-luc-4 sites mRNA and 80 ng F-luc mRNA each, and harvested at various time points. R-luc expression was not normalized to F-luc but instead was scaled to the total level of R-luc activity recovered. (c) Time course of miRNA-mediated repression. Cell cultures were transfected with R-luc-4 sites reporter, the F-luc control, either with or without the CXCR4 miRNA, and harvested at different time points. Transfections used either direct mRNA transfection ("cap&tail" transcripts) or conventional plasmid transfection. Repression by the miRNA was calculated as detailed in Methods. (d) Titration of CXCR4 miRNA concentration. Cells were transfected with a cocktail of 20 ng "cap&tail" R-luc-4 sites mRNA, 80 ng of F-luc mRNA, and varying concentrations of CXCR4 miRNA, followed by harvest and lysis after 16 h. Most data points in (a–d) represent averaged measurement from multiple experiments, the error bars represent standard deviation where appropriate. The figure displays data that were in part previously published (Humphreys *et al.*, 2005).

sites over time and compared it with the mRNA transfection data (Figure 6.2d). We note that plasmid transfection yields approximately 1.5-fold higher levels of miRNA-mediated repression compared to mRNA transfection.

Several conclusions can be drawn from these tests. First, analysis of the R-luc expression time course allows an estimate for the functional half-life of transfected R-luc-4 sites mRNA of 310 min in the absence of co-transfected miRNA (Figure 6.2b), similar to its half-life of 270 min in the presence of CXCR4 miRNA (Humphreys *et al.*, 2005). Second, the increase in repression over time (Figure 6.2d) may reflect a time lag in active miRNP complex formation relative to the onset of translation of the R-luc reporter mRNA. Third, transfection into HeLa cells, of 20 ng of R-luc mRNA in a transfection cocktail containing 2 nM CXCR4 miRNA, followed by cell harvest after 16 h of incubation yields robust and specific (Figure 6.3)

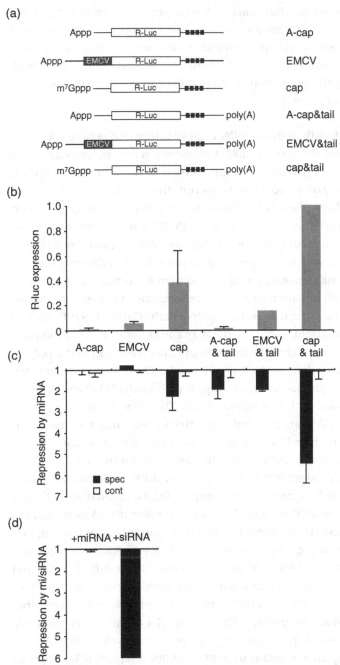

Figure 6.3. CXCR4 miRNA repression on modified R-luc mRNAs. (a) Schematic of the R-luc-4 sites mRNAs. The 5′ end of the mRNAs was modified with either a physiological m^{7}G-cap mRNA or with a "blocked" A-cap. Three of the mRNAs carry a poly(A) tail at the 3′ end as indicated. Further, two of the mRNAs exhibit the EMCV IRES sequence in their 5′ UTR. (b) Normalized R-luc activity of each mRNA transcript from transfections in the absence of CXCR4 miRNA. Expression from R-luc-4 sites "cap&tail" mRNA is set to 1.0. (c) Repression of the different R-luc-4 sites mRNAs by the specific CXCR4 miRNA (filled bars) or non-specific let-7 miRNA (open bars). Average results from three to five experiments are shown with standard deviation. (d) Repression of EMCV-R-luc-4 sites by either the CXCR4 miRNA or the perfectly complementary CXCR4 siRNA. Average results from two independent experiments are shown. Data in panels (b) and (c) were previously published (Humphreys *et al.*, 2005)

miRNA-mediated repression. Thus, nuclear transcription and processing of the target mRNA is not strictly required for miRNA-mediated repression. The increased magnitude of repression in plasmid-transfected cells may result from a more favorable timing of the onset of mRNA transcription with respect to miRNP formation, increasing the proportion of R-luc mRNA molecules that encounter functional CXCR4–miRNP complexes.

miCXCR4 blocks the functions of the mRNA cap structure and poly(A) tail

Having established that the CXCR4 miRNA efficiently represses translation of a m^7G-capped and polyadenylated ("cap&tail") version of the R-luc-4 sites mRNA, we then tested a derivative R-luc-4 sites transcript, the translation of which was driven by the IRES from the Cricket Paralysis Virus (CrPV) intergenic region (Humphreys *et al.*, 2005). Translation from the CrPV IRES, which does not require any canonical initiation factors (Hellen and Sarnow, 2001; Pestova and Hellen, 2003), was found to be no longer repressible by the miRNA, indicating that the initiation phase of translation was the major target of miRNA-mediated regulation.

Canonical translation initiation on a typical cellular mRNA is jointly promoted by the physiological cap structure and the poly(A) tail (Gallie, 1991; Preiss and Hentze, 1998; Preiss and Hentze, 2003; Sonenberg and Dever, 2003). We therefore asked if miRNA repression targets the function of either the 5′ cap or the poly(A) tail by preparing four versions of the R-luc "4-sites" mRNA: with no poly(A) tail and either an A-cap or a physiological m^7G-cap, or with a poly(A) tail and either an A-cap or a physiological m^7G-cap (Figure 6.3a). The non-physiological A-cap provides stability to RNA transcripts but is inactive in recruiting the translation initiation machinery to the mRNA. The four R-luc-4 sites mRNA versions display the typical functional synergy between the physiological cap and the poly(A) tail in that combining both end modifications on the same mRNA molecule has a more than additive effect on R-luc translation (Figure 6.3b). Functional and physical stability analyses confirmed that all versions of R-luc-4 sites mRNA were similarly stable in transfected cells (Humphreys *et al.*, 2005). Importantly, this transcript set displays markedly different responses to the CXCR4 miRNA (Figure 6.3c). Compared to the response of the "cap&tail" version of the mRNA (~5.5 fold repression), translation driven either solely by the physiological cap ("cap"), or solely by the poly(A) tail ("A-cap & tail"), was only partially responsive to the CXCR4 miRNA (~2-fold repression). The low level of translation from the mRNA having neither a physiological cap nor a poly(A) tail ("A-cap"; Figure 6.3b) was completely resistant to miRNA addition (Figure 6.3c). None of the R-luc-4 sites mRNA versions was affected by co-transfection of an unrelated control miRNA (*C. elegans let-7*, Figure 6.3c) confirming the target specificity of the observed effects. We conclude that within initiation, the functions of both the cap structure and the poly(A) tail are important for full miRNA-mediated repression.

miCXCR4 targets eIF4E function

To further investigate the targets of miRNA-mediated translational repression, we used the special mode of initiation by the encephalomyocarditis virus (EMCV)

IRES (Jackson, 2000; Hellen and Sarnow, 2001; Poyry *et al.*, 2004). The recruitment of eIF4G is a central step in early initiation that is jointly stimulated by the cap and poly(A) tail (Sachs, 2000; Preiss and Hentze, 2003; Sonenberg and Dever, 2003; Gebauer and Hentze, 2004). The cap structure interacts with the cap-binding protein eIF4E, while the poly(A) tail is bound to its binding protein PABP. Note that eIF4E and PABP concurrently bind to eIF4G and recruit it to the mRNA. The EMCV IRES is still responsive to poly(A)-mediated stimulation via the eIF4G-PAPBP interaction, but it recruits eIF4G without employing eIF4E (Bergamini *et al.*, 2000; Hellen and Sarnow, 2001; Svitkin *et al.*, 2001). Thus, it retains most features of canonical initiation without a need for eIF4E (Jackson, 2000; Hellen and Sarnow, 2001; Poyry *et al.*, 2004).

We prepared two additional R-luc "4 sites" mRNA versions carrying the EMCV IRES sequence in front of the R-luc reading frame; both were A-capped, one carried a poly(A) tail ("EMCV&tail"), and the other had no tail ("EMCV"; Figure 6.3a). These EMCV IRES mRNAs gave robust levels of R-luc expression (approximately 10% of the physiological m^7G-cap driven mRNA) and expression was enhanced by the poly(A) tail (Figure 6.3b). Translation driven solely by the EMCV IRES was found to be completely resistant to miRNA repression (Figure 6.3c), confirming the importance of the cap structure for miRNA-mediated repression. It is possible that the target sites may be inaccessible to the miRNA owing to some unforeseen secondary structure conformation induced by the presence of the EMCV IRES sequence. To address this, we transfected a perfectly complementary short interfering CXCR4 RNA variant (siCXCR4) with EMCV-R-luc-4 sites mRNA. siCXCR4 led to robust reduction of the EMCV-R-luc-4 sites mRNA (Figure 6.3d), indicating that the CXCR4 target sites were still accessible on the mRNA. Conversely, the lack of miRNA-mediated repression of the "EMCV" mRNA indicates that the miRNA is not indirectly inhibiting translation due to accelerated decay. This is consistent with other functional and physical stability analyses that we performed (Humphreys *et al.*, 2005), which also indicated the absence of appreciable miRNA-mediated target mRNA degradation in this system.

The "EMCV&tail" mRNA produced a partial response to the CXCR4 miRNA, comparable in level to the "A-cap&tail" mRNA (Figure 6.3c), providing further evidence that a miRNA targets the function of the poly(A) tail. Together, the findings with the EMCV-based constructs confirm that miRNA-mediated repression targets cap-structure and poly(A) tail function during the initiation phase of translation. They further indicate that the miRNA-mechanism targets the recruitment and/or function of eIF4E.

Similarities of the miRNA mechanism to other translational control paradigms

We have examined the sensitivity of different modes of translation initiation (canonical or IRES-driven) to miRNA-mediated repression and found that two different viral IRES elements, the CrPV IRES (which essentially bypasses all the translation initiation apparatus (Humphreys *et al.*, 2005), and the EMCV IRES (which selectively dispenses with the requirement for eIF4E during initiation), were not inhibited by a miRNA. This indicates that miRNAs target the canonical

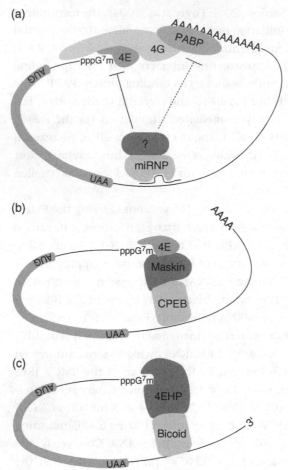

Figure 6.4. Working model of miRNA-mediated repression and its similarity to other translational control mechanisms. (a) A miRNP bound to the 3′ UTR represses translation initiation, by blocking the function of the cap structure and/or eIF4E. An unknown factor that bridges between the miRNP and the mRNA 5′ end may either preclude access of eIF4E to the cap structure or prevent productive association of cap-bound eIF4E with eIF4G. The miRNP further targets a function of the poly(A) tail, perhaps by interfering with a function of PABP. See main text for further explanations. (b) The *Xenopus* CPEB/maskin complex is an example of a 3′ UTR-tethered factor that forms a repressive interaction with cap-bound eIF4E. (c) Bicoid bound to a 3′ UTR motif in *Drosophila caudal* mRNA recruits a repressive cap-binding protein, termed 4EHP, that displaces eIF4E from the cap.

mechanism of translation initiation, and within that a role of the cap-binding protein eIF4E. Consistent with these observations, we found that the extent of miRNA-mediated repression of an mRNA with a generic 5′ UTR was lessened when the mRNA lacked either a functional cap structure or a poly(A) tail. As outlined above, both mRNA appendages – through their respective binding proteins eIF4E and PABP – stimulate the recruitment of eIF4G during early initiation (Sachs, 2000; Preiss and Hentze, 2003; Sonenberg and Dever, 2003; Gebauer and Hentze, 2004). A block of this process would thus be a plausible explanation for the observed miRNA effects (Figure 6.4a). The miRNP, perhaps through an additional bridging factor, could interfere with eIF4E function by either inhibiting its access to the cap structure, or by blocking eIF4E once bound to the cap. Full repression also requires the poly(A) tail indicating that miRNAs may also target the poly(A) tail function in eIF4G recruitment. Alternatively or perhaps additionally, they may target the less well understood roles of the poly(A) tail and PABP during the 60 S joining step of initiation (Kahvejian *et al.*, 2005) or possibly in mediating ribosome recycling on circular polysomes (Preiss and Hentze, 2003; Sonenberg and Dever, 2003).

The mechanistic aspects of miRNA-mediated translational repression as described here are reminiscent of other examples of mRNA-specific translational control. In *Xenopus* oocytes, the cytoplasmic polyadenylation control element-binding protein (CPEB) binds to the 3' UTR of maternal mRNAs and recruits the eIF4E-binding protein Maskin to block translation by establishing a repressive interaction between the 3' UTR-bound repressor complex and cap-bound eIF4E (Figure 6.4b) (Gebauer and Hentze, 2004; Richter and Sonenberg, 2005). CPEB-bound dormant mRNAs further have short poly(A) tails. During *Drosophila* embryogenesis, Bruno-mediated translational repression of *oskar* mRNA, and Smaug-mediated translational repression of *nanos* mRNA exhibits several similar features. Bound to elements in the 3' UTR, both proteins can recruit a repressive eIF4E-binding protein termed Cup to block initiation at the 5' end (Wilhelm *et al.*, 2003; Nakamura *et al.*, 2004; Nelson *et al.*, 2004; Zappavigna *et al.*, 2004). Repression of *Drosophila caudal* mRNA by Bicoid was shown to function through recruitment of the repressive cap-binding protein 4E-HP to displace eIF4E (Cho *et al.*, 2005) (Figure 6.4c). Future miRNA research will be informed by these similarities and will address, for instance, the identity of the putative factor that bridges between the miRNP and eIF4E or the cap structure.

Use of mRNA transfection to assess the generality of the miRNA mechanism
The work presented here has employed a particular synthetic miRNA duplex, exogenously added to the cells by transfection. Although previous work had shown that such RNA duplexes are functionally equivalent to endogenously expressed miRNAs (Zeng *et al.*, 2002; Doench *et al.*, 2003; Doench and Sharp, 2004), it is still relevant to ascertain whether the mechanism of action that we have uncovered is also used by other miRNAs and, in particular, by endogenous miRNAs. A partial answer to this question comes from work by Filipowicz and colleagues (Pillai *et al.*, 2005). These authors have independently studied the miRNA mechanism, using similar R-luc or F-luc reporter constructs, and direct mRNA transfection. Their constructs exhibited multiple target sites for the *let-7* miRNA in the 3' UTR. Translational repression by endogenous *let-7* miRNA was observed on both R-luc and F-luc reporter mRNAs containing the target sites. Their study further reported that the function of the cap structure and eIF4E during translation initiation was the likely molecular target of miRNA action, while no requirement for the poly(A) tail was noted. Thus, other miRNA/mRNA target pairings involving an endogenous miRNA trigger repression of translation that appears broadly compatible with the model proposed here. The two studies (Humphreys *et al.*, 2005; Pillai *et al.*, 2005) nevertheless differ with

regard to the poly(A) tail involvement. It will thus be interesting to investigate the origins of this apparent discrepancy. Several different miRNA/target pairings need to be tested to establish the generality or otherwise of a poly(A) tail contribution to the repressive mechanism. Further, the parallels to the maskin/ CPEB mechanism (Figure 6.4b) warrant systematic tests to establish whether miRNA-mediated repression affects, or is affected by, the poly(A)-tail length of the target mRNA. The direct mRNA transfection approach can be employed to address these questions.

The differential sensitivity of IRES-driven reporter mRNAs to translational repression versus mRNA cleavage (Figure 6.3d) may be used to classify the mode of action of a miRNA under study. More generally, sets of distinctly modified, *in vitro* transcribed reporter mRNAs similar to the ones reported here could be conveniently used to test what mechanism of action is employed by a given miRNA/ target mRNA pairing throughout a variety of cell types (e.g. representing different organisms, tissue types, or developmental stages).

The miRNA mechanism and processing bodies

Several recent studies have now established that miRNA function is linked to P-bodies (Jakymiw *et al.*, 2005; Liu *et al.*, 2005; Pillai *et al.*, 2005; Sen and Blau, 2005). P-bodies are microscopically discrete, cytoplasmic foci that contain non-translating mRNA and are enriched for proteins involved in mRNA decapping and 5' to 3' exonucleolytic decay (Coller and Parker, 2005; Marx, 2005). They are present in eukaryotic cells of different origin, where they exhibit generally similar protein compositions. Notably, they do not contain ribosomes or translation factors (see below). The function of P-bodies as sites of mRNA degradation is firmly established (see Sheth and Parker, 2003). More recent work, predominantly in *S. cerevisiae*, further described them as a major site of reversible storage of translationally inactive mRNA (Brengues *et al.*, 2005; Coller and Parker, 2005). A number of experimental manipulations leading to a block in translation initiation and polysome disassembly stimulate P-body formation, while a block in elongation leads to P-body reduction (Cougot *et al.*, 2004; Brengues *et al.*, 2005; Teixeira *et al.*, 2005). Two yeast P-body resident proteins, Dhh1p and Pat1p, act as broad repressors of translation and have critical roles in establishing the repressed mRNP state (Coller and Parker, 2005). These and other observations suggest that when mRNAs exit translation, they undergo a change in mRNP composition, which then allows sequestration in P-bodies, either for storage or degradation.

The mammalian homologue of Dhh1p, RCK/p54 is also found in P-bodies (Cougot *et al.*, 2004; Andrei *et al.*, 2005) as are the Ago proteins (Jakymiw *et al.*, 2005; Liu *et al.*, 2005; Pillai *et al.*, 2005; Sen and Blau, 2005). Further, mRNAs that are targeted by a miRNA collect in P-bodies (Liu *et al.*, 2005; Pillai *et al.*, 2005). How the miRNA-mechanism triggers the sequestration of the mRNA to P-bodies is not known at present. A plausible scenario is that miRNP-binding induces an mRNA-specific exit from translation, leading to assembly of a repressive mRNP and localization to P-bodies. As detailed earlier, the mRNA-specific events are likely to involve a bridging factor to block cap/eIF4E function. Intriguingly, several reports have demonstrated that, unlike their yeast counterparts, mammalian P-bodies do contain eIF4E (but not other translation factors), and also the 4E-Transporter (4E-T) protein. eIF4E-localization to P-bodies depends on an interaction with 4E-T (Andrei et al., 2005; Ferraiuolo et al., 2005), which also blocks eIF4E-association with eIF4G. It will thus be interesting to verify whether the miRNA-mechanism employs the eIF4E/4E-T interaction.

Finally, there are indications that, like the miRNA mechanism, other mRNA-specific translational control mechanisms might also feed into repression systems that are associated with cytoplasmic foci similar to P-bodies. For instance, the *Drosophila* Dhh1p-homologue, Me31B, forms granules containing multiple oocyte-localizing mRNAs. At least for two such mRNAs, *oskar* and *Bicaudal-D*, Me31B is furthermore required to repress their translation during transport (Nakamura et al., 2001). Maternal mRNA storage in *Xenopus* oocytes (Figure 6.4b) occurs in cytoplasmic granules, which contain the frog homologue of Dhh1p, Xp54 (Ladomery et al., 1997). Xp54 interacts with the CPEB protein and can mediate translational repression (Minshall and Standart, 2004). Overexpression of CPEB in mammalian cells increases P-body formation, and the protein accumulates within the foci (Wilczynska et al., 2005). Detailed comparisons of the composition of all these repressive cytoplasmic structures will be required to fully appreciate the relatedness of their function in post-transcriptional regulation.

Concluding remarks

There has been much progress over the past year in deciphering the mechanism by which miRNAs can regulate expression of partially mismatched target mRNAs. On the one hand, it was demonstrated that miRNAs interfere with the initiation phase of translation, by targeting functions of eIF4E, the mRNA cap structure and at least in one case also its poly(A) tail. On the other hand, it was revealed that the miRNA-repression leads to relocalization of the target mRNA to the repressive (and potentially degradative) environment of the P-bodies. Determining the causal and temporal relationships of these processes is undoubtedly a focus of intense ongoing research. In the interim, we are left with a reasonable working hypothesis that the miRNA-mediated initiation block represents the early stage of a process that ultimately leads to aggregation of the repressed mRNA and its transfer to P-bodies. Many mechanistic details of the mechanism remain to be determined. For instance, what is the bridging factor that mediates the effect of a 3′ UTR-bound miRNP on translation initiation at the 5′ end? Why does it usually require multiple (mismatched) miRNA binding sites within the 3′ UTR of a target mRNA to elicit clear repression? Is translational inhibition by a miRNA reversible? What determines whether the repressed mRNA is destined for stable storage or for degradation within the P-body environment? These and other questions will be the subject of miRNA research for some years to come.

ACKNOWLEDGEMENTS

This work was supported by grants from the National Health and Medical Research Council and the Australian Research Council to T.P. and D.I.K.M., and by the Victor Chang Cardiac Research Institute. T.P is supported by The Sylvia & Charles Viertel Charitable Foundation.

NOTE ADDED IN PROOF

The mechanism of miRNA-mediated repression of translation continues to be the focus of intense research. Several new studies have appeared since preparation of this article, supporting the notion of miRNAs acting on the initiation phase of translation. These include studies that have reconstituted the miRNA mechanisms in cell-free translation systems (Wang *et al.*, 2006; Thermann and Hentze, 2007), a report that argonaute proteins, core components of the miRNP, act as cap-binding proteins and compete with eIF4E similar to the model shown in Figure 6.4c (Kiriakidou *et al.*, 2007), and findings that the initiation factor eIF6 co-purifies with the miRNP complex and is required for miRNA-mediated translational repression (Chendrimada *et al.*, 2007). Contrasting with this trend, two papers have appeared supporting two post-initiation models, either premature ribosome drop-off (Petersen *et al.*, 2006) or nascent protein degradation (Notrott *et al.*, 2006).

REFERENCES

Ambros, V. (2004). The functions of animal microRNAs. *Nature*, **431**, 350–355.

Andrei, M. A., Ingelfinger, D., Heintzmann, R. *et al.* (2005). A role for eIF4E and eIF4E-transporter in targeting mRNPs to mammalian processing bodies. *RNA*, **11**, 717–727.

Bagga, S., Bracht, J., Hunter, S. *et al.* (2005). Regulation by let-7 and lin-4 miRNAs results in target mRNA degradation. *Cell*, **122**, 553–563.

Bartel, D. P. (2004). MicroRNAs: genomics, biogenesis, mechanism, and function. *Cell*, **116**, 281–297.

Bergamini, G., Preiss, T. and Hentze, M. W. (2000). Picornavirus IRESes and the poly(A) tail jointly promote cap-independent translation in a mammalian cell-free system. *RNA*, **6**, 1781–190.

Brengues, M., Teixeira, D. and Parker, R. (2005). Movement of eukaryotic mRNAs between polysomes and cytoplasmic processing bodies. *Science*, **310**, 486–489.

Brennecke, J., Hipfner, D. R., Stark, A., Russell, R. B. and Cohen, S. M. (2003). Bantam encodes a developmentally regulated microRNA that controls cell proliferation and regulates the proapoptotic gene hid in *Drosophila*. *Cell*, **113**, 25–36.

Brennecke, J., Stark, A., Russell, R. B. and Cohen, S. M. (2005). Principles of microRNA-target recognition. *Public Library of Science Biology*, **3**, e85.

Chen, C. Z., Li, L., Lodish, H. F. and Bartel, D. P. (2004). MicroRNAs modulate hematopoietic lineage differentiation. *Science*, **303**, 83–86.

Chendrimada, T. P., Gregory, R. I., Kumaraswamy, E. *et al.* (2005). TRBP recruits the Dicer complex to Ago2 for microRNA processing and gene silencing. *Nature*, **436**, 740–774.

Chendrimada, T. P., Finn, K. J., Ji, X. *et al.* (2007). MicroRNA silencing through RISC recruitment of eIF6. *Nature*, **447**, 823–829.

Cho, P. F., Poulin, F., Cho-Park, Y. A. *et al.* (2005). A new paradigm for translational control: inhibition via 5'-3' mRNA tethering by Bicoid and the eIF4E cognate 4EHP. *Cell*, **121**, 411–423.

Coller, J. and Parker, R. (2005). General translational repression by activators of mRNA decapping. *Cell*, **122**, 875–886.

Cougot, N., Babajko, S. and Seraphin, B. (2004). Cytoplasmic foci are sites of mRNA decay in human cells. *Journal of Cell Biology*, **165**, 31–40.

Cullen, B. R. (2004). Derivation and function of small interfering RNAs and microRNAs. *Virus Res*, **102**, 3–9.

Doench, J. G. and Sharp, P. A. (2004). Specificity of microRNA target selection in translational repression. *Genes & Development*, **18**, 504–511.

Doench, J. G., Petersen, C. P. and Sharp, P. A. (2003). siRNAs can function as miRNAs. *Genes & Development*, **17**, 438–442.

Dostie, J., Mourelatos, Z., Yang, M., Sharma, A. and Dreyfuss, G. (2003). Numerous microRNPs in neuronal cells containing novel microRNAs. *RNA*, **9**, 180–186.

Ferraiuolo, M. A., Basak, S., Dostie, J. *et al.* (2005). A role for the eIF4E-binding protein 4E-T in P-body formation and mRNA decay. *Journal of Cell Biology*, **170**, 913–924.

Gallie, D. R. (1991). The cap and poly(A) tail function synergistically to regulate mRNA translational efficiency. *Genes & Development*, **5**, 2108–2116.

Gebauer, F. and Hentze, M. W. (2004). Molecular mechanisms of translational control. *Nature Review of Molecular and Cell Biology*, **5**, 827–835.

Griffiths-Jones, S. (2004). The microRNA Registry. *Nucleic Acids Research*, **32**, D109–111.

He, L. and Hannon, G. J. (2004). MicroRNAs: small RNAs with a big role in gene regulation. *Nature Review Genetics*, **5**, 522–531.

Hellen, C. U. and Sarnow, P. (2001). Internal ribosome entry sites in eukaryotic mRNA molecules. *Genes & Development*, **15**, 1593–1612.

Humphreys, D. T., Westman, B. J., Martin, D. I. and Preiss, T. (2005). MicroRNAs control translation initiation by inhibiting eukaryotic initiation factor 4E/cap and poly(A) tail function. *Proceedings of the National Academy of Sciences USA*, **102**, 16961–16966.

Iizuka, N., Najita, L., Franzusoff, A. and Sarnow, P. (1994). Cap-dependent and cap-independent translation by internal initiation of mRNAs in cell extracts prepared from *Saccharomyces cerevisiae*. *Molecular Cell Biology*, **14**, 7322–7330.

Jackson, R. J. (2000). Comparative view of initiation site selection mechanisms. In *Translational Control of Gene Expression*, Sonenberg, N. and Mathews, M. B. (eds.). Cold Spring Harbor, pp. 127–184.

Jakymiw, A., Lian, S., Eystathioy, T. *et al.* (2005). Disruption of GW bodies impairs mammalian RNA interference. *Nature Cell Biology*, **7**, 1167–1174.

Jin, P., Zarnescu, D. C., Ceman, S. *et al.* (2004). Biochemical and genetic interaction between the fragile X mental retardation protein and the microRNA pathway. *Nature Neuroscience*, **7**, 113–117.

Johnston, R. J. and Hobert, O. (2003). A microRNA controlling left/right neuronal asymmetry in *Caenorhabditis elegans*. *Nature*, **426**, 845–849.

Kahvejian, A., Svitkin, Y. V., Sukarieh, R., M'Boutchou, M. N. and Sonenberg, N. (2005). Mammalian poly(A)-binding protein is a eukaryotic translation initiation factor, which acts via multiple mechanisms. *Genes & Development*, **19**, 104–113.

Kiriakidou, M., Tan, G. S., Lamprinaki, S. *et al.* (2007). An mRNA m7G Cap binding-like motif within human *Ago2* represses translation. *Cell*, **129**, 1141–1151.

Krek, A., Grun, D., Poy, M. N. *et al.* (2005). Combinatorial microRNA target predictions. *Nature Genetics*, **37**, 495–500.

Ladomery, M., Wade, E. and Sommerville, J. (1997). Xp54, the *Xenopus* homologue of human RNA helicase p54, is an integral component of stored mRNP particles in oocytes. *Nucleic Acids Research*, **25**, 965–973.

Lai, E. C. (2004). Predicting and validating microRNA targets. *Genome Biology*, **5**, 115.

Lewis, B. P., Shih, I. H., Jones-Rhoades, M. W., Bartel, D. P. and Burge, C. B. (2003). Prediction of mammalian microRNA targets. *Cell*, **115**, 787–798.

Liu, J., Valencia-Sanchez, M. A., Hannon, G. J. and Parker, R. (2005). MicroRNA-dependent localization of targeted mRNAs to mammalian P-bodies. *Nature Cell Biology*, **7**, 719–723.

Marx, J. (2005). Molecular biology. P-bodies mark the spot for controlling protein production. *Science*, **310**, 764–765.

Meister, G. and Tuschl, T. (2004). Mechanisms of gene silencing by double-stranded RNA. *Nature*, **431**, 343–349.

Meister, G., Landthaler, M., Patkaniowska, A. *et al.* (2004). Human Argonaute2 mediates RNA cleavage targeted by miRNAs and siRNAs. *Molecular Cell*, **15**, 185–197.

Meister, G., Landthaler, M., Peters, L. *et al.* (2005). Identification of novel argonaute-associated proteins. *Current Biology*, **15**, 2149–2155.

Minshall, N. and Standart, N. (2004). The active form of Xp54 RNA helicase in translational repression is an RNA-mediated oligomer. *Nucleic Acids Research*, **32**, 1325–1334.

Mourelatos, Z., Dostie, J., Paushkin, S. *et al.* (2002). miRNPs: a novel class of ribonucleoproteins containing numerous microRNAs. *Genes & Development*, **16**, 720–728.

Nakamura, A., Amikura, R., Hanyu, K. and Kobayashi, S. (2001). Me31B silences translation of oocyte-localizing RNAs through the formation of cytoplasmic RNP complex during *Drosophila* oogenesis. *Development*, **128**, 3233–3242.

Nakamura, A., Sato, K. and Hanyu-Nakamura, K. (2004). *Drosophila* cup is an eIF4E binding protein that associates with Bruno and regulates oskar mRNA translation in oogenesis. *Developmental Cell*, **6**, 69–78.

Nelson, M. R., Leidal, A. M. and Smibert, C. A. (2004). *Drosophila* Cup is an eIF4E-binding protein that functions in Smaug-mediated translational repression. *European Molecular Biology Organization Journal*, **23**, 150–159.

Nottrott, S., Simard, M. J. and Richter, J. D. (2006). Human *let-7a* miRNA blocks protein on actively translating polyribosomes. *Nature Structural & Molecular Biology*, **13**, 1108–1114.

Olsen, P. H. and Ambros, V. (1999). The lin-4 regulatory RNA controls developmental timing in *Caenorhabditis elegans* by blocking LIN-14 protein synthesis after the initiation of translation. *Developmental Biology*, **216**, 671–680.

Ostareck, D. H., Ostareck-Lederer, A., Shatsky, I. N. and Hentze, M. W. (2001). Lipoxygenase mRNA silencing in erythroid differentiation: The 3′ UTR regulatory complex controls 60 S ribosomal subunit joining. *Cell*, **104**, 281–290.

Pasquinelli, A. E. and Ruvkun, G. (2002). Control of developmental timing by microRNAs and their targets. *Annual Review of Cell and Developmental Biology*, **18**, 495–513.

Pestova, T. V. and Hellen, C. U. (2003). Translation elongation after assembly of ribosomes on the Cricket paralysis virus internal ribosomal entry site without initiation factors or initiator tRNA. *Genes & Development*, **17**, 181–186.

Petersen, C. P., Bordeleau, M., Pelletier, J. and Sharp, P. A. (2006). Short RNAs Repress Translation after Initiation in Mammalian Cells. *Molecular Cell*, **21**, 533–542.

Pillai, R. S., Artus, C. G. and Filipowicz, W. (2004). Tethering of human Ago proteins to mRNA mimics the miRNA-mediated repression of protein synthesis. *RNA*, **10**, 1518–1525.

Pillai, R. S., Bhattacharyya, S. N., Artus, C. G. *et al.* (2005). Inhibition of translational initiation by Let-7 MicroRNA in human cells. *Science*, **309**, 1573–1576.

Poyry, T. A., Kaminski, A. and Jackson, R. J. (2004). What determines whether mammalian ribosomes resume scanning after translation of a short upstream open reading frame? *Genes & Development*, **18**, 62–75.

Preiss, T. and Hentze, M. W. (1998). Dual function of the messenger RNA cap structure in poly(A)-tail-promoted translation in yeast. *Nature*, **392**, 516–520.

Preiss, T. and Hentze, M. W. (2003). Starting the protein synthesis machine: eukaryotic translation initiation. *Bioessays*, **25**, 1201–1211.

Richter, J. D. and Sonenberg, N. (2005). Regulation of cap-dependent translation by eIF4E inhibitory proteins. *Nature*, **433**, 477–480.

Sachs, A. (2000). Physical and functional interactions between the mRNA cap structure and the poly(A) tail. In *Translational Control of Gene Expression*, Sonenberg, N. and Mathews M. B. (eds.) Cold Spring Harbor, pp. 447–465.

Seggerson, K., Tang, L. and Moss, E. G. (2002). Two genetic circuits repress the *Caenorhabditis elegans* heterochronic gene lin-28 after translation initiation. *Developmental Biology*, **243**, 215–225.

Sen, G. L. and Blau, H. M. (2005). Argonaute 2/RISC resides in sites of mammalian mRNA decay known as cytoplasmic bodies. *Nature Cell Biology*, **7**, 633–636.

Sheth, U. and Parker, R. (2003). Decapping and decay of messenger RNA occur in cytoplasmic processing bodies. *Science*, **300**, 805–808.

Sonenberg, N. and Dever, T. E. (2003). Eukaryotic translation initiation factors and regulators. *Current Opinion in Structural Biology*, **13**, 56–63.

Svitkin, Y. V., Imataka, H., Khaleghpour, K. *et al.* (2001). Poly(A)-binding protein interaction with eIF4G stimulates picornavirus IRES-dependent translation. *RNA*, **7**, 1743–1752.

Teixeira, D., Sheth, U., Valencia-Sanchez, M. A., Brengues, M. and Parker, R. (2005). Processing bodies require RNA for assembly and contain nontranslating mRNAs. *RNA*, **11**, 371–382.

Thermann, R. and Hentzel, M. W. (2007). *Drosophila miR2* induces pseudo-polysomes and inhibits translation initiation. *Nature*, **447**, 875–979.

Wang, B., Love, T. M., Call, M. E., Doench, J. G. and Novina, C. D. (2006). Recapitulation of short RNA-directed translational gene silencing *in vitro*. *Molecular Cell*, **22**, 553–560.

Wilczynska, A., Aigueperse, C., Kress, M., Dautry, F. and Weil, D. (2005). The translational regulator CPEB1 provides a link between dcp1 bodies and stress granules. *Journal of Cell Science*, **118**, 981–992.

Wilhelm, J. E., Hilton, M., Amos, Q. and Henzel, W. J. (2003). Cup is an eIF4E binding protein required for both the translational repression of oskar and the recruitment of Barentsz. *Journal of Cell Biology*, **163**, 1197–1204.

Xu, P., Vernooy, S. Y., Guo, M. and Hay, B. A. (2003). The *Drosophila* microRNA Mir-14 suppresses cell death and is required for normal fat metabolism. *Current Biology*, **13**, 790–795.

Zappavigna, V., Piccioni, F., Villaescusa, J. C. and Verrotti, A. C. (2004). Cup is a nucleocytoplasmic shuttling protein that interacts with the eukaryotic translation initiation factor 4E to modulate *Drosophila* ovary development. *Proceedings of the National Academy of Sciences USA*, **101**, 14 800–14 805.

Zeng, Y. and Cullen, B. R. (2003). Sequence requirements for micro RNA processing and function in human cells. *RNA*, **9**, 112–123.

Zeng, Y., Wagner, E. J. and Cullen, B. R. (2002). Both natural and designed microRNAs can inhibit the expression of cognate mRNAs when expressed in human cells. *Molecular Cell*, **9**, 1327–1333.

Zhao, Y., Samal, E. and Srivastava, D. (2005). Serum response factor regulates a muscle-specific microRNA that targets Hand2 during cardiogenesis. *Nature*, **436**, 214–220.

7 *In situ* analysis of microRNA expression during vertebrate development

Diana K. Darnell, Stacey Stanislaw, Simran Kaur, Tatiana A. Yatskievych,
Sean Davey, Jay H. Konieczka and Parker B. Antin*

Introduction

A widespread class of non-coding, regulatory RNAs has been recently characterized. Because of their short length (21–22 nucleotides) they are called microRNA
or miRNA. These miRNAs have been identified in diverse organisms (prokaryote,
eukaryote, vertebrates, invertebrates, plants, fungi) and in viruses (Tuschl *et al.*,
1999; Elbashir *et al.*, 2001; Griffiths-Jones, 2004; Berezikov and Plasterk, 2005;
Griffiths-Jones *et al.*, 2006). Their apparently ancient function is to regulate
specific protein concentration by inhibiting the first step of translation or by
inducing specific mRNA degradation by 3′ UTR binding (He and Hannon, 2004;
Pillai, 2005; Valencia-Sanchez *et al.*, 2006). Both molecular and bioinformatics
tools have been used to identify candidate miRNAs and their target mRNAs. Based
on the numbers generated in these studies, it is estimated that vertebrate genomes
may contain hundreds of miRNA genes that may regulate stability or translation
of approximately one quarter of all mRNAs (Bentwich *et al.*, 2005; Berezikov and
Plasterk, 2005; Legendre *et al.*, 2005; Xie *et al.*, 2005).

Disruption of miRNA function often produces aberrations of important processes including organogenesis, and cell diversification, proliferation, and survival (Reinhart *et al.*, 2000; Brennecke *et al.*, 2003; Dostie *et al.*, 2003; Ambros, 2004;
Calin *et al.*, 2004; Alvarez-Garcia and Miska, 2005; Giraldez *et al.*, 2005). The
miRNA function has also been implicated in regulating stem cell renewal and
the onset of certain cancers (Hatfield *et al.*, 2005, Lu *et al.*, 2005). Therefore,
miRNAs regulate important processes in animal development, physiology and
disease. Identifying new miRNAs, determining their expression patterns and
demonstrating which mRNAs are their targets have become significant research
aims in many labs. Our lab has focused on demonstrating the temporal and spatial
expression of the known miRNAs in the chicken.

Their small size has made detection of mature miRNAs challenging. With modifications, standard Northern blot (Chen and Okayama, 1987; Valoczi *et al.*, 2004;

---

* Author to whom correspondence should be addressed.

MicroRNAs: From Basic Science to Disease Biology, ed. Krishnarao Appasani. Published by
Cambridge University Press. © Cambridge University Press 2008.

Watanabe *et al.*, 2005) and microarray (Barad *et al.*, 2004; Thomson *et al.*, 2004) protocols have been successfully used to demonstrate miRNA expression. However, neither of these techniques gives the tissue resolution of whole mount *in situ* hybridization. Unfortunately, although whole mount *in situ* hybridization with antisense RNA probes has been used for miRNA (Hornstein *et al.*, 2005; Lancman *et al.*, 2005), these traditional probes are typically not sensitive enough to detect these small transcripts reliably. Dependable detection is possible, however, using a new nucleic acid analog that exhibits superior hybridization kinetics, specificity and biostability (Wahlestedt *et al.*, 2000; Elmen *et al.*, 2004, 2005). This analog is called LNA, short for Locked Nucleic Acids (Koshkin *et al.*, 1998). LNAs are DNA oligos in which the introduction of a methylene bridge between the 2′O and the 4′C on the ribose ring of every third nucleic acid "locks" the structure into a high binding-affinity, RNA-mimicking conformation. On Northern blots, LNA probes are 10-fold more sensitive than similar DNA probes (Valoczi *et al.*, 2004). Similarly, LNA probes surpassed DNA oligonucleotide probes in the thermal stability of the hybrids they formed with target RNAs during *in situ* hybridization studies (Thomsen *et al.*, 2005). LNA probe chemistry and their applications in animal development are detailed by Kauppinen in Chapter 17. A comprehensive, *in situ* hybridization survey of miRNA expression in zebrafish using LNA probes detected the majority of miRNAs analyzed and demonstrated their dynamic spatial and temporal expression patterns (Weinholds *et al.*, 2005). In contrast, a similar *in situ* hybridization screen in mouse detected only a small number of the most highly expressed miRNAs (Kloosterman *et al.*, 2006). Thus, even using LNAs, the ability to detect miRNAs in vertebrate embryos has been variable and no comprehensive screen had successfully been done in an amniote model species.

Fortunately for *in situ* expression analysis of amniote miRNA, the chicken embryo provided an alternative to mouse. Chicken embryos are easily and inexpensively obtained. Their early development closely parallels the development of other amniotes, including humans. And whole mount *in situ* hybridization protocols have been optimized for chick to give outstanding sensitivity and low background (Nieto *et al.*, 1996; Bell *et al.*, 2004). Consequently, we undertook a comprehensive, whole mount *in situ* hybridization expression analysis of 112 mature miRNA sequences that are encoded by 134 distinct chicken miRNA genes. Owing to short probe length, modifications of traditional *in situ* protocols were employed to improve detection. These included substituting the standard antisense RNA probes with LNA oligos, changing the labeling strategy to incorporate terminal rather than internal DIG-UTP, and lowering the annealing temperatures to minimize melting of the short duplexes. Changes were also made in embryo handling to allow for increased numbers of embryos at one time in this high-throughput screen.

Methodology

Locked Nucleic Acid (LNA) probes

LNAs complementary to the mature miRNAs identified in chick were supplied by Exiqon A/S (Denmark). Each LNA was 22 nucleotides long. Each LNA oligo was identical in sequence to the reverse complement of one or more of the 134 proposed chick miRNA sequences. Because some miRNA genes arose from duplications of others, some miRNA mature sequences are identical to others. In such cases, our probes hybridized to miRNAs transcribed from more than one gene. Multigenic transcript hybridization is also possible but not assured for miRNAs differing by single nucleotides (Kloosterman *et al.*, 2006). Sequences for the miRNAs used can be found in the miRBase Sequence Database (http://microrna. sanger.ac.uk/sequences/index.shtml).

Probe labeling

Standard antisense RNA probes used to detect mRNAs are labeled with Digoxigenin-labeled UTP substituted into the RNA synthesis reaction. In contrast, for LNA probes a single Digoxigenin-labeled UTP was added to the 3′ end of each LNA-modified DNA oligonucleotide using a DIG-UTP 3′ end-labeling kit, (Roche, 2nd Generation; Cat. No. 03353575910). Labeling reactions were carried out according to the manufacturer's protocol with one exception. The labeling reaction was not cleaned up because embryos hybridized with cleaned and uncleaned probes gave similar results (data not shown). Probes were added at 1 µl/ml hybridization solution.

Embryo collection and preparation

Fertile chicken eggs (Hyline, Iowa) were incubated in a forced-draft, humidified incubator at 37.5 °C for 0.5–5 days, depending on the stages desired. Embryos were collected into chilled chick saline (123 mM NaCl in nanopure water), removed from the vitelline membrane and cleaned of yolk. Extra-embryonic membranes and large body cavities (brain vesicles, atria, allantois, eye) were opened to minimize trapping of the *in situ* reagents. Embryos were fixed in fresh, cold 4% paraformaldehyde in PBS overnight at 4 °C. Embryos were maintained at or near 4 °C during collection, because significant loss of labeling was correlated with increased time at room temperature during collection.

Embryos were rinsed in PBS, then in PBS plus 1% Tween 20 (PBTw), and dehydrated by steps (25%, 50%, 75%, and 100%) into methanol before being cooled to −20 °C overnight (or up to 10 days). Rehydration reversed this series. Embryos were rinsed 2 × in PBS and older embryos were treated with proteinase K: stages 8–13 and 14–18 at 10 µg/ml of proteinase K for 10 and 20 min, respectively; stages 19 and older at 20 µg/ml of proteinase K for 20 min. Embryos were rinsed repeatedly in PBTw to stop the digestion, and were moved into prehybridization (see below). Embryos were stored until use either at the methanol step or in prehybridization at −20 °C for fewer than 10 days. Embryos stored for more than 10 days showed considerable decrease in hybridization signal.

In situ hybridization

Prepared embryos were transferred to a standard prehybridization solution (Nieto *et al.*, 1996) in which Triton X-100 was replaced with Tween-20 (50% formamide, 5 × SSC, 2% blocking powder, 0.1% Tween-20, 0.1% CHAPS, 50 µg/ml yeast RNA, 5mM EDTA, 50 µg/ml heparin, DEPC water). Other modifications from the standard *in situ* protocol relating to the high-throughput nature of our screening are as follows. Hybridizations were carried out in 24-well plates in a shaking hybridization oven at a temperature between 21 °C and 23 °C below the reported melting temperature of the LNAs. Recent work indicates that up to a 5 °C spread in annealing temperature (20–25 °C below the melting temperature) is consistent with hybridization (Kloosterman *et al.*, 2006). Washes were carried out using 6- or 12-well plates containing 15 mm or 24 mm Netwell Inserts, respectively, with attached 74 μM polyester mesh bottoms (Corning, Inc., Cat. No. 3477, 3479). Inserts helped to maximize wash volume and minimize embryo handling and damage for high-throughput screening. Prewarming the wash solutions to the hybridization temperature before washing was crucial. Embryos in the Netwell inserts could be moved quickly into plates filled with pre-warmed wash buffer, minimizing cooling for high-throughput processing. Anti-DIG antibody binding was carried out in 24-well plates at 4 °C on a nutator. Color reactions (NBT/BCIP) were for 1–6 h at room temperature on a nutator until signal or background became visible, followed by overnight washing in KTBT. A second or third round of color reaction followed until each probe had yielded strong signal, or until the negative control began to show background label. Reactions were stopped with KTBT and embryos were then washed in PBS, dehydrated by a methanol series to remove background and enhance signal, then stored in PBS plus 0.1% sodium azide.

Results

Chick miRNA expression analysis

Of the 112 unique miRNA sequences analyzed, 107 are listed in miRBase version 8.0 (Griffiths-Jones *et al.*, 2006) and five were discovered by sequence homology comparisons with other species and are not yet listed there. Results of our screen showed that the majority (76%) of screened miRNA genes were expressed and detectable using single-labeled probes during the first four days of embryogenesis.

Few miRNAs were expressed during gastrulation and early neurulation (stages 3–12), more during morphogenesis and patterning stages (stages 13–16), and additional miRNA transcripts were expressed in a variety of tissues during organogenesis, limb and pharyngeal arch development (stages 17–25). Labeling patterns and their associated miRNA transcripts could be divided into five categories: those in one or related tissues (Table 7.1, Figure 7.1A–C); those in multiple, apparently unrelated cell types or structures (Table 7.2, Figure 7.1D–H); those nearly ubiquitous (Table 7.3, Figure 7.1I–K); those ubiquitous, and those lacking detectable expression (Table 7.4). All four categories with detectable hybridization contained miRNAs that showed temporal changes in expression in both region and intensity. Even most ubiquitously expressed miRNAs showed differential expression between tissues at some stage during development.

Tissue specific expression was observed in cells from all three germ layers. Muscle specific transcripts (mesoderm) in the body (somitic myotome) and in the heart (cardiac myotome) bound several probes related to miR-1, including

Table 7.1. Probes hybridizing to miRNA expressed in discrete tissues or regions

miRNA	Tissue
let-7c, -7k; miR-9, -100, -124a/b, -153, -219, -449	CNS
miR-1 and 1b, 133a	myotome and myocardium
miR-7, -23b, -126, -137, -138, -142–3p, -144	cardiac/vascular/blood
miR-15a, -18b, -217	limb mesoderm
miR-21	amnion
miR-34b, -222a, -455, -490	notochord
miR-183	cranial and spinal ganglia
miR-204	pigmented retina
miR146, -196, -2058b, -383	surface ectoderm
miR-199a	mesoderm

Table 7.2. Probes that hybridized to miRNA expressed in multiple tissues

let-7a	dorsal spinal cord, hindbrain, limb mesoderm
let-7b	early surface ectoderm and endoderm, notochord, wing, hindbrain, spinal cord
let-7f	hindbrain, endocardium
miR-10b	neural tube/spinal cord, limb buds, epidermis w/o apical ectodermal ridge
miR-20a	hindbrain, spinal cord, forebrain, face (mes & ecto), ubiquitous w/o heart
miR-30a-5p	notochord, mesonephric duct, surface ectoderm w/o AER, arches, trunk mesoderm
miR-30a-3p	notochord, mesonephric duct, aorta
miR-30e	mesonephric duct, surface ectoderm
miR-33	amnion, extraembryonic ectoderm, atrium
miR-34a	atria, mesonephric ducts
miR-99a	atria, motor horns of spinal cord, lungs
miR-106	spinal cord, hindbrain, pituitary rudiment, limb, arch muscles and mesoderm
miR-125b	surface ectoderm, CNS, sinus venosus, arches, gonad, limb bud
miR-128a/b	telencephalon, pharynx/pharyngeal arches, heart tube, limb bud
miR-130b	epiblast, pharyngeal arches, posterior limb bud mesenchyme
miR-135a	hindbrain floor, midbrain roof, telencephalon, wingbud mesoderm, arches
miR-137	sinus venosus, cardinal veins, surface ectoderm
miR-140	notochord, face and arch mesoderm, dorsal hindbrain
miR-184	lens, cranial ganglia, lateral plate mesoderm, pharyngeal arches
miR-187	brain closure/roofplate, thoracic mesoderm
miR-200a	surface ectoderm, limb, arches, midbrain roof, mesonephric duct
miR-200b	surface ectoderm, limb, mesonephric duct
miR-206	myotome of somite, telencephalon
miR-216	limb, surface ectoderm, pharyngeal arches
miR-218	spinal cord motor horns, hindbrain, atria
miR-222b	pharyngeal arches, mesonephric tubules
miR-302	notochord, floor plate of midbrain, gut endoderm
miR-367	epiblast, aorta, notochord
miR-375	pancreas, pituitary rudiment

miR-1.1 – 1.2 (duplication with identical sequence, Figure 7.1A), miR-1b (duplication with one central nucleotide mutated) and -133a-2, 133a-1, 133a-3 (a near neighbor, also duplicated and polycystronic to the miR-1 family). Additionally, miR-206 (a duplication of miR-1 with three mutations) was expressed in the

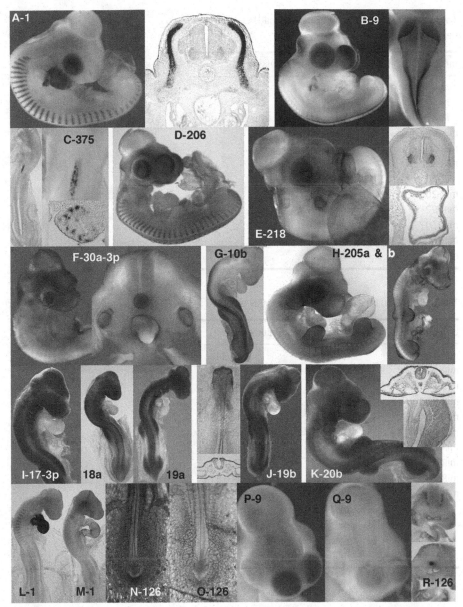

Figure 7.1 Expression of micro RNA in chick detected by whole mount *in situ* hybridization. (A–C) Expression in single tissue types. (A) miR-1 expression in the somitic myotome and cardiac myotome, stage 23 lateral view. Transverse section at the level of the spinal cord showing somitic myotome. (B) miR-9 expression in the telencephalic vesicles, midbrain, hindbrain and spinal cord, stage 24 lateral view, and dorsal view showing expression in the hindbrain with rhombomeric variability. (C) miR-375 expression in endodermal cells that will give rise to the pancreas in whole mount at stage 17, close up and in cross section. (D–H) Expression in multiple tissue types. (D) miR-206 expression in somitic myotome (mesoderm) and forebrain vesicles (ectoderm) at stage 24 lateral view. (E) miR-218 expression in the atria (mesoderm) at stage 25 lateral view, and in cross section in the spinal cord motor horns (ectoderm; top), and atria (bottom). (F) miR-30a-3p shown in whole mount at stage 19, and transverse cut showing notochord, mesonephric ducts and dorsal aorta (axial and intermediate mesoderm). (G) miR-10b expression in the entire length of the spinal cord (ectoderm) and limb buds (mesoderm). (H) miR-205a and b expression in the surface ectoderm shown at stages 23 and 21. (I–K) Nearly ubiquitous expressers, polycistronic, showing considerable

Table 7.3. Probes with hybridization to miRNAs expressed nearly ubiquitously

miR	tissues not expressing
miR-15b	heart
miR-17-3p	heart
miR-17-5p	heart and apical ectodermal ridge
miR-18	heart
miR-19a	heart
miR-19b	heart
miR-20	heart
miR-20b	heart
miR-92	heart
miR-107	heart
miR-130a	heart
miR-363	heart

Table 7.4. Ubiquitous and Not Detected miRNAs

Ubiquitous	Not Detected	
miR-16	let-7d	miR-34c
miR-22	let-7g	miR-101
miR-26a	miR-7b	miR-103
miR-30d	miR-9*	miR-122a
miR-194	miR-16	miR-133b
miR-203	miR-24	miR-142-5p
miR-215	miR-27b	miR-148a
miR-221	miR-29a	miR-155
miR-223	miR-29b	miR-181a
miR-365	miR-29c	miR-181b
	miR-30b	miR-190
	miR-30c	miR-213
	miR-31	miR-301
	miR-32	

Figure 7.1 (cont.)
expression at older stages in the body but much reduced in the heart. (I) miR-17-3p, -18a and -19a expression at stage 18/19, lateral view. (J) miR-19b expression at younger stages is restricted to the ectoderm and endoderm (whole mount and transverse section, stage 10), but later expands to near ubiquity, excepting heart. (K) A similar pattern seen for miR-20b, which is also not expressed in the apical ectodermal ridge of the limb (lower transverse section). (L) Stage 17 embryo strongly labeled with miR-1 double-labeled probe. (M) Stage 17 embryo weakly labeled with miR-1, 3′ single-labeled probe (same time in color reaction as (L)). (N) Stage 15 embryo tail, strongly labeled with miR-126 double-labeled probe. (O) Stage 15 embryo tail weakly labeled with miR-126 3′ single-labeled probe (same time in color reaction as (N)). (P) Stage 20 embryo strongly labeled with miR-9 double-labeled probe. (Q) Stage 20 embryo weakly labeled with miR-9 3′ single-labeled probe (same time in color reaction as (P)). (R) Stage 19 embryo transected across the lumbar trunk labeled with miR-126 3′ single-labeled probe (top, no notochord label detected) vs. miR-126 tailed (bottom, strong notochord label detected). (See color plate 5)

somitic myotome but not in the heart (see Figure 7.1D), likely owing to a muta- tion in a Mef-2 regulatory element upstream of this gene that is required for heart development (Zhao *et al.*, 2005; Darnell *et al.*, 2007). The nervous system (ecto- derm) expressed the largest number of miRNAs (41), contributing to both multi- tissue and tissue-specific expression patterns. Neural-specific expression was documented for let-7c (hindbrain, spinal cord), -7i (forebrain) and -7k (hindbrain and forebrain); miR-9 (all CNS levels, Figure 7.1B), miR-100 (spinal cord and hindbrain), -124a and b (spinal cord, hindbrain, midbrain and pituitary rudi- ment), -153 (all levels), -183 (cranial and spinal ganglia), -204 (pigmented retina) and -449 (hindbrain). Endoderm-specific expression was detected for miR-375 in the pancreatic precursors (Figure 7.1C).

Multiple-tissue expression (excluding ubiquitous and nearly ubiquitous) was detected for 36 miRNAs, including miR-206 (myotome of somite and forebrain, Figure 7.1D), miR-218 (atrium of heart and motor horns of spinal cord, Figure 7.1E), miR-30a-3p (notochord, nephric ducts and dorsal aorta, Figure 7.1F), miR-10b (limbs and spinal cord, Figure 7.1G) and miR-205a and b (surface ectoderm, limb and pharyngeal arch, Figure 7.1H). In addition, some miRNAs were first expressed in a few tissues or regions, and later become ubiquitously expressed, for example, let-7b, mir-138, -144 and -146 (Darnell *et al.*, 2007).

Expression of another group of miRNAs (nearly ubiquitous) was observed in discrete tissues at early embryonic stages, and later in nearly all tissues, while remaining conspicuously absent or restricted in one or a few tissues. Many of these are located in one of two miRNA polycistrons. miR-17 (5p and 3p), -18a, -19a, -20a -19b, and -92 are expressed from a polycistron (Lee *et al.*, 2002) on chromosome 1 and expressed strongly in nearly all tissues with the exception of heart (Figure 7.1I, J) and apical ectodermal ridge (AER) of the limb. miR-106, -20b (Figure 7.1K), -18b and -363 are expressed from a duplicated polycystron on chro- mosome 4, and show a similar expression pattern. At younger stages (Figure 7.1J, K) these same miRNAs are expressed specifically in the surface ectoderm and endoderm (inserts) and then in the limb mesenchyme but not the AER (Figure 7.1K insert).

Thus many individual miRNAs are expressed in multiple tissues, whereas expression patterns of others are more limited (e.g. solely expressed in the somitic myotome). Two general groups are evident from the pool of non-ubiquitously expressing miRNA: those that are expressed discretely in one or a few locations or tissues (e.g. only in the CNS), and those that are expressed broadly, but fail to express in isolated locations or tissues (e.g. nearly ubiquitous, but with greatly reduced expression in heart and some surface ectoderm). Detection in a wide variety of temporally and spatially distinct patterns implicates miRNAs in impor- tant developmental processes and suggests novel regulatory functions. Additional images can be found at our web site (GEISHA URL), the online repository for chicken embryo *in situ* hybridization data, and in Darnell *et al.* (2007).

Sensitivity

Approximately one quarter of the chick miRNAs assessed gave no visible expres- sion pattern using single-labeled probes. In addition, *in situ* hybridization analysis of mouse miRNA detected only the most abundantly expressed miRNA transcripts (Kloosterman *et al.*, 2006). These observations indicate that more sensitive probes might expand the number of miRNAs whose expression is detectable by *in situ* hybridization. In a pilot study on sensitivity, we compared standard probes with a single end-labeled UTP (3'-DIG) with probes pre-labeled during synthesis (3' and 5' double-DIG; Exiqon, Denmark) and with tailed probes (multiple-3' DIG-labeled UTP interspersed with unlabeled ATP spacers). We tested the sensitivity of these labeling differences for several expressing miRNAs: miR-126 (vasculature) using 3' single-, double-labeled and tailed probes and miR-9 (neural), miR-183 (cranial ganglia) and miR-1 (muscle) using 3' single- and double-labeled probes and miR-124 (CNS) and -206 (muscle) for single-labeled and tailed probes. The four double-labeled probes were visibly more sensitive (shorter time in color reaction, stronger signal when optimized) than the single-labeled probes (or for the miR-126 tailed probe) for *in situ* reactions (Figure 7.1L–Q). The tailing reaction hybridized with 3.3 × to 10 × greater sensitivity on dot blots (data not shown), but all three tailed probes gave strong signal in notochord instead of or in addition to the expression pattern detected using standard single-labeled probes (Figure 7.1R). The tailing kit (DIG Oligonucleotide Tailing Kit, 2nd generation, Roche cat # 03353583910) was designed to add spacer ATPs with the DIG-labeled UTPs. This may create a synthetic oligonucleotide string, which hybridizes artifactually to a different RNA expressed in the notochord, eradicating the specificity of the original miRNA template. Based on this pilot study, double-labeled probes gave the most sensitive *in situ* signal and may be able to detect miRNA transcripts at levels currently undetectable with single-labeled probes.

Relevance

Although information on miRNA target evolution and regulatory strategies of miRNAs were not the focus of this work, temporal and spatial expression data such as these are required to advance both of these fields. For example, a comparison of expression patterns between chick and fish (Ason *et al.*, 2007) demonstrated that although miRNA ortholog sequences are often conserved, their expression patterns (and therefore possibly their targets and functions) are not as frequently conserved. Expression patterns also illuminate the current debate about the developmental "strategy" of miRNA function. In one view, targets are regulated by miRNA temporally and/or quantitatively within tissues where expression of those targets is desirable to create a developmental switch. In these cases, miRNAs post-transcriptionally down-regulate their targets as cells switch from one set of proteins to another during a differentiation step. This mechanism has been demonstrated in *C. elegans* for seam cell maturation and was instrumental in the identification of the first miRNA and its target (Lee *et al.*, 1993, Wightman *et al.*, 1993, reviewed by Rougvie, 2005). A second strategy involves targets that are inhibited qualitatively in tissues in which expression of those targets is undesirable. This has been called "mutual exclusion" (Stark *et al.*, 2005) and would also aid in transitions between cell fates by preventing leaky transcription in neighboring cells. Although every expression pattern could represent this strategy, the nearly ubiquitous (excepting heart) miRNAs seems like ideal candidates for mutual exclusion. At least one putative target RNA for a member of this group codes for a protein found only in the heart. According to miRBase target prediction (Griffiths-Jones *et al.*, 2006), miR-20b has the heart gene *myosin regulatory light chain 2 A, cardiac muscle isoform* as a target although it is expressed strongly everywhere but heart (Figure 7.1K). As targets are proposed and tested, information on the expression patterns of the miRNAs will be important for assessing both specific developmental function, and larger issues of global regulatory strategy.

Our results provide a valuable foundation for future studies of miRNA function and regulation. For example, many miRNAs are found in intergenic regions or on the antisense strand of known genes. These miRNAs presumably have their own promoters and our expression data will allow the grouping of similarly expressed genes, which can be used for bioinformatics work on promoter function. For example, in a promoter comparison between two miRNA genes with overlapping but non-identical expression and sequence, miR-1 (muscle and heart, Figure 7.1A) and -206 (muscle alone), we identified promoter element mutations likely involved in this regulatory difference (Darnell *et al.*, 2007). Other miRNAs are imbedded in genes (on the sense strand) and presumably they are co-transcribed with their host gene from its promoter (see Ohler *et al.*, 2004; Figure 7.1G).

Finally, some miRNAs can behave directly as oncogenes (Brennecke *et al.*, 2003; He *et al.*, 2005) or as tumor suppressors (Johnson *et al.*, 2005), whereas others can be up-regulated by an oncogene, *myc* (O'Donnell *et al.*, 2005). Knowing the expression patterns of genes, including miRNAs should speed the process of identifying genetic causes of neoplastic transformation. For example, miR-205a (Figure 7.1H), is expressed in the epidermis and exists in two copies imbedded in TRPM1 and 3 (ensembl). In humans, a loss of TRPM1 expression is correlated with cutaneous metastatic melanoma (Duncan *et al.*, 1998). TRPM1 in humans also contains an imbedded miRNA. It is not clear at present whether the cancerous transformation down regulates the gene or vice versa. If the latter is true, it will be important to consider whether the loss of the principle gene, or the imbedded microRNA causes the cancer. Knowledge of gene expression patterns and their correlation with cell types involved in disease will inform research into the mechanisms of disease. Until overlapping expression patterns are demonstrated for relevant genes in humans, expression patterns from chick orthologs may inform research in human disease.

Concluding remarks

Our investigation into the embryonic expression patterns of most known microRNAs in the chick has yielded data that will be useful in comparative genomics, bioinformatics, developmental biology, cancer biology and many other disciplines. The expression patterns presented here likely represent only a fraction of the total miRNAs in the chick genome. Thus our work provides a starting point for investigations of miRNA regulation and function.

REFERENCES

Alvarez-Garcia, I. and Miska, E. A. (2005). MicroRNA functions in animal development and human disease. *Development*, **132**, 4653–4662.

Ambros, V. (2004). The functions of animal microRNAs. *Nature*, **431**, 350–355.

Ason, B., Darnell, D. K., Wittbrodt, B. *et al.* (2007). Differences in vertebrate microRNA expression. (In press.)

Barad, O., Meiri, E., Avniel, A. *et al.* (2004). MicroRNA detection by oligonucleotide microarrays: system establishment and expression profiling of human disease. *Genome Research*, **14**, 2486–2494.

Bell, G. W., Yatskievych, T. A. and Antin, P. B. (2004). GEISHA, A whole-mount *in situ* hybridization gene expression screen in chicken embryos. *Developmenta Dynamics*, **229**, 677–687.

Bentwich, I., Avniel, A., Karov, Y. *et al.* (2005). Identification of hundreds of conserved and nonconserved microRNAs. *Nature Genetics*, **37**, 766–770.

Berezikov, E. and Plasterk, R. H. A. (2005). Camels and zebrafish, viruses and cancer: a microRNA update. *Human Molecular Genetics*, **14**, R183–R190.

Brennecke, J., Hipfner, D. R., Stark, A., Russel, R. B. and Cohen, S. M. (2003). Bantam encodes a developmentally regulated microRNA that controls cell proliferation and regulates the proapoptotic gene hid in *Drosophila*. *Cell*, **113**, 25–36.

Calin, G. A., Sevignani, C., Dumitru, C. D. *et al.* (2004). Human micro RNA genes are frequently located at fragile sites and genomic regions involved in cancers. *Proceedings of the National Academy of Sciences USA*, **101**, 2999–3004.

Chen, C. and Okayama, H. (1987). High efficiency transformation of mammalian cells by plasmid DNA. *Molecular Cell Biology*, **7**, 2745–2752.

Darnell, D. K., Kaur, S., Stanislaw, S., Konieczka, J. K., Yatskievych, T. A. and Antin, P. B. (2007). MicroRNA expression during chick embryo development. (In press.)

Dostie, J., Mourelatos, Z., Yang, M., Sharma, A. and Dreyfus, G. (2003). Numerous microRNPs in neuronal cell containing novel microRNAs. *RNA*, **9**, 180–186.

Duncan, L. M., Deeds, J., Hunter, J. *et al.* (1998). Down-regulation of the novel gene melastatin correlates with potential for melanoma metastasis. *Cancer Research*, **58**, 1515–1520.

Elbashir, S. M., Lendeckel, W. and Tuschl, T. (2001). RNA interference is mediated by 21- and 22-nucleotide RNAs. *Genes & Development*, **15**, 188–200.

Elmen, J., Zhang, H. Y., Zuber, B. *et al.* (2004). Locked nucleic acid containing antisense oligonucleotides enhance inhibition of HIV-1 genome dimerization and inhibit virus replication. *Federation of the European Biochemical Society Letters*, **578**, 285–290.

Elmen, J., Thonberg, H., Ljungberg, K. *et al.* (2005). Locked nucleic acid (LNA) mediated improvements in siRNA stability and functionality. *Nucleic Acids Research*, **33**, 439–447.

Giraldez, A. J., Cinalli, R. M., Glasner, M. E. *et al.* (2005). MicroRNAs regulate brain morphogenesis in zebrafish. *Science*, **308**, 833–838.

Griffiths-Jones, S. (2004). The microRNA registry. *Nucleic Acids Research*, **32**, D109–D111.

Griffiths-Jones, S., Grocock, R. J., van Dongen, S., Bateman, A. and Enright, A. J. (2006). miRBase: microRNA sequences, targets and gene nomenclature. *Nucleic Acids Research*, **34**, D140–D144.

Hatfield, S. D., Scherbata, H. R., Fischer, K. A. *et al.* (2005). Stem cell division is regulated by the microRNA pathway. *Nature*, **435**, 974–978.

He, L. and Hannon, G. J. (2004). MicroRNAs: small RNAs with a big role in gene regulation. *Nature Reviews in Genetics*, **5**, 522–532.

He, L., Thomson, J. M., Hemann, M. T. *et al.* (2005). A microRNA polycistron as a potential human oncogene. *Nature*, **435**, 828–833.

Hornstein, E., Mansfield, J. H., Yekta, S. *et al.* (2005). The microRNA miR-196 acts upstream of Hoxb8 and Shh in limb development. *Nature*, **438**, 671–674.

Johnson, S. M., Grosshans, H., Shingara, J. *et al.* (2005). RAS is regulated by the let-7 microRNA family. *Cell*, **120**, 635–647.

Kloosterman, W. P., Wienholds, E., de Bruijn, E., Kauppinen, S. and Plasterk, R. H. A. (2006). *In situ* detection of miRNAs in animal embryos using LNA-modified oligonucteotide probes. *Nature Methods*, **3**, 27–29.

Koshkin, A. A., Singh, S. K., Nielsen, P. *et al*. (1998). LNA (Locked Nucleic Acids): synthesis of the adenine, cytosine, guanine, 5-methylcytosine, thymine, and uracil bicyclonucleoside monomers, oligomerization, and unprecedented nucleic acid recognition. *Tetrahedron*, **54**, 3607–3630.

Lancman, J. J., Caruccio, N. C., Harfe, B. D. *et al*. (2005). Analysis of the regulation of lin-41 during chick and mouse limb development. *Developmental Dynamics*, **234**, 948–960.

Lee, J. H., Koyano-Nakagawa, N., Naito, Y. *et al*. (1993). cAMP activates the IL-5 promoter synergistically with phorbol ester through the signaling pathway involving protein kinase A in mouse thyoma line El-4. *Journal of Immunity*, **151**, 6135–6142.

Lee, Y., Jeon, K., Lee, J. T., Kim, S. and Kim, V. N. (2002). MicroRNA maturation: stepwise processing and subcellular localization. *European Molecular Biology Organization Journal*, **21**, 4663–4670.

Legendre, M., Lambert, A. and Gautheret, D. (2005). Profile-based detection of microRNA precursors in animal genomes. *Bioinformatics*, **21**, 841–845.

Lu, J., Getz, G., Miska, E. A. *et al*. (2005). MicroRNA expression profiles classify human cancers. *Nature*, **435**, 834–838.

Nieto, M. A., Patel, K. and Wilkinson, D. G. (1996). *In situ* hybridization analysis of chick embryos in whole mount and tissue sections. In *Methods in Cell Biology*, vol. 51. New York: Academic Press, Inc.

O'Donnell, K. A., Wentzel, E. A., Zeller, K. I., Dang, C. V. and Mendell, J. T. (2005). c-Myc-regulated microRNAs modulate E2F1 expression. *Nature*, **435**, 839–843.

Ohler, U., Yekta, S., Lim, L. P., Bartel, D. P. and Burge, C. B. (2004). Patterns of flanking sequence conservation and a characteristic upstream motif for microRNA gene identification. *RNA*, **10**, 1309–1322.

Pillai, R. S. (2005). MicroRNA function: multiple mechanisms for a tiny RNA? *RNA*, **11**, 1753–1761.

Reinhart, B. J., Slack, F. J., Basson, M. *et al*. (2000). The 21-nucleotide let-7 RNA regulates developmental timing in *Caenorhabditis elegans*. *Nature*, **403**, 901–906.

Rougvie, A. E. (2005). Intrinsic and extrinsic regulators of developmental timing: from miRNAs to nutritional cues. *Development*, **132**, 3787–3798.

Stark, A., Brennecke, J., Bushati, N., Russell, R. B. and Cohen, S. M. (2005). Animal microRNAs confer robustness to gene expression and have a significant impact on 3' UTR evolution. *Cell*, **123**, 1133–1146.

Thomsen, R., Nielsen, P. S. and Jensen, T. H. (2005). Dramatically improved RNA *in situ* hybridization signals using LNA-modified probes. *RNA*, **11**, 1745–1748.

Thomson, J. M., Parker, J., Perou, C. M. and Hammond, S. M. (2004). A custom microarray platform for analysis of microRNA gene expression. *Nature Methods*, **1**, 47–53.

Tuschl, T., Zamore, P. D., Lehmann, R., Bartel, D. P. and Sharp, P. A. (1999). Targeted mRNAs degradation by double-stranded RNA *in vitro*. *Genes & Development*, **13**, 3191–3197.

Valencia-Sanchez, M. A., Liu, J. D., Hannon, G. J. and Parker, R. (2006). Control of translation and mRNA degradation by miRNAs and siRNAs. *Genes & Development*, **20**, 515–524.

Valoczi, A., Hornyik, C., Varga, N. *et al*. (2004). Sensitive and specific detection of microRNAs by northern blot analysis using LNA-modified oligonucleotide probes. *Nucleic Acids Research*, **32**, e175.

Wahlestedt, C., Salmi, P., Good, L. *et al*. (2000). Potent and nontoxic antisense oligonucleotides containing locked nucleic acids. *Proceedings of the National Academy of Sciences USA*, **97**, 5633–5638.

Watanabe, T., Takeda, A., Mise, K. *et al*. (2005). Stage specific expression of microRNAs during *Xenopus* development. *Federation of the European Biochemical Society Letter*, **579**, 318–324.

Weinholds, E., Kloosterman, W. P., Miska, E. *et al.* (2005). MicroRNA expression in zebrafish embryonic development. *Science*, **309**, 310–311.

Wightman, B., Ha, I. and Ruvkun, G. (1993). Posttranscriptional regulation of the heterochronic gene lin-14 by lin-4 mediates temporal pattern formation in *C. elegans*. *Cell*, **75**, 855–862.

Xie, X., Lu, J., Kulbokas, E. J. *et al.* (2005). Systematic discovery of regulatory motifs in human promoters and 3′ UTRs by comparison of several mammals. *Nature*, **434**, 338–345.

Zhao, Y., Samal, E. and Srivastava, D. (2005). Serum response factor regulates a muscle-specific microRNA that targets Hand2 during cardiogenesis. *Nature*, **436**, 214–220.

8 MicroRNA function in the nervous system

Gerhard Schratt* and Michael Greenberg

Introduction

It is fair to say that the nervous system is among the most complex of all the organ systems in animals. In a typical mammal, billions of nerve cells (neurons) form a highly organized, intricate network of connections. The creation of neuronal networks, and their support by surrounding glial cells, is a prerequisite for the correct functioning of the nervous system. Given the complexity of the mature nervous system and the underlying processes that lead to its proper development, it is not surprising that microRNA-dependent regulation of gene expression plays an especially important role in the developing and mature nervous system. Processes in which microRNAs have been implicated in the nervous system include the initial specification of neuronal cell identity, the formation and refinement of synaptic connections between individual neurons during development and the ongoing plasticity of synapses in the adult. The functions of microRNAs during these different stages of nervous system development will be discussed, followed by a review of our knowledge about the role of microRNAs in diseases caused by neuronal dysfunction. It should be noted that microRNA function in glial cells, the other major cell type in the nervous system, will not be discussed in this chapter. This omission is not because of the absence of microRNAs from glial cells, but it simply reflects our limited knowledge about microRNAs in this cell type.

Our first insight into a possible role of microRNAs in the nervous system came from studies that cloned small RNAs from RNA preparations of various tissues, including brain tissues such as the cortex (Lagos-Quintana *et al.*, 2002). These initial experiments provided some important clues that had significant impact onto subsequent research. First, a surprisingly large number of microRNAs were identified, which suggested a widespread role for microRNAs in the developing and mature nervous system. Second, some of the identified microRNAs showed a

---

* Author to whom correspondence should be addressed.

MicroRNAs: From Basic Science to Disease Biology, ed. Krishnarao Appasani. Published by Cambridge University Press. © Cambridge University Press 2008.

high degree of sequence conservation, sometimes ranging from primates down to lower organisms such as the nematodes. For example, the let-7 microRNA, which is involved in developmental timing in the nematode *Caenorhabditis elegans* (Reinhart *et al.*, 2000), is highly expressed in a variety of brain tissues, suggesting that microRNAs such as let-7 might regulate cell differentiation in the nervous system of higher vertebrates as well. On the other extreme, some microRNAs are restricted to higher organisms or even primates, raising the intriguing possibility that microRNAs might also be implicated in the regulation of nervous system functions that are unique to mammalians that are capable of higher cognitive function (Berezikov *et al.*, 2005). A large number of microRNAs were found enriched in polyribosome fractions of cultured neurons (Kim *et al.*, 2004). Polyribosomes assemble on highly translated mRNAs and serve as a hallmark of ongoing translation. This finding therefore provided the first evidence that microRNAs in neurons, as in a variety of other cell types, are involved in the regulation of mRNA translation.

More recently, large-scale expression profiling of microRNAs was performed using microRNA gene arrays (Krichevsky *et al.*, 2003; Miska *et al.*, 2004; Sempere *et al.*, 2004). These arrays contain oligonucleotide probes for all microRNAs registered in the microRNA database and therefore allow for the rapid and quantitative expression analysis of all known microRNAs in RNA isolated from a given tissue such as the brain. However, it should be kept in mind that microarrays are inherently limited to known microRNAs and do not lead to the discovery of novel microRNAs, as is the case for cloning-based approaches. At any rate, expression profiling of microRNAs at different stages of brain development revealed that the expression of a large number of microRNAs is rather dynamic, e.g. showing high expression in early progenitor cells and not in differentiated neurons, or vice versa. These studies therefore gave a first glimpse of the widespread and versatile roles of microRNAs at different stages of nervous system development, from neuronal differentiation in the embryo to plasticity in the adult. In the following sections, these various functions of microRNAs will be illustrated with examples from the recent literature.

MicroRNAs in neuronal cell fate determination

Differentiated cell types are characterized by the expression of a unique set of gene products. Therefore, once a cell commits itself to adopt a specialized phenotype, e.g. a highly polarized neuronal phenotype, it has to dramatically change its gene expression pattern, both by activating the expression of cell type-specific genes and by repressing the expression of genes that are not compatible with the respective phenotype. Studies over the last few decades have shown that a large proportion of these decisions are made at the level of gene transcription. Classical examples in the nervous system include the proneural bHLH transcription factor family (NeuroD, Neurogenin, Mash), and REST, a global repressor that helps to prevent the expression of neuronal genes in non-neuronal tissues (Ballas and Mandel, 2005; Lee, 1997). Given that microRNAs were initially discovered in

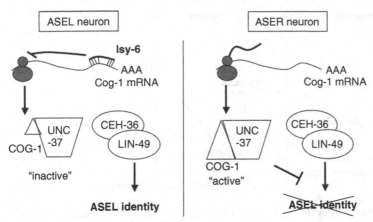

Figure 8.1. Signaling pathway for the specification of sensory neuron asymmetry in *C. elegans*. In left sensory neurons (ASEL; left), the lsy-6 microRNA binds to cog-1 mRNA, thereby inhibiting cog-1 translation and accumulation of cog-1 protein. This results in inactivity of the cog-1/unc-37 repressor complex, allowing ceh-36/lin-49 to promote ASEL identity. In right sensory neurons (ASER; right), cog-1 protein can accumulate owing to the absence of lsy-6, leading to the repression of ceh-36/lin-49 activity and ASEL identity.

the worm as regulators of cell fate determination during developmental timing, the cloning of microRNAs from neuronal tissues gave rise to the hypothesis that neuronal microRNAs might by analogy serve as regulators of the neuronal cell fate. Experimental support for this hypothesis first came from the analysis of *C. elegans* mutants that displayed defects in bilateral asymmetry of sensory neurons (Johnston and Hobert, 2003). In wild-type worms, chemosensory neuron pairs are bilaterally symmetrical based on anatomical considerations, but nevertheless display asymmetrical gene expression patterns. However, a screen for worms in which asymmetry was lost yielded several candidate genes for specifying sensory neuron asymmetry.

One of the mutant alleles, *lsy-6*, mapped to a genomic locus that did not contain a protein-coding gene, but instead a microRNA. In wild-type worms, *lsy-6* inhibits the translation of a zinc finger transcription factor, *cog-1*, specifically in the prospective left sensory neurons (ASEL, Figure 8.1). When *lsy-6* function is lost, cog-1 is inappropriately expressed, leading to the expression of right-specific genes in ASEL and loss of the asymmetrical gene expression patterns. Therefore, the *lsy-6* microRNA plays an instructive role in the specification of left–right neuron identity by controlling the expression of a regulatory transcription factor. Subsequently, the same group has identified another microRNA, miR-273, that functions within the sensory neuron left-right asymmetry pathway (Chang *et al.*, 2004). MiR-273 represses expression of the *die-1* transcription factor, which in turn is a positive regulator of *lsy-6* expression. This mechanism helps to keep *lsy-6* levels low in right sensory neurons (ASER). Thus, genetic evidence from the worm indicates that microRNAs can function as regulatory switches by inhibiting the expression of cell-type specific transcription factors.

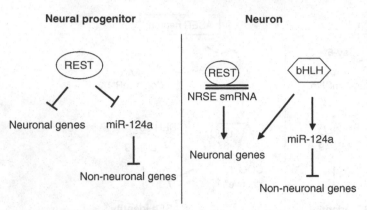

Figure 8.2. Signaling pathways for neuronal gene expression in vertebrates. In neural progenitors (left), the transcription factor REST represses the expression of neuronal genes including miR-124a. In neurons (right), binding of the NRSE small modulatory RNA to REST activates REST, which, together with pro-neural bHLH transcription factors, leads to the expression of neuronal genes, including miR-124a. The miR-124a mediated repression of non-neuronal genes helps to maintain the neuronal phenotype.

A role for microRNAs in neuronal development of vertebrates was first demonstrated in zebrafish. Fish deficient for mature microRNAs owing to the lack of the microRNA processing enzyme Dicer display specific developmental defects, including abnormal brain morphogenesis (Giraldez *et al.*, 2005). These defects could be attributed to defects in neuronal differentiation rather than effects on early patterning or fate specification in the embryonic nervous system. Consistent with this observation, it was found that microRNAs that are highly expressed in mammalian neurons, such as miR-124a and miR-9, help to maintain an already specified neuronal phenotype by preventing inappropriate expression of non-neuronal genes (Figure 8.2).

This mechanism is reminiscent of the transcription factor NRSF/REST (Neuronal Restricted Silencing Factor/RE-1 Silencing Transcription factor), which acts as a global repressor of the transcription of neuronal genes in non-neuronal tissues. In fact, recent results indicate that NRSF/REST and miR-124a might act in the same pathway (Conaco *et al.*, 2006). NRSF/REST appears to inhibit the expression of miR-124a in neuronal stem cells. Upon neuronal differentiation, NRSF/REST is degraded, leading to the expression of miR-124a and the down-regulation of non-neuronal genes. In contrast to *lsy-6*, which targets one crucial transcription factor, *cog-1*, to regulate neuronal specification, miR-124a appears to down-regulate a large number of target genes to exert its effect. When miR-124a is expressed in HeLa cells, it down-regulates a large number of non-neuronal transcripts as judged by microarray analysis, thereby conferring a neuron-like expression profile to this transformed cell line (Lim *et al.*, 2005). The action of miR-124a is also mechanistically quite distinct from the worm counterparts, in that it seems to affect mRNA levels rather than inhibiting the translation of otherwise stable mRNAs. Given recent evidence that microRNAs interact with the cellular mRNA degradation machinery within processing bodies (P-bodies), this effect could be easily explained by delivery of target mRNAs to P-bodies subsequent to translational inhibition (Liu *et al.*, 2005). Such a mechanism could be important to eliminate erroneously transcribed mRNAs from neurons and maybe other cell types as well. In the future, it will be interesting to know whether microRNAs in higher organisms function solely as permissive regulators, or whether some might constitute regulatory switches analogous to the *lsy-6* microRNA in *C. elegans*.

Yet another twist to the role of small RNAs during neuronal differentiation comes from the recent identification of a small modulatory dsRNA, NRSE (Kuwabara *et al.*, 2004). NRSE expression is induced during neuronal differentiation, although the genomic origin of the non-coding RNA remains unknown. In contrast to known microRNAs or siRNAs, NRSE seems to regulate neuronal differentiation at the level of transcription. NRSE interacts directly with the transcriptional repressor NRSF/REST in the nucleus. The NRSE/REST interaction switches the activity of NRSF/REST from a repressor to an activator of transcription by an unknown mechanism. The result is an increase in the transcription of neuronal genes concomitant with the commitment of neuronal progenitors to adopt a neuronal cell fate. A role of small non-coding RNAs at the chromatin level is well established in a number of organisms, including yeast and plants (Grewal and Rice, 2004). However, in these examples, RNAi-like molecules work in a more global fashion through the silencing of large regions of the genome ("heterochromatin silencing"). It remains an open question as to whether microRNAs and other small non-coding RNAs work through similar mechanisms in the nervous system of higher organisms.

MicroRNAs in activity-dependent synapse formation and refinement

Once neurons are specified and differentiated, they have the fascinating ability to form synaptic connections with hundreds of neighboring neurons, leading to the formation of intricate neuronal networks that lie at the heart of nervous system function. Synaptic connections are by no means static, but they can undergo intense remodeling and refinement in response to extra-cellular cues. The ability of synapses to undergo these long-lasting changes in structure and strength, known as synaptic plasticity, likely underlies higher cognitive functions such as learning and memory. Furthermore, synaptic dysfunction is responsible for several neurological diseases, including mental retardation, epilepsy and autism. Studies over the last few decades have highlighted an important role for activity-dependent translation of pre-existing mRNAs for long-lasting forms of synaptic plasticity. The first hint that microRNAs might be involved in activity-dependent translation during synaptic development came from the observation that the protein mutated in fragile-X mental retardation, fragile-X mental retardation protein (FMRP), is part of the RISC multiprotein complex (Caudy *et al.*, 2002; Jin *et al.*, 2004).

RISC (RNAi-induced Silencing Complex) mediates RNAi- and microRNA-dependent post-transcriptional gene silencing. FMRP is an RNA-binding protein that regulates the translation of dendritic mRNAs, and neurons deficient for FMRP function display aberrant translation of synaptic proteins and defective development of post-synaptic dendritic spines. Taken together, these observations suggest that FMRP may exert its effect on translation via the recruitment of microRNAs and the associated RISC complex to the respective target mRNAs. (The FMRP/miRNA connection is extensively covered in Chapter 27). More recently, other effectors of small RNA function, namely the RNAseIII-like enzyme dicer and the Argonaute family member eiF2C2, were detected in dendritic spines and postsynaptic densities (PSDs) (Lugli *et al.*, 2005). These findings further support the idea that a small RNA-based mechanism, possibly involving microRNAs, could operate locally at the synapse and participate in synaptic plasticity.

Another line of evidence that microRNAs are involved in synaptic development is provided by expression studies using microarrays and Northern blot analysis. Intriguingly, a number of microRNAs are absent from neural progenitors or newly differentiated neurons, but are highly expressed at times when synapse formation and remodeling takes place. However, the function of specific microRNAs during post-mitotic neuronal morphogenesis and synaptic development is just beginning to emerge. One of these microRNAs, miR-132, was identified in a genome-wide screen that was conducted to elucidate *in vivo* targets for the activity-dependent transcription factor CREB (cAMP Response Element Binding protein) (Vo *et al.*, 2005). CREB-dependent transcription has been widely implicated in neuronal morphogenesis, plasticity and survival, and the identification of a set of CREB-regulated microRNAs raised the possibility that some of the CREB effects might be actually mediated by microRNAs. In fact, miR-132 expression induces the formation of neurites, precursors of axonal and dendritic processes that will ultimately engage in the formation of synaptic connections. MiR-132 mediates neurite outgrowth by inhibiting the translation of a negative regulator in the neurite outgrowth pathway, the GTPase activating protein p250GAP. MiR-132 expression itself is under the control of neuronal activity and neurotrophins, thereby providing a link between extrinsic cues and the control of neuronal morphogenesis. Taken together, this study provides a possible mechanism for activity-dependent repression of protein synthesis in neurons. At present, it is unknown whether miR-132-mediated repression of translation is also important during other activity-regulated processes such as neuronal survival, synapse formation and plasticity.

In neurons, mRNAs are not only translated into proteins in the somatic cytoplasm, but also in dendritic processes close to synapses. This so called local translation has become an increasingly attractive model during the last decade, since it potentially provides a mechanism for synapse-specific changes that are known to occur when individual synapses undergo long-lasting changes in synaptic strength. Neuronal mRNAs subject to local translational control at synapses are initially transported into the synaptodendritic compartment in the form of high molecular weight ribonucleoprotein particles (RNP) (Kiebler and DesGroseillers, 2000).

RNPs contain motor proteins (kinesins and dyneins) and RNA-binding proteins that regulate the trafficking of the mRNAs into dendrites and prevent premature

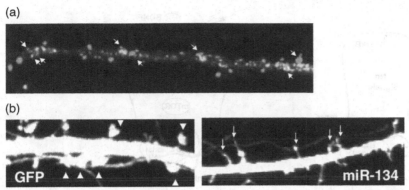

Figure 8.3. Expression of miR-134 in dendrites and its role in dendritic spine morphogenesis. (a) miR-134 (green; *in situ* hybridization signal) is present near post-synaptic sites (red; PSD-95 immunostaining) in dendrites of rat hippocampal neurons. Arrows denote sites of miR-134/PSD-95 co-localization. (b) Expression of miR-134 together with green fluorescent protein (GFP) in hippocampal neurons (right) leads to reduced volume of dendritic spines as compared to neurons that only express GFP (small arrows vs. arrowheads). (See color plate 6)

translation. It is believed that dendritic mRNAs are stored as RNPs at synapses in a translationally dormant state until an incoming presynaptic stimulus relieves the translational blockade. Recently, work from our laboratory showed that a brain-specific microRNA, miR-134, is part of the local translation machinery at synapses (Schratt *et al.*, 2006). By *in situ* hybridization, miR-134 is found throughout the cytoplasm but also frequently in close proximity to post-synaptic specializations known as dendritic spines (Figure 8.3). Functionally, miR-134 levels control the size of dendritic spines, presumably by regulating the local synthesis of important synaptic proteins. The only *bona fide* miR-134 target mRNA identified so far is *Limk1*, an mRNA that encodes for a kinase that regulates actin polymerization and thereby promotes dendritic spine growth. *Limk1* mRNA and miR-134 co-localize within neuronal dendrites, and miR-134 negatively regulates the translation of the *Limk1* mRNA. Thus, similar to other neuronal microRNAs, miR-134 prevents inappropriate accumulation of proteins important for neuronal function. However, in contrast to the classical microRNA mechanism, the repression of Limk1 translation at synapses by miR-134 is not constitutive, but can be relieved by brain-derived neurotrophic factor (BDNF), a neurotrophin that is released in response to synaptic stimulation. Thus, the miR-134-mediated translational blockade is rapidly reversible, a feature that is required to allow for rapid bi-directional changes in dendritic spine structure during synaptic plasticity (Figure 8.4).

Taken together, these examples illustrate an additional function of microRNAs in the mature nervous system, namely in activity-dependent morphogenesis and synaptic development. Moreover, a role for microRNAs in activity-dependent processes also challenges the prevailing dogma that microRNAs function solely as rheostats that ensure the homeostasis of cellular gene expression programs. Instead, these findings establish microRNAs as part of highly dynamic regulatory pathways that are responsive to extracellular cues such as synaptic stimulation.

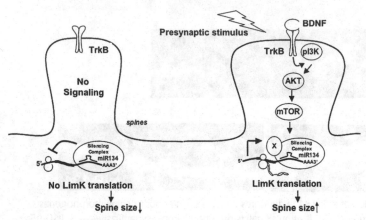

Figure 8.4. Model for the role of miR-134 in dendritic protein synthesis. (left) In the absence of BDNF signaling, Limk1 mRNA translation is inhibited by a miR-134 associated silencing complex. Low Limk1 protein levels prevent spine growth. (right) Presynaptic stimulus leads to the release of BDNF and activation of a TrkB/PI3K/Akt/mTOR signaling pathway and subsequent unknown modification (x) of the silencing complex. Translational block of Limk1 mRNA is relieved, leading to new Limk1 protein synthesis and the growth of dendritic spines.

How can microRNAs integrate extracellular stimuli to achieve a dynamic regulation of gene expression? Based on the few examples described in the literature and our preliminary results, two models, not mutually exclusive, seem plausible.

First, the activity of microRNAs might be regulated at the level of transcription of the microRNA precursor. This model is especially attractive given the well-established role of activity-dependent transcription in neurons. We have already discussed this point in the context of the miR-132 microRNA. In the same screen, a number of additional microRNAs were found to be regulated by CREB, suggesting that CREB might be a global regulator of microRNA expression downstream of neuronal activity and trophic signals. Work from our laboratory recently demonstrated that another activity-regulated transcription factor, MEF2 (Myocyte Enhancer Factor 2), plays a critical role in synaptogenesis (Flavell *et al.*, 2006). RNAi-mediated knockdown of MEF2 leads to increased synapse formation, suggesting that MEF2 works as a negative regulator of synapses. Although the targets of MEF2 that mediate this effect are largely unknown at present, large-scale genomic mapping of MEF2-binding sites identified many microRNAs as potential targets, raising the intriguing possibility that MEF2, like CREB, might use microRNA regulation as one way to regulate synaptic function. Not surprisingly, expression of the miR-134 microRNA that negatively regulates dendritic spine development is also under the control of neuronal activity (Schratt *et al.*, 2006). MiR-134 is part of a large cluster of more than 50 microRNAs within the imprinted *Dlk1-Gtl2* locus (Seitz *et al.*, 2004). There are reasons to believe that the whole cluster is regulated via a common enhancer element, although the nature of this element is unclear at present. In the future, it will be interesting to determine the function of other microRNAs within the cluster. Taken together, activity-dependent regulation of microRNAs is a powerful mechanism by which neurons can dynamically respond to changes in network activity on a cell-wide level.

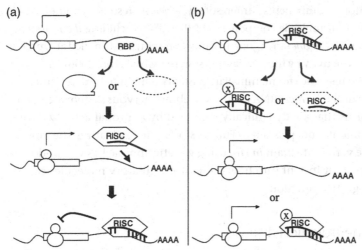

Figure 8.5. Models for the regulation of local microRNA function by neuronal activity. (a) Neuronal activity leads to the release of an RNA-binding protein (RBP) from the dendritic mRNA, either because of degradation (dashed line) or post-translational modification of the RBP. The microRNA associated RISC complex is now able to bind to its cognate site in the 3'UTR of the mRNA and subsequently inhibits productive translation of the mRNA. (b) Neuronal activity leads to post-translational modification (X) or degradation (dashed line) of RISC component(s). Both mechanisms then relieve the translational block and allow for dendritic synthesis of the mRNA.

Alternatively or in addition, microRNA activity might be regulated locally at or near synapses. This model, in contrast to regulation at the transcriptional level, allows for independent regulation of individual synapses in response to presynaptic activation, a hallmark of synaptic plasticity. Two models for local microRNA action at synapses, not mutually exclusive, have gained popularity recently and will be discussed here (Figure 8.5). Both models postulate delivery of microRNA target mRNAs into the synaptodendritic compartment, as already discussed for the dendritic miR-134 microRNA. In one model (a), the microRNA target is translated, probably as a result of basal synaptic activity. Synaptic stimulation locally increases the accessibility of the target mRNA for the cognate microRNA, e.g. by the release of an RNA-binding protein (RBP), leading to microRNA-target-interaction and subsequent repression of translation. The net result would be activity-dependent local repression of mRNA translation, as postulated for miR-132. In another model (b), the microRNA is already associated with its target in mutiprotein-RNA-complexes, so called miRNPs, ensuring basal repression of the target mRNA during transport and at the synaptic storage site. The translational blockade is subsequently relieved in response to synaptic stimulation.

Our results indicate that at least miR-134 might operate by the latter mode. However, important questions remain, including: what are the proteins within the microRNA-associated complexes that integrate extra cellular signals, and how can specific post-translational modifications of these proteins alter the translational competency of the complex as a whole? For example, inactivation of the silencing complex could be achieved by the activity-dependent modification or release of inhibitory components. Interestingly, recent results obtained from *Drosophila* indicate that multiple components of the RISC, including *dicer, armitage* (an RNA helicase) and *aubergine* (an Argonaute family protein) are degraded at synapses in response to neural activity by the proteasome system (Ashraf *et al.*, 2006). The exact mechanism and the relevance for mammalian plasticity remain to be determined.

In summary, a subset of microRNAs in the vertebrate nervous system are part of regulatory platforms that are dynamically regulated by neuronal activity during synaptic development and plasticity. Future studies investigating the role of microRNAs in the vertebrate brain *in vivo* using genetic approaches will undoubtedly provide important insight into how these tiny regulatory molecules contribute to higher cognitive function.

MicroRNAs in neurological disease

Given the emerging role of microRNAs in nervous system development, it comes as no surprise that the disruption of microRNA function might contribute to a number of neurological disorders. However, the potential role of microRNAs in neurological diseases is just beginning to be explored and our current knowledge about the role of specific microRNAs in neurological diseases is presently correlative at best. Nevertheless, this section will try to summarize what is known about potential connections between microRNA function and neurological diseases, and will also speculate concerning areas that warrant further investigation.

Several recent observations point to a role for microRNAs in the development of several forms of cancer, including cancers of the nervous system (Hammond, 2005). For example, microarray experiments identified a set of microRNAs that correlates with a glioblastoma phenotype, suggesting that these microRNAs might be related to the etiology or progression of the cancerous state (Ciafrè *et al.*, 2005; see also Chapter 26 by Ciafrè in this book). Interestingly, it was found that the expression pattern of microRNAs in cancer cells of different origin represents a molecular fingerprint that is unique to the history of the individual cancer (Lu *et al.*, 2005). This observation gives hope that microRNA expression profiles might in the near future be used to diagnose certain forms of malignancies.

Other neurological diseases in which microRNAs are likely to play an important role are diseases that are caused by defects in post-transcriptional RNA metabolism, including splicing, RNA transport and translation. Fragile-X mental retardation is a genetic disorder that is caused by mutations in the FMRP protein, an RNA-binding protein that regulates RNA transport into and translation within neuronal dendrites. We have already presented the rather intriguing connection between microRNAs and the fragile-X mental retardation syndrome, and an in depth discussion of the recent literature can be found in Chapter 27 in this book. Defective RNA metabolism also likely underlies another neurological disease, spinal muscular atrophy (SMA). SMA is a neurodegenerative disease that affects primarily motor neurons and is among the primary causes of infant death (Monani, 2005). In SMA, the RNA-binding protein small motor neuron protein

(SMN) is mutated. SMN is a rather ubiquitously expressed protein that seems to play a role in mRNA splicing as well as in axonal transport and possibly local mRNA translation in axons. It is still puzzling why mutations in a ubiquitously expressed gene such as SMN can lead to the rather specific degeneration of motor neurons. To gain more insight into the molecular basis of SMA, Dreyfuss and colleagues biochemically purified RNAs associated with the SMN multiprotein complex and thereby identified multiple small non-coding RNAs, including microRNAs (Mourelatos *et al.*, 2002). It is therefore tempting to speculate that microRNAs might be involved in the splicing/trafficking and/or translation of cargo mRNA that are associated with the SMN complex in the axons of motor neurons. It will be interesting to determine whether microRNA function is required for SMN-associated activities, such as mRNA splicing and transport, as well as axonal protein synthesis in motor neurons.

Given that microRNAs fulfill such diverse functions, it can be anticipated that the role of microRNAs in neurological disease is far more widespread than so far recognized. Large-scale cloning and expression analysis of microRNAs in different disease conditions, including Alzheimer's and Parkinson's disease, is underway and will undoubtedly yield more candidates for microRNAs that are relevant for disease. In addition, the mapping of genetic loci that are linked to multifactorial diseases of the nervous system, such as autism, promises to reveal important contributions from microRNA-dependent regulation of gene expression. Finally, the identification of disease genes as microRNA targets will aid in the delineation of the underlying mechanisms. Rett-syndrome (RTT) is a neurodevelopmental disorder caused by mutations in the MeCP2 gene that encodes a methylated DNA binding protein. Recently, it was reported that translation of the MeCP2 mRNA appears to be under the control of the activity-regulated microRNA miR-132 (R. Goodman, unpublished results). It is therefore conceivable that deregulated MeCP2 expression at the level of mRNA translation might contribute to Rett Syndrome. Post-transcriptional regulation of MeCP2 might be even more relevant given that aberrant MeCP2 expression in transgenic mice can cause a Rett-like phenotype. MiR-134, the other well-characterized activity-regulated microRNA, is also potentially linked to neurological disease. The only identified miR-134 target, Limk1, is one of the candidate genes for Williams Syndrome (WS), a genetic disorder that, among other non-neuronal features, is characterized by severe defects in cognitive function and visuospatial cognition (Frangiskakis *et al.*, 1996). The genetic lesion in WS is rather large and comprises several genes, and it is still controversial which of the genes within the deleted region is in fact responsible for the neurological symptoms. However, since the phenotypes of Limk1 mutant mice are not dissimilar from WS patients, Limk1 can be considered one of the most likely disease candidates identified to date. By inference, one might expect that perturbation of the Limk1 upstream regulator miR-134 would have similar consequences for higher cognitive function. Answers can be expected in the near future from the generation of brain-specific miR-134 knockout mice.

Finally, the recent discovery that microRNAs are involved in imprinting might have implications for other neurological disorders. Imprinting is defined as the

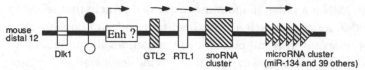

Figure 8.6. Schematic of the imprinted Dlk1-Gtl2 locus at mouse distal chromosome 12. Arrows denote putative transcripts originating from an unknown common Enhancer (Enh) element upstream of the GTL2 (gene trap locus 2) gene. Dlk1: Delta-like 1. RTL1: retrotransposon-like 1. Filled and open circle represent hyper- and hypomethylated state of the imprinting center, respectively. Maternally and paternally expressed genes are shown by striped and white boxes, respectively.

mechanism whereby a genetic region on one of the parental chromosomes is transcriptionally silenced, mostly through methylation of stretches of genomic DNA. Interestingly, misexpression of imprinted genes is often associated with striking defects of nervous system function, as observed for example in Angelman Syndrome, Prader–Willi Syndrome, and a role for small non-coding RNAs in these diseases has been recently proposed. As mentioned earlier, the miR-134 gene is located within a large cluster of microRNAs (>50) on mouse distal chromosome 12 (Figure 8.6). The nearby *Dlk1-Gtl2* locus is a classical example of an imprinted region, and a recent study demonstrated that all of the nearby microRNAs tested are subject to imprinting as well. Since small non-coding RNAs mediate DNA methylation and heterochromatin silencing in lower organisms such as yeast, it is conceivable that expression of the microRNA cluster on mouse distal chromosome 12 might contribute to establish the imprinting status of this genomic locus. This microRNA-based mechanism also has implications for disease, since deregulated gene expression at the *Dlk1-Glt2* domain is associated with the Callipyge phenotype, a form of muscular hypertrophy. Whether deregulated expression of this domain has also consequences for neuronal development is unknown.

In summary, multiple lines of evidence point to a rather widespread role for microRNAs in various forms of neurological disease, presumably as a consequence of the diverse mechanisms of microRNA action. Therefore, interfering with microRNA function, either by delivery of exogenous microRNA or blockade of endogeneous microRNAs (e.g. by antisense oligonucelotides or RNAi-mediated knockdown of microRNA precursors), may represent a way to restore physiological levels of critical microRNA targets and thereby contribute to the treatment of neurological diseases such as Rett, Williams, and Angelman Syndrome. Silencing of microRNAs *in vivo* has recently been achieved, although the approach in the present form seems not to be applicable to interfere with microRNA function in the brain (Krutzfeldt *et al.*, 2005).

Concluding remarks and perspectives

MicroRNA research in the nervous system is still very young and a plethora of important questions remain unanswered. Above all, it is still unclear what role, if any, microRNAs play in the development and function of the nervous system in higher vertebrates *in vivo*. On the other hand, *in vitro* studies have yielded several important insights into cellular processes that potentially involve microRNAs. These include neuronal differentiation, maintenance of the neuronal phenotype, neuronal morphogenesis and synaptic development/plasticity. During these processes, microRNAs employ a surprising variety of molecular mechanisms, ranging from the control of mRNA translation/stability to epigenetic regulation of chromatin structure and transcription. It can be expected that these findings represent just the tip of the proverbial iceberg, and we should keep an open mind towards novel microRNA-based mechanisms in the developing and mature nervous system. To appreciate the full repertoire of microRNA function in the nervous system, a combined effort from many different areas of biology is required. These include not only the bioinformatics prediction of microRNA target mRNAs, but also the biochemical purification of *in vivo* assembled microRNA complexes in the brain, cell biological detection of the spatio-temporal expression of microRNAs and their targets, and, last but not least, genetic approaches to elucidate the *in vivo* function of microRNAs in the context of a living organism. In addition, large-scale functional microRNA screens are starting to become feasible in vertebrate systems. For example, in an effort to obtain a more global picture of microRNA function during synaptic development, our laboratory has recently embarked on a large-scale loss-of-function screen in rat hippocampal neurons using 2′ O-methylated antisense oligonucleotides directed against microRNAs that are enriched at the synapse. Results from this and similar approaches will undoubtedly keep microRNA researchers busy for the foreseeable future.

REFERENCES

Ashraf, S. I., McLoon, A. L., Sclarsic, S. M. and Kunes, S. (2006). Synaptic protein synthesis associated with memory is regulated by the RISC pathway in *Drosophila*. *Cell*, **124**, 191–205.

Ballas, N. and Mandel, G. (2005). The many faces of REST oversee epigenetic programming of neuronal genes. *Current Opinion in Neurobiology*, **15**, 500–506.

Berezikov, E., Guryev, V., van de Belt, J. *et al.* (2005). Phylogenetic shadowing and computational identification of human microRNA genes. *Cell*, **120**, 21–24.

Caudy, A. A., Myers, M., Hannon, G. J. and Hammond, S. M. (2002). Fragile X-related protein and VIG associate with the RNA interference machinery. *Genes & Development*, **16**, 2491–2496.

Chang, S., Johnston, R. J., Jr., Frokjaer-Jensen, C., Lockery, S. and Hobert, O. (2004). MicroRNAs act sequentially and asymmetrically to control chemosensory laterality in the nematode. *Nature*, **430**, 785–789.

Ciafrè, S. A., Galardi, S., Mangiola, A. *et al.* (2005). Extensive modulation of a set of microRNAs in primary glioblastoma. *Biochemical Biophysical Research Communications*, **334**, 1351–1358.

Conaco, C., Otto, S., Han, J. J. and Mandel, G. (2006). Reciprocal actions of REST and a microRNA promote neuronal identity. *Proceedings of the National Academy of Sciences USA*. (In press.)

Flavell, S. W., Cowan, C. W., Kim, T. K. *et al.* (2006). Activity-dependent regulation of MEF2 transcription factors suppresses excitatory synapse number. *Science*. (In press.)

Frangiskakis, J. M., Ewart, A. K., Morris, C. A. *et al.* (1996). LIM-kinase1 hemizygosity implicated in impaired visuospatial constructive cognition. *Cell*, **86**, 59–69.

Giraldez, A. J., Cinalli, R. M., Glasner, M. E. *et al.* (2005). MicroRNAs regulate brain morphogenesis in zebrafish. *Science*, **308**, 833–838.

Grewal, S. I. and Rice, J. C. (2004). Regulation of heterochromatin by histone methylation and small RNAs. *Current Opinion in Cell Biology*, **16**, 230–238.

Hammond, S. M. (2005). MicroRNAs as oncogenes. *Current Opinion in Genetics Development*, **16**, 4–9.

Jin, P., Zarnescu, D. C., Ceman, S. *et al.* (2004). Biochemical and genetic interaction between the fragile X mental retardation protein and the microRNA pathway. *Nature Neuroscience*, **7**, 113–117.

Johnston, R. J. and Hobert, O. (2003). A microRNA controlling left/right neuronal asymmetry in *Caenorhabditis elegans*. *Nature*, **426**, 845–849.

Kiebler, M. A. and DesGroseillers, L. (2000). Molecular insights into mRNA transport and local translation in the mammalian nervous system. *Neuron*, **25**, 19–28.

Kim, J., Krichevsky, A., Grad, Y. *et al.* (2004). Identification of many microRNAs that copurify with polyribosomes in mammalian neurons. *Proceedings of the National Academy of Sciences USA*, **101**, 360–365.

Krichevsky, A. M., King, K. S., Donahue, C. P., Khrapko, K. and Kosik, K. S. (2003). A microRNA array reveals extensive regulation of microRNAs during brain development. *RNA*, **9**, 1274–1281.

Krutzfeldt, J., Rajewsky, N., Braich, R. *et al.* (2005). Silencing of microRNAs in vivo with 'antagomirs'. *Nature*, **438**, 685–689.

Kuwabara, T., Hsieh, J., Nakashima, K., Taira, K. and Gage, F. H. (2004). A small modulatory dsRNA specifies the fate of adult neural stem cells. *Cell*, **116**, 779–793.

Lagos-Quintana, M., Rauhut, R., Yalcin, A. *et al.* (2002). Identification of tissue-specific microRNAs from mouse. *Current Biology*, **12**, 735–739.

Lee, J. E. (1997). Basic helix-loop-helix genes in neural development. *Current Opinion in Neurobiology*, **7**, 13–20.

Lim, L. P., Lau, N. C., Garrett-Engele, P. *et al.* (2005). Microarray analysis shows that some microRNAs downregulate large numbers of target mRNAs. *Nature*, **433**, 769–773.

Liu, J., Valencia-Sanchez, M. A., Hannon, G. J. and Parker, R. (2005). MicroRNA-dependent localization of targeted mRNAs to mammalian P-bodies. *Nature Cell Biology*, **7**, 719–723.

Lu, J., Getz, G., Miska, E. A. *et al.* (2005). MicroRNA expression profiles classify human cancers. *Nature*, **435**, 834–838.

Lugli, G., Larson, J., Martone, M. E., Jones, Y. and Smalheiser, N. R. (2005). Dicer and eIF2c are enriched at postsynaptic densities in adult mouse brain and are modified by neuronal activity in a calpain-dependent manner. *Journal of Neurochemistry*, **94**, 896–905.

Miska, E. A., Alvarez-Saavedra, E., Townsend, M. *et al.* (2004). Microarray analysis of microRNA expression in the developing mammalian brain. *Genome Biology*, **5**, R68.

Monani, U. R. (2005). Spinal muscular atrophy: a deficiency in a ubiquitous protein; a motor neuron-specific disease. *Neuron*, **48**, 885–895.

Mourelatos, Z., Dostie, J., Paushkin, S. *et al.* (2002). miRNPs: a novel class of ribonucleoproteins containing numerous microRNAs. *Genes & Development*, **16**, 720–728.

Reinhart, B. J., Slack, F. J., Basson, M. *et al.* (2000). The 21-nucleotide let-7 RNA regulates developmental timing in *Caenorhabditis elegans*. *Nature*, **403**, 901–906.

Schratt, G. M., Tuebing, F., Nigh, E. A. *et al.* (2006). A brain-specific microRNA regulates dendritic spine development. *Nature*, **439**, 283–289.

Seitz, H., Royo, H., Bortolin, M. L., Lin, S. P., Ferguson-Smith, A. C. and Cavaille, J. (2004). A large imprinted microRNA gene cluster at the mouse Dlk1-Gtl2 domain. *Genome Research*, **14**, 1741–1748.

Sempere, L. F., Freemantle, S., Pitha-Rowe, I. *et al.* (2004). Expression profiling of mammalian microRNAs uncovers a subset of brain-expressed microRNAs with possible roles in murine and human neuronal differentiation. *Genome Biology*, **5**, R13.

Vo, N., Klein, M. E., Varlamova, O. *et al.* (2005). A cAMP-response element binding protein-induced microRNA regulates neuronal morphogenesis. *Proceedings of the National Academy of Sciences USA*, **102**, 16 426–16 431.

9 MicroRNA expression that controls the amount of branched chain α-ketoacid dehydrogenase in mitochondria of human cells

Dean J. Danner*, John Barnes and Benjamin Mersey

Introduction

More than 50 years ago a family was studied to explain the repeated loss of their children with death shortly after birth. Careful characterization of their newborn child found sweet smelling urine that was likened to maple syrup (Menkes *et al.*, 1954). Examination of blood and urine from these patients eventually identified an increased concentration of the branched chain amino acids (BCAA) and their α-ketoacids formed by transamination reactions. It was the elevated presence of these six compounds in body fluids that gave the sweet smell akin to maple syrup (Menkes, 1959). Continued work demonstrated that increased leucine, isoleucine, and valine and their branched chain ketoacids (BCKA) resulted from a defective ability of the patient to decarboxylate the BCKA. Investigations further showed that BCAA catabolism occurred in the mitochondria of all mammalian cells. Since the children within the family that were dying did not have this activity, but the parents had normal conditions, it was reasoned that an autosomal recessive inherited trait was causing the disease and each parent carried a single mutant gene. An affected individual must have acquired two mutant alleles for the same gene although it took 20 plus years to identify the genes and their mutations to cause the disease (Chuang and Shih, 2001).

Starting in the 1960s most of the studies to characterize the cause of the disease, now named maple syrup urine disease (MSUD), were focused on protein(s) since the DNA and gene structure was not yet to the level of analysis. Information gained by these studies confirmed that the disease was inherited as an autosomal recessive trait and the inhibited reaction was with a mitochondrial multienzyme complex. Purification of this complex, defined as the branched chain α-ketoacid dehydrogenase (BCKD) complex, did not succeed until the late 1970s (Pettitt *et al.*, 1978; Danner *et al.*, 1979). In the middle of the 1980s we were able to isolate a cDNA that represented the gene product that is the core of the BCKD complex (Litwer and Danner, 1985).

---

* Author deceased; correspondence should be addressed to Peng Jin.

MicroRNAs: From Basic Science to Disease Biology, ed. Krishnarao Appasani. Published by Cambridge University Press. © Cambridge University Press 2008.

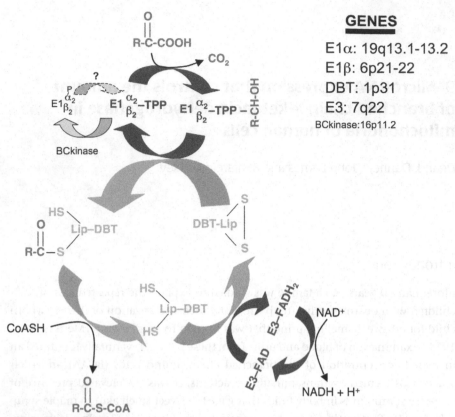

GENES

E1α: 19q13.1-13.2
E1β: 6p21-22
DBT: 1p31
E3: 7q22
BCkinase:16p11.2

Figure 9.1. BCKD reactions.

When the three genes that encode the unique components of the BCKD complex were identified, analysis began to define nucleic acid changes in gene alleles that cause MSUD (Peinemann and Danner, 1994; Chuang and Shih, 2001). The known mutations in each of the three genes are equal in number and all the different types of mutations causing disease were identified (Nellis and Danner, 2001). However, the specific mutation in many of the patients is not known and why some individuals express the disease later in life than others has not been defined at the gene level. Figure 9.1 shows the basic understanding of BCKD.

Among the most important roles for the BCAA (leucine, isoleucine, and valine) include: (a) serving as major components of most proteins, (b) crossing the blood brain barrier more efficiently than other amino acids and thereby being used to form glutamate, and (c) at least leucine stimulates insulin release in pancreatic cells. Mammals cannot synthesize these amino acids so BCAA are obtained from diet intake or endogenous protein breakdown. When not used for protein synthesis or the other tissue specific functions, the BCAA are metabolized in a pathway that exists in all cells with mitochondria. After transamination of the amino acid, the branched chain α-ketoacid (BCKA) is decarboxylated by the activity BCKD. The activity of BCKD commits the BCAA to their degradation. The first two enzymes, branched chain aminotransferase and BCKD act on all three BCAA but after decarboxylation, each branched chain acyl-CoA produced enters a separate pathway (Figure 9.2).

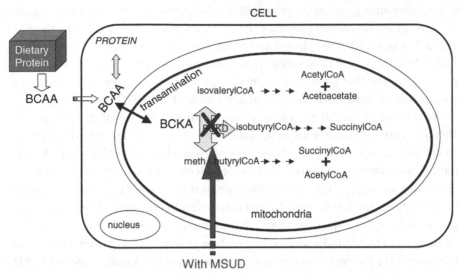

Figure 9.2. Metabolism of the branched chain amino acids and the human inherited defect that blocks the pathway.

BCKD is present in mitochondria of all cells and in mammals it was found that the activity state of BCKD is controlled by expression of a complex-specific kinase (BCkinase) (Odessey, 1982). This enzyme adds a phosphate to the E1α subunit that blocks the binding site for the BCKA substrate (Ævarsson *et al.*, 2000). Genes encoding the catalytic components of BCKD have been found in genomes of *C. elegans, Drosophila*, and fish but the BCkinase gene is only found in all the mammalian genomes. Activity and functional studies on BCKD have been done only in mice, rats, cows, and humans.

Controls for amount of BCKD activity

During the past ten years, the expression of microRNA (miR) has gained interest and their expression, although first noted in worms, is now reported to occur in plants, flies and even humans among many other species. Recently it was reported that a miR in *Drosophila* was used to control BCKD activity in response to dietary protein intake by the fly (Stark *et al.*, 2003). The fly miR277 was reported to match the mRNA for components of BCKD and when the miR was expressed the formation of BCKD was reduced. An hypothesis was made for the flies that when they eat a low protein diet miR277 expression is used to block formation of BCKD to control the amount of BCAA the flies need. However, it has not been demonstrated what controls the expression of this miR gene in the fly and no other investigation of BCKD expression affected by miRs has been reported in flies.

Recently more than 200 miRs have been detected for potential expression and function in humans. Many of the miRs appear to have the potential of binding to various mRNAs from different genes (Du and Zamore, 2005). It is not understood how the miR selects their target mRNA and whether that changes with cell type, development state, or metabolites. The coding sequences for miRs are found either as separate genes or are produced from an intronic sequence within a gene that encodes a functional protein. Current data suggest that in humans most of the miRs block mRNA translation and do not destroy the mRNA (Filipowicz *et al.*, 2005; Humphreys *et al.*, 2005; Pillai *et al.*, 2005; Smirnova *et al.*, 2005). Reports are being presented that show human miRs appear to be expressed only in certain tissues and are most active during cell development (Miska, 2005).

Our interest in human miRs was initiated from two sides. First, the report that *Drosophila* used miR expression to regulate the amount of BCKD needed in response to dietary protein (Stark *et al.*, 2003). Second, some human patients with MSUD have a less devastating condition than other patients and often do not demonstrate sensitivity to dietary protein until they are over the age of one year. For many MSUD patients the genetic basis has not been determined and this is especially true for those that don't display the MSUD clinical phenotype until later in life. Likewise the activity of BCKD and a western blot of these proteins has not been determined in many of the patients. Based on the report that flies use a miR to control BCKD expression in response to dietary protein, it raised the question of whether miR expression in humans affects the expression level of BCKD. miRNAs would provide a means of controlling the amount of BCKD in mammals while the BCkinase expression controls the activity state of the BCKD present within a cell. The increasing number of reports that miR expression is related to a number of different human diseases makes investigation of the MSUD patients for the function of miRNA expression reasonable (Alvarez-Garcia and Miska, 2005). We began to investigate if any of the human miRs that potentially bind the mRNA for components of BCKD can influence the amount of BCKD within a cell.

All the catalytic BCKD proteins are reported to have a 22–26 h lifetime, although turnover of mitochondrial proteins is difficult to define and is thought to be longer than most other cell proteins (Honda *et al.*, 1991). Data suggest that if the BCKD proteins are not associated with one another and specifically associated with the core dihydrolipoamide branch chain transacylase (DBT) protein then they are likely to be degraded. The one protein that may be present in excess is the E1αsubunit that is the substrate for the BCkinase (Yeamen *et al.*, 1984; Sitler *et al.*, 1998). However, it appears that the amount of BCKD within a cell depends on the amount of DBT that is being produced. Expression of BCkinase level still controls the fraction of the existing BCKD that is catalytically active. Since human miRs seem to be associated mainly with mRNA for DBT, it is important to determine whether DBT expression in different cell types is affected by the expression level of an miR in that cell type. If expression of DBT is decreased the other subunits of the complex would also decrease and the total amount of BCKD differs with cell type. We do know that the amount of BCKD varies with different tissues but have assumed this is caused by gene expression levels for component proteins. Therefore determining if the amount of BCKD within a cell type reflects the miR expression that controls the amount of DBT protein could provide another means of causing MSUD.

Two human miRs were found with the potential to bind to DBT mRNA within the terminal exon 11. Exon 11 for DBT holds the TGA stop codon and all of the 3′ UTR nucleotide sequence. One miR matches mRNA sequence within the coding

5' CUGACCUCUGAAUUGACAGCC 3'   has miR192
3' GACUGGAUA mRNA for DBT

5' UAGCACCAUUUGAAAUCAGU 3'   has miR29b
3' AUCGUGGUA mRNA for DBT

Figure 9.3. Predicted binding sites for the microRNAs with the mRNA for DBT.

region (miR29b-2) and the other (miR192) matches bases in the 3' UTR (Figure 9.3). Only the first 7–9 bases for each miR bind directly while the later sequences appear to bind to sequences further upstream in the mRNA. This condition is what is proposed to occur when the miR blocks translation of the mRNA (Hammond, 2005). The findings align with what is most common for other miRs that have been reported in humans. To investigate whether these identified miRs with the potential to bind human DBT mRNA will control expression in humans as found with *Drosophila*, human cell lines were transfected with a synthesized miR29b and miR192. As seen in Figure 9.3, the first nine bases of each miR tested pair with sequence in the mRNA for DBT. This finding suggested that translation of the mRNA would be reduced. The initial study focused on miR29b expression in HEK293 cells (Mersey *et al.*, 2005). A western blot of various concentrations of mitochondrial protein showed that the amount of DBT protein decreased after the cells were exposed to the miR and yet the analysis of the mRNA for DBT did not decrease during the time course after miR addition. What did happen was BCKD complex total activity decreased during this time course. The amount of DBT protein decreased and in fact all components of the complex decreased during this time period. Therefore since the amount of DBT mRNA remained constant while the protein amount decreased, the assumption is that translation of the mRNA was decreased. The exact mechanism for this action has not been well characterized in mammals. Data with miR29b was published and work with miR192 is now being conducted (Mersey *et al.*, 2005). Initial studies suggest that both miRs being studied by us have similar function (Tables 9.1–9.3).

Although transfection of cells with a putative miR did influence the amount of DBT and BCKD complex within a cell, it says nothing of endogenous expression of these miR molecules in that cell. For the miR29b there are two different gene sites, one on chromosome 1 and the other on chromosome 7. Expression of miR29b-2 from its gene on the long arm of chromosome 1 is detected in different human cell types while expression of miR29b-1 located on chromosome 7 was not expressed in the same cells tested. Interestingly the gene for DBT is also located on chromosome 1 but is located on the short arm so it is unlikely their presence on the same chromosome has any relation. Conditions for expression of the miR29b-2 gene need investigation.

Endogenous expression of miR29b was tested by assessment of the cells that were transfected with antisense RNA to potentially block the function of endogenous miR29b-2. This condition produced small increases in the amount of DBT in the mitochondria that also resulted in an increase in the total BCKD complex activity. The results suggest that the endogenous miR29b-2 expression does work at least in

Table 9.1. BCKD activity (pmol CO_2/µg protein/min) after treating with random sequence RNA and specific miR 29b for the indicated time

	24 h	48 h	96 h	144 h
Random miR	45.33	46.03	44.96	41.26
miR29b	39.72	32.09	33.21	23.69

The data are representative of at least three repeated analyses and each value is the average of at least four individual assays.

Table 9.2. Decrease in BCKD after cell transfection with O-methyl RNA to block endogenous miR29b (results are as described in Table 9.1)

	24 h	48 h	72 h
Random miR	63.71	42.54	69.51
O-me-29b	65.44	50.78	76.12

Table 9.3. Effect of transfection of HEK293 cells with two concentrations of miR192 on BCKD total activity 48 h after addition

Random miR	50 pmol miR192	100 pmol miR192
81.21	53.00	51.48
100%	65%	63%

part on the DBT mRNA translation. Our studies have not sought the binding of miR29b-2 to other mRNAs. BCKD expression levels need to be compared in cell types expressing low and high levels of miR29b-2. Further investigations will compare effects of the two miRs we have found to bind DBT mRNA. They do not overlap in binding sites but initial studies of cells transfected with both miRs were not additive so preferential binding by one may alter the binding of the other.

Addition of antisense RNA to block endogenous miR192 expression in cells resulted in an increase in DBT protein near to that we found with antisense miR29b. The gene for miR192 is located on chromosome 11 and expression will likely vary with cell type. In a recent report, it was shown that inhibition of miR192 expression decreased cell growth for two different cell types, HeLa and A549 lung carcinomas (Cheng *et al.*, 2005). This suggests preferential binding of miR192 may not be for DBT mRNA. Recall that we did not find effects on other mRNAs for miR29b-2. Although transfection of cells with miR192 did decrease DBT protein amounts, but when cells were transfected with both miR192 and miR29b-2 an additive effect was not found. This may reflect a preference of each miR for their preferred target mRNA.

Potential for MSUD disease with miR expression

Expression of miR29b-2 or miR192 in cell lines from patients with MSUD has not been investigated. One might expect that if BCKD is not functioning the miR expression would be low since the amount of BCAA is high. Since *Drosophila* use the miR277 expression to control BCAA concentration the same may be expected with mammals. Cells from MSUD patients with the intermittent clinical condition are of special interest as are those where the second mutant gene allele has not been defined. A higher expression of one or both of these miRs could decrease the total amount of BCKD found in cells and the cell type may have specific effects to reduce the amount of BCKD in certain cells. This would mimic mutations in genes for the other components of BCKD formation. Recall that protein expression for BCKD proteins in *Drosophila* was determined by dietary intake of protein that apparently induced miR277 expression.

REFERENCES

Ævarsson, A., Chuang, J. L., Wynn, R. M. *et al.* (2000). Crystal structure of human branched-chain α-ketoacid dehydrogenase and the molecular basis of multienzyme complex deficiency in maple syrup urine disease. *Structure Fold. Des.*, **8**, 277–291.

Alvarez-Garcia, I. and Miska, E. A. (2005). MicroRNA functions in animal development and human disease. *Development*, **132**, 4653–4662.

Cheng, A. M., Byrom, M. W., Shelton, J. and Ford, L. P. (2005). Antisense inhibition of human miRNAs and indications for an involvement of miRNA in cell growth and apoptosis. *Nucleic Acids Research*, **33**, 1290–1297.

Chuang, D. T. and Shih, V. E. (2001). Maple syrup urine disease (branched chain ketoaciduria). In *The Metabolic and Molecular Basis of Inherited Disease*, 8th edn, Scriver, C. R., Beaudet, A. L., Sly, W. S. and Valle, D. (eds.), Vol. 2. New York: McGraw-Hill, pp. 1971–2005.

Danner, D. J., Lemmon, S. K., Besharse, J. C. and Elsas, L. J. (1979). Purification and characterization of branched chain alpha-ketoacid dehydrogenase from bovine liver mitochondria. *Journal of Biological Chemistry*, **254**, 5522–5526.

Du, T. and Zamore, P. D. (2005). microPrimer: the biogenesis and function of microRNA. *Development*, **132**, 4645–4652.

Filipowicz, W., Jaskiewicz, L., Kolb, F. A. and Pillai, R. S. (2005). Post-transcriptional gene silencing by siRNAs and miRNAs. *Current Opinion in Structural Biology*, **15**, 331–341.

Hammond, S. M. (2005). Dicing and slicing. The core machinery of the RNA interference pathway. *Federation of the European Biochemical Sciences Letters*, **579**, 5822–5829.

Honda, K., Ono, K., Mori, T. and Kochi, H. (1991). Both induction and activation of the branched-chain 2-oxo acid dehydrogenase complex in primary-cultured rat hepatocytes by clofibrate. *Journal of Biochemistry (Tokyo)*, **109**, 822–827.

Humphreys, D. T., Westman, B. J., Martin, D. I. and Preiss, T. (2005). MicroRNAs control translation initiation by inhibiting eukaryotic initiation factor 4E/cap and poly(A) tail function. *Proceedings of the National Academy of Sciences USA*, **102**, 16 961–16 966.

Litwer, S. and Danner, D. J. (1985). Identification of a cDNA clone in λgt11 for the transacylase component of branched chain ketoacid dehydrogenase. *Biochemical Biophysical Research Communications*, **131**, 961–967.

Menkes, J. H. (1959). Maple syrup disease. Isolation and identification of organic acids in the urine. *Pediatrics*, **23**, 348–353.

Menkes, J. H., Hurst, P. L. and Craig, J. M. (1954). A new syndrome: progressive familial infantile cerebral dysfunction associated with an unusual urinary substance. *Pediatrics*, **14**, 462–467.

Mersey, B. D., Jin, P. and Danner, D. J. (2005). Human microRNA (miR-29b) expression controls the amount of branched chain α-ketoacid dehydrogenase complex in a cell. *Human Molecular Genetics*, **14**, 3371–3377.

Miska, E. A. (2005) How microRNAs control cell division, differentiation and death. *Current Opinion in Genetics and Development*, **15**, 563–568.

Nellis, M. M. and Danner, D. J. (2001). Gene preference in maple syrup urine disease. *American Journal of Human Genetics*, **68**, 232–237.

Odessey, R. (1982). Purification of rat kidney branched chain oxo acid dehydrogenase complex with endogenous kinase activity. *Biochemical Journal*, **204**, 353–356.

Peinemann, F. and Danner, D. J. (1994). Maple syrup urine disease 1954 to 1993. *Journal of Inheritable and Metabolic Diseases*, **17**, 3–15.

Pettit, F. H., Yeaman, S. J. and Reed, L. J. (1978). Purification and characterization of branched chain α-ketoacid dehydrogenase complex of bovine kidney. *Proceedings of the National Academy of Sciences USA*, **75**, 4881–4886.

Pillai, R. S., Bhattacharyya, S. N., Artus, C. G. *et al.* (2005). Inhibition of translational initiation by Let-7 MicroRNA in human cells. *Science*, **309**, 1573–1576.

Sitler, T. L., McKean, M. C., Peinemann, F., Jackson, E. and Danner, D. J. (1998). Import rate of the E1b subunit of human branched chain a-ketoacid dehydrogenase is a limiting factor in the amount of complex formed in the mitochondria. *Biochimica Biophysica Acta*, **1404**, 385–392.

Smirnova, L., Grafe, A., Seiler, A. *et al.* (2005). Regulation of miRNA expression during neural cell specification. *European Journal of Neuroscience*, **21**, 1469–1477.

Stark, A., Brennecke, J., Russell, R. B. and Cohen, S. M. (2003). Identification of Drosophila microRNA targets. *Public Library of Science Biology*, **1**, 1–13.

Yeaman, S. J., Cook, K. G., Boyd, R. W. and Lawson, R. (1984). Evidence that the mitochondrial activator of phosphorylated branched chain 2 oxoacid dehydrogenase complex is the dissociated E1 component of the complex. *Federation of the European Biochemical Sciences Letters*, **172**, 38–42.

10 MicroRNAs and the regulation of leaf shape

Ramiro E. Rodriguez, Carla Schommer and Javier F. Palatnik*

Introduction: microRNAs meet plant development

The conspicuous group of 20–21 nt small RNAs designated as microRNAs (miRNAs) was first discovered in *Caenorhabditis elegans*. In 1993, *lin-4*, the first identified miRNA, was shown to recognize several partially complementing sites in the 3' UTR of the *lin-14* mRNA and to inhibit its translation (Lee *et al.*, 1993). At that time it was thought, however, that miRNAs were an oddity restricted to worms. Their participation in biological pathways in other animals was vastly underestimated and any role in plant development was therefore even more unexpected.

The situation changed with the arrival of the new millennium when it was recognized that a second worm miRNA, *let-7* (Reinhart *et al.*, 2000), was conserved in many animals with bilateral symmetry (Pasquinelli *et al.*, 2000). Several dozens of new miRNAs were then readily identified through cloning efforts in *Drosophila*, *Caenorhabditis* and humans (Lagos-Quintana *et al.*, 2001; Lau *et al.*, 2001; Lee and Ambros, 2001). By 2002 there were already four small RNA libraries made from *Arabidopsis thaliana* allowing a first systematic scan for miRNAs in the fully sequenced model plant (Llave *et al.*, 2002a; Mette *et al.*, 2002; Park *et al.*, 2002; Reinhart *et al.*, 2002).

The knowledge on miRNA biogenesis and action burst when a connection with an apparently unrelated process called cosuppression in plants or RNA interference in animals was discovered (Zamore and Haley, 2005). In this RNA-mediated silencing pathway, a long double stranded RNA (dsRNA) is processed into small RNAs named short interfering RNAs (siRNAs) (Baulcombe, 2005). The key enzyme in this process is a ribonuclease III called DICER which cleaves dsRNA. This enzyme was also recognized to be involved in miRNA biogenesis, by processing the fold-back structures of the miRNA precursors (Bartel, 2004). However, despite sharing biogenesis and effector components,

---

* Author to whom correspondence should be addressed.

MicroRNAs: From Basic Science to Disease Biology, ed. Krishnarao Appasani. Published by Cambridge University Press. © Cambridge University Press 2008.

Table 10.1. *Arabidopsis* microRNA families and their targets

microRNA Family	Targets	No. Targets[a]	Targets' function
	I. Transcription factors		
miR156	SBP[1–3]	11/16	Flowering time
miR159	MYB,[2,4–6]	8/203	Anther development
miR319	TCP[4]	5/24	Leaf shape
miR160	ARF[7,8]	3/23	Auxin signaling
miR164	NAC[9–11]	6/113	SAM establishment, organ boundaries definition and root development
miR166	HD-ZIP[12]	5/94	SAM establishment and organ asymmetry
miR167	ARF[2,3]	2/23	Auxin signaling
miR169	HAP2[2,13]	8/10	Unknown
miR171	SCR[3,14,15]	3/32	Unknown
miR172	AP2[1,3,16,17]	6/146	Flowering time
miR396	GRF[13]	7/9	Leaves and cotyledons growth and development
	II. RNA metabolism		
miR162	DICER-LIKE (DCL)[18]	1	microRNA biogenesis
miR168	Argonaute (AGO1)[19]	1	microRNA action
miR173	trans-acting siRNA1 and 2 (TAS1, TAS2)[20]	4	ta-siRNA biogenesis
miR390	trans-acting siRNA3 (TAS3)[20]	1	ta-siRNA biogenesis
miR403	Argonaute (AGO2)[20]	1	microRNA action
	III. Others		
miR161	Pentatricopeptide repeat protein[2,21,22]	17	Unknown
miR163	S-adenosylmethionine-dependent methyltransferase (SAMT)[21]	5	Metabolism
miR397	Laccase/Cu oxidase[13,23]	3	Unknown – redox reactions
miR393	TIR1/F-Box[13]	4	Auxin receptor
miR395	ATP sulfurylase[13]	4	Sulfate assimilation
	Sulfate transporter[20]	1	Sulfate assimilation
miR398	Cu/Zn superoxide dismutase[13]	2	Stress response
	Cytchrome c oxidase[13,23]	1	Mitochondrial energetic metabolism
miR399	E2-ubiquitin conjugating enzyme[20,24]	1	Ubiquitin conjugation – response to low phosphate
miR408	Laccase/diphenol oxidase[1]	3	Unknown – redox reactions
	Plantacyanin[23,25]	1	Unknown – redox reactions
miR447	2 phosphoglycerate kinase[20]	1	Unknown

[a] Regulated transcription factors/total members in the corresponding transcription factor family according to the "Database of *Arabidopsis* Transcription Factors" (http://datf.cbi.pku.edu.cn/index, *Bioinformatics* **21**, 2568 (2005)).

[1] Schwab, R. *et al*. *Developmental Cell*, **8**, 517–527 (2005).
[2] Rhoades, M. W. *et al*. *Cell*, **110**, 513–520 (2002).
[3] Kasschau, K. D. *et al*. *Developmental Cell*, **4**, 205–217 (2003).
[4] Palatnik, J. F. *et al*. *Nature*, **425**, 257–263 (2003).
[5] Millar, A. A. and Gubler, F. *Plant Cell*, **17**, 705–721 (2005).
[6] Achard, P. *et al*. *Development*, **131**, 3357–3365 (2004).
[7] Wang, J. W. *et al*. *Plant Cell*, **17**, 2204–2216 (2005).

[8] Mallory, A. C. *et al*. *Plant Cell*, **17**, 1360–1375 (2005).
[9] Laufs, P. *et al*. *Development*, **131**, 4311–4322 (2004).
[10] Guo, H. S. *et al*. *Plant Cell*, **17**, 1376–1386 (2005).
[11] Mallory, A. C. *et al*. *Current Biology*, **14**, 1035–1046 (2004a).
[12] Bowman, J. L. *Bioessays*, **26**, 938–942 (2004).
[13] Jones-Rhoades, M. W. and Bartel, D. P. *Molecular Cell*, **14**, 787–799 (2004).
[14] Llave, C. *et al*. *Science*, **297**, 2053–2056 (2002b).
[15] Parizotto, E. A. *et al*. *Genes and Development*, **18**, 2237–2242 (2004).
[16] Aukerman, M. J. and Sakai, H. *Plant Cell*, **15**, 2730–2741 (2003).
[17] Chen, X. *Science*, **303**, 2022–2025 (2004).
[18] Xie, Z. *et al*. *Current Biology*, **13**, 784–789 (2003).
[19] Vaucheret, H. *et al*. *Genes and Development*, **18**, 1187–1197 (2004).
[20] Allen, E. *et al*. *Cell*, **121**, 207–221 (2005).
[21] Allen, E. *et al*. *Nature Gentics*, **36**, 1282–1290 (2004).
[22] Vazquez, F. *et al*. *Molecular Cell*, **16**, 69–79 (2004).
[23] Sunkar, R. and Zhu, J. K. *Plant Cell*, **16**, 2001–2019 (2004).
[24] Fujii, H. *et al*. *Current Biology*, **15**, 2038–2043 (2005).
[25] Sunkar, R. *et al*. *Plant Cell*, **17**, 1397–1411 (2005).

there is a major difference in the mechanism of action between animal miRNAs and siRNAs: while animal miRNAs with only limited complementarity to their target mRNAs inhibit translation, siRNAs perfectly match to their targets and guide them to degradation (Bartel, 2004).

Shortly after the first plant miRNAs were identified, Rhoades *et al.* (2002) realized that they have a very good complementarity to potential target mRNAs, much better than their animal counterparts. Typical plant miRNA-regulated genes contain one target site in the coding sequence, while in animals there are usually several target sites in the 3' UTR (Bartel, 2004). Allowing only three mismatches between miRNA and target mRNAs it is already possible to identify many, albeit not all, targets of a given miRNA (Rhoades *et al.*, 2002). To make things even more compelling, most plant miRNAs and their target genes are conserved in evolution; and it is easy enough to compare potential miRNA regulated genes from different plant species (e.g. *Arabidopsis* and rice) to find an almost perfect conservation of the target site (Palatnik *et al.*, 2003; Rhoades *et al.*, 2002; Sunkar and Zhu, 2004). From the information gained in these initial studies it was already clear that a major category of plant miRNA targets are transcription factors suggesting a direct link between miRNAs and plant development (see Table 10.1 for a summary of validated plant miRNA targets).

Using a transient assay, where miRNAs and their targets are co-infiltrated in tobacco leaves, Llave *et al.* (2002b) demonstrated that plant miRNAs can cause degradation of their regulated mRNAs. They also showed that the miRNA directed cleavage can be mapped to position 10 of the target site (Llave *et al.*, 2002b) in a similar or identical way to siRNAs (Elbashir *et al.*, 2001). A modification of a 5' RACE PCR (Llave *et al.*, 2002b) especially designed to map this cleavage position *in vivo* is nowadays used as a hallmark for experimental validation of potential target RNAs (Jones-Rhoades and Bartel, 2004; Kasschau *et al.*, 2003; Llave *et al.*, 2002b; Palatnik *et al.*, 2003; Sunkar *et al.*, 2005).

The general role of miRNAs in plant development is also evident from the consequences of disrupted DICER activity. *Arabidopsis* has four *DICER LIKE* (*DCL*) genes, which display a functional specialization in different small RNA pathways (Xie *et al.*, 2004), *DCL1* being responsible for miRNA processing (Park *et al.*, 2002; Reinhart *et al.*, 2002). Interestingly enough, *dcl1* mutant alleles were isolated several times in the past from different genetic screens for defects in embryo, ovule or flower development in *Arabidopsis*, indicating the multiple roles of the miRNA pathway in plant development (Schauer *et al.*, 2002). Mutations in other genes encoding components required for miRNA biogenesis and function such as *HEN1* (Boutet *et al.*, 2003; Park *et al.*, 2002), *HYL1* (Han *et al.*, 2004; Vazquez *et al.*, 2004; Yu *et al.*, 2005), *HASTY* (Park *et al.*, 2005) and *ARGONAUTE1* (*AGO1*) (Bohmert *et al.*, 1998; Kidner and Martienssen, 2004; Kidner and Martienssen, 2005; Vaucheret *et al.*, 2004) also have pleiotropic effects on plant development, the most notorious being defects in flowers and leaves.

Control of leaf morphogenesis by microRNAs

The *jaw-D* mutant

The discovery of miRNAs as main regulators of gene expression prompted to re-evaluate older genetic results (Palatnik *et al.*, 2003; Tang *et al.*, 2003), with a special interest in finding miRNAs acting in specific developmental processes. One example was the activation tagged mutant *jaw-D*, that displayed drastic changes in leaf morphology (Figure 10.1) (Palatnik *et al.*, 2003; Weigel *et al.*, 2000).

Leaves are determinate organs supporting the majority of photosynthetic processes in land plants. There is a wide variety of leaf morphologies and shapes that essentially aid the success of plants in nature. Although a good amount of information has been accumulated on how organs acquire their particular fate, only little is known about the way they are shaped. Wild-type (wt) *Arabidopsis* rosette leaves are nearly flat organs. Not so in the *jaw-D* mutant, where they become crinkly with a pronounced negative curvature (Palatnik *et al.*, 2003; Weigel *et al.*, 2000). Interestingly, *jaw-D* has also other phenotypic defects including a moderate delay in flowering time, cotyledon epinasty, changed petal color and fruit shape (Palatnik *et al.*, 2003; Weigel *et al.*, 2000).

Studies on leaf development in different species have shown that a front of cell cycle arrest moves gradually from the tip to the base of the leaf primordia. Therefore, cell proliferation is arrested at the tip of the leaf, while it still proceeds at the base (Tsukaya, 2005). Recently, Nath *et al.* (2003) demonstrated that making a leaf flat is not a trivial task. Snapdragon leaves are normally flat, but they become crinkly in plants lacking CINCINNATA (CIN) (Figure 10.1). The mutant leaves cannot be flattened without folds or cuts. A comparative study suggested that the cell cycle arrest in the leaf margins is delayed in the *cin* mutant compared to wild type (Nath *et al.*, 2003). As a result, cell proliferation is prolonged in the leaf margins in relation to the central regions of the *cin* mutant leaf blades, causing the leaves to become crinkly (Nath *et al.*, 2003).

CIN encodes a transcription factor of the TCP family (named after the first identified members: TEOSINTE BRANCHED1, CYCLOIDEA and PCFs) (Cubas *et al.*, 1999; Nath *et al.*, 2003). TCP transcription factors have been implicated in the control of cell division and various morphogenetic traits such as floral asymmetry in snapdragon and the fate of axillary meristems in maize. Although TCPs

Figure 10.1. Control of leaf morphogenesis by TCPs and miR319. (a) Leaves of wild-type (wt) *Arabidopsis* (left) and *jaw-D* mutant overexpressing miR319a. (b) Leaves of wt snapdragon (left) and *cin* mutant, lacking a *TCP* gene with target sequence to miR319. (c) Scheme representing the growth arrest front in developing wt leaf (left) and plants with reduced TCP activity; dots indicate proliferating regions.

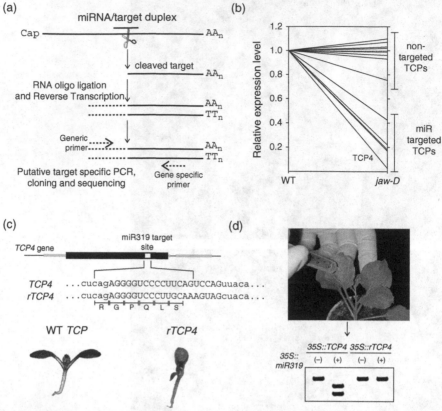

Figure 10.2. Tools to study miRNAs regulating plant development. (a) Scheme showing a modified 5′ RACE-PCR used to map the cleavage site on target mRNAs. (b) Microarray experiments reveal targeted transcripts. Levels of the transcription factors belonging to the TCP family in *jaw-D* mutant: the five TCP genes with a miRNA target sequence are the only that are significantly affected. (c) Disruption of miRNA-mediated regulation by introducing silent mutations on the miRNA target sequence. Bottom, *Arabidopsis* transgenic plants expressing wt or miRNA resistant *TCP4*. (d) Transient assay in *Nicotiana benthamiana* plants. *Agrobacterium* strains harboring different vectors with miRNAs and targets are co-infiltrated in leaves (top). The effects of the miRNA on the targets are detected by Northern blot (bottom).

are plant specific they display some similarities to the bHLH family of transcription factors present in both plants and animals (Cubas *et al.*, 1999).

The *jaw-D* mutant was isolated from an activation tagging screen which is based on the random insertion of viral transcriptional enhancers into the genome activating the transcription of nearby genes (Weigel *et al.*, 2000). As the activated region responsible for the *jaw-D* phenotype does not contain any protein coding gene, it was originally suspected that a potential non-coding RNA was responsible for the phenotype (Weigel *et al.*, 2000).

Global expression profiles using microarrays revealed that five *TCP*s were among the top affected genes in the *jaw-D* mutant (Palatnik *et al.*, 2003). Actually, *TCP4*, the closest *Arabidopsis* relative to *CIN* was the most downregulated gene in the mutant compared to wild-type plants (Figure 10.2). Sequence analysis of the five downregulated *TCP* transcription factors revealed that in addition to homology in the DNA-binding TCP domain they shared a short RNA motif that was absent in the other non-affected *Arabidopsis TCPs*. This sequence was also found in *CIN* and certain *TCPs* from nearly 20 different species, suggesting the existence of a miRNA target site (Palatnik *et al.*, 2003). A closer inspection of the region producing the non-coding RNA in *jaw-D* revealed the presence of a fold-back structure that contains a potential miRNA at the base of one of the stems with complementarity to the conserved region in the *TCPs* (Palatnik *et al.*, 2003).

A biochemical characterization revealed that this miRNA, called miR-JAW or miR319a according to the current nomenclature, exists in wild type plants and was overexpressed more than 20 times in the *jaw-D* mutant (Palatnik *et al.*, 2003). The fact that only recently miR319a has been cloned a few times from small RNA libraries (Gustafson *et al.*, 2005; Sunkar and Zhu, 2004), suggests that the miRNA is normally expressed at low levels or in specific tissues. In the *jaw-D* mutant the viral enhancers are inserted on chromosome 4 between genes At4g23710 and At4g23720, where they activate the transcription of the miRNA precursor. This causes the over-production of miR319a and the concomitant clearing of *TCP* transcripts (Palatnik *et al.*, 2003; Weigel *et al.*, 2000).

The impact of miRNA resistant genes

The first plant miRNA identified through a genetic screen was miR319a (Palatnik *et al.*, 2003). A similar screen had previously contributed to the identification of the *Drosophila* miRNA *bantam* (Brennecke *et al.*, 2003). Other plant miRNA mutants such as *eat-D*, *jabba-1D* and *men1* that overexpress miR172a-2 (Aukerman and Sakai, 2003), miR166g (Williams *et al.*, 2005) and miR166a (Kim *et al.*, 2005) respectively, were also generated by activation tagging screens, indicating that gain of function screens are a powerful genetic approach to isolate miRNA mutants.

In general, the overexpression of a miRNA provides information on the biological function of its targets as it produces a multiple target-gene knock down. In good agreement, disruption of the snapdragon *CIN* gene which contains a miR319 target sequence (Nath *et al.*, 2003) or the combination of individual insertional knock-outs in the *Arabidopsis* miRNA targeted *TCP* genes (C. Schommer *et al.*, unpublished results) produces changes in leaf morphology resembling *jaw-D* and miR319 overexpression (Palatnik *et al.*, 2003; Weigel *et al.*, 2000).

The preferred way to gain insight into the biological role of specific miRNAs would be to study loss of function mutations. Recently, the inactivation of miR164 family members was described (Baker *et al.*, 2005; Guo *et al.*, 2005; Mallory *et al.*, 2004a). However, owing to the fact that miRNAs are usually encoded by small families of potentially redundant members and that the small miRNA precursors are not susceptible to nonsense and frameshift mutations it is usually a difficult task to completely disrupt a specific miRNA activity.

An alternative approach to study miRNA loss of function is the introduction of silent mutations in a target gene that abolish the interaction with the miRNA but do not change the encoded amino acids, generating miRNA-resistant genes

Figure 10.3. Effects of miRNA manipulation on plant development. (a) Wt, miR319 overexpressor and transgenic plants expressing low levels of microRNA resistant TCP4. (b, c) Overexpression of miR164, (b) Cotyledon fusions in seedlings (right); (c) stem–leaf and leaf–leaf fusions (right). (d, e) Overexpression of miR156, (d) faster generation of leaves in the overexpressors (right); (e) delay in flowering time in miR156 overexpressors leading to a bushy plant (right). (f, g) Overexpression of miR172, (f) early flowering time in the transgenics (right); (g) patterning deffects on flowers overexpressing miR172 (right). (See color plate 7)

(Figure 10.2). This is ideally done in the genomic context of the target gene in order to study the endogenous contribution of miRNA-mediated regulation. Transformation of *Arabidopsis* with a miRNA-resistant *TCP4* (*rTCP4*) resulted in transgenic plants with severe embryo patterning defects and lethality, whereas transformation with the wild-type version of *TCP4* had no effect on the plants at all (Figure 10.2) (Palatnik *et al.*, 2003). Transgenic lines with weaker phenotypes, expressing miRNA-resistant *TCP4* or *TCP2* are affected in leaf development, displaying smaller leaves (Figure 10.3) (Palatnik *et al.*, 2003).

Back to leaf morphogenesis
The miR319/TCP system is one of the required factors for the rather complex process of leaf morphogenesis. The precise coordination of growth distribution in the developing leaf, crucial to the final size and shape of the organ, involves the concerted action of various transcription factor networks and hormone signaling pathways, recently reviewed by Piazza *et al.* (2005) and Tsukaya (2005). Some of the identified components include transcription factors AINTEGUMENTA, JAGGED, BLADE ON PETIOLE and GRFs (Horiguchi *et al.*, 2005; Kim *et al.*, 2003; Piazza *et al.*, 2005; Tsukaya, 2005).

Several lines of evidence indicate the importance of plant hormone auxin in leaf growth. The auxin resistant *axr1* and *axr2* mutants have smaller leaves than wild-type plants and the auxin inducible gene *ARGOS* has been also shown to regulate the cell cycle rate in leaves (Hu *et al.*, 2003; Tsukaya, 2005). Interestingly, the auxin pathway seems to be repeatedly controlled by small RNAs. The auxin receptor TIR1 is controlled by miR393, and two miRNAs miR160 and miR167 control several Auxin Response Factors (ARFs) (see Table 10.1). Transgenic plants expressing a miRNA-resistant version of ARF16 or ARF17, controlled by miR160, display abnormal leaf development (Mallory *et al.*, 2005; Wang *et al.*, 2005). Moreover, several other ARFs are controlled by transacting siRNAs, a recently discovered small RNA pathway in plants (Allen *et al.*, 2005).

Comparative studies in several species pointed out that the KNOXI family of homeodomain containing transcription factors can contribute to the different leaf morphologies seen in nature (Tsiantis and Hay, 2003). KNOXI transcription factors are required for stem cell maintenance as part of the shoot apical meristem (SAM) function and several different repressor systems act to exclude KNOXI activity from *Arabidopsis* simple leaves (Tsiantis and Hay, 2003). However, KNOXI expression is seen in leaf primordia of plants with lobed or dissected leaves such as tomato (Bharathan *et al.*, 2002; Hareven *et al.*, 1996). Conversely, ectopic KNOXI expression in leaf primordia can produce leaf lobes in *Arabidopsis* or increase the dissection of a tomato leaf (Tsiantis and Hay, 2003).

Despite the widespread conservation of the miR319/TCP system throughout the plant kingdom there is still no evidence that a variation in this system (e.g. an increase in miR319 level) can contribute to the different leaf morphologies seen in nature. Comparison of the transcriptome profiles of KNOXI overexpressors and *jaw-D* revealed that the two pathways act independently (C. Schommer *et al.*, unpublished results). It may be interesting to analyze the potential participation of miR319 and the TCPs in natural variation or their role, if any, in the formation of a dissected leaf.

Establishment of organ polarity and organ boundaries by miRNAs

In general, microRNAs regulate key biological processes in plants (Table 10.1) and changes in their expression levels usually have profound consequences (Figure 10.3). Apart from miR319, two well studied miRNAs, miR165/166 and miR164 have been shown to participate in leaf development.

The first one is involved in the establishment of organ dorsoventrality, regarded as the abaxial/adaxial axis (Bowman, 2004). The adaxial side (which faces the shoot apical meristem) is specialized in light absorbtion, and contains palisade cells, whereas the abaxial side is specialized in gas exchange and generally contains spongy cells. The class III homeodomain/leucine zipper (HD-ZIPIII) transcription factors *REVOLUTA(REV)*, *PHABULOSA(PHB)* and *PHAVOLUTA(PHV)* are specifically expressed in the adaxial part of the leaves to confer adaxial identity (Bowman, 2004). Several dominant gain-of-function mutations in these genes cause an expanded expression domain, promoting adaxialization and radialization of leaves (Jover-Gil *et al.*, 2005).

Interestingly, the first identified mutations resulted in amino acid changes in a conserved sterol binding domain of the transcription factors and it was thought that a change in protein activity was causing the observed phenotypes (McConnell *et al.*, 2001). After the identification of miR165/166, it was recognized that the miRNA target site was affected in the dominant mutants (Rhoades *et al.*, 2002), which prompted to reevaluate the previous results (Tang *et al.*, 2003). Actually, it was shown that the *phd-1d* allele is resistant to miRNA-cleavage in wheat germ extracts (Tang *et al.*, 2003) and, furthermore, that silent mutations in the miRNA target site can also cause similar gain of function phenotypes (Emery *et al.*, 2003; Zhong and Ye, 2004).

The miR164 system participates in the determination of organ boundaries. Five transcription factors belonging to the NAC gene family are regulated by miR164 (Jones-Rhoades and Bartel, 2004; Laufs *et al.*, 2004; Mallory *et al.*, 2004a). Mutations in two of the miR164 targets, *CUC1* and *CUC2*, result in cotyledon fusion and loss of the apical meristem (Aida *et al.*, 1997). In good agreement, overexpression of miR164 results in similar phenotypes as seen in *cuc1 cuc2* double mutants and some additional phenotypes such as stem–leaf fusions (Figure 10.3), again pointing to the role of these genes in the establishment of organ boundaries (Mallory *et al.*, 2004a; Schwab *et al.*, 2005). Disruption of miR164 regulation, by creating miRNA restistant *CUC1* (Mallory *et al.*, 2004a) or *CUC2* (Laufs *et al.*, 2004) genes leads to a severe reduction of leaf petioles and broadened leaf shape.

miRNAs are highly specific regulators of plant development

Once miRNAs have been identified, either by direct cloning, genetic screens or bioinformatic predictions, the next challenge is to understand how they are integrated into biological processes. A key intermediate step to achieve this goal is the comprehensive identification of targets for each specific miRNA.

In a thorough study, Mallory et al. (2004b) investigated the mechanistic aspects of miRNA:target interaction in Arabidopsis, using miR165/166 and their target PHABULOSA (PHB). A systematic mutation scan revealed that target mismatches between position 3 and 11 are the most relevant, producing at least a ten-fold decrease in miRNA-guided cleavage activity in vitro. Sequence analysis of other miRNA-target duplices confirmed that the 5′ regions of miRNAs match their targets with less mismatches than the 3′ portions (Mallory et al., 2004b). In this regard, plant miRNAs resemble animal miRNAs: the so called seed region in the 5′ end of animal miRNAs is the driving force for target mRNA identification, and mutations in this region disproportionately affect miRNA activity (Brennecke et al., 2005; Doench and Sharp, 2004).

Although animals and plants use miRNAs to regulate gene expression, there are important differences in the outcome of the control. Animal miRNAs have been predicted to have dozens or even up to a hundred target mRNAs that correspond to multiple gene categories (Lewis et al., 2005). At a variance, plant miRNAs have only a few target genes that usually belong to the same gene family (Allen et al., 2005; Jones-Rhoades and Bartel, 2004).

Global transcript profiling of plants and animals supports this differential concept in mRNA regulation. Lim et al. (2005) investigated the consequences of increased miR1 and miR124 levels in HeLa cells, following changes in the transcriptome by microarrays. They found that around one hundred genes were significantly downregulated and most of them contained sites in their 3′ UTR complementary to the miRNAs 5′ seed (Lim et al., 2005).

Schwab et al. (2005) analyzed the expression profiles of transgenic plants overexpressing five different miRNAs: miR319, miR159, miR156, miR164 and miR172. In agreement with the computational predictions (Jones-Rhoades and Bartel, 2004) and in contrast to the effects of high miRNA levels in animals, the only downregulated genes with similarity to the corresponding miRNAs were the expected few conserved targets. The only exception was the OPT1 (Oligo Peptide Transporter 1) mRNA that was degraded by ectopic miR159a (Schwab et al., 2005).

The degradation of the conserved miRNA targets and OPT1 in Arabidopsis was caused by the direct action of the miRNAs and the slicer activity of AGO, while the downregulation of animal target transcripts is probably a secondary consequence of the miRNA-target interaction such as the recruitment of the complexes into P-bodies (Liu et al., 2005b; Sen and Blau, 2005). The intriguingly different concept of target regulation between animals and plants supports the idea that plant and animal miRNAs evolved independently.

The activity of miR159a and miR319a is an example for the exquisite miRNA specificity in plants (Figure 10.4). These two miRNAs share 17 of the 21 nucleotides of their mature sequences and together have homology to 5 TCP and 8 MYB transcription factors (Jones-Rhoades and Bartel, 2004; Palatnik et al., 2003). However, the experimental data indicates that these two miRNAs have non-overlapping sets of targets. First, overexpression of miR319a, as seen in jaw-D, causes

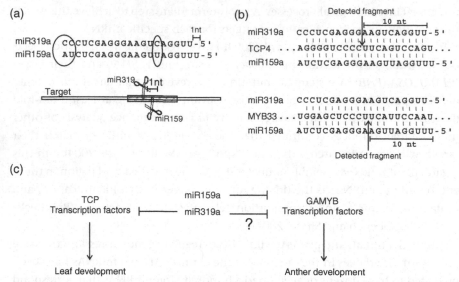

Figure 10.4. Specific effects of two nearly identical plant miRNAs. (a) Sequences of *Arabidopsis* miR319a and miR159a. Differences are highlighted with open circles. As the two miRNAs are offset in one nucleotide, they cut at different positions of a target sequence (bottom). (b) Interaction of miR319a and miR159a with the potential targets *TCP4* and *MYB33*. The arrow indicates the position of the cleavage site detected *in vivo* (Palatnik *et al.*, 2003). (c) Proposed model for the biological roles of miR319a and miR159a in plants.

the degradation of the *TCPs* and changes leaf morphology (Palatnik *et al.*, 2003), whereas overexpression of miR159a affects the *MYBs* and causes defects in stamen development (Achard *et al.*, 2004; Millar and Gubler, 2005; Schwab *et al.*, 2005). Second, as miR319a and miR159a are offset by one nucleotide, the actual miRNA executing target degradation *in vivo* can be identified by the sequence of the cleaved target fragments (Figure 10.4). The experimental results are consistent with miR319a guiding the cleavage of the *TCPs*, and miR159a acting on the *MYBs* (Palatnik *et al.*, 2003).

Interestingly, while miR159a is a very abundant miRNA cloned thousands of times, miR319a has been cloned only a few times (Gustafson *et al.*, 2005), suggesting different biological roles for these miRNAs. A systematic mutation analysis revealed that each difference between miR319a and miR159a can be important for target selection *in vivo* (Palatnik *et al.*, 2007). Plant miRNAs are usually encoded by small RNA families that often contain a few differences in their mature sequences. Therefore it is possible that miRNA family members could have a functional specialization towards specific target genes.

The lack of leaf phenotype in miR159 overexpressors suggests that its limited complementarity to the *TCPs* is not sufficient to trigger an efficient translational repression. Overall, it is currently not known whether plant miRNAs with limited complementarity to mRNAs, like animal miRNAs would in general be able to produce translational repression. The only plant miRNA reported so far that can inhibit translation is miR172 (Aukerman and Sakai, 2003; Chen, 2004), although it also seems capable of producing some transcript cleavage (Lauter *et al.*, 2005;

Schwab *et al.*, 2005). The interaction between miR172 and its targets, the AP2 transcription factors, is not different from other plant miRNA:target duplices, therefore it is still not clear why miR172 is able to work at least partially on translational level. At the same time, a few animal miRNAs with better complementarity to their targets have been reported to act through a cleavage mechanism similar to plant miRNAs (Davis *et al.*, 2005; Yekta *et al.*, 2004).

Conservation of miRNA pathways in plants

Several lines of evidence, including the lack of common miRNA-target systems, suggest that plant and animal miRNAs have evolved independently (Allen *et al.*, 2004). Among plants, however, miRNAs are largely conserved. Most of the validated miRNAs from *Arabidopsis* are predicted to exist in the distantly related rice, based on the conservation of the mature miRNA sequence and the ability of the surrounding sequence to form a fold-back structure (Bonnet *et al.*, 2004; Reinhart *et al.*, 2002; Wang *et al.*, 2004). Small RNA libraries from rice have also confirmed their expression (Liu *et al.*, 2005a; Sunkar *et al.*, 2005).

Homologs of the *Arabidopsis* miRNA target genes can also be found in rice with an almost perfect conservation of their miRNA-target sequences (Jones-Rhoades and Bartel, 2004; Sunkar *et al.*, 2005). A systematic expression profiling of *Arabidopsis* miRNAs has shown that they can be detected even in gymnosperms and ferns (Axtell and Bartel, 2005). Moreover, the conservation of target sequences has been used as a tool to find homologs of miRNA-regulated genes in other species (Axtell and Bartel, 2005).

Analysis of expressed sequence tag (EST) datasets has been proven to be a useful tool to identify miR319 precursors and its *TCP* targets in different species (Palatnik *et al.*, 2003). Similar results have been obtained for several other miRNA-target systems (Dezulian *et al.*, 2005; Jones-Rhoades and Bartel, 2004; Sunkar *et al.*, 2005; Sunkar and Zhu, 2004; Zhang *et al.*, 2005). Small RNA libraries from mosses have revealed the conservation of several *Arabidopsis* miRNAs such as miR319, miR156, miR160 and miR390 (Arazi *et al.*, 2005; Axtell and Bartel, 2005). An EST containing a putative miR319 precursor has also been identified in mosses (Arazi *et al.*, 2005).

The evolutionary conservation of the miR166/class III HD-Zip system is also noteworthy. The miRNA target site in the HD-Zip transcription factors is conserved even in liverworts and hornworts (Floyd and Bowman, 2004). Dominant mutations in the target site of these transcription factors that disrupt the interaction with the miRNA have been isolated from *Arabidopsis*, maize and tobacco (Emery *et al.*, 2003; Juarez *et al.*, 2004; Kim *et al.*, 2005; Mallory *et al.*, 2004b; McConnell *et al.*, 2001; McHale and Koning, 2004; Zhong and Ye, 2004). In all cases the phenotypes show changes in organ polarity, suggesting a functional conservation of the system. Moreover, miR166 has similar expression patterns in *Arabidopsis* and maize (Juarez *et al.*, 2004; Kidner and Martienssen, 2004) and a miR166 precursor has even been identified in a lycopod (Floyd and Bowman, 2004). Taken together, all these data suggest that this conserved miRNA-target system predates to more than 400 million years ago, similar to the *let-7* regulation in metazoans (Floyd and Bowman, 2004).

Perspectives

Although much is known about the biogenesis and mechanistic aspects of miRNA action, many aspects remain only poorly understood. One of them is how miRNAs themselves are regulated. Some of them, such as miR159, seem to be very abundant and widely expressed as judged by their cloning efficiency; others seem to be expressed in specific tissues as seen using miRNA reporters, such as miR171 (Parizotto *et al.*, 2004) and even some of them may respond to environmental stimuli, such as miR395 (Jones-Rhoades and Bartel, 2004). Although the current evidence indicates that most plant miRNAs may be transcribed by RNA pol II like regular protein coding genes (Aukerman and Sakai, 2003; Kurihara and Watanabe, 2004; Xie *et al.*, 2005), the control of their expression is still not characterized.

What are the roles of miRNA mediated regulation of gene expression? (i) From a mechanistic point of view there are at least two scenarios: it is possible that in some cases they may contribute to the repressed state of a target gene, while in others they may regulate the fine-tuning and quantitative level of expression of certain targets. (ii) From a biological perspective, miRNAs have added an extra level of complexity to known pathways and may bring some unexpected twists to them. On the other hand, there are several miRNAs regulating genes of unknown function. It is possible that the involvement of miRNAs may help to understand their function. As mentioned before, miRNA overexpression and miRNA resistant targets are valuable tools to uncover new biological functions. (iii) From a morphogenetic point of view it will be interesting to see if variations in the miRNA-target systems could account for some of the wide diversity seen in nature.

ACKNOWLEDGEMENTS

We would like to thank Utpal Nath and Rebecca Schwab for pictures of the snapdragon leaves and plants overexpressing miRNAs respectively. J. P. is supported by Fundacion Antorchas, ANCYP/PICT2004, Howard Hughes Medical Institute (International Scholar), and CDA007/2005-C from the Human Frontier Science Program.

REFERENCES

Achard, P., Herr, A., Baulcombe, D. C. and Harberd, N. P. (2004). Modulation of floral development by a gibberellin-regulated microRNA. *Development*, **131**, 3357–3365.

Aida, M., Ishida, T., Fukaki, H., Fujisawa, H. and Tasaka, M. (1997). Genes involved in organ separation in *Arabidopsis*: an analysis of the cup-shaped cotyledon mutant. *Plant Cell*, **9**, 841–857.

Allen, E., Xie, Z., Gustafson, A. M. *et al.* (2004). Evolution of microRNA genes by inverted duplication of target gene sequences in *Arabidopsis thaliana*. *Nature Genetics*, **36**, 1282–1290.

Allen, E., Xie, Z., Gustafson, A. M. and Carrington, J. C. (2005). microRNA-directed phasing during trans-acting siRNA biogenesis in plants. *Cell*, **121**, 207–221.

Arazi, T., Talmor-Neiman, M., Stav, R. *et al.* (2005). Cloning and characterization of micro-RNAs from moss. *Plant Journal*, **43**, 837–848.

Aukerman, M. J. and Sakai, H. (2003). Regulation of flowering time and floral organ identity by a MicroRNA and its APETALA2-like target genes. *Plant Cell*, **15**, 2730–2741.

Axtell, M. J. and Bartel, D. P. (2005). Antiquity of microRNAs and their targets in land plants. *Plant Cell*, **17**, 1658–1673.

Baker, C. C., Sieber, P., Wellmer, F. and Meyerowitz, E. M. (2005). The early extra petals1 mutant uncovers a role for microRNA miR164c in regulating petal number in *Arabidopsis. Current Biology*, **15**, 303–315.

Bartel, D. P. (2004). MicroRNAs: genomics, biogenesis, mechanism, and function. *Cell*, **116**, 281–297.

Baulcombe, D. (2005). RNA silencing. *Trends Biochemical Sciences*, **30**, 290–293.

Bharathan, G., Goliber, T. E., Moore, C. *et al.* (2002). Homologies in leaf form inferred from KNOXI gene expression during development. *Science*, **296**, 1858–1860.

Bohmert, K., Camus, I., Bellini, C. *et al.* (1998). AGO1 defines a novel locus of *Arabidopsis* controlling leaf development. *European Molecular Biology Organization Journal*, **17**, 170–180.

Bonnet, E., Wuyts, J., Rouze, P. and Van de Peer Y. (2004). Detection of 91 potential conserved plant microRNAs in *Arabidopsis thaliana* and *Oryza sativa* identifies important target genes. *Proceedings of the National Academy of Sciences USA*, **101**, 11 511–11 516.

Boutet, S., Vazquez, F., Liu, J. *et al.* (2003). Arabidopsis HEN1: a genetic link between endogenous miRNA controlling development and siRNA controlling transgene silencing and virus resistance. *Current Biology*, **13**, 843–848.

Bowman, J. L. (2004). Class III HD-Zip gene regulation, the golden fleece of ARGONAUTE activity? *Bioessays*, **26**, 938–942.

Brennecke, J., Hipfner, D. R., Stark, A., Russell, R. B. and Cohen, S. M. (2003). bantam encodes a developmentally regulated microRNA that controls cell proliferation and regulates the proapoptotic gene hid in *Drosophila. Cell*, **113**, 25–36.

Brennecke, J., Stark, A., Russell, R. B. and Cohen, S. M. (2005). Principles of microRNA-target recognition. *Public Library of Science-Biology*, **3**, e85.

Chen, X. (2004). A microRNA as a translational repressor of APETALA2 in *Arabidopsis* flower development. *Science*, **303**, 2022–2025.

Cubas, P., Lauter, N., Doebley, J. and Coen, E. (1999). The TCP domain: a motif found in proteins regulating plant growth and development. *Plant Journal*, **18**, 215–222.

Davis, E., Caiment, F., Tordoir, X. *et al.* (2005). RNAi-mediated allelic trans-interaction at the imprinted Rtl1/Peg11 locus. *Current Biology*, **15**, 743–749.

Dezulian, T., Remmert, M., Palatnik, J. F., Weigel, D. and Huson, D. H. (2005). Identification of plant microRNA homologs. *Bioinformatics*. (In press.)

Doench, J. G. and Sharp, P. A. (2004). Specificity of microRNA target selection in translational repression. *Genes & Development*, **18**, 504–511.

Elbashir, S. M., Lendeckel, W. and Tuschl, T. (2001). RNA interference is mediated by 21- and 22-nucleotide RNAs. *Genes & Development*, **15**, 188–200.

Emery, J. F., Floyd, S. K., Alvarez, J. *et al.* (2003). Radial patterning of *Arabidopsis* shoots by class III HD-ZIP and KANADI genes. *Current Biology*, **13**, 1768–1774.

Floyd, S. K. and Bowman, J. L. (2004). Gene regulation: ancient microRNA target sequences in plants. *Nature*, **428**, 485–486.

Fujii, H., Chiou, T.-J., Lin, S.-I., Aung, K. and Zhou, J. K. (2005). A miRNA involved in phosphate-starvation response in *Arabidopsis. Current Biology*, **15**, 2038–2043.

Guo, H. S., Xie, Q., Fei, J. F. and Chua, N. H. (2005). MicroRNA directs mRNA cleavage of the transcription factor NAC1 to downregulate auxin signals for arabidopsis lateral root development. *Plant Cell*, **17**, 1376–1386.

Gustafson, A. M., Allen, E., Givan, S. *et al.* (2005). ASRP: the *Arabidopsis* Small RNA Project Database. *Nucleic Acids Research*, **33**, D637–D640.

Han, M. H., Goud, S., Song, L. and Fedoroff, N. (2004). The *Arabidopsis* double-stranded RNA-binding protein HYL1 plays a role in microRNA-mediated gene regulation. *Proceedings of the National Academy of Sciences USA*, **101**, 1093–1098.

Hareven, D., Gutfinger, T., Parnis, A., Eshed, Y. and Lifschitz, E. (1996). The making of a compound leaf: genetic manipulation of leaf architecture in tomato. *Cell*, **84**, 735–744.

Horiguchi, G., Kim, G. T. and Tsukaya, H. (2005). The transcription factor AtGRF5 and the transcription coactivator AN3 regulate cell proliferation in leaf primordia of *Arabidopsis thaliana*. *Plant Journal*, **43**, 68–78.

Hu, Y., Xie, Q. and Chua, N. H. (2003). The *Arabidopsis* auxin-inducible gene ARGOS controls lateral organ size. *Plant Cell*, **15**, 1951–1961.

Jones-Rhoades, M. W. and Bartel, D. P. (2004). Computational identification of plant microRNAs and their targets, including a stress-induced miRNA. *Molecular Cell*, **14**, 787–799.

Jover-Gil, S., Candela, H. and Ponce, M. R. (2005). Plant microRNAs and development. *International Journal of Developmental Biology*, **49**, 733–744.

Juarez, M. T., Kui, J. S., Thomas, J., Heller, B. A. and Timmermans, M. C. (2004). microRNA-mediated repression of rolled leaf1 specifies maize leaf polarity. *Nature*, **428**, 84–88.

Kasschau, K. D., Xie, Z., Allen, E. *et al.* (2003). P1/HC-Pro, a viral suppressor of RNA silencing, interferes with *Arabidopsis* development and miRNA unction. *Developmental Cell*, **4**, 205–217.

Kidner, C. A. and Martienssen, R. A. (2004). Spatially restricted microRNA directs leaf polarity through ARGONAUTE1. *Nature*, **428**, 81–84.

Kidner, C. A. and Martienssen, R. A. (2005). The role of ARGONAUTE1 (AGO1) in meristem formation and identity. *Developmental Biology*, **280**, 504–517.

Kim, J., Jung, J. H., Reyes, J. L. *et al.* (2005). microRNA-directed cleavage of ATHB15 mRNA regulates vascular development in *Arabidopsis* inflorescence stems. *Plant Journal*, **42**, 84–94.

Kim, J. H., Choi, D. and Kende, H. (2003). The AtGRF family of putative transcription factors is involved in leaf and cotyledon growth in *Arabidopsis*. *Plant Journal*, **36**, 94–104.

Kurihara, Y. and Watanabe, Y. (2004). *Arabidopsis* micro-RNA biogenesis through Dicer-like 1 protein functions. *Proceedings of the National Academy of Sciences USA*, **101**, 12 753–12 758.

Lagos-Quintana, M., Rauhut, R., Lendeckel, W. and Tuschl, T. (2001). Identification of novel genes coding for small expressed RNAs. *Science*, **294**, 853–858.

Lau, N. C., Lim, L. P., Weinstein, E. G. and Bartel, D. P. (2001). An abundant class of tiny RNAs with probable regulatory roles in *Caenorhabditis elegans*. *Science*, **294**, 858–862.

Laufs, P., Peaucelle, A., Morin, H. and Traas, J. (2004). MicroRNA regulation of the CUC genes is required for boundary size control in *Arabidopsis* meristems. *Development*, **131**, 4311–4322.

Lauter, N., Kampani, A., Carlson, S., Goebel, M. and Moose, S. P. (2005). microRNA172 down-regulates glossy15 to promote vegetative phase change in maize. *Proceedings of the National Academy of Sciences USA*, **102**, 9412–9417.

Lee, R. C. and Ambros, V. (2001). An extensive class of small RNAs in *Caenorhabditis elegans*. *Science*, **294**, 862–864.

Lee, R. C., Feinbaum, R. L. and Ambros, V. (1993). The *C. elegans* heterochronic gene lin-4 encodes small RNAs with antisense complementarity to lin-14. *Cell*, **75**, 843–854.

Lewis, B. P., Burge, C. B. and Bartel, D. P. (2005). Conserved seed pairing, often flanked by adenosines, indicates that thousands of human genes are microRNA targets. *Cell*, **120**, 15–20.

Lim, L. P., Lau, N. C., Garrett-Engele, P. *et al.* (2005). Microarray analysis shows that some microRNAs downregulate large numbers of target mRNAs. *Nature*, **433**, 769–773.

Liu, B., Li, P., Li, X. *et al.* (2005a). Loss of function of OsDCL1 affects microRNA accumulation and causes developmental defects in rice. *Plant Physiology*, **139**, 296–305.

Liu, J., Valencia-Sanchez, M. A., Hannon, G. J. and Parker, R. (2005b). MicroRNA-dependent localization of targeted mRNAs to mammalian P-bodies. *Nature Cell Biology*, **7**, 719–723.

Llave, C., Kasschau, K. D., Rector, M. A. and Carrington, J. C. (2002a). Endogenous and silencing-associated small RNAs in plants. *Plant Cell*, **14**, 1605–1619.

Llave, C., Xie, Z., Kasschau, K. D. and Carrington, J. C. (2002b). Cleavage of Scarecrow-like mRNA targets directed by a class of *Arabidopsis* miRNA. *Science*, **297**, 2053–2056.

Mallory, A. C., Dugas, D. V., Bartel, D. P. and Bartel, B. (2004a). MicroRNA regulation of NAC-domain targets is required for proper formation and separation of adjacent embryonic, vegetative, and floral organs. *Current Biology*, **14**, 1035–1046.

Mallory, A. C., Reinhart, B. J., Jones-Rhoades, M. W. *et al.* (2004b). MicroRNA control of PHABULOSA in leaf development: importance of pairing to the microRNA 5′ region. *EMBO Journal*, **23**, 3356–3364.

Mallory, A. C., Bartel, D. P. and Bartel, B. (2005). MicroRNA-directed regulation of *Arabidopsis* AUXIN RESPONSE FACTOR17 is essential for proper development and modulates expression of early auxin response genes. *Plant Cell*, **17**, 1360–1375.

McConnell, J. R., Emery, J., Eshed, Y. *et al.* (2001). Role of PHABULOSA and PHAVOLUTA in determining radial patterning in shoots. *Nature*, **411**, 709–713.

McHale, N. A. and Koning, R. E. (2004). MicroRNA-directed cleavage of *Nicotiana sylvestris* PHAVOLUTA mRNA regulates the vascular cambium and structure of apical meristems. *Plant Cell*, **16**, 1730–1740.

Mette, M. F., van der Widen, J., Matzke, M. and Matzke, A. J. (2002). Short RNAs can identify new candidate transposable element families in *Arabidopsis*. *Plant Physiology*, **130**, 6–9.

Millar, A. A. and Gubler, F. (2005). The *Arabidopsis* GAMYB-like genes, MYB33 and MYB65, are microRNA-regulated genes that redundantly facilitate anther development. *Plant Cell*, **17**, 705–721.

Nath, U., Crawford, B. C., Carpenter, R. and Coen, E. (2003). Genetic control of surface curvature. *Science*, **299**, 1404–1407.

Palatnik, J. F., Allen, E., Wu, X. *et al.* (2003). Control of leaf morphogenesis by microRNAs. *Nature*, **425**, 257–263.

Palatnik, J. F., Wollmann, H., Schommer, C. *et al.* (2007). Differential targeting of *MYB* and *TCP* transcription factor genes by two related microRNAs in *Arabidopsis*. *Developmental Cell*, (in press).

Parizotto, E. A., Dunoyer, P., Rahm, N., Himber, C. and Voinnet, O. (2004). In vivo investigation of the transcription, processing, endonucleolytic activity, and functional relevance of the spatial distribution of a plant miRNA. *Genes & Development*, **18**, 2237–2242.

Park, M. Y., Wu, G., Gonzalez-Sulser, A., Vaucheret, H. and Poethig, R. S. (2005). Nuclear processing and export of microRNAs in *Arabidopsis*. *Proceedings of the National Academy of Sciences USA*, **102**, 3691–3696.

Park, W., Li, J., Song, R., Messing, J. and Chen, X. (2002). CARPEL FACTORY, a Dicer homolog, and HEN1, a novel protein, act in microRNA metabolism in *Arabidopsis thaliana*. *Current Biology*, **12**, 1484–1495.

Pasquinelli, A. E., Reinhart, B. J., Slack, F. *et al.* (2000). Conservation of the sequence and temporal expression of let-7 heterochronic regulatory RNA. *Nature*, **408**, 86–89.

Piazza, P., Jasinski, S. and Tsiantis, M. (2005). Evolution of leaf developmental mechanisms. *New Phytologist*, **167**, 693–710.

Reinhart, B. J., Slack, F. J., Basson, M. *et al.* (2000). The 21-nucleotide let-7 RNA regulates developmental timing in *Caenorhabditis elegans*. *Nature*, **403**, 901–906.

Reinhart, B. J., Weinstein, E. G., Rhoades, M. W., Bartel, B. and Bartel, D. P. (2002). MicroRNAs in plants. *Genes & Development*, **16**, 1616–1626.

Rhoades, M. W., Reinhart, B. J., Lim, L. P. *et al.* (2002). Prediction of plant microRNA targets. *Cell*, **110**, 513–520.

Schauer, S. E., Jacobsen, S. E., Meinke, D. W. and Ray, A. (2002). DICER-LIKE1: blind men and elephants in *Arabidopsis* development. *Trends in Plant Science*, **7**, 487–491.

Schwab, R., Palatnik, J. F., Riester, M. *et al.* (2005). Specific effects of microRNAs on the plant transcriptome. *Developmental Cell*, **8**, 517–527.

Sen, G. L. and Blau, H. M. (2005). Argonaute 2/RISC resides in sites of mammalian mRNA decay known as cytoplasmic bodies. *Nature Cell Biology*, **7**, 633–636.

Sunkar, R. and Zhu, J. K. (2004). Novel and stress-regulated microRNAs and other small RNAs from *Arabidopsis*. *Plant Cell*, **16**, 2001–2019.

Sunkar, R., Girke, T., Jain, P. K. and Zhu, J. K. (2005). Cloning and characterization of microRNAs from rice. *Plant Cell*, **17**, 1397–1411.

Tang, G., Reinhart, B. J., Bartel, D. P. and Zamore, P. D. (2003). A biochemical framework for RNA silencing in plants. *Genes & Development*, **17**, 49–63.

Tsiantis, M. and Hay, A. (2003). Comparative plant development: the time of the leaf? *Nature Reviews Genetics*, **4**, 169–180.

Tsukaya, H. (2005). Leaf shape: genetic controls and environmental factors. *International Journal of Developmental Biology*, **49**, 547–555.

Vaucheret, H., Vazquez, F., Crete, P. and Bartel, D. P. (2004). The action of ARGONAUTE1 in the miRNA pathway and its regulation by the miRNA pathway are crucial for plant development. *Genes & Development*, **18**, 1187–1197.

Vazquez, F., Gasciolli, V., Crete, P. and Vaucheret, H. (2004). The nuclear dsRNA binding protein HYL1 is required for microRNA accumulation and plant development, but not posttranscriptional transgene silencing. *Current Biology*, **14**, 346–351.

Wang, J. W., Wang, L. J., Mao, Y. B. *et al.* (2005). Control of root cap formation by microRNA-targeted auxin response factors in *Arabidopsis*. *Plant Cell*, **17**, 2204–2216.

Wang, X. J., Reyes, J. L., Chua, N. H. and Gaasterland, T. (2004). Prediction and identification of *Arabidopsis thaliana* microRNAs and their mRNA targets. *Genome Biology*, **5**, R65.

Weigel, D., Ahn, J. H., Blazquez, M. A. *et al.* (2000). Activation tagging in *Arabidopsis*. *Plant Physiology*, **122**, 1003–1013.

Williams, L., Grigg, S. P., Xie, M., Christensen, S. and Fletcher, J. C. (2005). Regulation of *Arabidopsis* shoot apical meristem and lateral organ formation by microRNA miR166g and its AtHD-ZIP target genes. *Development*, **132**, 3657–3668.

Xie, Z., Kasschau, K. D. and Carrington, J. C. (2003). Negative feedback regulation of *Dicer-Like 1* in *Arabidopsis* by microRNA-guided mRNA degradation. *Current Biology*, **13**, 784–789.

Xie, Z., Johansen, L. K., Gustafson, A. M. *et al.* (2004). Genetic and functional diversification of small RNA pathways in plants. *Public Library of Science-Biology*, **2**, E104.

Xie, Z., Allen, E., Fahlgren, N. *et al.* (2005). Expression of *Arabidopsis* MIRNA genes. *Plant Physiology*, **138**, 2145–2154.

Yekta, S., Shih, I. H. and Bartel, D. P. (2004). MicroRNA-directed cleavage of HOXB8 mRNA. *Science*, **304**, 594–596.

Yu, L., Yu, X., Shen, R. and He, Y. (2005). HYL1 gene maintains venation and polarity of leaves. *Planta*, **221**, 231–242.

Zamore, P. D. and Haley, B. (2005). Ribo-gnome: the big world of small RNAs. *Science*, **309**, 1519–1524.

Zhang, B. H., Pan, X. P., Wang, Q. L., Cobb, G. P. and Anderson, T. A. (2005). Identification and characterization of new plant microRNAs using EST analysis. *Cell Research*, **15**, 336–360.

Zhong, R. and Ye, Z. H. (2004). Amphivasal vascular bundle 1, a gain-of-function mutation of the IFL1/REV gene, is associated with alterations in the polarity of leaves, stems and carpels. *Plant & Cell Physiology*, **45**, 369–385.

III

Computational
biology of
microRNAs

11 miRBase: a database of microRNA sequences, targets and nomenclature

Anton J. Enright* and Sam Griffiths-Jones

Introduction

The miRBase database (formerly entitled the *microRNA Registry*) is the primary online repository for microRNA (miRNA) sequences and annotation (Griffiths-Jones, 2004; Griffiths-Jones *et al.*, 2006). When laboratories first began to clone and sequence increasing numbers of miRNAs (Lagos-Quintana *et al.*, 2002) it became apparent that a single resource for the naming, annotation and dissemination of published miRNAs was urgently required. With no predefined nomenclature or central repository for miRNA sequences there was a distinct danger that these sequences would appear in journals with inconsistent names. Furthermore, if miRNAs were independently identified by multiple laboratories and had multiple ambiguous identifiers this could hamper subsequent analysis. Previously, the RFAM project (Griffiths-Jones *et al.*, 2003) at the Wellcome Trust Sanger Institute had been cataloging and identifying RNAs and their evolutionary relationships and hence already had much of the expertise required. A number of laboratories involved in miRNA research discussed these issues and published a collaborative document detailing the agreed nomenclature for miRNAs and announcing the miRNA registry as their main repository (Ambros *et al.*, 2003).

Recently the resource has expanded to include miRNA annotations and automatically predicted targets for animal miRNAs and has been renamed miRBase (Griffiths-Jones *et al.*, 2006). This resource currently attracts a very large number of visitors and registered over 1.5 million page hits in July 2006 alone, illustrating the growing scientific interest in these important regulatory molecules. miRBase has three primary functions: firstly, to provide researchers with consistent names for novel miRNA gene discoveries prior to their publication; secondly, to provide a searchable database of all published miRNA sequences and annotation; and finally, to provide a pipeline for the prediction of miRNA target genes, and to make up-to-date high quality predictions available to the community.

---

* Author to whom correspondence should be addressed.

MicroRNAs: From Basic Science to Disease Biology, ed. Krishnarao Appasani. Published by Cambridge University Press. © Cambridge University Press 2008.

The sequence repository and gene nomenclature roles of the microRNA Registry are continued by the *miRBase Sequence* database and the *miRBase Registry* respectively. These sequence and annotation data are now also coupled to a target prediction system provided by a genomic pipeline that scans known miRNAs against genomic sequence obtained from the Ensembl database (Birney *et al.*, 2006) and currently uses the miRanda algorithm (Enright *et al.*, 2003). In the future we hope to use the target prediction framework to bring together other computational methods and experimentally validated targets into a combined resource that will better enable researchers to determine whether their gene or miRNA of interest has biologically valid miRNA binding sites.

The database is available online at http://microrna.sanger.ac.uk/. All data can be downloaded from the FTP site (ftp://ftp.sanger.ac.uk/pub/mirbase/) in a variety of formats, including FASTA sequences and relational database dumps, for local installation and querying (see miRBase Data Availability below). In this chapter we will describe the various facets of the miRBase database and how researchers interested in the biology of miRNAs can utilize them.

miRBase Registry

The initial goal of miRBase was to provide a primary repository for assigning and maintaining consistent gene nomenclature and nucleotide sequence for both published and unpublished miRNA sequences. Unpublished miRNA sequences are normally in the final stages of publication and hence are confidential until publication of the manuscript that describes them. Hence, this service means that all miRNAs will have a consistent name with respect to other miRNAs from their first mention in the literature. As new miRNA genes are discovered, each and every miRNA gene is submitted to the Registry for official naming. This can be a single miRNA sequence or a set of hundreds discovered by high-throughput approaches. The miRBase Registry communicates extensively with the Human and Mouse Genome Nomenclature Committees to ensure miRNA gene names are both consistent and sensible.

In order to avoid unnecessary gaps in this scheme and to take advantage of the peer review process to validate miRNA sequences, final miRBase names are assigned to novel miRNA sequences only *after* a publication describing their discovery is accepted for publication. Official names are then usually incorporated into a final version of the accepted manuscript, and it is this version that appears in print to the scientific community. Homologs of previously validated miRNA sequences do not require an *in press* publication to be included in the database. Computational miRNA predictions, in the absence of experimental validation in any related organism, are not currently assigned official gene names. The community has also agreed the guidelines and minimum criteria for experimental miRNA verification (Ambros *et al.*, 2003).

Assigned miRNA gene names are of the form XXXX-mir-YYYY, where "XXXX" represents a three or four-letter species code and "YYY" is a sequential numeric identifier. A newly discovered miRNA in any species is given the next numeric

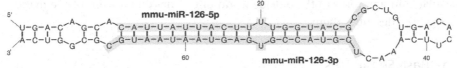

Figure 11.1. Predicted base-paired structure of mouse pre-mir-126. The two derived mature miRNA products are highlighted: miR-126-5p from the 5′ arm, and miR-126-3p from the 3′ arm.

identifier available from all organisms. Hence, if the last novel miRNA deposited were *hsa-mir-663* then the next novel miRNA in any other species would be *mir-664*. Other miRNA genes deposited that are deemed to encode the same miRNA, but at different loci are assigned a numeric suffix to distinguish them (e.g. *hsa-mir-521-1* and *hsa-mir-521-2*).

Traditionally in both miRBase and in the literature, it is common practice to refer to both the gene locus and precursor miRNA (pre-miRNA) of an miRNA as "*mir,*" while the mature miRNA product is designated as "*miR.*" When mature miRNAs are closely related in terms of their sequence, they are given additional suffixes that use an alphabetic character to distinguish them (e.g. hsa-miR-520a to hsa-miR-520d). Of course, combinations of these two classifications also occur (e.g. hsa-miR-519a-1 and has-miR-519a-2).

A small number of exceptions to this sequential naming scheme exist. These mainly arise from miRNAs that were discovered prior to miRBase and already had names in common use frequently from earlier mutation screens (e.g. *let*-7 and *lin*-4 from *Caenorhabditis elegans*). Such non-sequential identifiers are discouraged for novel miRNAs. One other important point of note is that plant and viral nomenclature schemes differ subtly, once again for historical reasons. Plant precursors are given names of the form MIR169a, and lettered suffixes are used to designate between both identical and non-identical mature miRNA sequences.

It has been shown recently, for some animal miRNAs, that two mature products may be formed with relatively high frequency from the same pre-miRNA. In many cases one mature sequence appears to be of much higher abundance than the other (which may be a non-functional by-product of biogenesis). For these cases, the alternative sequence (often termed the "star" sequence) is identified using an asterisk in miRBase (e.g. hsa-miR-520d & hsa-miR-520d*). Where cloning frequencies or other information are insufficient to determine which mature sequence is functional, or if both miRNAs appear to be functional, then suffixes are used to denote the hairpin arm of origin. For example, hsa-miR-30a-5p is expressed from the 5′ arm of the hsa-mir-30a hairpin, while hsa-miR-30a-3p originates from the 3′ arm (see Figure 11.1).

The process of submission of miRNA sequences to the database is straightforward. The *miRBase Sequences* section of the website contains a submission form for miRNAs. This form takes details of the accepted manuscript that will describe the miRNAs, the authors of the manuscript, the organism in question and finally a FASTA file containing nucleotide sequences for the miRNAs. Sequences are reviewed and assigned identifiers that should then be used in the accepted manuscript. Such sequences are held strictly privately and are not available in any form to the outside world until final publication of the manuscript that describes them.

miRBase Sequences

The goal of miRBase is not only to provide nomenclature for miRNAs, but also to act as a repository for miRNAs, their sequences and annotations. To this end, the miRBase Sequence database provides comprehensive access to all published miRNAs that have been curated as described above. Currently this database (release 8.2, July 2006) contains 4039 entries (miRNA hairpin precursors) that encode 3834 mature miRNA products, in primates, rodents, birds, fish, worms, flies, plants and viruses (Table 11.1). The growth of miRBase sequence entries has increased rapidly since the first release in 2002 (Figure 11.2).

Structure of a miRBase Sequence Entry

A screenshot illustrating the database web page for a typical miRNA sequence entry is shown (Figure 11.3). These pages aim to provide a quick and concise view of all information held about a particular miRNA of interest. A summary of the miRBase fields displayed for entries is described below (Table 11.3).

The fundamental unit stored by miRBase as an entry is the stem-loop portion of the primary transcript (pri-miRNA) for a given miRNA. This should hence also include the precursor products derived from Drosha mediated cleavage (pre-miRNA).

Unique identifiers are attached to each of these stem-loop sequences as described in the previous section, together with accession numbers (e.g. the accession for cel-mir-1 is MI0000001). The main reason accession numbers are also used, is that the miRNA identifiers may be subject to change across database releases, while accession numbers are guaranteed to be unique and stable. This is especially important for miRNAs as new experimental information frequently becomes available that sheds new light on the evolutionary relationships between existing miRNAs and may call for changes to the identifiers of existing entries.

As stated above, the core unit of an entry is the pri-miRNA stem-loop. Currently, little is known about the nature and biology of the primary transcripts of miRNAs. As such, no information can currently be represented in the database regarding these transcripts. Attached to stem-loop entries are associated mature miRNA sequences derived from each stem-loop. Each of these entries also consists of an identifier (as previously described) and once again, a stable unique accession number (e.g. for cel-miR-1 the accession is MIMAT0000003).

One further division exists between entries in *miRBase Sequences*. This division is because of the fact that while most mature miRNA sequences have been experimentally confirmed in the organism in which they are described, a large number of miRNAs can also be identified relatively trivially in closely related species by virtue of direct homology. The highly conserved nature of most miRNAs has illustrated that an miRNA confirmed in one species, e.g. human, will in many cases also be present in other species such as mouse and rat, especially when BLAST

Table 11.1. Breakdown of the number of miRNA entries deposited in miRBase across taxonomic groups and species

Taxonomic group	Species	No. of miRNAs
Arthropoda	*Drosophila melanogaster*	78
	Drosophila pseudoobscura	73
	Anopheles gambiae	38
	Apis mellifera	25
Nematoda	*Caenorhabditis elegans*	114
	Caenorhabditis briggsae	79
Vertebrata	*Homo sapiens*	462
	Mus musculus	358
	Danio rerio	337
	Rattus norvegicus	234
	Xenopus tropicalis	177
	Gallus gallus	152
	Tetraodon nigroviridis	132
	Fugu rubripes	131
	Bos taurus	98
	Pan paniscus	89
	Gorilla gorilla	86
	Pongo pygmaeus	84
	Pan troglodytes	83
	Macaca nemestrina	75
	Macaca mulatta	71
	Sus scrofa	54
	Lagothrix lagotricha	48
	Ateles geoffroyi	45
	Saguinus labiatus	42
	Lemur catta	16
	Xenopus laevis	7
	Canis familiaris	6
	Ovis aries	4
Viridiplantae	*Populus trichocarpa*	213
	Oryza sativa	182
	Arabidopsis thaliana	118
	Zea mays	96
	Sorghum bicolor	72
	Medicago truncatula	30
	Glycine max	22
	Physcomitrella patens	18
	Saccharum officinarum	16
Viruses	Epstein Barr virus	23
	Rhesus lymphocryptovirus	16
	Kaposi sarcoma-associated herpesvirus	13
	Human cytomegalovirus	11
	Mouse gammaherpesvirus 68	9
	Herpes Simplex Virus 1	1
	Simian virus 40	1

Growth of miRBase Sequences

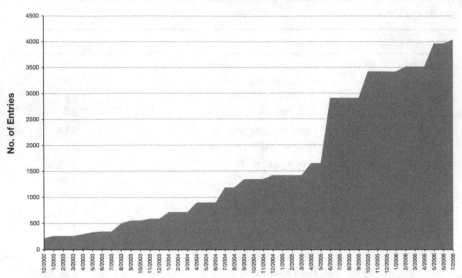

Figure 11.2. Growth of the number of entries (stem-loops) in the miRBase database since the first release in December 2006 until the current release.

searches and RNA folding algorithms can confirm the presence of highly homologous sequences to the miRNA locus in these related species. For example, 370 of the 454 human miRNA sequences in the database are annotated with experimental evidence, while the remaining 84 were linked to validated sequences in mouse, rat and fish. The database does not currently include computationally predicted miRNAs without some form of laboratory based experimental validation in a related organism. Those entries that represent experimentally validated miRNAs (Figure 11.3) are annotated together with their literature references and controlled vocabulary tags (Table 11.2) describing the experimental approach that was used to verify their presence.

The genomic location of an miRNA can have important implications for its biological function and for the design of experiments, which may, for instance, attempt to disrupt that particular miRNA. For this reason, the genomic coordinates for each miRNA precursor entry are provided where possible. Obviously, for organisms without a publicly released or assembled genome this is not possible. However, for the majority of organisms and entries in miRBase it is possible to provide genomic coordinates of miRNA loci, together with links to Ensembl.

The genomic context of an miRNA precursor is also of particular interest. Some miRNAs are derived from intergenic regions, either individually or in clusters; some from introns of protein-coding or non-coding transcripts; and a small few from overlapping exons of annotated transcripts (Rodriguez *et al.*, 2004; Weber, 2005). The Ensembl database (Birney *et al.*, 2006) of gene structures is used as the primary resource for determining the genomic context of miRNA precursors and hence contextual information displayed for the entry (Figure 11.3) is provided using the Ensembl coordinate system. These data show that as many as 35% of mammalian miRNAs overlap annotated genes. The vast majority of these are

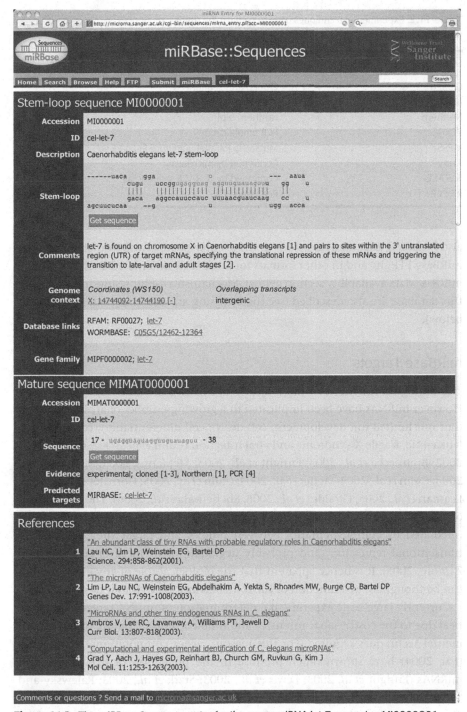

Figure 11.3. The *miRBase Sequences* entry for the worm miRNA let-7, accession MI0000001.

located in introns of protein coding genes. Approximately, only 14% of fly and worm miRNAs are intronic.

The *miRBase Sequences* pages also provide links to other (external or internal) web pages (e.g. Ensembl) that may provide further relevant information. It should

Table 11.2. Experimental techniques used to confirm the miRNA sequences present in the miRBase

Experiment tag	Description
Cloned	Cloned and sequenced
In situ	*In situ* hybridization
Northern	Northern blot
PCR	PCR amplification
MPSS	Massively Parallel Signature Sequenced
Array	Confirmed by microarray
5′ RACE	Confirmed by 5′ RACE sequencing
RT-PCR	Real-time PCR

also be noted, that all data shown in these entry pages is also available from the miRBase FTP site and in other formats for large-scale automated analysis (see the miRBase data availability section below). Mechanisms for browsing and searching this database are also described (see the Browsing and searching miRBase section below).

miRBase Targets

The biological functions of the majority of miRNAs are completely unknown. However miRNAs have been implicated in a growing number of areas, including (but not limited to): development; cell death; cell differentiation; colon cancer, leukaemia, fragile X syndrome and viral infection (Calin *et al.*, 2002; Caudy *et al.*, 2002; Brennecke *et al.*, 2003; Enright *et al.*, 2003; McManus, 2003; Michael *et al.*, 2003; Calin *et al.*, 2004; Pfeffer *et al.*, 2004; Calin *et al.*, 2005; Hornstein *et al.*, 2005; Leaman *et al.*, 2005; Giraldez *et al.*, 2006; Shcherbata *et al.*, 2006). The mechanism of miRNA action currently appears to be direct binding of target transcripts (mediated by complementarity between miRNA and target site) followed by translational repression (and probably degradation) of these targets (Du and Zamore, 2005). It is hence apparent that the biological role of any given miRNA can be thought of as a function of the roles of those genes that it targets.

Currently large-scale experimental determination of miRNA targets is impractical due to the costly, labor-intensive and time-consuming nature of the experimental techniques required. For this reason, recent computational approaches (Lai, 2004) have shown much promise in the prediction of likely targets of miRNAs (Enright *et al*, 2003; Lewis *et al.*, 2003; Stark *et al.*, 2003; Rajewsky and Socci, 2004; Rehmsmeier *et al.*, 2004). Most of these methods exploit the complementarity signal between miRNA and a 3′ UTR target site as the key signal for initial determination of candidate sites. Predicted sites may then be filtered by additional signals, such as thermodynamic stability of the predicted duplex and/or conservation of miRNA/target pairs in related species. A more detailed discussion of computational approaches can be found in Section III by other contributors.

Although computational miRNA target predictions are still far from perfect, they represent our first glimpse at the potential biological functions of miRNAs. For this reason we aim to provide the users of the miRBase database with both systematically predicted miRNA targets for the majority of animal miRNAs and externally provided miRNA target predictions from other computational or experimental techniques. Systematic miRNA target prediction in miRBase is a particularly computationally intensive task and currently involves searching over 2500 animal miRNAs against over 400 000 3′ UTRs from 17 species for potential target sites. For this reason, we currently use only one algorithm for core miRBase target predictions (miRanda v3.0), although external target predictions for miRNAs in miRBase are made available as external links in *miRBase Sequences*. We intend to include other methods, algorithms and improvements to this system as they become available in the future.

A computational pipeline performs the large amount of computational analysis required in parallel on a supercomputing cluster. This pipeline works as follows. Initially, the pipeline takes mature miRNA sequences from *miRBase Sequences* and also 3′ UTR sequences from all genomes available in the Ensembl database and scans each miRNA in each species against every available 3′ UTR. The current algorithm used (miRanda 3.0) utilizes a statistical model based on work by Rehmsmeier and his colleagues (Rehmsmeier *et al.*, 2004). The pipeline first uses the miRanda algorithm to scan each miRNA against control sets of 3′ UTRs and fit an Extreme Value Distribution (EVD) to the results of these control scans. Next, each miRNA is scanned against all 3′ UTRs from the same (or related) species using miRanda, and the previously derived EVD parameters are used to deduce a *P*-value from the score obtained for each potential site detected. Scores are normalized to correct for differences in the lengths of both 3′ UTRs and miRNAs. Multiple sites for a given miRNA on a 3′ UTR are combined in the pipeline using Poisson statistics. Finally, those sites which appear to be conserved among orthologous 3′ UTRs for an miRNA are assigned modified *P*-values based on the likelihood of those 3′ UTRs sharing significant blocks of conservation across their full length. One advantage of this approach is that while conservation of a target site will boost its P-value, it is not essential. Target sites with significant *P*-values, which are not conserved, will still be displayed, this is particularly important for miRNAs which may have recently evolved in certain species (Giraldez *et al.*, 2005).

The result of this analysis is a ranked list of target predictions for each mature animal miRNA in miRBase in any organism contained within the Ensembl genomic framework. All target predictions are ranked according to P-value and predictions where $P < 0.05$ is displayed on the web site. For each predicted target transcript we retrieve annotation information including its description, annotations and Gene Ontology (GO) terms (Harris *et al.*, 2004). The core data view in *miRBase Targets* is a *hit-list* page (Figure 11.4) which displays a summary of transcripts which are predicted to be targeted by a given miRNA or set of miRNAs. This view (Figure 11.4) also displays the function and GO terms associated with each predicted target so that interesting genes or functions may be rapidly identified by the user.

Figure 11.4. The *miRbase Targets* hit-list summary page for the worm miRNA cel-lin-4. Columns contain miRNA and predicted target gene information and summary of the scores, energies and P-values for each hit. Experimentally verified target genes (where known) are highlighted.

For more detailed information about any given prediction the user can access the *transcript detail* page. This page (Figure 11.5) shows the sequence of the 3′ UTR of a predicted target transcript, with the identified binding site highlighted. If that 3′ UTR was deemed have orthologs in other species, then their sequences are also displayed in a multiple alignment (Figure 11.5) where conserved binding sites can be shown. A landscape schematic of the 3′ UTR provides a reference diagram that conveys the relative positions of individual miRNAs predicted to bind that sequence and are color coded according to degree of sequence conservation across the 3′ UTR alignment. Finally, a detailed base-pairing view of each predicted miRNA:target duplex is shown together with the *P*-value associated with those sites. In each case the single-site, multiple-site and cross-species *P*-values are summarized together with the original miRanda score and thermodynamic

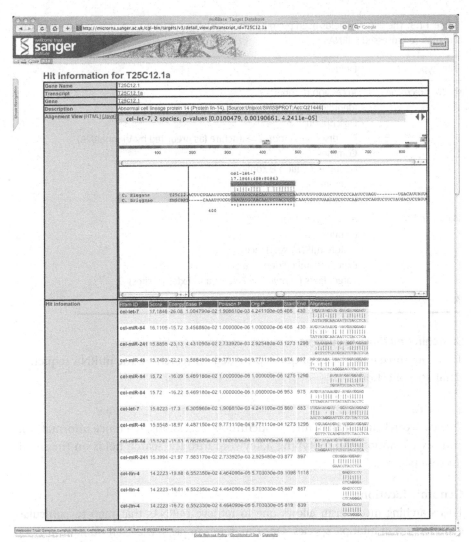

Figure 11.5. The *miRbase Targets* detailed transcript view of the worm lin-14 transcript (T25C12.1a). Predicted binding sites along the 3′ UTR for multiple miRNAs are shown (top) across a 3′ UTR schematic, which is coloured according to sequence conservation. The next pane (middle) shows an alignment (if available) between this transcript and its orthologs. In this case a predicted target site for cel-let-7 is highlighted and its conservation between *C. elegans* and *C. briggsae* is shown in orange. Finally, details for each individual site are shown (bottom) which include a full description of P-values obtained, relative coordinates of the site position in the 3′ UTR and a predicted duplex arrangement for the site. (See color plate 8)

stability (ΔG). Mechanisms for searching this resource for miRNAs or genes of interest are described below, together with details of data accessibility.

Browsing and searching miRBase

The miRBase web pages allow the user to both browse and search entries by organism, miRNA or gene. Both *miRBase Sequences* and *miRBase Targets* have independent search mechanisms to allow users to retrieve data.

Table 11.3. The different fields used by miRBase to describe both precursor miRNAs and mature processed miRNAs

Entry type	Fields
miRNA precursor	Accession number
	Identifier
	Description
	Precursor stem-loop 2° Structure (as predicted by ViennaRNA)
	Comments
	Genome location & context
	External database links
	Gene family
Mature miRNA	Accession number
	Identifier
	Mature miRNA sequence
	Experimental evidence tags
	Target links (to *miRBase Targets* and External sites)
Both	Literature references

Searching *miRBase Sequences*

There are three main types of query provided for searching the miRNA sequence database collection.

Keyword

One can search any or all of miRNA identifiers, accessions, reference information, and textual annotation by typing a keyword into the search fields on the main (or sub) pages.

Genomic location

This searching mechanism allows one to retrieve miRNAs that map to particular genomic loci in multiple species. One can, for example, retrieve all human miRNAs that map to chromosome 1, between sequence coordinates 1,000,000 and 1,200,000.

Sequence

Finally, it is straightforward to search any nucleotide sequence against the miRBase sequence library. Both precursor and mature sequences will be scanned, and alternative methods are provided to search a long sequence for homologs of known miRNAs, or to search the miRNA database for occurrences of a short sequence motif (see Table 11.3). The output from an example search for miRNA homologs in a query sequence is shown in Figure 11.6.

Searching *miRBase Targets*

For querying of systematically predicted miRNA targets, the *miRBase Targets* pages include a detailed *search* page. This page allows the user to search all predicted miRNA targets according to miRNA or gene of interest, keywords, gene names, and GO terms of gene descriptions. Any miRNA targets found from a search are shown on the more detailed *hit-list* page (see Figure 11.4).

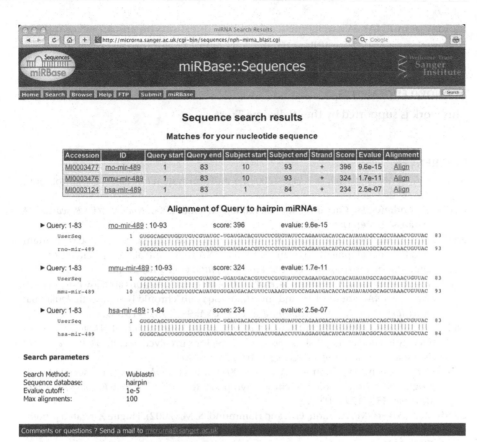

Figure 11.6. Example output from a *miRBase Sequence* search query for homologs of known miRNA precursors. Each hit is summarized in the table, followed by an alignment of the query to each database hit.

miRBase data availability

All data contained in miRBase are freely available. The web interface (http://microrna.sanger.ac.uk) is the primary mechanism for data access, but an FTP site also exists (ftp.sanger.ac.uk). The site also provides links for downloading flat-file dumps of the data in raw text formats or relational-database dumps. Distributed Annotation System (DAS) sources are also available to provide easy access to miRNA genomic locations, and these data are available as DAS tracks through both the Ensembl and UCSC genome browsers respectively (Birney *et al.*, 2006; Hinrichs *et al.*, 2006).

Summary

The miRBase database provides a comprehensive resource for miRNA gene annotation, nomenclature and systematic target prediction. These functions are essential to the continued growth of the miRNA community. We hope that miRBase will prove useful to the community and enhance our ability to understand the biology of miRNAs. We welcome feedback and questions on any aspect of miRNA nomenclature and the miRBase database by email to microrna@sanger.ac.uk.

ACKNOWLEDGEMENTS

This work is supported by the Wellcome Trust.

REFERENCES

Ambros, V., Bartel, B., Bartel, D. P. *et al.* (2003). A uniform system for microRNA annotation. *RNA*, **9**, 277–279.

Birney, E., Andrews, D., Caccamo, M. *et al.* (2006). Ensembl 2006. *Nucleic Acids Research*, **34** (Database issue), D556–D561.

Brennecke, J., Hipfner, D. R., Stark, A., Russell, R. B. and Cohen, S. M. (2003). bantam encodes a developmentally regulated microRNA that controls cell proliferation and regulates the proapoptotic gene hid in *Drosophila*. *Cell*, **113**, 25–36.

Calin, G. A., Dumitru, C. D., Shimizu, M. *et al.* (2002). Frequent deletions and down-regulation of micro-RNA genes miR15 and miR16 at 13q14 in chronic lymphocytic leukemia. *Proceedings of the National Academy of Sciences USA*, **99**, 15 524–15 529.

Calin, G. A., Sevignani, C., Dumitru, C. D. *et al.* (2004). Human microRNA genes are frequently located at fragile sites and genomic regions involved in cancers. *Proceedings of the National Academy of Sciences USA*, **101**, 2999–3004.

Calin, G. A., Ferracin, M., Cimmino, A. *et al.* (2005). A microRNA signature associated with prognosis and progression in chronic lymphocytic leukemia. *New England Journal of Medicine*, **353**, 1793–1801.

Caudy, A. A., Myers, M., Hannon, G. J. and Hammond, S. M. (2002). Fragile X-related protein and VIG associate with the RNA interference machinery. *Genes & Development*, **16**, 2491–2496.

Du, T. and Zamore, P. D. (2005). microPrimer: the biogenesis and function of microRNA. *Development*, **132**, 4645–4652.

Enright, A. J., John, B., Gaul, U. *et al.* (2003). MicroRNA targets in *Drosophila*. *Genome Biology*, **5**, R1.

Giraldez, A. J., Cinalli, R. M., Glasner, M. E. *et al.* (2005). MicroRNAs regulate brain morphogenesis in zebrafish. *Science*, **308**, 833–838.

Giraldez, A. J., Mishima, Y., Rihel, J. *et al.* (2006). Zebrafish MiR-430 promotes deadenylation and clearance of maternal mRNAs. *Science*, **312**, 75–79.

Griffiths-Jones, S. (2004). The microRNA Registry. *Nucleic Acids Research*, **32**, D109–111.

Griffiths-Jones, S., Bateman, A., Marshall, M., Khanna, A. and Eddy, S. R. (2003). Rfam: an RNA family database. *Nucleic Acids Research*, **31**, 439–441.

Griffiths-Jones, S., Grocock, R. J., van Dongen, S., Bateman, A. and Enright, A. J. (2006). miRBase: microRNA sequences, targets and gene nomenclature. *Nucleic Acids Research*, **34** (Database issue), D140–144.

Harris, M. A., Clark, J., Ireland, A. *et al.* (2004). The Gene Ontology (GO) database and informatics resource. *Nucleic Acids Research*, **32** (Database issue), D258–261.

Hinrichs, A. S., Karolchik, D., Baertsch, R. *et al.* (2006). The UCSC Genome Browser Database: update. *Nucleic Acids Research*, **34** (Database issue), D590–598.

Hornstein, E., Mansfield, J. H., Yekta, S. *et al.* (2005). The microRNA miR-196 acts upstream of Hoxb8 and Shh in limb development. *Nature*, **438**, 671–674.

Lagos-Quintana, M., Rauhut, R., Yalcin, A. *et al.* (2002). Identification of tissue-specific microRNAs from mouse. *Current Biology*, **12**, 735–739.

Lai, E. C. (2004). Predicting and validating microRNA targets. *Genome Biology*, **5**, 115.

Leaman, D., Chen, P. Y., Fak, J. *et al.* (2005). Antisense-mediated depletion reveals essential and specific functions of microRNAs in *Drosophila* development. *Cell*, **121**, 1097–1108.

Lewis, B. P., Shih, I. H., Jones-Rhoades, M. W., Bartel, D. P. and Burge, C. B. (2003). Prediction of mammalian microRNA targets. *Cell*, **115**, 787–798.

McManus, M. T. (2003). MicroRNAs and cancer. *Seminars in Cancer Biology*, **13**, 253–258.

Michael, M. Z., O'Connor, S. M., van Holst Pellekaan, N. G., Young, G. P. and James, R. J. (2003). Reduced accumulation of specific microRNAs in colorectal neoplasia. *Molecular Cancer Research*, **1**, 882–891.

Pfeffer, S., Zavolan, M., Grasser, F. A. *et al.* (2004). Identification of virus-encoded microRNAs. *Science*, **304**, 734–736.

Rajewsky, N. and Socci, N. D. (2004). Computational identification of microRNA targets. *Developmental Biology*, **267**, 529–535.

Rehmsmeier, M., Steffen, P., Hochsmann, M. and Giegerich, R. (2004). Fast and effective prediction of microRNA/target duplexes. *RNA*, **10**, 1507–1517.

Rodriguez, A., Griffiths-Jones, S., Ashurst, J. L., and Bradley, A. (2004). Identification of mammalian microRNA host genes and transcription units. *Genome Research*, **14**, 1902–1910.

Shcherbata, H. R., Hatfield, S., Ward, E. J. *et al.* (2006). The microRNA pathway plays a regulatory role in stem cell division. *Cell Cycle*, **5**, 172–175.

Stark, A., Brennecke, J., Russell, R. B. and Cohen, S. M. (2003). Identification of *Drosophila* microRNA targets. *Public Library of Science Biology*, **1**, E60.

Weber, M. J. (2005). New human and mouse microRNA genes found by homology search. *Federation of European Biochemical Society Letters*, **272**, 59–73.

12 Computational prediction of microRNA targets in vertebrates, fruitflies and nematodes

Dominic Grün and Nikolaus Rajewsky*

Introduction

MicroRNAs are a novel class of small endogenous non-coding RNAs, in their mature form typically of length 21–23 nt. They suppress protein production by binding to the mRNA of their target genes. The apparent fundamental role of microRNAs in nematode development, together with the outstanding degree of conservation of *let-7*, triggered large-scale searches for microRNAs in various organisms. In the order of at least 100–200 microRNAs per species have already been identified experimentally in *C. elegans*, *D. melanogaster*, *D. rerio*, *M. musculus* and *H. sapiens*, many of them being conserved over large evolutionary distances. Recent computational searches for microRNA genes in humans indicate that the true number of different microRNAs per species in mammals could exceed 500, including a substantial fraction of species or lineage specific microRNAs (Berezikov *et al.*, 2005). After the discovery of hundreds of microRNAs in a variety of plants and animals this class of small non-coding RNAs turns out to be an important player in post-transcriptional gene regulation. Mature microRNAs display complex expression patterns during development and in adult tissues. Some were specifically expressed at high numbers, others appear to be expressed in almost all tissues (Barad *et al.*, 2004; Baskerville and Bartel, 2005), suggesting substantial microRNA regulation in various physiological processes. The discovery and biogenesis of microRNAs is presented in other chapters of this book. Although it has been demonstrated that microRNAs, by imperfect binding to their target sites, can strongly inhibit translation of their targets, it has also become clear that they can also degrade their targets (for a review, see Rajewsky (2006)). Both processes are likely to be coupled, but there might also be cases where microRNAs virtually exclusively inhibit translation (or induce mRNA degradation, respectively).

* Author to whom correspondence should be addressed. Permanent address: Max Delbruck Centrum for Molecular Medicine, Robert Rossle Strasse 10, 13092 Berlin, Germany.

MicroRNAs: From Basic Science to Disease Biology, ed. Krishnarao Appasani. Published by Cambridge University Press. © Cambridge University Press 2008.

Although the first targets of microRNAs were discovered by genetic screens, experimental techniques are still too expensive and time-consuming to validate targets of microRNAs on a larger scale. Hence, computational predictions based on sequence properties of a microRNA/mRNA target duplex serve as an important first step to filter out likely candidates of microRNA target genes. A significant fraction of the reported confirmed target genes has been predicted computationally before experimental validation. However, current computational target prediction methods still rely on a rather limited amount of information that can be deduced from common features of only a handful of experimentally validated target sites for only a few microRNAs. Even examination of these few examples anticipates different modes of target recognition. Existing target prediction algorithms are thus likely to predict only particular classes of target sites or may operate with high false positive/negative rates. A particular danger is to "overtrain," i.e. to "learn" too many parameters from few examples. Thus, at present, algorithms that use simple features of microRNA/target site duplexes to predict target sites have an intuitive appeal over complicated methods.

After summarizing the common properties of validated target sites and an introduction of a basic model of target site recognition, some prominent examples of target prediction algorithms will be discussed.

How to detect a microRNA target site

What is known about recognition of microRNA target sites?

The strategy of choice to computationally predict microRNA genes is based on machine learning. The idea is to train an algorithm with a "training set" of known microRNAs such that it can assign a score to a new candidate to assess the likelihood of being a true microRNA. This technique obviously requires enough instances of validated microRNA genes with somewhat overlapping properties to provide sufficient training data. Owing to the large number of known microRNA genes, machine learning based algorithms became established as a prediction method of microRNA genes. As opposed to microRNAs themselves, prediction of their target genes is complicated by a rather small training set, comprising the few instances of validated target sites for only a few microRNAs. Most of the current target prediction algorithms use an empirical model of target site recognition based on experimental evidence and statistical properties deduced from the training set. The most prominent feature common to nearly all validated sites is the "binding nucleus" (Rajewsky and Socci, 2004), also called "seed" (Lewis *et al.*, 2003), a stretch of consecutive Watson–Crick base-pairings between the microRNA and the 3′ UTR of the targeted transcript, in most cases 6–8 nucleotides long and located at the 5′ end of the microRNA. The term "nucleus" was originally not restricted to the 5′ end of the microRNA. The idea of such nuclei had already been suspected after the discovery that *lin-4* partially complementary sites in the 3′ UTR of *lin-14* show a core region of perfect complementarity (Wightman *et al.*, 1993). Strong evidence for this target site feature came with Eric Lai's observation that known posttranscriptional elements in 3′ UTRs were perfectly complementary to the 5′ ends of several microRNAs (Lai, 2002). A more explicit model for binding site recognition was deduced after scoring a large number of possible ways to align microRNAs to their known target sites to obtain the maximum discrimination between microRNAs binding to known targets versus random background sequence (Rajewsky and Socci, 2004). The best method by far simply assigned a score to a stretch of consecutive base pairings between the microRNA and its target site, defined by the sum of single scores for each basepair belonging to the nucleus.

The best nucleus score discrimination between microRNAs binding to known sites versus background sequence yielded single basepair scores that were consistent with the known relative free energies of AU- and CG-basepairs. Interestingly, G:U wobble basepairs turned out to be strongly disfavored. The nucleus could thus be interpreted as a locus of favorable free energy where a rapid zip-up between the microRNA and the mRNA is guaranteed by perfect complementarity. This quick zip-up helps to overcome thermal diffusion and allows for stabilizing, but potentially unspecific, binding of the remaining 3′ portion of the microRNA to its target site. For most target sites, the overall free energy of the microRNA/target site duplex was a bad predictor. However, it is well known that perfect basepairing over the entire length of the microRNA to the target can induce target cleavage. Thus, very low free energies of target site duplexes can still be used to discover functional sites.

Experimental *in vitro* work by Doench and Sharp (2004) suggested deeper insight into the dependence of the repression efficiency of the target gene imposed by the microRNA on the presence of a minimal nucleus. Assuming synergistic effect of multiple sites (Doench *et al.*, 2003), the authors implanted two target sites subject to single nucleotide mutations flanked by two original sites of a particular microRNA into a reporter mRNA and measured the repression efficacy for each mutant in HeLa cells transfected with the reporter construct. First, only the nucleus region was mutated in order to examine the mutability of the nucleus without significant loss of repression. The results suggest a dependence of the repression level on the free energy of the nucleus section of the miRNA:mRNA duplex such that single point mutations of the nucleus region are only allowed if the free energy of the nucleus does not increase substantially. It was also shown that G:U wobble basepairs in the nucleus region hinder repression. Mutating the region of the target site predicted to bind the 3′ end of the microRNA revealed that for a binding site with 7-mer nuclei of perfect complementarity (perfect nuclei) single point mutations in this region only mildly affect repression efficiency. Furthermore, the authors examined how far target sites are allowed to overlap and observed that for sites with perfect nuclei repression remains unaffected as long as the nuclei regions of both sites do not overlap.

Along these lines, a recent study distinguished different classes of microRNA target sites based on an *in vivo* assay in the *Drosophila* wing imaginal disc (Brennecke *et al.*, 2005). A single target site was implanted in the 3′ UTR of a fluorescent reporter gene expressed in the imaginal disc, and the degree of repression was compared for microRNA-expressing and adjacent microRNA-non-expressing cells. The authors tested various types of mutations and came up with at least two different classes of target sites. The 5' dominant sites have sufficient complementarity in the nucleus region and need nearly no supports of the microRNA 3' end binding to the mRNA. Target sites with weak nuclei, i.e. nuclei shorter than seven basepairs or nuclei containing mismatches require compensatory binding of the microRNA 3' end to its target site above the background level. As opposed to 5' dominant sites, 3' compensatory sites form microRNA:mRNA duplexes of free energies lower than expected by matching the microRNA 3' end to randomly chosen background sequence. The contribution of this class of sites leads to a higher specificity, when minimal free energy requirements are included in target prediction algorithms as it is the case for nearly all existing ones.

As a prominent example of 3' compensatory binding, *lin-41* was shown to be bound by *let-7* in nematodes *in vivo* via two neighboring sites with imperfect nuclei residing in the 3' UTR of this gene (Vella *et al.*, 2004a, b, and references therein). In accordance with former experiments (Doench *et al.*, 2003), which demonstrated that increasing the number of miRNA or siRNA target sites increases the degree of repression of the target gene, the authors observed that only a single *let-7* binding site is insufficient to mediate repression of the target gene. The synergistic effect of both sites proved necessary to repress translation of *lin-41*. In case of weak 3' compensatory binding sites the synergistic effect is presumably even more crucial than for multiple sites comprising binding sites with perfect nuclei. However, merely multimerizing single miRNA binding sites does not necessarily suffice to acquire repression. For the two *lin-41* target sites, the authors showed that also the spacer sequence in between these sites plays an important role. Repression is significantly diminished if the spacer segment is mutated or if it is shortened too much. This observation suggests a more complicated architecture of multiple sites and thus anticipates the involvement of interaction between different sites mediated by additional proteins similar to the case of multiple transcription factor binding sites in gene regulatory modules. These details appear to be specific for each instance of multiple sites residing in a single transcript and are thus difficult to include in a general model of target recognition. Beyond this observation it is not unlikely that microRNA target sites divide in even more subclasses where target duplexes of each subclass are characterized by specific features, e.g. loops of particular sizes.

It has been attempted to determine more elaborate characteristics of microRNA:mRNA duplexes by mutation experiments of a single *let-7* site in human and mouse cell lines (Kiriakidou *et al.*, 2004; Chapter 13, this volume). From these experiments a set of rules was deduced, that was subsequently applied to predict targets of arbitrary microRNAs. However, it remains unclear to what extent specific rules derived for a single particular microRNA hold for other microRNAs or different classes of targets. Moreover, one is at risk of losing many true target sites which do not abide by these rules. Based on the finding that multiple sites could act synergistically (Doench *et al.*, 2003) the specificity of target prediction algorithms can be increased by searching for multiple target sites of a particular microRNA or the co-occurrence of binding sites for various microRNAs in the same transcript (Enright *et al.*, 2003; John *et al.*, 2004; Krek *et al.*, 2005; Grün *et al.*, 2005). In particular, searching for the co-occurrence of binding sites for co-expressed microRNAs led to the prediction and subsequent validation of the common regulation of the insulin-secretion regulating gene MTPN by *miR-375*, *miR-124* and *let-7b* in mouse, providing evidence for coordinate microRNA control in mammals (Krek *et al.*, 2005).

It is still an open question as to how far the folding structure of the mRNA influences the accessibility of microRNA target sites. It has been attempted in two studies to incorporate the secondary structure of the potential target into target prediction algorithms (Robins *et al.*, 2005; Zhao *et al.*, 2005). In both works it was assumed that the nucleus part of a microRNA binding site residing in the 3' UTR of the target mRNA requires a minimal overlap with single stranded sub-stretches of the folded transcript. For example, hairpin-loops or free terminal ends, since the pairing probability between the microRNA and double stranded elements is strongly reduced compared to the pairing probability between the microRNA and such single stranded elements.

In order to incorporate structural features, either the entire 3' UTR of the putative target mRNA (Robins *et al.*, 2005) or a rather small segment of sequence surrounding the pre-selected microRNA binding site (Zhao *et al.*, 2005) was folded. Since mRNA folding software becomes unreliable at a sequence length comparable to the average 3' UTR length of the organisms under consideration (perhaps with the exception of nematodes), it is questionable if predictions should be based on folding of whole 3' UTRs. On the other hand, if the mRNA is folded only locally it is unclear whether this reflects the true local folding structure, since inclusion of surrounding sequence or even worse of RNA binding proteins could entirely change the folding of a local element. In fact, microRNA target sites could even actively compete or facilitate the binding of RNA binding proteins. In both studies, the authors provide validations of some of their target predictions providing supporting evidence for their methods. However, both methods impose additional filters to extract candidates for target genes. In particular, both algorithms require the presence of a 7-mer binding nucleus which is known to be the consensus feature for a large class of target sites.

The current amount of information inferred from examination of validated sites is still too small to search the genomic sequence of a single species for target sites of microRNAs, with the rare exception of 3' UTRs that harbor several sites for a single microRNA (such as the fly 3' UTR for hid) or 3' UTRs with a microRNA/mRNA duplex of exceedingly low free energy). For many of the current prediction algorithms the first step consists in searching the 3' UTR of all genes for binding nuclei of microRNAs, i.e. sub-stretches of seven particular nucleotides. Since a given 7-mer will occur roughly every 16 000 bases by chance, this counting presumably comprises many false positives. Under the commonly accepted assumption that evolutionarily conserved sites are likely to be functional, cross-species comparison can significantly reduce the number of false positives. Cross-species comparison assumes enhanced conservation of true target sites due to selective pressure, while random 3′ UTR background sequence is assumed to mutate at a higher rate. Randomly chosen 7-mers are thus expected to have on average significantly lower conservation levels than 7-mers complementary to the nucleus region of microRNAs. This expectation was independently confirmed by several works on a genome-wide scale in vertebrates, fruitflies and nematodes (Grün et al., 2005; Krek et al., 2005; Lewis et al., 2005; Xie et al., 2005; Lall et al., 2006). However, cross-species comparisons allow only for target predictions of microRNAs conserved between the species under consideration. Furthermore, species or lineage specific target genes even of conserved microRNAs are missed by this technique. However, the sequencing of many relatively closely related metazoans, such as the recently completed sequencing of 12 fly species, will allow exploring lineage specific target predictions (N. Rajewsky, unpublished data).

Overview of different prediction methods

In this section, a number of target prediction algorithms is introduced in order to exemplify the variety of approaches being used to tackle the problem of microRNA target prediction.

The prediction software *miRanda*, a Smith–Waterman type approach with several heuristically determined parameters, was first applied to predict targets in *Drosophila* and later on in human (Enright et al., 2003; John et al., 2004). The authors provide lists of target genes ranked by the score, and the algorithms are validated by the recovery of nearly all previously known target sites, but no experimental support for a newly predicted target is provided. Since the target detection rules and the cut-off values were derived from known target sites their recovery was to be expected. A false-positive rate of 35% for a single binding site was derived by comparing the number of true microRNA targets to predicted numbers for shuffled microRNAs. At this false positive rate, the authors predict 535 target genes for 73 microRNAs in *Drosophila* and roughly 4400 target genes for ~150 conserved mammalian microRNAs.

Stark et al. (2003) present an algorithm for target prediction in *Drosophila melanogaster* that also relies on cross species comparison to *D. pseudoobscura*. All 3′ UTRs are screened for sequences complementary to the first eight residues of a microRNA allowing G:U wobble base pairings. For instances of conserved binding sites the free energy of the microRNA:mRNA duplex was computed and a Z-value score was assigned by comparing to the free-energy distribution for random-sequences. The Z-value cut-off was chosen such that only 0.3% of random matches are expected to exceed this threshold. The algorithm detects all previously known targets. Its predictive power was statistically evaluated by comparing the number of known sites at a given score to the expected number of random matches with the same scores. While single sites were observed to have a low statistical significance, multiple sites turn out to be highly significant. Experimental support was provided by verifying three targets each for the *miR-7* and the *miR-2* family with an *in vivo* reporter system in transgenic flies, expressing the microRNA of interest in the *imaginal disc*.

Based on their experimental analysis of structural requirements of microRNA target sites (Brennecke *et al.*, 2005), Stark *et al.* (2005) came up with predicted targets conserved between *D. melanogaster* and *D. pseudoobscura*. Essentially, either a conserved perfect binding nucleus of at least seven basepairs or a nucleus-length dependent minimal free energy of compensatory binding of the microRNA 3' end was required. The authors predict on average 179 target sites per microRNA, and by comparing to the number of conserved targets of shuffled microRNAs a signal to noise ratio of 2.4:1 was estimated. Eight out of nine tested to ranking targets were found be repressed by the predicted microRNA.

Genome-wide predictions of mammalian targets were published by Lewis *et al.* (2003, 2005). The first version of the prediction algorithm, *TargetScan*, combines thermodynamics-based modeling of the microRNA:mRNA duplex with cross-species comparison between human, mouse, rat and puffer fish. Perfect binding nuclei of at least seven nucleotides starting at position two of the microRNA and allowing for G:U wobble basepairs are searched in one species. The microRNA is subsequently hybridized to each target site in a UTR, yielding free-energy parameters that are used to derive a Z-score for each UTR and a ranked list of targeted UTRs for every microRNA. This procedure is repeated for each species and instances where a minimal Z-score and a minimal rank are exceeded in all species are considered a target. A false-positive rate of 22%–31% was inferred by comparing the predicted number of targets for true microRNAs and sets of carefully randomized microRNAs. Using reporter assays in HeLa cells, experimental support was provided for 11 out of 15 tested targets.

The second version, *TargetScanS*, published two years later, extended the cross-species comparison, requiring conserved binding nuclei in human, mouse, rat, dog and chicken. The minimal length was reduced to six and only Watson–Crick basepairings are allowed. Enhanced specificity was inferred by requiring either a conserved adenosine flanking the 3' end of the mRNA nucleus sequence or an increased nucleus length of seven in all species. Using these settings, the authors report a signal to noise ratio of 2.2 for conservation in mammals which corresponds to a false-positive rate of 29%. Above noise, the authors report 13 044 regulatory relationships between human genes and 148 human microRNAs, corresponding to an average of over 200 targets per microRNA. In total, 5300 out of 17 850 orthologous human genes (30%) are predicted targets of microRNAs.

The computational prediction method *DIANA-microT* developed by Kiriakidou *et al.* (2004) takes a different approach to predict targets in human. As already discussed in the previous section the method is based on rules that describe structural requirements of the whole microRNA:mRNA duplex beyond the presence of a binding nucleus. These rules are derived from mutational experiments of a single *let-7* binding site. Additionally, all duplexes have to pass a free-energy filter. A number of 5031 human targets for 94 microRNAs was reported. Requiring conservation of the target sites also in mouse reduces this number to 222 at a signal to noise ratio of 2:1 inferred by comparison to the number of conserved hits of shuffled microRNAs. Experimental evidence for seven predicted targets and successful recovery of known *C. elegans* targets was reported to validate the method.

Rehmsmeier *et al.* (2004) introduced fast folding software, *RNAhybrid*, that finds the energetically most favorable binding site of a small RNA in a large RNA. To account for the binding nucleus the algorithm can enforce basepairing of the microRNA 6-mer starting at position two to the mRNA. Statistical significance is assessed based on extreme value statistics for length normalized minimum free energies, a Poisson approximation for multiple sites and on orthologous targets in multiple organisms. *RNAhybrid* was used to generate target predictions in *Drosophila* and recovered some of the previously known target sites. For 78 microRNAs the authors come up with 227 targets of good E-values at an estimated signal to noise ratio of 2.9:1.

The algorithm *MovingTarget* was developed by Burgler and Macdonald (2005) as a tool for prediction of targets subject to flexible biological constraints. They derive a set of adjustable parameters from instances of experimentally validated microRNA targets: the number of target sites in the UTR, the hybridization energy, the number of consecutive basepairings of the mRNA to the microRNA 5' end, the total number of microRNA 5' nucleotides involved in basepairings, and the number of G:U wobble basepairs in the microRNA 5' region. The authors chose strict parameters to predict conserved targets in *Drosophila*. Requiring a target site to reside in UTR regions of a minimal conservation level of 80% between *D. melanogaster* and *D. pseudoobscura* they predict 83 high-likelihood targets for 78 microRNAs. In order to validate the algorithm three predictions were tested and experimental evidence for microRNA dependent down-regulation of the predicted targets was found. However, no shuffling tests were performed to estimate the false-positive rate.

Our own target prediction algorithm *PicTar* (Krek *et al.*, 2005; Grün *et al.*, 2005) and its improved version *PicTar2.0* (Lall *et al.*, 2006) also relies substantially on cross-species conservation. As a first step, all 3' UTRs alignments for a set of species are screened for all possible binding nuclei of a given set of microRNAs conserved among the same set of species. Binding nuclei comprise both perfect 7-mer nuclei, starting at position one or two of the microRNA, and imperfect nuclei allowing for one mismatch, i.e. a base substitution or insertion/deletion event inside of the mRNA target sequence. However, only nuclei with a non-zero conservation score are accepted. The conservation score is defined by the inverse ratio of the number of instances of a particular nucleus in a reference species and the number of conserved instances of the same nucleus. This conservation score reflects the evolutionary conservation of a particular nucleus due to selection pressure. Nearly all perfect nuclei of all conserved microRNA have high conservation scores, but only a few imperfect nuclei acquire a non-zero conservation score. For example, in a cross-species comparison of *C. elegans*, *C. briggsae* and *C. remanei*, *let-7* is predicted to have only four imperfect nuclei with non-zero conservation score. Remarkably, three of them are observed to reside in the conserved *lin-41* target sites. If a binding nucleus is detected, the corresponding microRNA is hybridized to the mRNA at this position using *RNAhybrid*. If the free energy of this duplex exceeds a certain threshold, the nucleus is kept and otherwise discarded. This microRNA-specific threshold is defined as a certain fraction of the free energy of a perfectly complementary microRNA:mRNA duplex. Perfect nuclei are currently not subject to free energy filtering, whereas the threshold for imperfect nuclei is 66% to account for the requirement of extended compensatory binding. A binding site is considered conserved and termed anchor site if arbitrary nuclei of the same microRNA are found in all species at overlapping positions in the same UTR. A user-defined minimal number of anchor sites for the set of microRNAs under consideration has to be identified in order to consider a particular UTR a target.

Every target UTR sequence (i.e. for each species) is subsequently assigned a score using a hidden Markov model (HMM) that was adopted from a HMM that was previously used to search for combinations of transcription factor binding sites in non-coding DNA. At first, each species is treated separately. The individual scores are combined in a phylogenetic scoring scheme to avoid trivial correlations based on phylogenetic correlations and to obtain a global score reflecting the likelihood of the transcript to be a conserved target. The states of the hidden Markov model are either single background nucleotides with emission probabilities equal to the single letter frequencies or microRNA binding nuclei. In the latter case, the conservation score defines the emission probability. The probabilistic model accounts for length effects, e.g. it scores a single binding nucleus higher if it occurs in a shorter UTR. Furthermore, occurrences of multiple sites have a significant impact on the probability of each site to be real and thus boost the score. The probabilistic model also treats competition of overlapping sites appropriately. We believe that this approach is especially useful to predict target genes under coordinate control of several co-expressed microRNAs. Another strength of the algorithm is its probabilistic nature. For example, it would be relatively straightforward to also incorporate mRNA and microRNA expression levels. The algorithm was used to predict targets in vertebrates, *Drosophila* and nematodes. The signal to noise ratio was also inferred by comparing to target numbers of randomized microRNAs. It shows a strong dependence on the score cut-off, e.g. for targets conserved in mammals it equals 2:1 without any score cut-off and increases to roughly 4:1 at a score cut-off of 6. At the same time, the average number of predicted targets per microRNA drops down by a factor of around three from 330 to 100. Without any score cut-off, 8000 human genes are predicted to be targets of 164 microRNAs conserved in mammals. In *Drosophila*, *PicTar* predicts more than 100 targets per microRNA at a minimal signal to noise ratio of 2.5:1. Approximately 3300 *Drosophila* genes are potential targets of 79 *Drosophila* microRNAs. In nematodes, 73 microRNAs are predicted to target 2000 genes, each microRNA having on average 55 targets at a minimal signal to noise ratio of 2.7:1.

PicTar2.0 recovers nearly all previously known targets with experimental evidence for binding site dependent regulation in *C. elegans* and *Drosophila*. *PicTar* target predictions led to the experimental validation of seven new microRNA-target gene interactions in mammals, twelve novel *let-7* targets in nematodes and to the experimental validation of the first instance of coordinate regulation of a mammalian gene by three different microRNAs. Although the discussed algorithms share some essential features, e.g. the detection of the binding nucleus, the overlap between the different prediction methods is in many instances surprisingly small. A representative comparison yielded overlaps of 50% or less for pairwise comparison of most algorithms with some exceptions, e.g. *PicTar* and *TargetScanS* which overlap by more than 70%. An experimental test of target prediction methods for *Drosophila* was performed by Stark *et al.* (2005). The authors compiled a list of 133 microRNA target pairs that have been experimentally tested by different groups (see references in Stark *et al.* (2005) and Rajewsky (2006)). Of all tested pairs, 71 were observed to be functional, while 62 did not show any microRNA dependent down-regulation. For these 133 interactions, the number of correctly predicted targets was compared to the number of false positives for each available genome wide *Drosophila* target prediction method, separately. Only three methods, their own algorithm, *PicTar* and *MovingTarget* had an accuracy of roughly 90% (percentage of predictions that were functional). The remaining methods by Enright *et al.* (2003), Rehmsmeier *et al.* (2004) (see also Chapter 14 in this book) and by Robins *et al.* (2005) have accuracy levels around 70%. The sensitivity of each algorithm was assessed by the percentage of functional sites that was predicted by each method. The predictions by Stark *et al.* (2005) recovered 81% followed by *PicTar* (69%) and *miRanda* (69%). However, assessment of the sensitivity is complicated by the small overlap between the predictions of some methods and the tested set of interactions. Finally, the authors computed the global overlap of all target predictions of their own methods with each different method, separately. While *PicTar* overlaps with 71% of the new predictions by Stark *et al.* (2005), the overlap with all remaining methods was substantially lower. However, it should be stressed that obviously (a) convergence of target prediction methods does not prove that they are more correct than others, and (b) the experimental assay uses over-expression of microRNAs, and thus might not detect true (endogenous) regulatory relationships. Hence, the ability of target prediction methods to correctly identify and exhaustively find endogenous, functional microRNA targets is still not resolved.

Insights from analyzing genome-wide target predictions

Although large scale experimental validation of computationally predicted microRNA targets is still missing, analysis of the prediction data together with specific experiments leads to increasing insights into microRNA biology, and vice versa has lent more credibility to target predictions. Comparison of the number of binding nuclei of real and randomized microRNAs conserved between related species revealed that at least 10%–40% of all genes are subject to microRNA regulation in vertebrates, insects and nematodes (Lewis *et al.*, 2005; John *et al.*, 2004; Krek *et al.*, 2005; Grün *et al.*, 2005; Xie *et al.*, 2005; Lall *et al.*, 2006). Furthermore, it seems likely that many of them could be coordinately regulated by co-expressed microRNAs (Krek *et al.*, 2005; Grün *et al.*, 2005; Enright *et al.*, 2003; John *et al.*, 2004), suggesting a cell-specific expression signature of microRNAs that is read out by the mRNA content of a cell (Hobert, 2004).

The statistical significance accompanied by experiments (Doench *et al.*, 2003; Brennecke *et al.*, 2005) indicates that a perfect 7-mer binding nucleus, encompassing positions 2–7 of the mature microRNA, could presumably be sufficient to mediate microRNA regulation and that this mode of regulation seems to be the most abundant one. However, it is possible that other classes of target sites are missed due to the incapability of current algorithms to infer specificity for these sites. Irrespective of this caveat, it became apparent that microRNA regulation affects a considerable fraction of all genes.

Evidence for widespread down-regulation of transcript levels by microRNAs is given by the observation of Lim *et al.* (2005) those mRNAs that were down-regulated upon transfection of HeLa cells with *miR-124* or *miR-1* had 3′ UTRs enriched with motifs complementary to the 5′ end of the respective microRNA. The mRNAs down-regulated in these experiments were found to be expressed in the tissue of endogenous microRNA expression but significantly down-regulated compared to their expression in *c.*80 different tissues, suggesting that micro-RNAs play a role in tissue identity. Combining target prediction data with expression profiles of microRNAs and mRNAs further reveals that tissue-specific up-regulation of a gene correlates with depletion of its 3′ UTR for microRNAs expressed in this particular tissue. Whereas enrichment of 3′ UTR target sites for tissue-specifically expressed microRNAs enhances the probability of the transcript to be down-regulated in the same tissue (Krützfeld *et al.*, 2005; Farh *et al.*, 2005, Stark *et al.*, 2005, Sood *et al.*, 2005). The reason for this correlation could be either direct microRNA regulation of the transcript or evolutionary avoidance of target sites for expressed microRNAs if the gene has to be spatially and temporarily co-expressed. Genes selectively depleted for target sites of certain microRNAs were previously postulated and called "anti-targets" (Bartel, 2004; Bartel and Chen, 2004). Notably, in Krützfeldt *et al.* it was shown, by combining *in vivo* knockdown of the murine liver specific miR-122 with microarray analysis before and after knockdown, that anti-targets for miR-122 were highly statistically enriched in genes in a specific functional category (cholesterol biosynthesis). Indeed, the only observable knockdown phenotype was an in-vivo reduction of 40% in cholesterol levels, underscoring the potential importance in identifying anti-targets. The computational identification of anti-targets for a specific 3′ UTR sequence is difficult; however functionally related groups of genes can be used to show that they are or are not depleted in microRNA target sites (Krützfeld *et al.*, 2005; Farh *et al.*, 2005; Stark *et al.*, 2005; Sood *et al.*, 2005). Estimates for the number of anti-targets range up to thousands of genes.

Analysis of the distribution of predicted microRNA targets among Gene Ontology (GO) categories (The Gene Ontology Consortium, 2000) can yield further insights into involvement of microRNAs in particular processes. For instance, Stark *et al.* (2003) observed functionally related targets for several microRNAs in *Drosophila*: *miR-7* was predicted to regulate the Notch signaling pathway (as predicted by Lai, 2002), putative targets of the *miR-2* family comprise at least three proapoptotic genes and *miR-277* potentially regulates the valine, leucine, and isoleucine catabolic pathway. The last example was recovered by our own

GO-term analysis in *Drosophila* (Grün *et al.*, 2005). Among the enriched GO categories in *Drosophila* we identified several metabolic processes, regulation of transcription, morphogenesis, organogenesis and development.

For nematodes, the distribution of enriched GO-terms (Lall *et al.*, 2006) also comprises several metabolic categories but another abundant class includes various kinds of transport regulation, which does not overlap with the results predicted for *Drosophila*. In addition, the enrichment of developmental categories and morphogenic terms is not observed in nematodes to the same extent as in flies. This finding indicates considerable evolution of microRNA regulation between different metazoan clades. To address this issue in more detail, we compared our target predictions for flies and vertebrates (Grün *et al.*, 2005). For each microRNA present in both clades and all orthologous genes we calculated the overlap of predicted microRNA/mRNA regulatory relationships. This overlap was close to what one would expect by chance and thus consistent with a scenario in which many regulators (microRNAs) are evolutionarily conserved while their targets have significantly changed, suggesting major rewiring of post-transcriptional control mediated by microRNAs during evolution. A more exhaustive analysis, comparing microRNAs, their targets and 3′ UTR motifs in general between flies, nematodes, and vertebrates is consistent with this analysis (K. Chen and N. Rajewsky, unpublished results). For flies and vertebrates, we also examined the number of targets of those microRNAs that have homologs in flies and in vertebrates. In addition we also found that the number of targets per microRNA in flies correlates with the number of targets in vertebrates, indicating that while specific microRNA/mRNA regulatory relationships have diverged between both clades, some global properties of the microRNA target "network" have been preserved. However, for some microRNAs the number of targets seems to have evolved in a species-specific way. In these cases we observed a correlation between enhanced abundance of targets in one clade and ubiquitous expression in the same clade whereas the overall expression level for the homologous microRNA in the other clade was low. One of the three microRNAs with a substantially increased relative number of targets in flies versus vertebrates, *miR-210* was observed to have a set of target genes strongly enriched with the GO category "female gamete generation," a likely clade-specific process. In view of the growing number of sequenced and annotated genomes and subsequent availability of microRNA target predictions for these organisms, the co-evolution of microRNAs and their targets will be addressed in growing detail, perhaps one of the most exciting aspects of this novel field.

Conclusions

Although considerable progress has been made to confidently predict a large number of microRNA targets and to gain insights into microRNA function, many caveats and the open questions remain. First, it is possible that we have not yet discovered all classes of microRNA targets. PicTar, for example, predicts only a handful of targets for certain microRNAs, and it is entirely possible that these microRNAs target sequences other than 3' UTRs (such as genomic DNA, or 5' UTRs, or coding sequences) or that they recognize targets by unknown mechanisms. Second, many of the published results are essentially statistical trends (for example the relative down-regulation of mRNA targets in the corresponding microRNA expression domain) and although these trends are statistically significant, it is dangerous to infer general principles based on significances since significance *per se* is not predictive. Third, microRNA targets could be context dependent; for example, a microRNA target could be functional only if certain other regulatory factors such as RNA binding proteins are present. It is noteworthy to remark that the human genome alone encodes thousands of proteins with RNA binding domains. Fourth, at least 30% but more likely up to 70% of all human genes have at least two different 3' UTR isoforms (defined by different polyadenylation sites). Since the majority of microRNA target predictions seem to fall in the alternative (longer) isoforms, and microRNAs have been implicated in mRNA deadenylation, there could be complicated regulatory mechanisms tied to microRNAs and 3' UTR isoforms (see Rajewsky (2006) and references therein). Fifth, current microRNA target predictions cannot predict if the microRNA is likely to inhibit translation, or induce degradation, or both. These differences in gene regulation could be related to differences in the target sequences, but are also already known to be tied to the particular argonaute protein present in the RISC complex that is guided by the microRNA to the target site. Sixth, none of the current target prediction methods incorporates expression levels of mRNAs and microRNAs. Since a single microRNA can have hundreds of targets (all with different mRNA expression levels), this is a very challenging computational problem, especially since most mRNA expression data are measured in the presence of microRNAs. Seventh, it will be necessary to develop better methods for identifying functional sites when no or only limited cross-species comparisons are available. Measuring mRNA expression levels before and after microRNA expression perturbation has shown to help in this regard (Krützfeld *et al.*, 2005, Sood *et al.*, 2005), but it might be also possible to use single nucleotide polymorphism data and population genetic methods in some cases (K. Chen and N. Rajewsky, unpublished results). In summary, predicting microRNA targets is very likely to remain a fast moving and exciting field with many surprises yet to be discovered.

REFERENCES

Barad, O., Meiri, E., Avniel, A. *et al.* (2004). MicroRNA expression detected by oligonucleotide microarrays: system establishment and expression profiling in human tissues. *Genome Research*, **14**, 2486–2494.

Bartel, D. P. (2004). MicroRNAs: genomics, biogenesis, mechanism, and function. *Cell*, **116**, 281–297.

Bartel, D. P. and Chen, C. Z. (2004). Micromanagers of gene expression:the potentially widespread influence of metazoan microRNAs. *Nature Review Genetics*, **5**, 396–400.

Baskerville, S. and Bartel, D. P. (2005). Microarray profiling of microRNAs reveals frequent coexpression with neighboring miRNAs and host genes. *RNA*, **11**, 241–247.

Berezikov, E., Guryev, V., van de Belt, J. *et al.* (2005). Phylogenetic shadowing and computational identification of human microRNA genes. *Cell*, **120**, 21–24.

Brennecke, J., Stark, A., Russell, R. B. and Cohen, S. M. (2005). Principles of microRNA-target recognition. *Public Library of Science Biology*, **3**, e85.

Burgler, C. and Macdonald, P. M. (2005). Prediction and verification of microRNA targets by MovingTargets, a highly adaptable prediction method. *BMC Genomics*, **6**, 88.

Doench, J. G. and Sharp P. A. (2004). Specificity of microRNA target selection in translational repression. *Genes & Development*, **18**, 504–511.

Doench, J. G., Petersen, C. P. and Sharp, P. A. (2003). siRNAs can function as miRNAs. *Genes & Development*, **17**, 438–442.

Enright, A. J., John, B., Gaul, U. *et al.* (2003). MicroRNA targets in Drosophila. *Genome Biology*, **5**, R1.

Farh, K. K., Grimson, A., Jan, C. *et al.* (2005). The widespread impact of mammalian microRNAs on mRNA repression and evolution. *Science*, **310**, 1817–1821.

Grün, D., Wang, Y. L., Langenberger, D., Gunsalus, K. C. and Rajewsky, N. (2005). microRNA target predictions across seven Drosophila species and comparison to mammalian targets. *Public Library of Science Computational Biology*, **1**, e13.

Hobert, O. (2004). Common logic of transcription factor and microRNA action. *Trends in Biochemical Sciences*, **29**, 462–468.

John, B., Enright, A. J., Aravin, A. *et al.* (2004). Human microRNA targets. *Public Library of Science Biology*, **2**, e363.

Kiriakidou, M., Nelson, P. T., Kouranov, A. *et al.* (2004). A combined computational-experimental approach predicts human microRNA targets. *Genes & Development*, **18**, 1165–1178.

Krek, A., Grün, D., Poy, M. N. *et al.* (2005). Combinatorial microRNA target predictions. *Nature Genetics*, **37**, 495–500.

Krützfeld J., Rajewsky, N., Braich, R. *et al.* (2005). Silencing of microRNAs *in vivo* with "antagomirs". *Nature*, **438**, 685–689.

Lai, E. C. (2002). Micro RNAs are complementary to 3′ UTR sequence motifs that mediate negative post-transcriptional regulation. *Nature Genetics*, **30**, 363–364.

Lall, S., Grün, D., Krek, A. *et al.* (2006). A genome wide map of conserved microRNA targets in *C. elegans*. *Current Biology*, **16**, 460–471.

Lewis, B. P., Shih, I. H., Jones-Rhoades, M. W., Bartel, D. P. and Burge, C. B. (2003). Prediction of mammalian microRNA targets. *Cell*, **115**, 787–798.

Lewis, B. P., Burge, C. B., Bartel, D. P. (2005). Conserved seed pairing, often flanked by adenosines, indicates that thousands of human genes are microRNA targets. *Cell*, **120**, 15–20.

Lim, L. P., Lau, N. C., Garrett-Engele, P., Grimson, A. and Schelter, J. M. (2005). Microarray analysis shows that some microRNAs downregulate large numbers of target mRNAs. *Nature*, **433**, 769–773.

Rajewsky, N. (2006). Computational microRNA target predictions in animals. *Nature Genetics*, **38**, S8–13.

Rajewsky, N. and Socci, N. D. (2004). Computational identification of microRNA targets. *Developmental Biology*, **267**, 529–535.

Rehmsmeier, M., Steffen, P., Hochsmann, M. and Giegerich, R. (2004). Fast and effective prediction of microRNA/target duplexes. *RNA*, **10**, 1507–1517.

Robins, H., Li, Y. and Padgett, R. W. (2005). Incorporating structure to predict microRNA targets. *Proceedings of the National Academy of Sciences USA*, **102**, 4006–4009.

Sood, P., Krek, A., Zavolan, M., Macino, G. and Rajewsky, N. (2005). Cell-type specific signatures of microRNAs on target mRNA expression. *Proceedings of the National Academy of Sciences USA*, **103**, 2746–2751.

Stark, A., Brennecke, J., Russell, R. B. and Cohen, S. M. (2003). Identification of Drosophila microRNA targets. *Public Library of Science Biology*, **1**, 397–409.

Stark, A., Brennecke, J., Bushati, N., Russell, R. B. and Cohen, S. M. (2005). Animal microRNAs confer robustness to gene expression and have a significant impact on 3′UTR evolution. *Cell*, **123**, 1133–1146.

The Gene ontology Consortium (2000). Gene ontology: tool for the unification of biology. *Nature Genetics*, **25**, 25–29.

Vella, M. C., Choi, E. Y., Lin, S. Y., Reinert, K. and Slack F. J. (2004a). The *C. elegans* microRNA let-7 binds to imperfect complementary sites from the *lin-41* 3'UTR. *Genes & Development*, **18**, 132–137.

Vella, M. C., Reinert, K. and Slack F. J. (2004b). Architecture of a validated microRNA::target interaction. *Chemistry & Biology*, **11**, 1619–1623.

Wightman, B., Ha, I. and Ruvkun, G. (1993). Posttranscriptional regulation of the heterochronic gene lin-14 by lin-4 mediates temporal pattern formation in *C. elegans*. *Cell*, **75**, 855–862.

Xie, X., Kulbokas, E. J., Golub T. R. *et al.* (2005). Systematic discovery of regulatory motifs in human promoters and 3' UTRs by comparison of several mammals. *Nature*, **434**, 338–345.

Zhao, Y., Samal, E. and Srivastava, D. (2005). Serum response factor regulates a muscle-specific microRNA that targets Hand2 during cardiogenesis. *Nature*, **436**, 214–220.

13 Computational approaches to elucidate miRNA biology

Praveen Sethupathy, Molly Megraw and Artemis G. Hatzigeorgiou*

Introduction

Research in the past decade has revealed that microRNAs (miRNAs) are widespread and that they are likely to underlie an appreciably larger set of disease processes than is currently known. The first miRNAs and their functions were determined via classical genetic techniques. Soon after, a number of miRNAs were discovered experimentally (Lagos-Quintana *et al.*, 2001). However, the characterization of miRNA function remained elusive owing to low-throughput experiments and often indeterminate results, most notably for those miRNAs which have multiple roles in multiple tissues. High-throughput experimental methods for miRNA target identification are the ideal solution, but such methods are not currently available. As a result, computational methods were developed, and are still regularly used, for the purpose of identifying miRNA targets.

Most current target prediction programs require the sequences of known miRNAs. Currently, there are 332 known miRNAs in the human genome. The estimation of the total number of miRNAs varies from publication to publication (Lim *et al.*, 2003; Bentwich *et al.*, 2005). In a recent paper, Bentwich *et al.* contended that there are at least 500 more miRNAs that are yet to be identified (Bentwich *et al.*, 2005). Despite the number of unknown miRNAs, computational approaches based on features of known miRNAs have been instrumental in the discovery of as-of-yet-unknown miRNAs in the genome. The past few years have witnessed an explosion in information regarding the genomic organization of miRNAs, the biogenesis of miRNAs, the targeting mechanisms of miRNAs, and the regulatory networks in which miRNAs are involved. Computational methods for the efficient storage and retrieval of this information have been critical.

In this chapter we will describe computational approaches that we have utilized in order to elucidate miRNA biology. More specifically, we will discuss the role of

---

* Author to whom correspondence should be addressed.

MicroRNAs: From Basic Science to Disease Biology, ed. Krishnarao Appasani. Published by Cambridge University Press. © Cambridge University Press 2008.

bioinformatics in the prediction of miRNAs and their targets, in the design and construction of databases, and web resources related to miRNA biology.

MicroRNA target prediction

Background

The first target prediction programs for human (Lewis *et al.*, 2003, Kiriakidou *et al.*, 2004) and for fruitfly (Enright *et al.*, 2003; Stark *et al.*, 2003; Rajewsky and Socci, 2004) were published in late 2003/early 2004. Since then, at least four others for human (John *et al.*, 2004; Lewis *et al.*, 2005; Krek *et al.*, 2005; Rusinov *et al.*, 2005) and five others for fruitfly (Rehmsmeier *et al.*, 2004; Burgler and Macdonald, 2005; Grün *et al.*, 2005; Robins et al., 2005; Saetrom *et al.*, 2005) have been published. A majority of these fourteen programs are available for public use (Sethupathy *et al.*, 2006). At a high level, most of these programs search for potential miRNA targets by scanning the genome for short segments that have features very similar to a set of experimentally supported miRNA targets. Here we describe the basic idea behind one of the first of these computational methods, DIANA-microT (Kiriakidou *et al.*, 2004).

The miRNA target prediction program DIANA-microT

The first computational approaches for identifying miRNA recognition elements (MREs) were based on the very few MREs that were known at that time. These approaches followed the hypothesis that functional miRNA:MRE interactions are guided by high affinity interactions, based on binding energies, between a given miRNA and its cognate MRE. One of the first of these approaches, DIANA-microT, identified putative miRNA:MRE interactions using a modified dynamic program-ming algorithm that calculated binding energies between two imperfectly paired RNAs. At a high level, this algorithm can be described as a modified global alignment algorithm between two RNA segments that allows both canonical (Watson–Crick) and G:U wobble base pairing. The energy score of an alignment was calculated by summing the binding energy of each consecutive dinucleotide in the alignment (Tinoco *et al.*, 1973). Bulges, loops, and mismatches were allowed in the alignment, but were assigned penalties.

To identify putative MREs in the human genome, the dynamic programming algorithm was used to align a sliding window of 38 nucleotides (nt), that "slid" over all human 3′ UTRs, with all known miRNAs. The 38 nt window was chosen based on the average length of a miRNA (19–24 nt) plus additional bases that were allowed for loops and bulges in the alignment (Kiriakidou *et al.*, 2004).

In the second step, all MREs of each miRNA found from the previous step were sorted based on the free energy score. From among the top scoring MREs, those that were conserved between human and mouse were further considered for experimental verification. Experimental verification of an MRE was conducted by cloning the MRE into the 3′ UTR of Renilla Luciferase (RL) and then measuring the activity of RL in the presence and absence of the corresponding miRNA. Luciferase assays of a dozen predicted MREs led to the identification of three

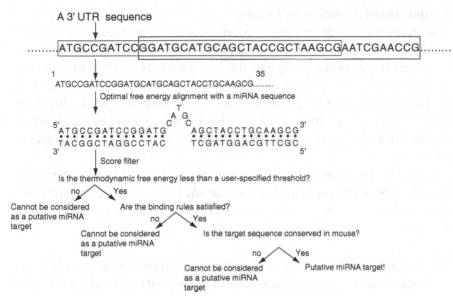

Figure 13.1. An outline of DIANA-microT's methodology for miRNA target prediction.

novel MREs, including one for the let-7b miRNA within the 3′ UTR of LIN-28, the human homolog of a *C. elegans* gene known to operate in the same pathway as let-7b. In order to identify key structural elements of the binding between a miRNA and its MRE, a series of mutated variants of the let-7b miRNA and the LIN-28 MRE were developed and tested in additional luciferase reporter silencing assays. This analysis allowed for the description of a specific set of binding rules for miRNA:MRE interactions:

(1) the 5′ end of the miRNA must form at least seven base pairs with the target site, with up to two G:U wobble pairs allowed if flanked by two canonical bindings, and with up to one nucleotide bulge interrupting the base pairing,

(2) there must be either a central loop of 2–3 nucleotides or a central bulge of 2–5 nucleotides on the miRNA or 6–9 nucleotides on the target mRNA, and

(3) the 3′ end of the miRNA must form at least five base pairs with the target, with G:U wobble pairs allowed and with single or dinucleotide bulges allowed (Kiriakidou *et al.*, 2004).

These new rules were incorporated into the original algorithm to generate the final DIANA-microT algorithm (Figure 13.1). In summary, the basic elements of the program are as follows.

For each miRNA:

(a) find all sites within all known 3′ UTRs that bind to the miRNA with an energy score that exceeds a user-specified threshold,

(b) apply a filter on the predicted target sites based on the specific binding rules mentioned above,

(c) apply a final filter on the remaining predicted target sites based on sequence and position conservation between human and mouse.

Validation of target prediction programs

The performance of target prediction programs can be examined in one of three manners: *in vitro, in vivo* and *in silico*. A popular method for *in vitro* examination is the reporter (i.e. luciferase) assay which is mentioned above. Depending on the organism, reporter assays are also popular in *in vivo* systems. Other *in vivo* techniques range from traditional phenotypic analysis of MRE or miRNA mutants to miRNA under/over expression assays. For a more comprehensive listing of various experimental techniques for target validation, we refer the reader to Sethupathy *et al.* (2006).

The *in silico* validation is currently accomplished by calculating the ratio of the number of targets predicted for real miRNAs to the number of targets predicted for shuffled miRNAs. There are several different ways to produce shuffled miRNAs:

(a) By simple random shuffling of the real miRNA sequence. Resulting sequences have the same length and same nucleotide distribution as real miRNAs (Enright *et al.*, 2003).

(b) By selecting from step (a) only those shuffled sequences that have the same ability to bind to the human 3′ UTR sequences as the real miRNAs. The calculation of binding ability is based on the dinucleotide distribution of the 3′ UTR sequences.

(c) By selecting from step (a) only those shuffled sequences that have the same number of hits in a single genome as real miRNAs do, and by testing whether predicted sites of real miRNAs are preferentially conserved across a set of species (Lewis *et al.*, 2003; Lewis *et al.*, 2005).

Status of target prediction

Since there are very few experimentally supported miRNA targets, it is very difficult to discern which programs perform better than others. As a result, the choice of which program to use is best determined by the specific nature of the question at hand. For example, if one wanted to minimize the number of predictions to test experimentally, it would be most beneficial to choose a program with few predictions per miRNA. Rather, if one wanted to get a sense of the many pathways possibly affected by miRNAs, it would be more advantageous to choose a program that achieves high sensitivity. Table 13.1 provides a summary of how a sampling of current target prediction programs perform on experimentally verified miRNA targets recorded in TarBase (Sethupathy *et al.*, 2006). There is emerging evidence that there are different classes of miRNA targets (Brennecke *et al.*, 2005). The current set of experimentally verified targets is still small and possibly biased towards the class(es) of targets that are predicted by certain programs. Because the targets available in TarBase may not be representative of all target classes, the reader is encouraged to interpret the results presented in Table 13.1 not as a direct comparison of programs but as a means of ascertaining the relative strengths of each program.

Table 13.1. TarBase based comparison of five recent target prediction programs

	Percentage of experimentally supported targets predicted	Number of additional targets predicted
TargetScan (MIT)	0%	~280
DIANA-microT (UPenn)	3%	~480
miRanda (Sloan-Kettering)	51%	~13 600
TargetScanS (MIT)	60%	~9200
PicTar (NYU)	55%	~11 100

Results are as of 2006. This table is intended to show the wide disparity in the predictions of different programs and the relative strengths of each (sensitivity vs. specificity).

MicroRNA gene prediction

Background

The first computational methods for miRNA identification in worm and fruitfly emerged in 2003 (Ambros *et al.*, 2003; Grad *et al.*, 2003; Lai *et al.*, 2003; Lim *et al.*, 2003) and are still emerging for application in other organisms such as plant, human and various viruses (Adai *et al.*, 2005; Lindow and Krogh, 2005; Pfeffer *et al.*, 2005; Sewer *et al.*, 2005; Zhang *et al.*, 2005). In all of these methods, the structure used for the computational identification of miRNAs is the RNA hairpin. The mature miRNA is located on one or both arms (i.e. sides) of the hairpin stem. Computational identification methods define a set of features that characterize the structure of the miRNA hairpin. Different approaches are then used to classify potential miRNA hairpins based on the specific feature values of the hairpins. For each RNA hairpin, this specific set of feature values is often called a feature "vector." Here we describe the most recent computational approach to classify RNA hairpin feature vectors based on a method derived from the machine learning field called "support vector machines" (SVMs).

Feature extraction

Which features describe miRNAs, and how are these features combined? There are several biological factors that determine whether a miRNA is processed and functional: thermodynamic stability of the miRNA hairpin within the cytoplasm, structural accessibility for enzymes to cleave the mature miRNA from the hairpin, and the ability of a mature miRNA to be loaded into the protein complex which carries it to a target. The features used by most gene prediction programs to capture these biological factors include: free-energy of the miRNA hairpin, number of paired bases in the stem of the hairpin, sizes of loops and bulges in a hairpin, and sequence conservation of the hairpin with other species. The most prominent approach to combining these features is to score a prediction based on a weighted linear combination of the features, with weights determined according to presumed biological importance or according to the individual discriminative power of each feature between real miRNAs and control sequences.

Machine learning technique for miRNA prediction

A heretofore less-explored approach to feature combination is the use of machine learning techniques. The idea behind using machine learning is that if the collection of known miRNAs is truly representative of yet-undiscovered miRNAs, feature score combinations can be "learned" from the set of known miRNAs in a way that maximizes total discriminative power between real miRNAs and control sequences. Depending on the technique chosen, these score combinations will not necessarily be a linear weighting of features. For example, suppose that the requirement for overall miRNA hairpin stability increases dramatically for hairpins with small end-loops. This trend can be exploited by a technique which allows a non-linear relationship between free-energy requirement and end-loop size. Furthermore, some machine learning methods are able to "pick out" beneficial feature combinations even if the features are inter-dependent, by modeling the same underlying concept. For example, the number of bulges and loops and the total hairpin free energy are not mutually exclusive features. However, both features are functionally impor-tant *in vivo* for enzyme recognition and cleavage.

Support vector machines for miRNA prediction

The machine learning technique called a "support vector machine" (SVM) classi-fier is one method which allows for both feature dependence and non-linear relationships among features. To understand how an SVM classifier works, first think of the task of separating two classes of points in space. If two classes are linearly separable, we can define separating hyperplanes for their separation (see Figure 13.2). Think of a "separating hyperplane" as a plane which divides two sets of points in a multi-dimensional space. There are many ways to choose a separat-ing hyperplane, and the SVM method specifically chooses the plane which maxi-mizes the distance between this plane and the nearest points from either class (Figure 13.2c). A similar concept is used to pick the hyperplane boundary even if the data sets overlap – the hyperplane is drawn so as to minimize the total "error" for points which are on the "wrong" side of the hyperplane (Figure 13.2d). The total error for these mis-classified points is calculated as the sum of the distance from each of these points to the margin boundary. If the classes overlap, as is the case with miRNA gene features, we need to produce nonlinear boundaries to separate them. SVMs allow for this by constructing a linear boundary in a large, transformed version of the feature space (Figure 13.2a, b). This transformation of the space is achieved through a feature map called the kernel function. A complete technical description of SVM learning is available in Cortes and Vapnik (1995), and for accessible further reading we suggest the text by Hastie *et al.* (2001).

In order to use an SVM classifier for miRNA gene prediction, one must first train the classifier using a set of "positive" examples – typically hairpins asso-ciated with known miRNAs – and a set of "negative" examples. Choices for negative examples are less obvious, but usually they are random subsequences from intergenic regions, tRNAs, rRNAs and mRNAs. After positive and negative hairpin data sets are chosen, the feature value vectors are computed for each

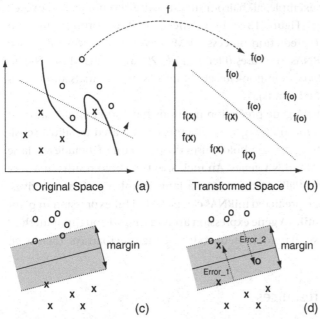

Figure 13.2. This figure depicts two important aspects of SVMs: the kernel function and the margin. Parts (a) and (b) show how a feature map using a kernel function can simplify the classification task. Part (c) shows a picture of the maximum margin separating two classes; there are three "support points" on the margin boundary. Part (d) shows the margin with errors in the case where two classes are not separable in the space shown.

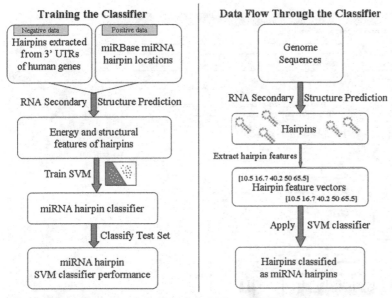

Figure 13.3. A diagram of the miRNA hairpin classification procedure.

hairpin. An SVM classifier is then trained on the positive and negative sets of feature vectors. To better understand how the classifier will perform on "unseen" data, this classifier can be applied to a "test set" of hairpins which has not been included in the training data. Finally, the classifier is applied to the set of hairpins of interest – for example, all hairpins in an entire genome which exceed a chosen folding energy. Figure 13.3 shows the data flow through an SVM classifier to illustrate the prediction process. SVMs have been successfully used for the prediction of miRNAs in viruses (Pfeffer *et al.*, 2005), clustered miRNAs in mammals (Sewer *et al.*, 2005), and non-clustered miRNAs in mammals and viruses (DIANA-group, unpublished results).

All computational miRNA gene prediction methods have some limits on their accuracy. The thresholds for miRNA gene predictors can be used to adjust sensitivity and specificity. A stringent threshold causes the predictor to include less false positives, but also less real miRNA genes. An inclusive threshold results in a very sensitive prediction, potentially at the cost of a large number of false positives. After a threshold is chosen, predicted miRNAs can be tested for expression in plant or animal tissues using a miRNA gene expression array, a high-throughput method which is conceptually similar to protein-coding gene microarrays (Bentwich *et al.*, 2005).

Databases and web resources

TarBase

We discuss here a recently published database called TarBase (Sethupathy *et al.*, 2006). TarBase describes at least fifty human/mouse genes, thirty fruitfly genes, ten worm genes, and one zebrafish gene which have been experimentally validated as miRNA targets (Figure 13.4). The total number of target sites recorded in TarBase

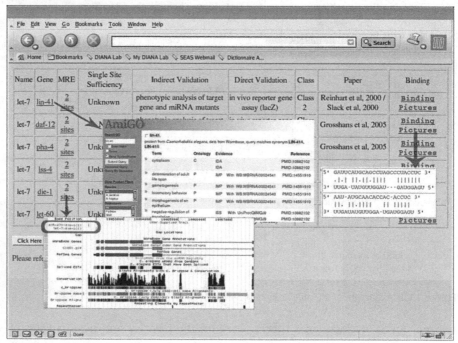

Figure 13.4. A snapshot of the results page showing the various functionalities of the database. Target genes are linked to Gene Ontology (GO), miRNA Recognition Elements (MREs) are linked to UCSC Genome Browser via custom tracks, and binding pictures are viewable in a new window.

for human/mouse, fruitfly, worm, and zebrafish is ~550 – a much larger data set for the next generation of target prediction programs. Furthermore, TarBase describes each supported target site by the miRNA which binds it, the gene in which it occurs, the location within the 3′ UTR where it occurs, the nature of the experiments that were conducted to validate it, and the sufficiency of the site to induce translational repression. Such a comprehensive description of each target site will be useful for focused bioinformatic and experimental studies to further understand the features of miRNA targeting, the mechanisms of miRNA based translational repression, and the roles of miRNAs in various biological networks. TarBase can be accessed at http://www.diana.pcbi.upenn.edu/tarbase.

TarBase also includes functional links to other databases such as Gene Ontology (GO) (Ashburner *et al.*, 2000) and UCSC Genome Browser (Kent *et al.*, 2002). A target gene with experimental support is linked to GO in order to provide a clearer picture of what kinds of biological pathways the targeting miRNA may regulate. The specific target sites within the target gene are linked to UCSC Genome Browser via custom tracks for facile viewing of their genic location and sequence composition. Finally, each target site is also linked to an in-house database of all known miRNA:target site base pairing diagrams. All of this centralized data are valuable either for inquiries regarding a specific gene/miRNA/target site/experimental technique or for extracting general features of miRNA targeting or miRNA based mechanisms for translational repression.

To ensure that TarBase is up-to-date, two semi-automated means of submission are provided. Readers are directed to http://www.diana.pcbi.upenn.edu/tarbase for submission instructions.

MicroRNA Genomics Resource and Cluster Server

MicroRNAs may be located either in the introns of protein-coding genes, outside of protein-coding genes entirely, or more rarely in a gene exon or UTR. There is some evidence that miRNAs located within the introns of protein-coding genes may be processed from the mRNAs of these "host genes" (Baskerville and Bartel, 2005). The Sanger Institute's miRBase provides the locations and descriptions of all known miRNAs in plants, animals, and viruses. Currently about one third of the approximately 330 known human miRNAs are located within the introns of known protein-coding genes, and the vast majority of the remaining miRNAs are considered "intergenic" with respect to protein-coding genes. The DIANA microRNA Genomics Resource and Cluster Server (http://www.diana.pcbi.upenn.edu/mirGen/) provides a resource for detailed and up-to-date analysis of the genomic locations of miRNAs in many organisms by interfacing miRBase miRNA coordinates with UCSC genome annotation data.

The Genomics Resource allows the user to explore where whole-genome collections of microRNAs are located with respect to UCSC Known Genes, Refseq Genes, Genscan predicted genes, and to other current UCSC genome annotation sets such as CpG islands and pseudogenes. This resource addresses all organisms for which the UCSC Genome Browser Database (Hinrichs *et al.*, 2006) and miRBase both provide data – this includes the most commonly accessed genomes in vertebrates, insects, and nematodes. The Cluster Server allows the user to discover which miRNAs fall into clusters at any user-defined inter-miRNA distance, using any eukaryotic genome miRNA collection available at miRBase. The Cluster Server resource also connects miRNAs to predicted targets by DIANA-microT, experimentally supported targets in TarBase, and target GO function.

Together, the tools provided by the DIANA microRNA Genomics Resource and Cluster Server allow users to perform their own unique investigations into miRNA genomics-related questions using the most recent data available.

Discussion

Bioinformatics has played a critical role in the understanding of miRNA genomic organization, miRNA targeting, and the functions of miRNAs in various biological networks. Furthermore, bioinformatics has also been instrumental in the efficient storage and retrieval of information related to miRNA biology through the design and construction of databases. However, despite the advances made in understanding miRNA function, there are still many open questions.

- Have we exploited all the features of miRNA targeting in our target prediction algorithms? For example, recent studies provide an example of at least one *C. elegans* target gene in which the binding sites of two different miRNAs must be a certain distance from each other in order to induce translational repression. If this phenomenon is widespread in miRNA targeting, then perhaps more extensive experimental studies will provide the data necessary for incorporation of this and other related features into new target prediction programs.
- Why do miRNAs tend to be clustered in the human genome? Could they have evolved in this manner in order to preserve functional miRNA units? How can this bias for clustering be used in the prediction of as-of-yet-unknown miRNAs?
- How are miRNAs themselves regulated, both transcriptionally and post-transcriptionally? Insight into this area will provide a greater understanding of how miRNAs fit into biological networks.

Continued crosstalk between the computational and experimental realms will provide a more swift and efficient discovery of why miRNAs are critical for life and how they can be utilized for therapeutics.

REFERENCES

Adai, A., Johnson, C., Mlothshwa, S. *et al.* (2005). Computational prediction of miRNAs in *Arabidopsis thaliana*. *Genome Research*, **15**, 78–91.

Ambros, V., Lee, R. C., Lavanway, A., Williams, P. T. and Jewell, D. (2003). MicroRNAs and other tiny endogenous RNAs in *C. elegans*. *Current Biology*, **13**, 807–818.

Ashburner, M., Ball, C. A., Blake, J. A. *et al.* (2000). Gene Ontology: tool for the unification of biology. *Nature Genetics*, **25**, 25–29.

Baskerville, S. and Bartel, D. P. (2005). Microarray profiling of microRNAs reveals frequent coexpression with neighboring miRNAs and host genes. *RNA*, **11**, 241–247.

Bentwich, I., Avniel, A., Karov, Y. *et al.* (2005). Identification of hundreds of conserved and nonconserved human microRNAs. *Nature Genetics*, **37**, 766–770.

Brennecke, J., Stark, A., Russell, R. B., and Cohen, S. M. (2005). Principles of microRNA-target recognition. *Public Library of Science Biology*, **3**, 404–418.

Burgler, C. and Macdonald, P. M. (2005). Prediction and verification of microRNA targets by MovingTargets, a highly adaptable prediction method. *BMC Genomics*, **6**, 88.

Cortes, C. and Vapnik, V. (1995). Support vector networks. *Machine Learning*, **20**, 1–25.

Enright, A. J., John, B., Gaul, U. *et al.* (2003). MicroRNA targets in *Drosophila*. *Genome Biology*, **5**, R1.

Grad, Y., Aach, J., Hayes, G. D. *et al.* (2003). Computational and experimental identification of *C. elegans* microRNAs. *Molecular Cell*, **11**, 1253–1263.

Grün, D., Wang, Y., Langenberger, D., Gunsalus, K. C. and Rajewsky, N. (2005). MicroRNA target predictions across seven Drosophila species and comparison to mammalian targets. *Public Library of Science Computational Biology*, **1**, 51–66.

Hastie, T., Tibshirani, R. and Friedman, J. (2001). *The Elements of Statistical Learning*. New York: Springer-Verlag.

Hinrichs, A. S., Karolchik, D., Baertsch, R. *et al.* (2006). The UCSC Genome Browser Database: update 2006. *Nucleic Acids Research*, **34**, D590–598.

John, B., Enright, A. J., Aravin, A. *et al.* (2004). Human microRNA targets. *Public Library of Science Biology*, **2**, 1862–1879.

Kent, W. J., Sugnet, C. W., Furey, T. S. *et al.* (2002). The Human Genome Browser at UCSC. *Genome Research*, **12**, 996–1006.

Kiriakidou, M., Nelson, P., Kouranov, A. *et al.* (2004). A combined computational-experimental approach predicts human miRNA targets. *Genes & Development*, **18**, 1165–1178.

Krek, A., Grun, D., Poy, M. N. *et al.* (2005). Combinatorial microRNA target predictions. *Nature Genetics*, **37**, 495–500.

Lagos-Quintana, M., Rauhut, R., Lendeckel, W. and Tuschl, T. (2001). Identification of novel genes coding for small expressed RNAs. *Science*, **194**, 797–799.

Lai, E. C., Tomancak, P., Williams, R. W. and Rubin, G. M. (2003). Computational identification of *Drosophila* microRNA genes. *Genome Biology*, **4**, **R42**, 1–20.

Lewis, B. P., Shih, I., Jones-Rhoades, M. W., Bartel, D. P. and Burge, C. B. (2003). Prediction of mammalian microRNA targets. *Cell*, **115**, 787–798.

Lewis, B. P., Burge, C. B. and Bartel, D. P. (2005). Conserved seed pairing, often flanked by adenosines, indicates that thousands of human genes are microRNA targets. *Cell*, **120**, 15–20.

Lim, L. P., Lau, N. C., Weinstein, E. G. *et al.* (2003). The microRNAs of *Caenorhabditis elegans*. *Genes & Development*, **17**, 991–1008.

Lindow, M. and Krogh, A. (2005). Computational evidence for hundreds of non-conserved plant microRNAs. *BMC Genomics*, **6**, 119.

Pfeffer, S., Sewer, A., Lagos-Quintana, M. *et al.* (2005). Identification of microRNAs of the herpesvirus family. *Nature Methods*, **2**, 269–276.

Rajewsky, N. and Socci, N. D. (2004). Computational identification of microRNA targets. *Developmental Biology*, **267**, 529–535.

Rehmsmeier, M., Steffen, P., Hochsmann, M. and Giegerich, R. (2004). Fast and effective prediction of microRNA/target duplexes. *RNA*, **10**, 1507–1517.

Robins, H., Li, Y. and Padgett, R. W. (2005). Incorporating structure to predict microRNA targets. *Proceedings of the National Academy of Sciences USA*, **102**, 4006–4009.

Rusinov, V., Baev, V., Minkov, I. N. and Tabler, M. (2005). MicroInspector: a web tool for detection of miRNA binding sites in an RNA sequence. *Nucleic Acids Research*, **33**, 696–700.

Saetrom, O., Snove Jr., O. and Saetrom, P. (2005). Weighted sequence motifs as an improved seeding step in microRNA target prediction algorithms. *RNA*, **11**, 995–1003.

Sethupathy, P., Corda, B. and Hatzigeorgiou, A. G. (2006). TarBase: a comprehensive database of experimentally supported animal microRNA targets. *RNA*, **12**, 192–197.

Sewer, A., Paul, N., Landgraf, P. *et al.* (2005). Identification of clustered microRNAs using an ab initio prediction method. *BMC Bioinformatics*, **6**, 267.

Stark, A., Brennecke, J., Russell, R. B. and Cohen, S. M. (2003). Identification of *Drosophila* microRNA targets. *Public Library of Science Biology*, **1**, 1–13.

Tinoco Jr., I., Borer, P. N., Dengler, B. *et al.* (1973). Improved estimation of secondary structure in ribonucleic acids. *Nature New Biology*, **246**, 40–41.

Zhang, B. H., Pan, X. P., Wang, Q. L., Cobb, G. P. and Anderson, T. A. (2005). Identification and characterization of new plant microRNAs using EST analysis. *Cell Research*, **15**, 336–360.

14 The RNAhybrid approach to microRNA target prediction

Marc Rehmsmeier

Introduction

MicroRNAs (miRNAs) are 19–24 nt long RNAs that post-transcriptionally regulate their target genes. The regulation is effected by binding of the RISC-incorporated miRNA to the target mRNA. Upon near-perfect hybridization, the target mRNA is cleaved and subsequently degraded. Less strong hybridizations lead to translational repression of the mRNA or to its degradation. Besides the elucidation of mechanistic aspects of the miRNA pathway, the reconstruction of the miRNA regulatory network is of great interest. A fundamental part in the reconstruction process is the knowledge of miRNA targets. Since reverse-genetic approaches are limited by their time and cost-intensiveness, a reliable prediction of miRNA targets is indispensable. Indeed, while the total number of targeted genes is estimated to be one third of the whole human gene complement, 10 000 genes (Lewis *et al.*, 2005), the current number of experimentally validated targets, according to the Diana TarBase (Sethupathy *et al.*, 2006; see also Chapter 13 in this book), is 55 (at the time of writing). A number of prediction methods have contributed considerably to the generation of interesting hypotheses about possible animal miRNA/target relationships (John *et al.*, 2004; Kiriakidou *et al.*, 2004; Rehmsmeier *et al.*, 2004; Brennecke *et al.*, 2005; Lall *et al.*, 2006).

RNAhybrid is a method that offers a database of target predictions, a download version, and an easy-to-use web-interface for online target predictions. At the same time, RNAhybrid allows the user broad control of the search. One area of control is the structural requirements of a miRNA/target interaction. One such structural requirement is that the "seed" region of the miRNA, nucleotides 2 to 7 or 8, forms a perfect Watson–Crick hybridization with its target (Lewis *et al.*, 2005). While it is possible that the majority of functional miRNA/target interactions indeed fulfil this requirement, exceptions are entirely possible. For example, experimentally validated targets of the *let-7* and *lin-4* miRNAs in *Caenorhabditis elegans* appear to have unpaired nucleotides either in the seed region or the matching target part (Grosshans and Slack, 2002). Brennecke *et al.* (2005) have shown that much shorter seed regions of four nucleotides can be functional as

MicroRNAs: From Basic Science to Disease Biology, ed. Krishnarao Appasani. Published by Cambridge University Press. © Cambridge University Press 2008.

long as the miRNA 3′-part forms a compensatory hybridization. Thus by restricting predicted miRNA target interactions to those that contain seed matches, one might easily miss non-seed sites. For plants, the situation is altogether different, since it is generally assumed that only near-perfect hybridizations along the entire miRNA are functional. This heterogeneity of structural requirements is encompassed by RNAhybrid, and the user can, according to his or her domain of application, employ any of these.

The other area of control the user has over the search is an estimate of statistical significance. It is not sufficient to rank predictions merely by score, since scores lack the statistical interpretation that is necessary to express the reliability of the predictions. In the area of protein database searching, established methods like BLAST (Altschul *et al.*, 1990) annotate their search results with p- or E-values which allow for an estimate of the false discovery rate, the expected number of false positive predictions among all positive, i.e. reported, predictions. Such a statistical guidance is especially important in large-scale predictions of miRNA targets, where deceptively good-looking target sites can frequently occur by chance. The prediction of miRNA targets is a step-wise process, from the identification of individual binding sites, over multiple binding sites of one miRNA in the same target candidate, to the comparative analysis of orthologous targets over several species, and each of these steps has its own statistical intricacies.

The remainder of this chapter is divided into a review of the RNAhybrid method and the presentation of a number of use-cases of different complexity.

Methodology

Availability of RNAhybrid
RNAhybrid can be used in three ways: first, with a web-interface that allows the online prediction of miRNA targets; second, as a locally installed download version; and third, as a database of predicted miRNA targets, RNAhybridDB. The download version can be used as a command-line tool or from a graphical user interface (GUI). All versions of RNAhybrid can be accessed at http://bibiserv.techfak.uni-bielefeld.de/rnahybrid.

Brief description of RNAhybrid
The following paragraphs give a brief review of the RNAhybrid approach. For a more detailed description, see Rehmsmeier *et al.* (2004). Figure 14.1 shows a visualization of RNAhybrid's workflow.

Energy minimization
The algorithmic core of RNAhybrid finds energetically optimal binding sites of a short RNA (usually a miRNA) in a target candidate sequence. It can be seen as an extension of the classical algorithm for RNA secondary structure prediction (Zuker and Stiegler, 1981). However, while algorithmically, secondary structure prediction precedes inside-out, RNAhybrid proceeds from

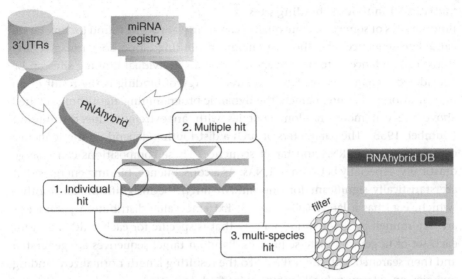

Figure 14.1. RNAhybrid workflow. RNAhybrid takes candidate target sequences, for example 3′ UTRs, and a set of miRNAs, and looks for energetically favourable binding sites (triangles). Statistical significance is evaluated in a step-wise fashion, from individual hits over multiple hits to multi-species hits. Results are filtered according to p-value thresholds and stored in a database of miRNA target predictions, RNAhybridDB. This database can be queried from a web interface.

left to right, zipping together target candidate and miRNA, allowing for some teeth of the zip not to interlock. While this resembles pair wise sequence alignment algorithms, the building blocks of alignments are different to those of hybridizations. The former are matches, mismatches, insertions, and deletions; the latter are stacked base pairs, bulges, internal loops, and end-loops. Thus (apart from what is common to all dynamic programming optimization algorithms), RNA secondary structure prediction, pair wise sequence alignment, and RNA hybridization as in RNAhybrid are rather different. The advantages of RNAhybrid as a dedicated miRNA target prediction algorithm are, first, effectiveness, in that the algorithm directly calculates the desired result without the necessity to adapt existing algorithms, and, second, speed, in that undesired calculations such as the consideration of branching structures are not performed.

Length normalization

The fact that miRNAs are short and made up of only four nucleotides frequently gives rise to the prediction of energetically favorable but random and probably non-functional hybridizations. The longer a target candidate sequence, the higher the probability of such chance occurrences of binding sites. While a predicted target site in a short candidate sequence might be convincing, it can be statistically insignificant if predicted in a long candidate sequence. RNAhybrid eliminates the dependence of binding energy on target length by normalizing binding energies accordingly.

Statistics of individual binding sites

Binding sites of some moderate quality can nearly always be found for any target candidate sequence, and the free energy of binding alone does not give any statistical guidance as to how likely it is that an individual binding site can be considered a chance event. Since the free energy of binding is the result of an optimization procedure, namely the dynamic programming algorithm sketched above, we can model random energies with an extreme value distribution (Gumbel, 1958). The parameters of such a distribution depend on the sequence compositions of miRNA and target sequence. These compositions can change drastically, especially between miRNAs. As a consequence, binding energies that are statistically significant for one miRNA might be insignificant for another (which, e.g., has a larger GC-content). Extreme value distribution parameters are determined by RNAhybrid in a way that is specific for each miRNA and for each set of target candidate sequences. Random target sequences are generated and then searched with the miRNA. To the resulting length normalized binding energies, an extreme value distribution is fitted. The fitted parameters can then be used in the search of the actual database of target candidates to calculate for each observed binding site its p-value, i.e. the probability that a random binding site would have been of the same or a better quality.

Statistics of multiple binding sites

Multiple binding sites of one miRNA in the same target candidate sequence typically increase the confidence one has in believing that the target is indeed regulated by the miRNA. Not only is this a biological consideration, but it is also a statistical one, since multiple independent chance occurrences are less likely than individual ones. RNAhybrid takes this into account by calculating p-values for multiple binding sites, based on the individual p-values. Larger numbers of binding sites are assigned better, i.e. smaller, p-values, thereby increasing the likelihood that the detected binding sites are biologically functional.

Comparative analysis of orthologous targets

Further evidence for functional binding sites comes from their evolutionary conservation across orthologous targets from multiple species. Although such conservation is not a necessity for a binding site to be functional, it gives additional support if present. One has to be careful, however, as to what extent binding site conservation is the result of a stabilizing selection of the binding site or merely an artifact of an overall sequence conversation. If two sequences from different species are very similar owing to a short evolutionary time scale, the presence of binding sites in both sequences is hardly more suprising than the presence in one. The extent of this surprise, which again is nothing else than statistical significance, is adjusted by RNAhybrid to the overall sequence conservation. This sequence conservation is measured as the effective number of targets. When the targets in hand are very similar, then their effective number is small. When they are diverse, sharing conserved binding sites only, their effective number is close to

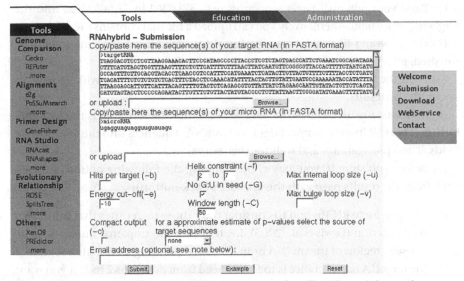

Figure 14.2. Screenshot of the RNAhybrid online web interface. The left panel shows other tools available on the Bielefeld Bioinformatics Server.

their actual number. Neglecting the statistical dependence between orthologous sequences could easily result in a huge over-estimate of significance and thus in a large number of false-positive predictions.

Use cases

The following sections describe use cases of increasing complexity. Each use case comprises a step-by-step protocol.

Target prediction on a per-gene basis

Frequently, the aspiring user of RNAhybrid will have in hand only a small number of genes he or she is interested in. The question will be whether these genes are possibly regulated by miRNAs. The easiest way to quickly check for potential, good binding sites is via the RNAhybrid web-interface for online target prediction.

(1) Go to the RNAhybrid submission page at http://bibiserv.techfak.uni-bielefeld.de/rnahybrid/submission.html. See Figure 14.2 for an example screenshot of the web interface.

(2) Paste the miRNA sequences in Fasta format into the miRNA sequence field. Alternatively you can upload a file that contains the sequences. In Fasta format, each sequence entry consists of a line beginning with the symbol " > " (the description line) and followed by the sequence's identifier and description, and one or more lines with the sequence itself. A Fasta file can contain multiple such sequence entries, in which case it is also called a multiple Fasta file. The sequence entries in such a file just follow each other, each one starting with a new line. For RNAhybrid, each entry has to have a unique identifier, and there must be no blanks between the " > " and the identifier.

(3) Paste your potential target sequences (e.g. 3′ UTRs) in Fasta format into the target sequence field. Alternatively you can upload a file that contains the sequences.

(4) If you would like to get an email notification once the calculations have finished, give your email address in the respective field.

(5) Press the Submit button. After a short while the web server will get back to you with the results.

(6) If you are unsure about what data to paste into the form, press the Example button. This will insert example target and miRNA sequences into the appropriate fields. Then press Submit and wait for the results.

(7) Before you submit, you may choose among the following options (just go back from the results page with the browser's back-button):

(I) You can force RNAhybrid to consider only those target sites that fulfill the seed criterion (Lewis *et al.*, 2005), i.e. that have no unpaired nucleotides in the seed region of the miRNA or in the corresponding seed-match region of the target. A useful choice is to force a seed from positions 2 to 7. If you want to be more restrictive, you can disallow any non-Watson–Crick base pairing (G:U pairs) in the seed. These options will reduce the number of spurious target sites, though at the risk of missing *bona fide* sites that do not have a seed. For plant miRNA target prediction the seed region might be set from 8 to 12, so that the expected cleavage site is covered.

(II) For the prediction of plant miRNA targets, where only few nucleotides in the miRNA are unpaired, the options for restricting loop sizes are especially useful. A useful choice might be one unpaired nucleotide on each side of an internal loop and one unpaired nucleotide in a bulge loop. Internal loops are regions in a hybridization duplex that consist of two base pairs that enclose unpaired nucleotides on both sides, i.e. in the target and in the miRNA. Bulge loops are similar, but have unpaired nucleotides on only one side.

(III) If you have chosen to restrict your search to target sites that contain a seed match, you can speed up the calculation by using the seed-match speed-up option. With that, RNAhybrid first searches for seed-matches, and only upon the detection of such a match performs the complete calculation of optimal hybridization in a window around the seed match. A window size of 50 nt is suggested. For seeds of length 6, with no G:U pairs allowed, this option accelerates the search by a factor of 8.

(IV) For statistical guidance, i.e. p-value calculation, choose the "set" option. In this option you can tell RNAhybrid from which organism your target candidate sequences are. This information is used by RNAhybrid to heuristically calculate p-values on the basis of the miRNA's "energy potential." The larger this potential, the better chance hybridizations will be, and p-values for actual binding energies are adjusted accordingly.

(8) You may also want to install RNAhybrid locally on your computer system. Especially, if you would like to get p-values while using any of the options (I) to (III), since the heuristic p-value calculation assumes that none of these options is chosen. Follow the "download" link on the RNAhybrid welcome page

http://bibiserv.techfak.uni-bielefeld.de/rnahybrid. The local RNAhybrid version can either be used as a command-line tool or in combination with a graphical user interface (GUI). To run the GUI, a java runtime environment has to be installed on your computer (see http://java.sun.com). For options, see the manual page and the following two use-cases.

Target prediction on a per-miRNA basis

In case you have a few new miRNAs, maybe from a previous miRNA prediction or cloning strategy, these miRNAs generally might target any genes from the organism they come from. Since this usually involves one to two orders of magnitude more comparisons than in the "per-gene basis" use-case described above, statistical guidance for the interpretation of results is necessary. The following steps (2) to (8) describe the use of the command-line version of RNAhybrid. All options in the GUI are marked with the corresponding command-line parameter letters.

(1) For "seed-free" analyses follow the steps from the previous use-case, including the p-value heuristics option (7)(IV) above. Note that the total amount of data that can be uploaded is limited.

(2) For structurally constrained analyses (including any kind of seed or restricted loop sizes), download RNAhybrid and install it on your local computer system (see (8) in the previous use-case).

(3) Save your miRNAs in Fasta format (see previous use-case). Let us assume that the file is called mirnas.fasta.

(4) Save the target candidate sequences in Fasta format. Such candidate sequences can, e.g., be downloaded from the ensembl database at http://www.ensembl.org. Either get a set of genes you are interested in or the whole set of genes from one organism. The easiest way to do this is to use ensembl BioMart; follow the respective link on the ensembl start page. Let us assume that your potential targets are saved in a file called targets.fasta.

(5) Start RNAhybrid with the following options:

RNAhybrid -F -b 5 -r -o 1 -p 0.01 -t targets.fasta -q mirnas.fasta

For every miRNA/target combination, RNAhybrid will look for up to 5 binding sites (-b 5), calculate a multiple-hit p-value (-r, see also "Statistics of muliple binding sites" above), sort the output by p-value (-o 1), and report only hits with p-values not larger than 0.01 (-p 0.01). P-values are calculated from parameters of extreme value distributions, which in turn are estimated on the results of the database search (induced by option -F).

(6) With option -F, RNAhybrid calculates p-values based on the results of the actual database search. This, however, only gives accurate estimates of statistical significance if the database contains at least a couple of hundred sequences. If this is not the case, you can tell RNAhybrid to generate random sequences and to calculate p-values on the basis of these, e.g.:

RNAhybrid -S 1000 -b 5 -r -o 1 -p 0.01 -t targets.fasta -q mirnas.fasta

Given the option -S 1000, RNAhybrid generates 1000 random sequences for each miRNA. The calibration is performed in the beginning and protocolled in the output.

(7) You can also use the p-value heuristic described above, but do not do this in combination with any structural constraint (seed or loop size restrictions). The corresponding option is -s, e.g.

RNAhybrid -s 3utr_human -b 5 -r -o 1 -p 0.01 -t targets.fasta -q mirnas.fasta

See also the RNAhybrid help (RNAhybrid -h) or the RNAhybrid manual for further arguments to the -s option.

(8) If you do not use the heuristics option -s, you can give RNAhybrid any of the structural options as described in steps (1)–(3) in the previous use-case.

Genome-wide multi-species target prediction

The most exciting, and at the same time computationally and statistically most demanding use-case, is the genome-wide target prediction across multiple species. The conservation of a miRNA regulatory network will have resulted in the conservation of miRNA/target relationships. In addition, chance occurrences of good binding sites in several orthologous transcripts are less likely than in a single transcript. Thus, the occurrence of binding sites in orthologous genes across species increases the likelihood that the predicted miRNA/target relationship is real, and with the likelihood our confidence in the prediction. However, as pointed out in the brief review of RNAhybrid, the overall similarities between orthologous transcripts have to be taken into account so that the statistical surprise of a conserved miRNA/target relationship is not over-estimated. For the following let us assume that we would like to predict targets for human and mouse miRNAs.

(1) Save your miRNA sequences in Fasta format (as described above). All human miRNAs go to one file, all mouse sequences to another. Let us assume that these files are called hsa-miRNA.fasta and mmu-miRNA.fasta, respectively. miRNAs have to be uniquely annotated according to their orthologies. miRNAs are usually called hsa-miR-18 and mmu-miR-18 or similar. Later you will have to tell RNAhybrid how to recognize orthologous miRNAs, and this could be done by focusing on the non-species part of the miRNA name, miR-18 in this example. Human and mouse miRNAs that are identical in this part are considered orthologous.

(2) Save the candidate target sequences in Fasta format. Again, these sequences might come from the ensembl database (see above). The target sequences also have to be annotated with respect to their orthologies. One way could be to number the genes like this:

>100_my_human_gene
CTCACCATGGATGATGATACTGCCGTGCTCGTCATTGACAACGGCTCTGG
CATGTGCAAGGCCGGCTTTGCAGGTGACGA . . .

and

>100_my_mouse_gene
AGGAAAGAAGACTCTCAGTTGTACAGTGCTGGAAGCACACACGCTTCTCT
GCACCCGACTGTGTCGTAGAGCTTGATTTC . . .

That these two transcripts are orthologous can be told from the fact that they are both annotated with the number 100. Let us assume that the target candidate sequences are saved in files called human_3utr.fasta and mouse_3utr.fasta, respectively.

(3) Start RNAhybrid as follows:

RNAhybrid -F -b 5 -o 1 -p 0.01 -t human_3utr.fasta,mouse_3utr.fasta

-q hsa-miRNAs.fasta,mmu-miRNAs.fasta -a '[0–9][0–9]*_','-.*'

(without line-break). Be careful that the comma-separated file lists do not contain any blank characters. Option -a gives the target and miRNA keys. These keys tell RNAhybrid how to recognize orthologous miRNAs or targets, respectively. These keys are regular expressions. The target key '[0–9][0–9]*_' means "a digit, followed by an arbitrary number (possibly zero) of digits, followed by an underscore." The miRNA key '-.*' means "a dash followed by an arbitrary (possibly zero) number of characters of any kind (including digits and other symbols)." RNAhybrid analyzes all human miRNAs against all human 3' UTRs, all mouse miRNAs against all mouse 3' UTRs, and combines the results on the basis of the orthology information with which the sequences are annotated.

(4) If you have just one miRNA set with which you would like to search multiple genomes, you can give just that (however, you also have to provide a miRNA key):

RNAhybrid -F -b 5 -o 1 -p 0.01 -t human_3utr.fasta,mouse_3utr.fasta

-q miRNA.fasta -a '[0–9][0–9]*_','-.*'

This takes the miRNAs from miRNA.fasta and with them analyzes both human and mouse 3' UTR sequences.

(5) It is recommended not to neglect the statistical dependence of orthologous target sequences, especially for closely related species. RNAhybrid accounts for this dependence by calculating, for each group of orthologous target candidate sequences, the effective number of these sequences (see RNAhybrid review above). This is done with the -k option:

RNAhybrid -k 1000 -F -b 5 -o 1 -p 0.01 -t human_3utr.fasta, mouse_3utr.fasta

-q hsa-miRNA.fasta,mmu-miRNA.fasta -a '[0–9][0–9]*_','-.*'

With -k 1000, the miRNA under consideration is shuffled 1000 times for each considered set of orthologous target sequences. With these shuffles, the target sequences are probed for their statistical dependence. Larger values for -k might give more accurate results, but will also be computationally more demanding.

Database of RNAhybrid target predictions

At the time this chapter is being written, a database interface for pre-calculated predictions of miRNA targets is being developed. This database, RNAhybridDB, will be accessible from the RNAhybrid start page. The user can choose the homology circle he or she is interested in, e.g. human/mouse/rat/dog, the kind of structural requirement he or she is interested in, e.g. a seed with no G:U basepairs allowed, and select or type in miRNA and gene names. Available predictions for these combinations will then be retrieved and displayed.

Selection of candidates

The results of the core task of RNAhybrid, the identification of energetically favorable binding sites for given miRNAs and target candidate sequences, do not themselves guide the user as to which predictions can be considered reliable. While in a small-scale analysis, such as target prediction on a per-gene basis, this reliability might be judged on by the user on the basis of binding energies and "believable" shapes of the predicted hybridization, such a judgment is impossible for large-scale analyses. The reason for this is not only one of manageability, but foremost a statistical one. The infinite-monkey theorem says that a monkey at a typewriter, given enough time, will eventually type one of Shakespeare's plays. Given less time, it might type out a shopping list. In the theorem, the typing monkey embodies a random process. For miRNA target prediction, the process is the analysis of potential targets for good-looking binding sites. The more data is processed, the more likely it is that good-looking binding sites will occur merely by chance. In the monkey picture, after a medium amount of time one would not expect the monkey to have written more than a shopping list, certainly not one of Shakespeare's plays. After a very long time, the appearance of one of Shakespeare's plays would not be surprising and thus statistically insignificant. Whenever we come across a shopping list in daily life, however, we can have strong confidence that this list is not the work of a randomly typing monkey. The credibility of a prediction is a matter of search-space. RNAhybrid outputs p-values which always refer to individual miRNA/target combinations. An easier interpretation is given by E-values, which are expected numbers of false-positive predictions of a certain quality. To get E-values from RNAhybrid p-values, multiply the p-values with the number of genes and the number of miRNAs. If the miRNAs fall into a smaller number of families that consist of similar sequences, take the number of families instead the number of miRNAs. This calculation allows the user to control the overall number of false-positive predictions. For example, in a search of 10 000 3' UTR sequences with 100 miRNAs, setting the miRNA/target p-value threshold to 1.0e-5 will result in an expected 10 false positive predictions (1.0e-5 times 10 000 times 100). This 10 is the E-value of the entire search.

Concluding remarks

RNAhybrid is a versatile tool for the prediction of miRNA targets. It can be used in a number of different situations, such as small-scale analyses on a per-gene basis or genome-wide multi-species target predictions. RNAhybrid can be used online or installed and run locally. A database of pre-calculated target predictions, RNAhybridDB, will complement the existing applications. RNAhybrid's major strength is the breadth of control the user has over the program's behavior, including a wide selection of structural requirement options and the calculation of p-values as a statistical guidance to the selection of confident predictions.

ACKNOWLEDGEMENTS

The author thanks Leonie Ringrose for her advice on the language of this chapter and for providing the idea of the shopping list.

REFERENCES

Altschul, S. F., Gish, W., Miller, W., Myers, E. W. and Lipman, D. J. (1990). Basic local alignment search tool. *Journal of Molecular Biology*, **215**, 403–410.

Brennecke, J., Stark, A., Russell, R. B. and Cohen, S. M. (2005). Principles of microRNA–target recognition. *Public Library of Science Biology*, **3**, e85.

Grosshans, H. and Slack, F. J. (2002). Micro-RNAs: small is plentiful. *Journal of Cell Biology*, **156**, 17–21.

Gumbel, E. J. (1958). *Statistics of Extremes*. New York: Columbia University Press.

John, B., Enright, A. J., Aravin, A. *et al.* (2004). Human microRNA targets. *Public Library of Science Biology*, **2**, e363.

Kiriakidou, M., Nelson, P. T., Kouranov, A. *et al.* (2004). A combined computational-experimental approach predicts human microRNA targets. *Genes & Development*, **18**, 1165–1178.

Lall, S., Grün, D., Krek, A. *et al.* (2006). A genome-wide map of conserved microRNA targets in *C. elegans*. *Current Biology*. (In press.)

Lewis, B. P., Burge, C. B. and Bartel, D. P. (2005). Conserved seed pairing, often flanked by adenosines, indicates that thousands of human genes are microRNA targets. *Cell*, **120**, 15–20.

Rehmsmeier, M., Steffen, P., Höchsmann, M. and Giegerich, R. (2004). Fast and effective prediction of microRNA/target duplexes. *RNA*, **10**, 1507–1517.

Sethupathy, P., Corda, B. and Hatzigeorgiou, A. G. (2006). TarBase: a comprehensive database of experimentally supported animal microRNA targets. *RNA*, **12**, 192–197.

Zuker, M. and Stiegler, P. (1981). Optimal computer folding of large RNA sequences using thermodynamics and auxiliary information. *Nucleic Acids Research*, **9**, 133–148.

15 Machine learning predicts microRNA target sites

Pål Sætrom* and Ola Snøve Jr.

Introduction

Ceanorhabditis elegans' *lin-4* and *let-7* were discovered seven years apart in the 1990s (Lee *et al.*, 1993; Wightman *et al.*, 1993; Reinhart *et al.*, 2000). Because of their importance for correct timing in post-embryonic larval development in worms, these non-protein-coding molecules were first referred to as small temporal RNAs, but following the discovery of numerous RNAs with similar characteristics (Lau *et al.*, 2001; Lee and Ambros, 2001; Lagos-Quintana *et al.*, 2001), *lin-4* and *let-7* have been recognized as the founding members of the microRNA (miRNA) family. The details of the discovery of miRNAs and their involvement in various pathways are described in earlier chapters in Part I of this book.

To enable functional inference, it is important to identify the miRNA targets, and many efforts have been made to solve this problem in the past few years. MicroRNAs are known to regulate gene expression on two levels, namely by translational suppression (Olsen and Ambros, 1999) and mRNA depletion (Yekta *et al.*, 2004). But despite massive resources invested in this problem, we have yet to find more than one human miRNA with assigned target and function (Hornstein *et al.*, 2005). Massive evidence does, however, support the crucial role of miRNAs in accurate and timely regulation of messages, which means that we have to develop new approaches to identify their genetic targets.

Current microRNA target prediction algorithms are filters that start with a relatively coarse seed step that identifies candidates based on sequence similarity between the miRNA and the potential target site. These candidate sites must subsequently pass additional steps before being classified as miRNA targets. For instance, a miRNA target with functional significance would be expected to be evolutionary conserved between species, and consequently, you may filter out any candidates that lack a predefined conservation in a set of species.

---

* Author to whom correspondence should be addressed.

MicroRNAs: From Basic Science to Disease Biology, ed. Krishnarao Appasani. Published by Cambridge University Press. © Cambridge University Press 2008.

We have argued that target prediction algorithms should be compared after the seeding step, as subsequent filtering steps, such as target site conservation, will be common to the different seeding algorithms. Thus, the quality of the initial seeding step will also determine the quality of the final results (Sætrom *et al.*, 2005a). Furthermore, we have shown that machine learning can be used for complex classification problems with limited validated data available, such as short interfering RNA efficacy predictions (Sætrom, 2004; Sætrom and Snøve Jr., 2004) and non-protein-coding RNA gene finding (Sætrom *et al.*, 2005b).

This chapter will present our machine learning method for miRNA target prediction (Sætrom *et al.*, 2005a). We show that our method, called TargetBoost, gives reliable predictions of miRNA target sites; in fact, we show that TargetBoost compares favorably with two methods whose seed step is based on sequence properties and thermodynamic stability calculations. We then illustrate and discuss the effect of requiring that target sites should be conserved in related species. Finally, we summarize and outline some future challenges of miRNA target prediction.

Methodology

The fundamental problem of predicting miRNA target sites is to automatically determine the target sites of a given miRNA. To solve this problem, one has to create a function or algorithm that reliably predicts whether or not a given miRNA targets a given RNA sequence. Our approach is to use machine learning to create a target site predictor that describes the common characteristics of a set of experimentally verified target sites. The following sections describe our machine learning approach, the methods we use to analyze its predictive accuracy, and the dataset we used to train our predictor.

Motifs describe microRNA target sites

The basic assumption underlying our method is that certain nucleotide positions in a miRNA are more likely to form base pairs with the miRNA's target sites. In other words, we assume that if the nucleotides at specific positions in the miRNA base-pair with a site in mRNA, the miRNA will target this site. Several experimental studies support these assumptions; for example, the first eight nucleotides in the 5' end of the miRNA are particularly important for target site recognition (Kiriakidou *et al.*, 2004; Brennecke *et al.*, 2005; Doench and Sharp, 2004; see also Chapter 13 in this book).

Such base-pairing requirements for target-site recognition can be expressed as sequence motifs. To illustrate this, consider Figure 15.1, which shows the *C. elegans* *lin-4* miRNA and one of its target sites in *lin-14*. The binding between the 5' end of *lin-4* and this target site can be expressed as the sequence motif CUCAGGGA, which represents the nucleotides in mRNA paired by the eight nucleotides in the 5' end of the miRNA.

Although this motif can be used to search for *lin-4*'s target sites in other genes, it has several shortcomings. First, all nucleotides in the motif must be matched for it

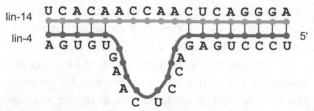

Figure 15.1. One of *lin-4*'s target sites in *lin-14*.

to recognize a target site, but some of *lin-4*'s target sites do not require a complete match of all eight nucleotides. Second, even though all eight nucleotides contribute to target site recognition, some of the positions may be more important than others (Lewis *et al.*, 2005). Third, the motif can only be used to find *lin-4*'s target sites, as it contains the nucleotides specific for *lin-4*'s 5' end. This is a major weakness, as we want the motifs to model the general properties of miRNA target sites and be applicable for all miRNAs.

To solve this major shortcoming, instead of using the miRNA's nucleotides in the motifs, we use the nucleotides' positions. Let the complement of the nucleotide at each position in the miRNA be represented by the special characters P_1, P_2, P_3, and so on, where the subscript indicates the nucleotide's position from the miRNA's 5' end. The above motif for the *lin-4/lin-14* target site can then be rewritten as $P_8P_7P_6P_5P_4P_3P_2P_1$. This motif can now be used to find target sites for any miRNA simply by replacing all the special characters with the appropriate nucleotide.

We can then extend this basic motif to allow incomplete matches between the nucleotides. One approach is to allow a certain number of mismatches in an ungapped alignment of the motif and potential target sites. Using the same example as before, we may write this as $\{P_8P_7P_6P_5P_4P_3P_2P_1 : p \geq x\}$, which translates to a motif where at least x of the eight nucleotides should match in an ungapped alignment.

Genetic programming creates target site motifs

Assuming that the motif-model outlined in the previous section can describe target sites in general, we now need a method that automatically creates such target site motifs from a training set. Genetic programming is one method that can do this. Genetic programming is a member of the family of optimization methods called evolutionary computation, which uses simulated evolution in a population of candidate solutions to try to find the optimal solution to a particular problem (Foster, 2001). In genetic programming, the candidate solutions are symbolic expressions (Koza, 1992). Here, the candidate solutions will be target site motifs; that is, we will use genetic programming to find motifs that can discriminate true target sites from random sequences.

Note that in our implementation, there are no prior constraints as to which nucleotide positions genetic programming should include in the motifs; that is, all nucleotide positions occurs with equal probability in the initial motifs. We do, however, expect that the final motifs will contain the most important nucleotide positions for target site recognition.

Boosting improves the predictive power of target site motifs

As outlined in the previous sections, given a training set of known target sites, genetic programming can automatically find target site motifs that describe the target sites' common properties. There are, however, two problems with this approach. First, the solutions produced by genetic programming may be suboptimal (see, for example, Sætrom, 2004). A single motif may, for example, not be able to completely capture all target site properties. Second, the motifs themselves can only give a *yes* or *no* answer as to whether a particular mRNA sequence is a potential target site. Lab experiments have, however, shown that the protein downregulation from different target sites varies, for example, with the number of base pairs in the 5′ region (Brennecke *et al.*, 2005). Thus, when creating a motif one is faced with the dilemma of whether the motif should be highly specific and only find the strongest and most likely target sites, or very sensitive and find all potential target sites at the expense of many false positive predictions. Instead, the target site model should give a more dynamic score of the site's strength or probability of being a true target site.

Both problems can, however, be solved. Boosting is a method that can combine single suboptimal solutions into a joint optimal model (Meir and Rätsch, 2003). This joint model is a weighted sum of the individual motifs' predictions, and gives a dynamic score in the range [− 1, 1]. To build the joint model, boosting iteratively trains the single solutions by assigning higher weights to difficult examples (Freund and Schapire, 1997). Our boosting algorithm combines boosting with genetic programming to create a weighted sequence motif model; Sætrom (2004) describes the algorithm in general and Sætrom *et al.* (2005a) describes how we use it to model miRNA target sites in particular.

ROC describes the overall performance of target site predictors

Most algorithms output a score that says something about how confident the user can be about a given prediction. Hence, it is often necessary to set a threshold that determines whether or not a given sequence is a miRNA target or not. You should select the performance tuning that best fits your needs: there is a huge difference between trying to figure out how many miRNA targets there are in the transcriptome, and determining which gene is the most likely target for your favorite miRNA.

Receiver operating characteristics (ROC) is a method that is used to compare the performance characteristics of different algorithms. It is a curve that plots an algorithm's ability to find true targets versus its ability to identify all targets for different score thresholds. At one end of the scale, you will have decreased the threshold to a point where the algorithm finds most true target sites, but in doing that, the algorithm has output a few extra false sites as well (sensitive predictions). At the other end of the scale, most of the targets you obtain will be true sites, but the algorithm may have left out many true sites as well (specific predictions).

The desired performance may be a compromise between the algorithm's sensitivity and specificity, and the ROC curve is therefore valuable since it contains information about an algorithm's performance across the whole spectrum of thresholds. Also, the area under the curve, which is referred to as the ROC score, shows the stability of the algorithm since a higher score means higher average performance for different thresholds.

If you are looking for all targets, you should therefore choose an algorithm that performs well in the high sensitivity area, whereas you would prefer algorithms with high performance in the high specificity area if you want to find targets for particular miRNAs. A practical application of the ROC curve in the latter case is to measure how many true targets an algorithm is able to identify before a predefined number of false predictions occurs. That is, you want to make sure that most true targets will be included before false predictions clutter your list.

Positive and negative training sets

To train and test the classifiers, we used a set of 36 experimentally verified target sites as positive data and a larger set of random sequences as negative data. The positive data were the target sites for the miRNAs *let-7*, *lin-4*, *miR-13a*, and *bantam* in *C. elegans* and *D. melanogaster* (Boutla *et al.*, 2003; Brennecke *et al.*, 2003; Rajewsky and Socci, 2004). We padded each site with its surrounding sequence, such that each positive site was 30 nt long; longer sites were discarded. The negative data were 3000 random 30 nt strings with the same nucleotide distribution as *D. melanogaster*'s 3' UTRs (Rajewsky and Socci, 2004; see also Chapter 12 in this book).

Results and discussion

Near-perfect binding in the miRNA 5' end is the most important target site motif

Any manual study of a few known target sites will show that miRNAs tend to have near-perfect complementarity between their 5' ends and their targets (Lee *et al.*, 1993). In an effort to find the most conserved motifs in the 3' UTRs of various species, Xie *et al.* (2005) reported that several octamers were significantly more conserved than would be expected by random. Lewis *et al.* (2005) found that conserved motifs in 3' UTRs are complementary to miRNAs' 5' ends. Specifically, they argue that hexamers are highly conserved, but admit that octamers give fewer false positives. This was later confirmed by Xie *et al.* (2005) when they searched for highly conserved motifs in the 3' UTRs of various species.

Our machine learning method confirms this tendency for near-perfect binding between the miRNAs' 5' ends and their targets (Sætrom *et al.*, 2005a). Several independently repeated analyses of the dataset produced the same target site motif in the first boosting iteration (Figure 15.2). This motif states that six of the eight bases at the miRNA's 5' end have to match for a sequence to be recognized as a potential target site. This indicates that most target sites in the training set demand a near-perfect match in the 5' end of the miRNA; in fact, the motif identified 34 of the 36 sites in the training set (data not shown). Thus, this motif may represent a minimal condition for miRNA target site recognition. As a comparison, the seed-motif, which requires perfect complementarity of nucleotides

(a)

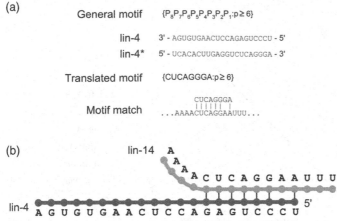

General motif $\{P_8 P_7 P_6 P_5 P_4 P_3 P_2 P_1 : p \geq 6\}$

lin-4   3' - AGUGUGAACUCCAGAGUCCCU - 5'

lin-4*   5' - UCACACUUGAGGUCUCAGGGA - 3'

Translated motif $\{CUCAGGGA : p \geq 6\}$

Motif match

```
          CUCAGGGA
          ||||||  |
...AAAACUCAGGAAUUU...
```

(b)

Figure 15.2. The motif for near-perfect binding in the miRNA 5′ end. Panel (a) shows the general motif, how the general motif translates into a specific motif for *lin-4*, and how the specific motif matches one of *lin-4*'s target sites in *lin-14*. Panel (b) shows the binding between *lin-4* and *lin-14*.

two through seven in the 5′ end (Lewis *et al.*, 2005), only identifies 19 of the 36 sites in the training set.

Weighted sequence motifs give good target site predictions

Most classification problems are sufficiently complex to prevent single character-istics from obtaining satisfactory performance. In our case, a predictor that includes only one motif is not likely to capture all miRNA target binding char-acteristics. We therefore use a technique called boosting, as described previously, to create a predictor that is a set of differentially weighted sequence motifs. We call this algorithm TargetBoost (Sætrom *et al.*, 2005a).

To get a good estimate of our algorithm's ability to predict new target sites – that is, target sites that were not used to train the predictor – we used two forms of cross-validation. The first, 10-fold cross-validation, usually gives a good estimate of a predictor's predictive accuracy (Stone, 1974; Kohavi, 1995). Here, it may be biased, however, as the number of verified target sites varied greatly for the four miRNAs, and the miRNA having the most target sites (*let-7*) would likely be present in both the training and test sets in many of the 10 folds. We therefore also used a second approach, called "leave-one-miRNA-out" cross-validation, which did not have this potential bias. In this approach, the training set consisted of all target sites from all miRNAs but one; the remaining miRNA's target sites were the test set. This gave four training and test sets.

Figure 15.3 compares TargetBoost's 10-fold and leave-one-miRNA-out cross-validation ROC curves to the ROC curve of a random predictor. Even though TargetBoost's performance varies on the different test sets, as indicated by the differ-ences in the ROC curves, the ROC curves have at least two things in common. First, the ROC curves show that TargetBoost's predictions are far from random. Second, TargetBoost makes very specific predictions and returns few false positive sites among the highest scoring true sites. To illustrate, when TargetBoost had identified 50% of the true sites, the relative number of false positive sites were 0.04%, 0.1%, 3%, 0.5%,

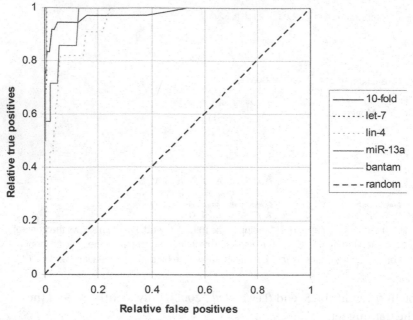

Figure 15.3. TargetBoost makes both good overall and highly specific target site predictions.

and 0.3% on the 10-fold, *let-7*, *lin-4*, *miR-13a*, and *bantam* test sets. What is more, on all test sets, the highest scoring target site was a true target site.

Weighted sequence motifs give more specific predictions than predictions based on sequence and thermodynamics alone

So how does our machine learning-generated weighted sequence motifs compare with other approaches? In Sætrom *et al.* (2005a), we suggested that filter approaches should be compared at the seed level, as any algorithm can add additional filters, such as target site conservation in related species, to improve its performance. As previously described, others have used sequence properties of the binding site or thermodynamic binding stability to calculate the score in the seed step. Rajewsky and Socci (2004) suggested an algorithm that used position-weighted sequence properties in its seed step, and Rehmsmeier *et al.* (2004) compute minimal free energy while forcing near-perfect complementarity in the miRNAs' 5′ ends. We compare our algorithm to these methods, called Nucleus and RNAhybrid, as they represent sequence properties and thermodynamic stability approaches, respectively.

Table 15.1 compares the overall performance of the three algorithms. The table shows that TargetBoost has the highest overall performance and gives the most stable predictions across all miRNAs; that is, both Nucleus and RNAhybrid have significantly lower performance than the best algorithm on at least one miRNA, but TargetBoost is either the highest performing algorithm or performs similar to the highest performing algorithm on all miRNAs.

Even though overall performance is important, when searching for target sites in the transcriptome, the negative sites will outnumber the true sites by far. We

Table 15.1. TargetBoost is the most stable algorithm: ROC scores that are not significantly different from the highest score on a particular miRNA are in boldface (90% confidence level)

Algorithm	let-7	lin-4	miR-13a	bantam	all
TargetBoost	**0.997**	**0.944**	**0.972**	**0.998**	0.979
RNAhybrid	**0.989**	0.931	**0.979**	**0.991**	0.967
Nucleus	0.988	**0.962**	0.928	**0.998**	0.973

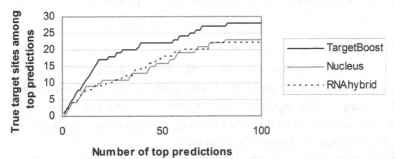

Figure 15.4. TargetBoost identifies more true target sites than do RNAhybrid and Nucleus. The figure shows the number and occurrence of true target sites among the top 100 predictions for TargetBoost, RNAhybrid, and Nucleus on the complete dataset.

therefore want a target site predictor to return as many positive sites as possible before the number of sites in the result set becomes too large. As Figure 15.4 shows, TargetBoost lists more true target sites among the top predictions on the complete dataset than do RNAhybrid and Nucleus. Statistical tests on the leave-one-miRNA-out cross-validation results confirmed these observations (Sætrom *et al.*, 2005a).

Target site conservation improves specificity

One of the most widely used filters is the target conservation requirement: functional target sites of miRNAs will tend to be conserved throughout evolution, so there is no doubt that this will improve the specificity of the algorithm (Lewis *et al.*, 2005). But if you want to find the most likely off-target sites of an siRNA, which is miRNAs' synthetic counterpart, you cannot rely on sequence conservation. This demonstrates why it is important to work on the seed step of algorithms to improve miRNA target prediction.

To illustrate the effect of adding a target conservation filter, we used TargetBoost and RNAhybrid to predict the target sites of 32 miRNAs in the 3′ UTRs of the human RefSeq genes. We used a cutoff of 0 on the TargetBoost predictions and a free energy cutoff of $-25 \, \Delta G°_{37}$ on the RNAhybrid predictions to get the initial list of predicted target sites. We then required that the target sites were conserved in both human, chimp, mouse, and rat. This conservation filter reduced the number of target sites predicted by TargetBoost from 37844 to 1466 and by RNAhybrid from 38681 to 1213. Thus, the filter removed about 96% of the initial predictions made by both algorithms.

Even though the conservation filter improves the specificity of the target site predictions, there are several drawbacks to relying on target site conservation. First, one will identify only the regulatory sites that are common between the different species. Thus, one risks missing the sites that make humans different from rodents, for example. Second, many miRNAs seem to be unique to humans (Bentwich *et al.*, 2005), and one can hardly expect the target sites of these miRNAs to be conserved in other species. Thus, if you are studying conserved regulatory targets of miRNAs, you can use the conservation filter to improve the specificity of the target site predictions. Otherwise, you cannot simply ignore unconserved target sites.

Concluding remarks

This chapter has (1) described the basics of our machine learning algorithm for predicting miRNA target sites, (2) shown that our algorithm has identified a basic motif for miRNA target site recognition, and (3) shown that our algorithm compares favorably to two other algorithms that use sequence properties and thermodynamic stability calculations to predict target sites.

In our opinion, miRNA target prediction is still in its infancy. Current approaches have focused mostly on the first question of how many targets there are or may be in various species (Lewis *et al.*, 2003; 2005), but few actual targets have been identified. Scientists working in bioinformatics should keep in mind that one important job of ours is to reduce the risk of lab experiments – that is, to suggest validation experiments that have the highest possible probability to succeed. We are therefore convinced that new approaches that can combine various data sources to improve the specificity of miRNA target prediction algorithms will become highly successful going forward.

Ultimately, miRNA target prediction is only one aspect of understanding microRNAs' role in gene regulation. To illustrate the complexity of this task, the *mir-17-92* polycistron was shown to have oncogenic potential (He *et al.*, 2005) about at the same time as it was reported that the same polycistron can function as a tumor suppressor (O'Donnell *et al.*, 2005; see also Chapter 22 in this book). The cancerous phenotype is therefore likely to be a consequence of an unstable regulatory circuit that depends on more than one interaction. In addition to the development of miRNA target prediction algorithms with higher specificity, we would also argue that development of new systems biology approaches that can handle the complexity of the regulatory mechanisms seems necessary.

REFERENCES

Bentwich, I., Avniel, A., Karov, Y. *et al.* (2005). Identification of hundreds of conserved and nonconserved human microRNAs. *Nature Genetics*, **37**, 766–770.
Boutla, A., Delidakis, C. and Tabler, M. (2003). Developmental defects by antisense-mediated inactivation of micro-RNAs 2 and 13 in *Drosophila* and the identification of putative target genes. *Nucleic Acids Research*, **31**, 4973–4980.
Brennecke, J., Hipfner, D. R., Stark, A., Russell, R. B. and Cohen, S. M. (2003). *bantam* encodes a developmentally regulated miRNA that controls cell proliferation and regulates the proapoptotic gene *hid* in *Drosophila*. *Cell*, **113**, 25–36.

Brennecke, J., Stark, A., Russell, R. B. and Cohen, S. M. (2005). Principles of microRNA-target recognition. *Public Library of Science Biology*, **3**, e85.

Doench, J. G. and Sharp, P. A. (2004). Specificity of microRNA target selection in translational repression. *Genes & Development*, **18**, 504–511.

Foster, J. A. (2001). Evolutionary computation. *Nature Reviews Genetics*, **2**, 428–436.

Freund, Y. and Schapire, R. E. (1997). A decision-theoretic generalization of on-line learning and an application to boosting. *Journal of Computer and System Sciences*, **55**, 119–139.

He, L., Thomson, M., Hemann, M. T. *et al.* (2005). A microRNA polycistron as a potential human oncogene. *Nature*, **435**, 828–833.

Hornstein, E., Mansfield, J., Yekta, S. *et al.* (2005). The microRNA miR-196 acts upstream of Hoxb8 and Shh in limb development. *Nature*, **438**, 671–674.

Kiriakidou, M., Nelson, P. T., Kouranov, A. *et al.* (2004). A combined computational-experimental approach predicts human microRNA targets. *Genes & Development*, **18**, 1165–1178.

Kohavi, R. (1995). A study of cross-validation and bootstrap for accuracy estimation and model selection. In *Proceedings of the 14th IJCAI*, pp. 1137–1143.

Koza, J. R. (1992). *Genetic Programming: On the Programming of Computers by Natural Selection*. Cambridge, Massachusetts: MIT Press.

Lagos-Quintana, M., Rauhut, R., Lendeckel, W. and Tuschl, T. (2001). Identification of novel genes coding for small expressed RNAs. *Science*, **294**, 853–858.

Lau, N. C., Lim, L. P., Weinstein, E. G. and Bartel, D. P. (2001). An abundant class of tiny RNAs with probable regulatory roles in *Caenorhabditis elegans*. *Science*, **294**, 858–862.

Lee, R. C. and Ambros, V. (2001). An extensive class of small RNAs in *Caenorhabditis elegans*. *Science*, **294**, 862–864.

Lee, R. C., Feinbaum, R. and Ambros, V. (1993). The *C. elegans* heterochronic gene *lin-4* encodes small RNAs with antisense complementarity to *lin-14*. *Cell*, **75**, 843–854.

Lewis, B. P., Hung, Shih, I., Jones-Rhoades, M. W., Bartel, D. P. and Burge, C. B. (2003). Prediction of mammalian microRNA targets. *Cell*, **115**, 787–798.

Lewis, B. P., Burge, C. B. and Bartel, D. P. (2005). Conserved seed pairing, often flanked by adenosines, indicates that thousands of human genes are microRNA targets. *Cell*, **120**, 15–20.

Meir, R. and Rätsch, G. (2003). An introduction to boosting and leveraging. In *Advanced Lectures on Machine Learning*, Mendelson, S. and Smola, A., eds., volume 2600, pp. 118–183. Springer-Verlag.

O'Donnell, K. A., Wentzel, E. A., Zeller, K. I., Dang, C. V. and Mendell, J. T. (2005). c-Myc-regulated microRNAs modulate E2F1 expression. *Nature*, **435**, 839–843.

Olsen, P. H. and Ambros, V. (1999). The *lin-4* regulatory RNA controls developmental timing in *Caenorhabditis elegans* by blocking LIN-14 protein synthesis after the initiation of translation. *Developmental Biology*, **216**, 671–680.

Rajewsky, N. and Socci, N. D. (2004). Computational identification of microRNA targets. *Developmental Biology*, **267**, 529–535.

Rehmsmeier, M., Steffen, P., Höchsmann, M. and Giegerich, R. (2004). Fast and effective prediction of microRNA/target duplexes. *RNA*, **10**, 1507–1517.

Reinhart, B. J., Slack, F. J., Basson, M. *et al.* (2000). The 21-nucleotide *let-7* RNA regulates developmental timing in *Caenorhabditis elegans*. *Nature*, **403**, 901–906.

Sætrom, P. (2004). Predicting the efficacy of short oligonucleotides in antisense and RNAi experiments with boosted genetic programming. *Bioinformatics*, **20**, 3055–3063.

Sætrom, P. and Snøve, Jr., O. (2004). A comparison of siRNA efficacy predictors. *Biochemical and Biophysical Research Communications*, **321**, 247–253.

Sætrom, O., Snøve, Jr., O. and Sætrom, P. (2005a). Weighted sequence motifs as an improved seeding step in microRNA target prediction algorithms. *RNA*, **11**, 995–1003.

Sætrom, P., Sneve, R., Kristiansen, K. I. *et al.* (2005b). Predicting non-coding RNA genes in *Escherichia coli* with boosted genetic programming. *Nucleic Acids Research*, **33**, 3263–3270.

Stone, M. (1974). Cross-validatory choice and assessment of statistical predictions. *Journal of the Royal Statistical Society, Series B (Methodological)*, **36**, 111–147.

Wightman, B., Ha, I. and Ruvkun, G. (1993). Posttranscriptional regulation of the heterochronic gene *lin-14* by *lin-4* mediates temporal pattern formation in *C. elegans*. *Cell*, **75**, 855–862.

Xie, X., Lu, J., Kulbokas, E. *et al.* (2005). Systematic discovery of regulatory motifs in human promoters and 3′ UTRs by comparison of several mammals. *Nature*, **434**, 338–345.

Yekta, S., Shih, I. and Bartel, D. P. (2004). MicroRNA-directed cleavage of *HOXB8* mRNA. *Science*, **304**, 594–596.

16 Models of microRNA–target coordination

Neil R. Smalheiser* and Vetle I. Torvik

Introduction

The number of microRNAs appears to be ever-growing, as more intensive sequencing of small RNAs reveals a large population of sequences that are expressed at low abundance, or in a tissue- or stage-specific manner. Computational studies have also predicted the existence of thousands of candidate microRNA precursor hairpin structures throughout mammalian genomes (reviewed in Bentwich, 2005). The number of predicted potential targets per microRNA is also steadily increasing, with the recognition that binding of a 7-mer seed at the 5'-end of a microRNA may be sufficient to regulate a target mRNA functionally (Doench and Sharp, 2004; Farh *et al.*, 2005; Lim *et al.*, 2005; Stark *et al.*, 2005; Sood *et al.*, 2006). But how are microRNAs and their targets coordinated – if at all?

A random model

One recent paper proposes that microRNAs arise whenever a RNA hairpin structure happens to be transcribed, that happens to be competent for processing by Drosha and Dicer (Svoboda and Cara, 2006). Most of these microRNAs will have no function at all, at least not initially: They will bind to a relatively large number of putative target regions at random (a 7-mer sequence will bind randomly every $4^7 = 16\,384$ bases on average), and those target regions that happen to be associated with a useful phenotypic response will tend to be retained over evolutionary time whereas those mRNAs that show deleterious responses will become relatively depleted in target sequences (Svoboda and Cara, 2006). Such a random model predicts that no coordination exists initially between microRNAs and their targets, but that coordinated sets of microRNAs and targets emerge over time via natural selection.

---

* Author to whom correspondence should be addressed.

MicroRNAs: From Basic Science to Disease Biology, ed. Krishnarao Appasani. Published by Cambridge University Press. © Cambridge University Press 2008.

Cases in which microRNAs and their targets have an intrinsic relationship

On the other hand, at least four situations have been described in which microRNA precursors do have a close intrinsic relationship with their target sequences. (a) In plants, several microRNA precursors arise from inverted duplicated repeat sequences, so that the microRNA naturally targets the same mRNA from whence it was derived (Allen *et al.*, 2004). (b) Several mammalian microRNAs are encoded on the opposite strand from their target mRNA, so that they comprise a perfect antisense match to the target (Smalheiser, 2003; Royo *et al.*, 2006). (c) Several human microRNAs have been proposed to arise from within processed pseudogenes, one of which arises from the antisense strand of a beta-5 tubulin gene and hence would be expected to target the normal cognate beta-5 tubulin mRNA (Devor, 2006). (d) A set of four human microRNA precursors were described that derived entirely from MIR/LINE2 genomic repeats (Smalheiser and Torvik, 2005). In these cases, two different segments of MIR repeats had become apposed in opposite orientations along a chromosome, forming a hairpin structure that was transcribed and processed as microRNAs (Figure 16.1). These microRNAs would be expected to target MIR/LINE2 repeats inserted into the 3'-UTR regions of mRNAs in the proper (opposite) orientation.

It is worth emphasizing that genomic repeats are well-represented among the sequences that are predicted to be good microRNA precursor hairpin candidates (Sewer *et al.*, 2005). Indeed, repeat sequences are routinely observed during small RNA cloning projects, but most, if not all, have been interpreted as rasiRNAs arising from dsRNA (Aravin *et al.*, 2003; Chen *et al.*, 2005). Any microRNAs derived from genomic repeats that form precursor hairpins would have been excluded from miRBase or other databases because they do not map unambiguously to one or a few distinct sites within the genome. Thus, it is possible that the contribution of genomic repeats to microRNA gene evolution has been underestimated. It is also conceivable that some "conventional" microRNA precursors have evolved originally from this pool of potential microRNA precursor hairpins formed by genomic repeats, but that progressive changes occurring over evolution have obscured their origins. As well, transposable elements can contribute promoter sequences that can affect transcription of nearby genes (see Nigumann *et al.*, 2002; Jordan *et al.*, 2003), possibly including microRNA precursors (Smalheiser, 2003; Borchert *et al.*, 2006). Thus, genomic repeats may influence the probability of a microRNA precursor hairpin being transcribed, even if the repeat sequences are not themselves incorporated within the microRNA precursor.

Cases in which microRNAs hit genomic repeats or regulatory elements within 3'-UTRs

Another way that microRNAs may interact with multiple mRNA targets in a coordinated fashion is by interacting with specific sequences within 3'-UTRs such as those that regulate mRNA transport, stability and polyadenylation.

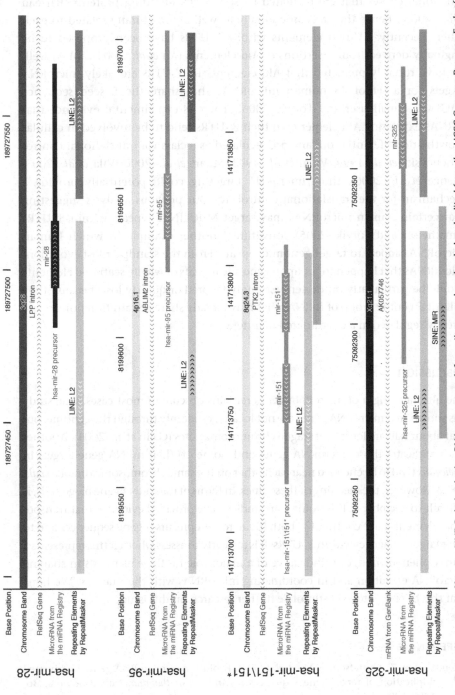

Figure 16.1. Genomic structure of human LINE-2 derived microRNA precursors. Information was downloaded and edited from the UCSC Genome Browser. Each of the precursors resides within a mRNA intron, and each flanks the junction of two L2 repeats in opposite orientation (darker shading indicates less divergence from the L2 consensus). (Reprinted from Smalheiser and Torvik, 2005, with permission.)

(a) For example, a recent report proposes that miR-15 binds to AU-rich (ARE) elements and thereby mediates functional inhibition of translation (Jing *et al.*, 2005). Messenger RNAs bearing ARE elements comprise one of several distinct functional classes that are regulated by sets of RNA-binding proteins (Barreau *et al.*, 2006); hence they are functionally as well as structurally related to each other. Recently, AU-rich elements within 3'-UTRs have been proposed to be originally derived from insertions of Alu elements (An *et al.*, 2004). (b) As well, we have recently predicted that Alu elements in 3'-UTRs are likely microRNA targets for a set of 28 human microRNAs that share the 5'-seed sequence AAGUGC (Smalheiser and Torvik, 2006). There is experimental evidence that mRNAs expressing Alu elements in their 3'-UTRs tend to be involved in cellular growth and differentiation, and are regulated as a class via translational control (Krichevsky *et al.*, 1999; Vidal *et al.*, 1993; Stuart *et al.*, 2000; Vila *et al.*, 2003; Spence *et al.*, 2006); thus, microRNA targeting could potentially provide a mechanism for the translational control. (c) Our previous analysis suggesting that certain human microRNAs may target MIR/LINE2 repeats within 3'-UTRs (Smalheiser and Torvik, 2005) constitutes another example in which human microRNAs appear to target genomic repeats. From the standpoint of evolution, microRNAs that happen to hit upon genomic repeats or widely scattered elements should be particularly prominent (strongly favored or quickly lost) because they will hit an entire group of mRNAs that may already share certain functions or are already regulated in a coordinated manner.

Discussion

The ultimate origin of microRNA genes remains obscure in most cases. One study has shown that microRNA genes are not located randomly through the genome, but that about half are located at fragile chromosomal sites (Calin *et al.*, 2004). Another study indicates those microRNA genes and (some of) their mRNA gene targets in *C. elegans* tend to be encoded near each other on the same chromosome (Inaoka *et al.*, 2006). However, the meaning of these clues, in terms of microRNA gene biogenesis, is difficult to decipher at this point. Our studies have provided evidence that transposable elements have contributed both to microRNA precursor gene sequences and to their target sequences within 3'-UTRs. It is too early to assess whether this represents a minor phenomenon, or is indicative of a major role for these elements in shaping microRNA evolution and in coordinating microRNAs with their targets. We hope that this brief chapter will stimulate further research on this issue.

REFERENCES

Allen, E., Xie, Z., Gustafson, A. M. *et al.* (2004). Evolution of microRNA genes by inverted duplication of target gene sequences in *Arabidopsis thaliana*. *Nature Genetics*, **36**, 1282–1290.

An, H. J., Lee, D., Lee, K. H. and Bhak, J. (2004). The association of Alu repeats with the generation of potential AU-rich elements (ARE) at 3' untranslated regions. *BMC Genomics*, **5**, 97.

Aravin, A. A., Lagos-Quintana, M., Yalcin, A. *et al.* (2003). The small RNA profile during *Drosophila melanogaster* development. *Developmental Cell*, **5**, 337–350.

Barreau, C., Paillard, L. and Osborne, H. B. (2006). AU-rich elements and associated factors: are there unifying principles? *Nucleic Acids Research*, **33**, 7138–7150.

Bentwich, I. (2005). Prediction and validation of microRNAs and their targets. *Federation of the European Biochemical Society Letters*, **579**, 5904–5910.

Borchert, G. M., Lanier, W. and Davidson, B. L. (2006). RNA polymerase III transcribes human microRNAs. *Nature Structural and Molecular Biology*, **13**, 1097–1101.

Calin, G. A., Sevignani, C., Dumitru, C. D. *et al.* (2004). Human microRNA genes are frequently located at fragile sites and genomic regions involved in cancers. *Proceedings of the National Academy of Sciences USA*, **101**, 2999–3004.

Chen, P. Y., Manninga, H., Slanchev, K. *et al.* (2005). The developmental miRNA profiles of zebrafish as determined by small RNA cloning. *Genes & Development*, **19**, 1288–1293.

Devor, E. J. (2006). Primate microRNAs miR-220 and miR-492 lie within processed pseudogenes. *Journal of Heredity*, **97**, 186–190.

Doench, J. G. and Sharp, P. A. (2004). Specificity of microRNA target selection in translational repression. *Genes & Development*, **18**, 504–511.

Farh, K. K., Grimson, A., Jan, C. *et al.* (2005). The widespread impact of mammalian microRNAs on mRNA repression and evolution. *Science*, **310**, 1817–1821.

Inaoka, H., Fukuoka, Y. and Kohane, I. S. (2006). Lower expression of genes near microRNA in *C. elegans* germline. *BMC Bioinformatics*, **7**, 112.

Jing, Q., Huang, S., Guth, S. *et al.* (2005). Involvement of microRNA in AU-rich element-mediated mRNA instability. *Cell*, **120**, 623–634.

Jordan, I. K., Rogozin, I. B., Glazko, G. V. and Koonin, E. V. (2003). Origin of a substantial fraction of human regulatory sequences from transposable elements. *Trends in Genetics*, **19**, 68–72.

Krichevsky, A. M., Metzer, E. and Rosen, H. (1999). Translational control of specific genes during differentiation of HL-60 cells. *Journal of Biological Chemistry*, **274**, 14295–14305.

Lim, L. P., Lau, N. C., Garrett-Engele, P. *et al.* (2005). Microarray analysis shows that some microRNAs downregulate large numbers of target mRNAs. *Nature*, **433**, 769–773.

Nigumann, P., Redik, K., Matlik, K. and Speek, M. (2002). Many human genes are transcribed from the antisense promoter of L1 retrotransposon. *Genomics*, **79**, 628–634.

Royo, H., Bortolin, M. L., Seitz, H. and Cavaille, J. (2006). Small non-coding RNAs and genomic imprinting. *Genome Research*, **113**, 99–108.

Sewer, A., Paul, N., Landgraf, P. *et al.* (2005). Identification of clustered microRNAs using an ab initio prediction method. *BMC Bioinformatics*, **6**, 267.

Smalheiser, N. R. (2003). EST analyses predict the existence of a population of chimeric microRNA precursor-mRNA transcripts expressed in normal human and mouse tissues. *Genome Biology*, **4**, 403.

Smalheiser, N. R. and Torvik, V. I. (2005). Mammalian microRNAs derived from genomic repeats. *Trends in Genetics*, **21**, 322–326.

Smalheiser, N. R. and Torvik, V. I. (2006). Alu elements within human mRNAs are probable microRNA targets. *Trends in Genetics*, **22**, 532–536.

Sood, P., Krek, A., Zavolan, M., Macino, G. and Rajewsky, N. (2006). Cell-type-specific signatures of microRNAs on target mRNA expression. *Proceedings of the National Academy of Sciences USA*, **103**, 2746–2751.

Spence, J., Duggan, B. M., Eckhardt, C., McClelland, M. and Mercola, D. (2006). Messenger RNAs under differential translational control in Ki-ras-transformed cells. *Molecular Cancer Research*, **4**, 47–60.

Stark, A., Brennecke, J., Bushati, N., Russell, R. B. and Cohen, S. M. (2005). Animal microRNAs confer robustness to gene expression and have a significant impact on 3′UTR evolution. *Cell*, **123**, 1133–1146.

Stuart, J. J., Egry, L. A., Wong, G. H. and Kaspar, R. L. (2000). The 3′ UTR of human MnSOD mRNA hybridizes to a small cytoplasmic RNA and inhibits gene expression. *Biochemical Biophysical Research Communications*, **274**, 641–648.

Svoboda, P. and Cara, A. D. (2006). Hairpin RNA: a secondary structure of primary importance. *Cell Molecular Life Sciences*, **63**, 901–908.

Vidal, F., Mougneau, E., Glaichenhaus, N. *et al.* (1993). Coordinated posttranscriptional control of gene expression by modular elements including Alu-like repetitive sequences. *Proceedings of the National Academy of Sciences USA*, **90**, 208–212.

Vila, M. R., Gelpi, C., Nicolas, A. *et al.* (2003). Higher processing rates of Alu-containing sequences in kidney tumors and cell lines with overexpressed Alu-mRNAs. *Oncology Reports*, **10**, 1903–1909.

IV

Detection and quantitation of microRNAs

17 Detection and analysis of microRNAs using LNA (locked nucleic acid)-modified probes

Sakari Kauppinen

Introduction

MicroRNAs (miRNAs) are an abundant class of short endogenous RNAs that act as post-transcriptional regulators of gene expression by base-pairing with their target mRNAs. To date more than 3500 microRNAs have been annotated in vertebrates, invertebrates and plants according to the miRBase microRNA database release 8.0 in February 2006 (Griffiths-Jones, 2004; Griffiths-Jones et al., 2006), and many miRNAs that correspond to putative genes have also been identified. The miRNAs identified to date represent most likely the tip of the iceberg, and the number of miRNAs might turn out to be very large. Recent bioinformatic predictions combined with array analyses, small RNA cloning and Northern blot validation indicate that the total number of miRNAs in vertebrate genomes is significantly higher than previously estimated and maybe as many as 1000 (Bentwich et al., 2005; Berezikov et al., 2005; Xie et al., 2005). An increasing body of research shows that animal miRNAs play fundamental biological roles in cell growth and apoptosis (Brennecke et al., 2003), hematopoietic lineage differentiation (Chen et al., 2004), life-span regulation (Boehm and Slack, 2005), photoreceptor differentiation (Li and Carthew, 2005), homeobox gene regulation (Yekta et al., 2004; Hornstein et al., 2005), neuronal asymmetry (Johnston and Hobert, 2003), insulin secretion (Poy et al., 2004), brain morphogenesis (Giraldez et al., 2005), muscle proliferation and differentiation (Chan et al., 2005; Kwon et al., 2005; Sokol and Ambros, 2005), cardiogenesis (Zhao et al., 2005) and late embryonic development in vertebrates (Wienholds et al., 2005). Most recent reports indicate that miRNAs may not function as developmental switches, but rather play a role in maintaining tissue identity by conferring accuracy to gene-expression programs (Giraldez et al., 2005; Lim et al., 2005; Stark et al., 2005; Farh et al., 2005; Wienholds et al., 2005).

The expanding inventory of human miRNAs along with their highly diverse expression patterns and high number of potential target mRNAs suggest that miRNAs are involved in a wide variety of human diseases, including cancer. A number of chapters on this topic are found in Part V of this book. It has become

MicroRNAs: From Basic Science to Disease Biology, ed. Krishnarao Appasani. Published by Cambridge University Press. © Cambridge University Press 2008.

apparent that human miRNAs would not only be highly useful as biomarkers for future cancer diagnostics, but are rapidly emerging as attractive targets for disease intervention.

The current view that miRNAs represent a newly discovered, hidden layer of gene regulation has resulted in high interest among researchers around the world in the discovery of miRNAs, and their target mRNAs. Detection and analysis of the miRNAs is, however, not trivial, and the small size and sometimes low level of expression of different miRNAs require the use of sensitive and specific research tools. The focus of this chapter is on detection and analysis of microRNAs enhanced by LNA (locked nucleic acid)-modified oligonucleotide probes.

LNA (locked nucleic acid) – a synthetic RNA mimic for high-affinity miRNA targeting

LNAs comprise a new class of bicyclic high-affinity RNA analogs in which the furanose ring in the sugar–phosphate backbone is chemically locked in an RNA mimicking N-type (C3'-endo) conformation by the introduction of a 2'-O,4'-C methylene bridge (Figure 17.1; Koshkin *et al.*, 1998). Several studies have demonstrated that LNA-modified oligonucleotides exhibit unprecedented thermal stability when hybridized with their RNA target molecules (Koshkin *et al.*, 1998; Braasch and Corey, 2001; Kurreck *et al.*, 2002). Consequently, an increase in melting temperature (T_m) of $+2$–$10\,°C$ per monomer against complementary RNA compared to unmodified duplexes has been reported Examples of melting temperatures for LNAs complexed with RNA are shown in Table 17.1. Importantly, LNA incorporation generally improves mismatch discrimination compared to unmodified reference oligonucleotides. It should be underlined that LNA mediates high-affinity hybridization without compromising base pairing selectivity, and that the standard Watson–Crick base pairing rules are obeyed.

Structural studies of different LNA–RNA and LNA–DNA heteroduplexes based on NMR spectroscopy and X-ray crystallography have shown that LNA is an RNA mimic, which fits seamlessly into an A-type Watson–Crick duplex geometry (Petersen *et al.*, 2000; Petersen *et al.*, 2002; Nielsen *et al.*, 2004) similar to that of dsRNA duplexes. Furthermore, in heteroduplexes between LNA oligonucleotides and their complementary DNA oligonucleotides, an overall shift from a B-type

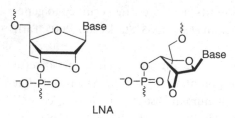

Figure 17.1. The structure of LNA monomers.

Table 17.1. Examples of melting temperatures (T_m values) for hybridization of LNA and DNA oligonucleotides to complementary RNA sequences

Probe sequence (5'→3')[a]	T_m (° C)	RNA target sequence (5'→3')[a]
ttttt	<10	aaaaaa
TTTTTt	40	aaaaaa
gtgatatgc	28	gcauaucac
gTGaTaTgc	58	gcauaucac
GTGATATG^{m}C	74	gcauaucac
tttttttttttttttttttt	40	aaaaaaaaaaaaaaaaaaaa
TtTtTtTtTtTtTtTtTtTt	71	aaaaaaaaaaaaaaaaaaaa
TTTTTTTTTTTTTTTTTTTT	>95	aaaaaaaaaaaaaaaaaaaa

[a] LNA nucleotides in uppercase, RNA and DNA nucleotides in lowercase. ^{m}C denotes LNA methylcytosine.
Source: From Singh *et al.* (1998), Jacobsen *et al.* (2004).

duplex towards an A-type duplex has been reported resulting in increased stability of the heteroduplexes. Another important observation is that LNA monomers are also able to twist the sugar conformation of flanking DNA nucleotides from an S-type (C2'-endo) towards an N-type sugar pucker in LNA-modified DNA oligo-nucleotides (Petersen *et al.*, 2002; Nielsen *et al.*, 2004).

The high thermal stability of short LNA oligonucleotides together with their improved mismatch discrimination has facilitated the design of highly accurate single nucleotide polymorphism (SNP) genotyping assays using allele-specific LNA probes (Jacobsen *et al.*, 2002a; Jacobsen *et al.*, 2002b; Mouritzen *et al.*, 2003). LNA substituted oligonucleotides have also been used to increase the sensitivity and specificity in gene expression profiling by spotted oligonucleotide microarrays (Tolstrup *et al.*, 2003) and, more recently, in efficient isolation of intact poly(A) + RNA from lysed cell and tissue extracts by LNA oligo(T) affinity capture (Jacobsen *et al.*, 2004). In addition, LNA oligonucleotides are readily transfected into cells using standard techniques, they are sequence-specific and non-toxic, and show improved nuclease resistance, which make them highly useful for potent and selective antisense-based gene silencing (Wahlestedt *et al.*, 2000; Braasch *et al.*, 2002; Fluiter *et al.*, 2003). Hence, LNA-modified probes are uniquely suited for mimicking miRNA structures, and for miRNA targeting *in vitro* or *in vivo*.

Use of LNA probes in miRNA Northern blot analysis

Most miRNA researchers use Northern blot analysis combined with polyacryla-mide gels to examine the expression of both the mature and precursor miRNAs, as well as to validate predicted miRNAs, since it allows both quantitation of the expression levels and miRNA size determination (Reinhart *et al.*, 2000; Lagos-Quintana *et al.*, 2001; Lee and Ambros, 2001). A major drawback of this method is its poor sensitivity, especially when monitoring expression of low-abundant miRNAs. Consequently, a large amount of total RNA per sample is required for Northern analysis, which is not feasible when the cell or tissue source is limited.

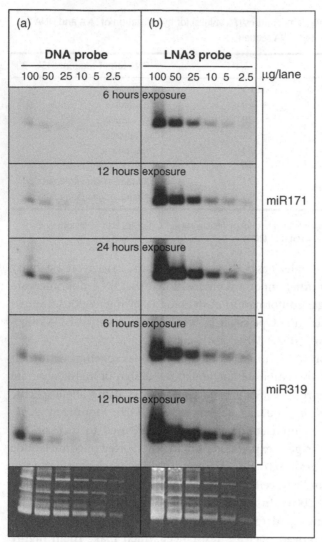

Figure 17.2. Northern blot analysis of two duplicate dilution series of *A. thaliana* total RNA hybridized with (a) ^{32}P-labeled DNA and (b) LNA probes, respectively, for *A. thaliana* miR171 and miR319. The gel loading controls are shown from ethidium bromide staining of the rRNAs (bottom panels). From Válóczi *et al.* (2004), reprinted by permission of Oxford University Press.

In a recent paper, Válóczi *et al.* (2004) describe highly efficient detection of miRNAs by Northern blot analysis using LNA-modified oligonucleotide probes and demonstrate their significantly improved sensitivity in detection of different miRNAs in mouse, *Arabidopsis thaliana* and *Nicotiana benthamiana*. The sensitivity in detecting mature microRNAs by Northern blots was increased by at least tenfold compared to end-labeled DNA probes of the same sequence (Figure 17.2), allowing the researcher to use less total RNA in the Northern analysis.

The discriminatory power of LNA-modified probes was assessed using three single mismatch and a double mismatch probe, respectively, for *A. thaliana* miR-171 alongside the perfect match LNA probe. Hybridization of the Northern blot

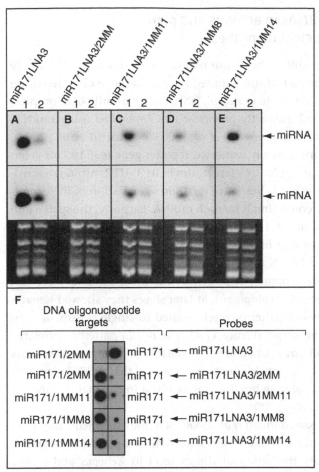

Figure 17.3. Assessment of the LNA probe specificity using (A) ^{32}P-labeled perfect match (miR171LNA3), (B) double mismatch (miR171LNA3/2MM) and (C-E) three single mismatch (miR171LNA3/MM11, miR171LNA3/MM8, miR171LNA3/MM14) probes, respectively, in the detection of miR171 in *A. thaliana* flowers (1) and leaves (2). The filters were washed at low stringency (upper panels) and high stringency (middle panels). (F) The perfect match and the different mismatched LNA probes were validated by hybridization to their respective perfect match DNA oligonucleotide targets as well as to a DNA oligo corresponding to the mature miR171 sequence. From Válóczi *et al.* (2004), reprinted by permission of Oxford University Press.

resulted in significantly decreased signals for the mature miR-171 by all three single mismatch probes compared to the perfect match probe, whereas no signals were detected by the double mismatch probe (Figure 17.3). This is in good agreement with previous reports for improved mismatch discrimination by LNA probes (Jacobsen *et al.*, 2002a; 2002b; Mouritzen *et al.*, 2003) and indicates that the sensitivity of miRNA detection by Northerns can be significantly improved using LNA probes, while simultaneously retaining or even improving specificity (Válóczi *et al.*, 2004). LNA probes have recently been used in validating novel candidate miRNAs by Northerns in *A. thaliana* (Xie *et al.*, 2005), and in detecting zebrafish miRNAs using digoxigenin (DIG)-labeled LNA probes in non-radioactive Northerns (Wienholds *et al.*, 2005).

In situ detection of miRNAs in animals and plants using LNA-modified oligonucleotide probes

The detection of mature miRNAs by *in situ* hybridization has been technically challenging because of their small size. Limited success has been obtained using plant sections (Chen *et al.*, 2004; Juarez *et al.*, 2004), whereas miRNA expression patterns can be determined indirectly in mouse and *Drosophila* using miRNA-responsive sensor constructs (Brennecke *et al.*, 2003; Mansfield *et al.*, 2004). Each sensor contains a constitutively expressed reporter gene (e.g. lacZ or green fluorescent protein) harbouring miRNA target sites in its 3′-UTR. Although sensitive, this approach is time-consuming, since it requires generation of the expression constructs and transgenic animals for each miRNA. Recently, the spatial and temporal expression patterns of 115 conserved vertebrate miRNAs were determined directly by whole-mount *in situ* hybridizations with LNA-modified DNA probes similar to those used for Northern blots (Válóczi *et al.*, 2004) on embryos and larvae of different developmental stages. While most miRNAs were not detected in the early stages of development, at later stages they showed remarkably tissue-specific expression patterns, often limited to single organs or even single cell layers within an organ (Figure 17.4). The *in situ* patterns correlated well with zebrafish miRNA array data and with available miRNA expression data for mammals. The temporal expression patterns suggested that miRNAs may not directly act as developmental switches, but are involved in the maintenance of tissue identity (Wienholds *et al.*, 2005).

In a more recent study, Kloosterman *et al.* (2006) describe a detailed analysis on the conditions for LNA probe-based *in situ* detection of miRNAs in zebrafish embryos, and demonstrate the utility of the method in *Xenopus* and mouse embryos. Using LNA-modified probes for miR-206, miR-124a and miR-122a in 72 h-old zebrafish embryos revealed expected patterns for all three miRNAs, whereas no *in situ* signals could be deteted with corresponding DNA probes, in accordance with the significantly increased affinity of LNA oligonucleotides toward complementary RNA molecules. The optimal miRNA *in situ* detection was obtained at hybridization temperatures of 20–25 °C below the calculated T_m value of the LNA probe, while the signal strength after 1 h of hybridization was comparable to *in situ* staining obtained after a standard overnight hybridization. For miR-122a and miR-206, specific *in situ* staining was lost upon introduction of a single central mismatch in the LNA probe, whereas two central mismatches were needed for adequate discrimination for miR-124a (Figure 17.5a). Importantly, specific detection of miR-206 and miR-124a could be achieved using shortened LNA versions complementary to a 14 nt region at the 5′-end of the miRNA (Figure 17.5b), thereby improving their specificity (Kloosterman *et al.*, 2006). The successful use of very short probes with T_m values in the range of body temperatures has significant implications for therapeutic applications, such as inhibition of disease-related miRNAs by LNA antago-mirs.

LNA-modified oligonucleotide probes are also well suited for sensitive and specific detection of mature miRNAs in plant tissues by *in situ* hybridization

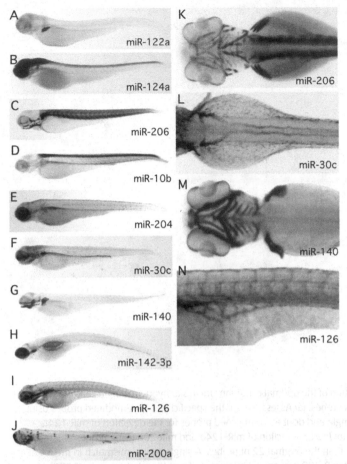

Figure 17.4. Highly diverse miRNA expression patterns in zebrafish embryos detected by whole-mount *in situ* hybridization using LNA probes. Representatives for miRNAs expressed in: (A) liver of the digestive system, (B) brain, spinal cord and cranial nerves/ganglia of the central and peripheral nervous systems, (C, M) muscles, (D) spinal cord, (E) pigment cells of the skin, (F, L) pronephros and presumably mucous cells of the excretory system, (G, M) cartilage of the skeletal system, (H) thymus, (I, N) blood vessels of the circulatory system, (J) lateral line system of the sensory organs. Embryos in (K, L, M, N) are higher magnifications of the embryos in (C, D, G, I), respectively. (A–J, N) are lateral views; (K–M) are dorsal views. All embryos are 72 h post-fertilization, except for (H), which is a 5-day old larva. From Wienholds *et al.* (2005), reprinted by permission of the American Association for the Advancement of Science. (See color plate 9)

(Válóczi *et al.*, unpublished results). The *in situ* results obtained with *N. benthamiana* indicate that miR160, miR164 and miR167 accumulate spatially and temporally in a highly restricted manner (Figure 17.6). These miRNAs are localized in cells and organs which show the most active division and differentiation activity, supporting the notion that plant miRNAs have a crucial role in cell fate determination and differentiation.

The spatial expression patterns of 38 miRNA genes were recently reported in *Drosophila* embryos using long riboprobes encompassing the pre-miRNA hairpin sequence (Aboobaker *et al.*, 2005). In another study, Sokol and Ambros (2005) reported on comparable expression patterns for miR-1 in *Drosophila* embryos

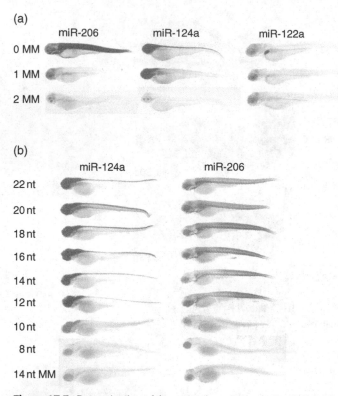

Figure 17.5. Determination of the optimal conditions for *in situ* hybridization in zebrafish embryos using LNA-modified DNA probes. (a) Assessment of the specificity of LNA-modified probes using perfectly matched and single and double mismatched probes for the detection of miR-124a, miR-122a and miR-206. (b) *In situ* detection of miR-124a and miR-206 using probes of 2, 4, 6, 8, 10, 12 and 14 nt shorter than the original 22 nt probes. A single central mismatch in the 14 nt probes for miR-124a and miR-206 prevents hybridization. From Kloosterman *et al.* (2006), reprinted by permission of Nature Publishing Group.

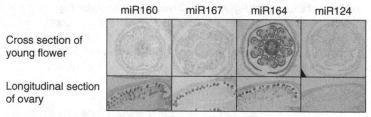

Figure 17.6. Detection of miR160, miR167, and miR164 expression patterns in developing flowers (upper panels) and ovaries (lower panels) of *N. benthamiana* by *in situ* hybridization using LNA probes. A probe for mouse miR-124a was used as negative control. From A. Válóczi, J. Burgyán and Z. Havelda, unpublished results.

obtained using both a full-length riboprobe for the primary transcript and an LNA probe for the mature miR-1, implying that LNA-enhanced *in situ* detection would also be useful as a complementary method, when direct detection of mature miRNAs is desirable. Another important finding is that use of LNA probes allows *in situ* detection of miRNAs in archival tissues, as recently reported for several brain-specific miRNAs in formalin-fixed paraffin-embedded sections (Nelson *et al.*, 2006). This will greatly facilitate future miRNA studies in human diseases, including miRNA detection in tumor sections.

LNA-enhanced miRNA expression profiling

DNA microarrays would appear to be a good alternative to Northern blot analysis to quantify miRNAs in a genome-wide scale, since arrays have excellent throughput (Fodor *et al.*, 1991; Schena *et al.*, 1995). Screening and detection of miRNAs by expression arrays has been described in several reports (Krichevsky *et al.*, 2003; Barad *et al.*, 2004; Liu *et al.*, 2004; Miska *et al.*, 2004; Thomson *et al.*, 2004). A major drawback of all DNA-based oligonucleotide array platforms regardless of the capture probe length is the difficulty to design T_m-normalized probe sets for genome-wide expression profiling due to the short nature of miRNAs. In addition, at least in some array platforms discrimination of highly homologous miRNA differing by just one or two nucleotides could not be achieved, although the 60-mer microarray by Barad *et al.* (2004) appears to have adequate specificity. Given their significantly improved sensitivity and high specificity in miRNA detection by Northerns and *in situ* hybridization, LNA-modified DNA oligonucleotides would present an attractive alternative to the currently used DNA oligoarray platforms. In a recent study, Castoldi *et al.* (2006) report on development of an LNA-based microarray, comprising a T_m-normalized LNA probe set for all human and mouse miRNAs annotated in the miRBase release 6.0 (Griffiths-Jones, 2004; Griffiths-Jones *et al.*, 2006). The substitution of DNA probes with LNA modifications allowed normalization of the probe set to a melting temperature of 72 °C. The resulting LNA arrays showed significantly increased sensitivity compared to DNA-based miRNA arrays, while simultaneously exhibiting high specificity enabling efficient discrimination between miRNA family members, at least when they differ in nucleotides close to the central position. Another important observation in this study was that the superior detection sensitivity of LNA arrays eliminated the need for RNA size selection and/or amplification, thereby allowing use of labeled total RNA directly for miRNA expression profiling (Castoldi *et al.*, 2006). A single-molecule method for highly specific and sensitive quantitation of miRNA expression using hybridization of fluoro-labeled LNA-modified DNA probes in solution to the miRNA of interest was recently developed and reported (Neely *et al.*, 2006; see also Chapter 19 in this book).

Functional microRNA analysis using LNA-antimiRs

The small size of miRNA genes makes it difficult to create loss-of-function mutants for functional analysis. Another potential problem is that many miRNA genes are present in several copies per genome occurring in different loci, which makes it even more difficult to obtain mutant phenotypes. An alternative approach to creating miRNA gene knock-outs has been reported by Hutvágner *et al.* (2004) and Leaman *et al.* (2005), in which 2'-O-methyl antisense oligonucleotides were used as potent and irreversible inhibitors of siRNA and miRNA function *in vitro* and *in vivo* in *Drosophila* and *C. elegans*, thereby inducing a loss-of-function phenotype. This method was recently applied to mouse studies, by conjugating 2'-O-methyl antisense oligonucleotides complementary to four different miRNAs with cholesterol (so called antago-mirs) for silencing miRNAs *in vivo* (Krützfedt *et al.*, 2005). Recent studies have reported that LNA-modified oligonucleotides can also mediate specific inhibition of miRNA function (Chan *et al.*, 2005; Lecellier *et al.*, 2005; Ørom *et al.*, 2006). Using the well-characterized interaction between the *Drosophila melanogaster* bantam miRNA and its target gene hid as a model, Ørom *et al.* (2006) have described the efficacy and specificity of the LNA-based silencing method. LNA-antimiRs can readily inhibit exogenously introduced miRNAs with high specificity, and furthermore inhibit endogenous bantam in *Drosophila melanogaster* cells, leading to up-regulation of its cognate target protein hid. The method showed stoichiometric and reliable inhibition of the targeted miRNA and would thus be applicable to functional analysis of miRNAs and validation of putative target genes (Ørom *et al.*, 2006). Since LNA antisense oligonucleotides have features that result in very high hybridization affinity towards complementary single stranded RNA without compromising specificity (Koshkin, 1998; Wahlestedt *et al.*, 2000; Braasch and Corey, 2001) and furthermore, show improved antisense efficacy and higher T_m toward complementary RNA compared to 2'-O methyl oligonucleotides of the same sequence (Kurreck *et al.*, 2002), LNAs could be used to improve *in vivo* efficacy of 2'-O methyl antago-mirs.

Conclusions

The remarkable hybridization properties of LNA, both with respect to affinity and specificity, position LNA as an enabling molecule for microRNA research. The facts that LNA probes are commercially available as miRCURY probes and arrays from Exiqon (www.exiqon.com), and that LNA nucleotides can be freely mixed with DNA nucleotides in oligonucleotide probes, make LNA a highly flexible tool. It is important to note that the increased cost of LNA oligonucleotides compared to DNA probes can be counterbalanced by the use of shorter, but more sensitive and specific LNA probes. Furthermore, as more laboratories begin to use LNA oligonucleotides in their research, the cost of LNA is expected to decrease, with simultaneous acceleration in the use of LNA probes in high-throughput genomics.

ACKNOWLEDGMENTS

The author wishes to thank Z. Havelda, W. Kloosterman and R. Plasterk for helpful comments on this manuscript, and A. Válóczi, J. Burgyán and Z. Havelda for sharing their unpublished results.

REFERENCES

Aboobaker, A. A., Tomancak, P., Patel, N., Rubin, G. M. and Lai, E. C. (2005). *Drosophila* microRNAs exhibit diverse spatial expression patterns during embryonic development. *Proceedings of the National Academy of Sciences USA*, **102**, 18017–18022.

Barad, O., Meiri, E., Avniel, A. *et al.* (2004). MicroRNA expression detected microarrays: system establishment profiling in human tissues. *Genome Research*, **14**, 2486–2494.

Bentwich, I., Avniel, A., Karov, Y. *et al.* (2005). Identification of hundreds of conserved and nonconserved human microRNAs. *Nature Genetics*, **37**, 766–770.

Berezikov, E., Guryev, V., van de Belt, J. *et al.* (2005). Phylogenetic shadowing and computational identification of human microRNA genes. *Cell*, **120**, 21–24.

Boehm, M. and Slack, F. (2005). A developmental timing microRNA and its target regulate life span in *C. elegans*. *Science*, **310**, 1954–1957.

Braasch, D. A. and Corey, D. R. (2001). Locked nucleic acid (LNA): fine-tuning the recognition of DNA and RNA. *Chemistry and Biology*, **8**, 1–7.

Braasch, D. A., Liu, Y. and Corey, D. R. (2002). Antisense inhibition of gene expression in cells by oligonucleotides incorporating locked nucleic acids: effect of mRNA target sequence and chimera design. *Nucleic Acids Research*, **30**, e 5160–5167.

Brennecke, J., Hipfner, D. R., Stark, A., Russel, R. B. and Cohen S. (2003). bantam encodes a developmentally regulated microRNA that controls cell proliferation and regulates the proapoptotic gene hid in *Drosophila*. *Cell*, **113**, 25–36.

Castoldi, M., Schmidt, S., Benes, V. *et al.* (2006). A sensitive array for microRNA expression profiling (mi-Chip) based on locked nucleic acids (LNA). *RNA*. (In press.)

Chan, J. A., Krichevsky, A. M. and Kosik, K. S. (2005). MicroRNA-21 is an antiapoptotic factor in human glioblastoma cells. *Cancer Research*, **65**, 6029–6033.

Chen, C. Z., Li, L., Lodish, H. F. and Bartel, D. P. (2004). MicroRNAs modulate hematopoietic lineage differentiation. *Science*, **303**, 83–86.

Farh, K. K., Grimson, A., Jan, C. *et al.* (2005). The widespread impact of mammalian microRNAs on mRNA repression and evolution. *Science*, **310**, 1817–1821.

Fluiter, K., ten Asbroek, A. L. M., de Wissel, M. B. *et al.* (2003). In vivo tumor growth inhibition and biodistribution studies of locked nucleic acid (LNA) antisense oligonucleotides. *Nucleic Acids Research*, **31**, 953–962.

Fodor, S. P., Read, J. L., Pirrung, M. C. *et al.* (1991). Light-directed, spatially addressable parallel chemical synthesis. *Science*, **251**, 767–773.

Giraldez, A. J., Cinalli, R. M., Glasner, M. E. *et al.* (2005). MicroRNAs regulate brain morphogenesis in zebrafish. *Science*, **308**, 833–838.

Griffiths-Jones, S. (2004). The microRNA Registry. *Nucleic Acids Research*, **32**, D109–D111.

Griffiths-Jones, S., Grocock, R. J., van Dongen, S., Bateman, A. and Enright, A. J. (2006). miRBase: microRNA sequences, targets and gene nomenclature. *Nucleic Acids Research* **34**, D140–D144.

Hornstein, E., Mansfield, J. H., Yekta, S. *et al.* (2005). The microRNA miR-196 acts upstream of Hoxb8 and Shh in limb development. *Nature*, **438**, 671–674.

Hutvágner, G., Simard, M. J., Mello, C. C. and Zamore, P. D. (2004). Sequence-specific inhibition of small RNA function. *Public Library of Science Biology*, **2**, 1–11.

Jacobsen, N., Fenger, M., Bentzen, J. *et al.* (2002a). Genotyping of the apolipoprotein B R3500Q mutation using immobilized locked nucleic acid capture probes. *Clinical Chemistry*, **48**(4), 657–660.

Jacobsen, N., Bentzen, J. Meldgaard, M. *et al.* (2002b). LNA-enhanced detection of single nucleotide polymorphisms in the apoli-poprotein E. *Nucleic Acids Research*, **30**, e100.

Jacobsen, N., Nielsen, P. S., Jeffares, D. C. *et al.* (2004). Direct isolation of poly(A)+ RNA from 4 M guanidine thiocyanate-lysed cell extracts using locked nucleic acid-oligo(T) capture. *Nucleic Acids Research*, **32**, e64.

Johnston, R. J. and Hobert, O. (2003). A microRNA controlling left/right neuronal asymmetry in *Caenorhabditis elegans*. *Nature*, **426**, 845–849.

Juarez, M. T., Kui, J. S., Thomas, J., Heller, B. A. and Timmermans, M. C. (2004). MicroRNA-mediated repression of rolled leaf 1 specifies maize leaf polarity. *Nature*, **428**, 84–88.

Kloosterman, W. P., Wienholds, E., de Bruijn, E., Kauppinen, S. and Plasterk, R. H. (2006). In situ detection of miRNAs in animal embryos using LNA-modified oligonucleotide probes. *Nature Methods*, **3**, 27–29.

Koshkin, A. A., Singh, S. K., Nielsen, P. *et al.* (1998). LNA (locked nucleic acids): synthesis of the adenine, cytosine, guanine, 5-methylcytosine, thymine and uracil bicyclonucleoside monomers, oligomerisation, and unprecedented nucleic acid recognition. *Tetrahedron*, **54**, 3607–3630.

Krichevsky, A. M., King, K. S., Donahue, C. P., Khrapko, K. and Kosik, K. S. (2003). A microRNA array reveals extensive regulation of microRNAs during brain development. *RNA*, **9**, 1274–1281.

Krützfeldt, J., Rajewsky, N., Braich, R. *et al.* (2005). Silencing of microRNAs in vivo with 'antagomirs'. *Nature*, **438**, 685–689.

Kurreck, J., Wyszko, E., Gillen, C. and Erdmann, V. A. (2002). Design of antisense oligonucleotides stabilized by locked nucleic acids. *Nucleic Acids Research*, **30**, 1911–1918.

Kwon, C., Han, Z., Olson, E. N. and Srivastava, D. (2005). MicroRNA1 influences cardiac differentiation in Drosophila and regulates Notch signaling. *Proceedings of the National Academy of Sciences USA*, **102**, 18986–18991.

Lagos-Quintana, M., Rauhut, R., Lendeckel, W. and Tuschl, T. (2001). Identification of novel genes coding for small expressed RNAs. *Science*, **294**, 853–858.

Leaman, D., Chen, P,Y., Fak, J. *et al.* (2005). Antisense-mediated depletion reveals essential and specific functions of microRNAs in Drosophila development. *Cell*, **121**, 1097–1108.

Lecellier, C. H., Dunoyer, P., Arar, K. *et al.* (2005). A cellular microRNA mediates antiviral defense in human cells. *Science*, **308**, 557–560.

Lee, R. C. and Ambros, V. (2001). An extensive class of small RNAs in *Caenorhabditis elegans*. *Science*, **294**, 862–864.

Li, X. and Carthew, R. W. (2005). A microRNA mediates EGF receptor signaling and promotes photoreceptor differentiation in the *Drosophila eye*. *Cell*, **123**, 1267–1277.

Lim, L. P., Lau, N. C., Garrett-Engele, P. *et al.* (2005). Microarray analysis shows that some microRNAs downregulate large numbers of target mRNAs. *Nature*, **433**, 769–773.

Liu, C.-G., Calin, G. A., Meloon, B. *et al.* (2004). An oligonucleotide microchip for genome-wide microRNA profiling in human and mouse tissues. *Proceedings of the National Academy of Sciences USA*, **101**, 9740–9744.

Mansfield, J. H., Harfe, B. D., Nissen, R. *et al.* (2004). MicroRNA-responsive 'sensor' transgenes uncover Hox-like and other developmentally regulated patterns of vertebrate microRNA expression. *Nature Genetics*, **36**, 1079–1083.

Miska, E. A., Alvarez-Saavedra, E., Townsend, M. *et al.* (2004). Microarray analysis of microRNA expression in the developing mammalian brain. *Genome Biology*, **5**, R68.

Mouritzen, P., Nielsen, A. T., Pfundheller, H. M., *et al.* (2003). Single nucleotide polymorphism genotyping using locked nucleic acid (LNA). *Expert Review of Molecular Diagnostics*, **3**, 27–38.

Neely, L. A., Patel, S., Garver, J. *et al.* (2006). A single-molecule method for the quantitation of microRNA gene expression. *Nature Methods*, **3**, 41–46.

Nelson, P. T., Baldwin, D. A., Kloosterman, W. P. *et al.* (2006). RAKE and LNA-ISH reveal microRNA expression and localization in archival human brain. *RNA*, **12**, 187–191.

Nielsen, K. E., Rasmussen, J., Kumar, R. *et al.* (2004). NMR studies of fully modified locked nucleic acid (LNA) hybrids: Solution structure of an LNA:RNA hybrid and characterization of an LNA:DNA hybrid. *Bioconjugate Chemistry*, **15**, 449–457.

Ørom, U. A., Kauppinen, S. and Lund, A. (2006). LNA-modified oligonucleotides mediate specific inhibition of microRNA function. *Gene*. (In press.)

Petersen, M., Nielsen, C. B., Nielsen, K. E. *et al.* (2000). The conformations of locked nucleic acids (LNA). *Journal of Molecular Recognition*, **13**, 44–53.

Petersen, M., Bondensgaard, K., Wengel, J. and Jacobsen, J. P. (2002). Locked nucleic acid (LNA) recognition of RNA: NMR solution structures of LNA:RNA hybrids. *Journal of American Chemical Society*, **124**, 5974–5982.

Poy, M. N., Eliasson, L., Krutzfeldt, J. *et al.* (2004) A pancreatic islet-specific microRNA regulates insulin secretion. *Nature*, **432**, 226–230.

Reinhart, B.J., Slack, F.J., Basson, M. *et al.* (2000). The 21 nucleotide let-7 RNA regulates developmental timing in *Caenorhabditis elegans*. *Nature*, **403**, 901–906.

Schena, M., Shalon, D., Davis, R.W. and Brown, P.O. (1995). Quantitative monitoring of gene expression patterns with a complementary DNA microarray. *Science*, **270**, 467–470.

Singh, S.K., Nielsen, P., Koshkin, A., Olsen, C.E. and Wengel, J. (1998). LNA (locked nucleic acids): synthesis and high-affinity nucleic acid recognition. *Chemical Communications*, **4**, 455–456.

Sokol, N.S. and Ambros, V. (2005). Mesodermally expressed Drosophila microRNA-1 is regulated by Twist and is required in muscles during larval growth. *Genes and Development*, **19**, 2343–2354.

Stark, A., Brennecke, J., Bushati, N., Russell, R.B. and Cohen, S.M. (2005). Animal microRNAs confer robustness to gene expression and have a significant impact on 3′UTR evolution. *Cell*, **123**, 1133–1146.

Thomson, J.M., Parker, J., Perou, C.M. and Hammond, S.M. (2004). A custom microarray platform for analysis of microRNA gene expression. *Nature Methods*, **1**, 1–6.

Tolstrup, N., Nielsen, P.S., Kolberg, J. *et al.* (2003). OligoDesign: optimal design of LNA (locked nucleic acid) oligonucleotide capture probes for gene expression profiling. *Nucleic Acids Research*, **31**, 3758–3762.

Válóczi, A., Hornyik, C., Varga, N. *et al.* (2004). Sensitive and specific detection of microRNAs by northern blot analysis using LNA-modified oligonucleotide probes. *Nucleic Acids Research*, **32**, e175.

Wahlestedt, C., Salmi, P., Good, L. *et al.* (2000). Potent and nontoxic antisense oligonucleotides containing locked nucleic acids. *Proceedings of the National Academy of Sciences USA*, **97**, 5633–5638.

Wienholds, E., Kloosterman, W.P., Miska, E. *et al.* (2005). MicroRNA expression in zebrafish embryonic development. *Science*, **309**, 310–311.

Xie, X., Lu, J., Kulbokas, E.J. *et al.* (2005). Systematic discovery of regulatory motifs in human promoters and 30 UTRs by comparison of several mammals. *Nature*, **434**, 338–345.

Yekta, S., Shih, I.-S. and Bartel, D. (2004). MicroRNA-directed cleavage of HOXB8 mRNA. *Science*, **304**, 594–596.

Zhao, Y., Samal, E. and Srivastava, D. (2005). Serum response factor regulates a muscle-specific microRNA that targets Hand2 during cardiogenesis. *Nature*, **436**, 214–220.

18 Detection and quantitation of microRNAs using the RNA Invader® assay

Hatim T. Allawi* and Victor I. Lyamichev

Introduction

MicroRNAs are short single-stranded RNA molecules ranging in length from 17 to 24 nucleotides. During the past few years, hundreds of miRNAs have been identified in plants, animals, and a number of viruses with their exact biological function not fully understood (Lagos-Quintana *et al.*, 2001; Lagos-Quintana *et al.*, 2002; Reinhart *et al.*, 2002; Hunter and Poethig, 2003; Lagos-Quintana *et al.*, 2003; Pfeffer *et al.*, 2004; Dunn *et al.*, 2005; Pasquinelli *et al.*, 2005; Wienholds and Plasterk, 2005; Wienholds *et al.*, 2005). It has been shown that miRNAs target messenger RNA (mRNAs) at specific sites inducing cleavage of the RNA or result in inhibition of translation (Bartel, 2004). MiRNAs also exhibit unique expression patterns in tumor cells and therefore maybe useful as molecular markers for cancer cells (Michael *et al.*, 2003; Calin *et al.*, 2004; Croce and Calin, 2005; Eis *et al.*, 2005). Moreover, microRNAs have been identified to be involved in regulation of cell and tissue development as well as several biological processes including cell proliferation and death, apoptosis, neuron development, DNA methylation and chromatin modification, and fat metabolism (Reinhart *et al.*, 2000; Pasquinelli and Ruvkun, 2002; Ambros, 2003; Brennecke *et al.*, 2003; Johnston and Hobert, 2003; Xu *et al.*, 2003; Bao *et al.*, 2004; Alvarez-Garcia and Miska, 2005; Croce and Calin, 2005; Miska, 2005).

Several methods of detecting and quantitating miRNAs have been developed. The size and sequence homology of some miRNAs makes their quantitation and differentiation challenging to conventional RT-PCR methods or standard microchip hybridization techniques. Northern blots were among the first techniques used for reliable miRNA detection and quantitation but are labor intensive, insensitive, requiring microgram quantities of total RNA, and are low throughput (Lim *et al.*, 2003). Rnase protection assays have also been used for reliable miRNA detection but are also labor intensive and offer lower throughput and less

---

* Author to whom correspondence should be addressed.

MicroRNAs: From Basic Science to Disease Biology, ed. Krishnarao Appasani. Published by Cambridge University Press. © Cambridge University Press 2008.

sensitive detection levels. Other methods also developed and applied successfully to miRNA detection and quantitation include the TaqMan® miRNA assay (Chen *et al.*, 2005), miRNA microarrays (Krichevsky *et al.*, 2003; Liu *et al.*, 2004), and fluorescently labeled hybridization miRNA probes. These methods offer easier alternatives to gel-based miRNA assays and are more specific and sensitive to miRNA detection and quantitation. However, the high cost of these assays and the need for synthesis of individually labeled probes for detecting different miRNA sequences render these methods suitable for detecting one miRNA at a time but cost prohibiting for analysis of several miRNAs in a high-throughput manner.

Here we describe the microRNA Invader® assay, which offers a sensitive and specific detection and quantitation method for miRNAs. The Invader miRNA assay has the ability to detect and quantitate as few as 20 000 molecules of an individual miRNA. It distinguishes between miRNAs and their precursors, as well as between closely related miRNA isotypes. The assay is rapid and can be performed in detergent lysates of cells without the need for total RNA isolation. Finally, the Invader assay does not require individually labeled probes or expensive instrumentation, thus making it the better choice for low-cost, high-throughput miRNA detection and quantitation.

Methodology

The Invader assay

The Invader assay uses a class of structure-specific 5′ nuclease enzymes, Cleavase®, that recognize the formation of an overlap flap structure between a probe oligonucleotide containing a 5′- single stranded flap and a 3′ invasive oligonucleotide that once hybridized to their targeted nucleic acid form an overlap by one base (Figure 18.1a) (Kaiser *et al.*, 1999; Lyamichev *et al.*, 1999; Hall *et al.*, 2000). Upon the formation of the overlap–flap structure, Cleavase cleaves the single-stranded 5′-flap rendering it free for detection or participation in a secondary signal amplification reaction. The secondary signal amplification reaction uses the released 5′-flap as an invasive oligonucleotide in a substrate consisting of a secondary reaction template (SRT) and a secondary probe oligonucleotide. The secondary probe contains a fluorescent dye (F) and a quencher (Q) which form a fluorescence resonance energy transfer (FRET) pair that is separated upon cleavage (de Arruda *et al.*, 2002; Eis *et al.*, 2001) (Figure 18.1b). An additional oligonucleotide, referred to as the arrestor, is added to the secondary reaction to capture any uncleaved primary probe and prevent it from hybridizing to the SRT. Separation of the fluorescent dye from the quencher generates fluorescent signal that is indicative of the presence of the targeted nucleic acid in the primary reaction. The fluorescent signal generated in the Invader assay is proportional to the amount of target present and can be used to calculate target concentration in unknown samples (Lyamichev *et al.*, 2000). Fluorescence is amplified by running the reaction at elevated temperatures resulting in multiple primary probe cleavages per target and multiple secondary FRET probe cleavages per primary cleavage event (Lyamichev *et al.*, 2000).

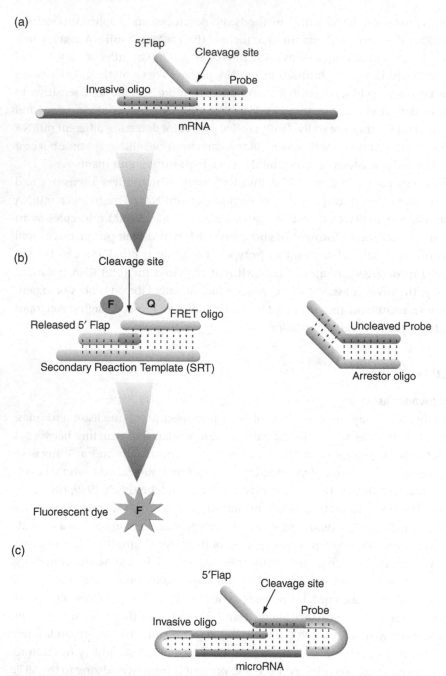

Figure 18.1. Schematic representation of the RNA invasive cleavage assay. (a) mRNA Invader® assay primary reaction: an overlap–flap substrate for the structure-specific 5′ nuclease, Cleavase® *substrate* is formed upon annealing of the invasive and probe oligonucleotides to the mRNA target. The mRNA target is shown in blue and the target-specific oligonucleotides are shown in green. The non-complementary 5′-flap of the probe oligonucleotide is in purple. The site of cleavage is indicated by an arrow. (b) Secondary reaction: a secondary overlap–flap structure is formed when a Secondary Reaction Template (SRT, purple) binds the 5′-flap cleaved in the primary reaction plus a FRET oligonucleotide (purple). The FRET oligonucleotide is labeled with a fluorophore (F) and a quencher (Q). Cleavage between the fluorophore and a quencher generates a fluorescence

Oligonucleotide synthesis

DNA oligonucleotides were synthesized using an Expedite 8909 synthesizer (PerSeptive Biosystems) following standard phosphoramidite chemistry. All phosphoramidites, including 2′-O-methyl phosphoramidites, were purchased from Glen Research (Sterling, VA). Synthesis and deprotection were performed according to the manufacturer's procedures. RNA oligonucleotides were purchased from Dharmacon Inc. (Lafayette, CO) and were deprotected and desalted according to the manufacturer's procedures. Denaturing polyacrylamide (20%) gel electrophoresis was used to purify all probe and RNA oligonucleotides used in this study. FRET and secondary reaction template (SRT) oligonucleotides were ion-exchange HPLC purified. Extinction coefficients for nucleosides and dinucleoside monophosphates were used to calculated the 260 nm extinction coefficients and concentrations for oligonucleotides (Cantor *et al.*, 1970).

Oligonucleotide designs for the miRNA Invader assay

In the microRNA Invader assay, the probe and invasive oligonucleotides are designed so that each hybridized to equal halves of the miRNA of interest (Figure 18.1c). In cases when the miRNA length is odd, the miRNA region complementary to the probe was one nucleotide longer than the region complementary to the invasive oligonucleotide. The probe and invasive oligonucleotides were modified by adding 2′-O-methylated stem-loops at their 3′ and 5′ ends, respectively. The addition of hairpin-forming stem-loops promotes stacking between the 3′ and 5′ ends of hybridized miRNA and the 5′ and 3′ ends of the invasive and probe oligonucleotides, respectively. This enhances the hybridization between the probe and invasive oligonucleotides and the target miRNA and makes the substrate thermodynamically more stable than it would without stacking (Walter and Turner, 1994; Walter *et al.*, 1994; Allawi *et al.*, 2004). Furthermore, it allows the assay to be performed at elevated temperatures resulting in higher stringency and more cleavage events in the assay (Lyamichev *et al.*, 2000).

The invasive oligonucleotide has a 3′ terminal overlapping nucleotide that is not complementary to the miRNA sequence, with optimal activity being in the order T = C > A > G. The probe oligonucleotide has a 5′-flap region 5′-AACGAGGCGCAC which, when cleaved from the probe in the primary reaction, is used in the secondary reaction (Figure 18.1b). In cases when the 5′-flap sequence was partially complementary to the miRNA sequence thus extending the length of the probe-miRNA heteroduplex, an alternative 5′-flap sequence (5′-CCGTCGCTGCGT) was used.

Figure 18.1. (cont.)
signal. A 2′-O-methyl arrestor oligonucleotide (red) complementary to the probe is added to the secondary reaction, to sequester the uncleaved probes and prevent their binding to the SRT.
(c) miRNA Invader assay primary reaction: the overall structure of the substrate resembles that shown in (a), except that the short size of the miRNA target requires the inclusion of extra stem-loop structures to the 5′ and 3′ ends of the invasive and probe oligonucleotides, respectively. The miRNA target is shown in blue, target-specific probe and invasive oligonucleotide sequences in green and 2′-O-methylated stem-loop hairpin regions are shown in red. The non-complementary 5′-flap is in purple and the arrow indicates the site of cleavage, which releases the 5′-flap. (See color plate 10)

Cell lysis and RNA preparation

HeLa and Hs 578 T cells were obtained from ATCC (catalog numbers: CCL-2, HTB-126, respectively) and grown as described (Allawi *et al.*, 2004). Approximately 1×10^6 of suspended cells were pelleted by centrifugation at 1000 g for 3 min, washed once with 1 ml PBS (no $MgCl_2$, no $CaCl_2$) (Invitrogen, Rockville, MD) and spun down at 1000 g for 3 min. The cell pellets were suspended in 100 µl of 10 mM MOPS (pH 7.5), 100 mM KCl, and 2 µl cell aliquots were removed and diluted with 98 µl of the lysis buffer containing 20 mM Tris-HCl, pH 8.5, 0.5% NP-40, 20 µg/ml tRNA and heated at 80 °C for 15 min. To remove cellular debris, cell lysates were then centrifuged at 1000 g for 3 min. Total RNA from HeLa and Hs 578 T cells were prepared using TRIZOL® (Invitrogen, Rockville, MD) according to the manufacturer's procedures. RNA concentration was determined from the absorption at 260 nm (assuming that 1 $A_{260} = 40$ µg/ml). Samples of total RNA isolated from specific human tissues were purchased from Clontech (Palo Alto, CA).

Invader microRNA assay

Invader reactions were performed in triplicate in 96-well microplates (MJ research, (now Bio-Rad) Waltham, MA) using the Invader miRNA assay generic reagent kit (Third Wave Technologies, Inc. Madison, WI). The optimal temperature of the primary reaction was determined by performing Invader reactions using 50 pM of synthetic miRNA in a gradient thermal cycler over a temperature range of 40 °C to 60 °C for 30 min, in 10 µl volumes containing 1 µM of each of the probe and invasive oligonucleotides. For no-target controls, 10 ng/µl yeast tRNA (Sigma) was substituted for samples or synthetic miRNA. Reactions were overlayed with 10 µl of clear Chill-Out™ 14 liquid wax (MJ research, (now Bio-Rad) Waltham, MA) to prevent evaporation.

Upon the completion of the primary reaction, 5 µl of a secondary reaction mixture was added to the primary reaction by pipetting directly below the liquid wax phase. The secondary reaction mixture was prepared by combining the arrestor oligonucleotide and the FRET cassettes to final concentration of 8 µM of arrestor and 1X FRET cassette. The secondary reaction was then performed at 60 °C for 15 min and the fluorescence signal detected using a CytoFluor® 4000 fluorescence plate reader (Applied Biosystems, Foster City, CA) using 485/20 nm excitation and 530/25 nm emission filters for the FAM dye. The optimal primary Invader reaction temperature was defined as the temperature at which the highest net signal (miRNA positive signal minus no-target signal) was observed. To determine the miRNA levels, unknown samples, 5 µl aliquots of cell lysate or 50–100 ng of total RNA (in 5 µl), were used in a 10 µl primary Invader reaction. The primary reaction was performed at the optimal predetermined temperature for each miRNA for 60–90 min and the secondary reaction was performed at 60 °C for 60–90 min. Samples of synthetic miRNA (5 µl) with known concentrations ranging from 5 fM to 5 pM were used in the Invader assay to obtain standard quantitation curves for miRNAs in unknown samples.

For biplex format Invader miRNA assays, Invader GAPDH mRNA and U6 RNA kits (Third Wave Technologies, Inc. Madison, WI; # 94–002) were used. The GAPDH and U6 Invader assays are configured to have their florescence signal reporting to Redmond Red™ dye (Epoch Biosciences, Bothell, WA) which is detected using 560/20 nm excitation and 620/40 nm emission filters.

Results and discussion

Specificity of the microRNA Invader assay

Let-7 was used as a model in this study to test the specificity of the miRNA Invader assay. Two microRNA Invader designs targeting the *a* and *c* variants of let-7 were made (Figure 18.2a,c) and tested for specificity against the *a*, *c*, *e*, and *f* variants of let-7. These let7 variants differed by one nucleotide at various positions of the let-7 miRNA. The let-7a and let-7c microRNAs differ by one nucleotide at position 19 of the miRNA. Therefore, the designs targeting *a* and *c* variants used the same probe oligonucleotide sequence but different invasive oligonucleotides (Figure 18.2a,c). The specificity of the Invader miRNA assay was investigated at several temperatures. For the assay targeting let-7a, addition of let-7a RNA resulted in much greater signal than did addition of the rest of the *c*, *e*, and *f* variant miRNAs (Figure 18.2b). Similarly, the let-7c assay generated significant signal in the presence of let-7c RNA but not the other let-7 variants (Figure 18.2d). The optimal signal generation temperature for both the let-7a and let-7c designs targeting their targeted miRNAs was 53 °C. Results show lower signal generation and temperature optima for non-target miRNAs (i.e. let-7c, e, and f for the let-7a-specific Invader assay and let-7a, e, and f for the let-7c-specific Invader assay). This shows that the Invader microRNA assay is capable of discrimination between miRNAs that differ in their sequence by as little as one nucleotide. Discrimination was greatest when the sequence differences between the let-7 variants were near the middle of the miRNA opposite overlap region of the probe and invasive oligonucleotide. Discrimination of let-7c Invader assay seemed to be higher than the let-7a Invader assay possibly due to the relative stabilities of the mismatched nucleotides (A–C versus G–T) in the invasive oligonucleotide region (Allawi and SantaLucia, 1997; Allawi and SantaLucia, 1998). This indicates that specificity of the miRNA Invader assay stems from a combination of both enzyme specificity requiring perfect Watson–Crick base pairing relative to the cleavage site and stability of hybridization of probe and invasive oligonucleotide in which mismatch formation with other than targeted miRNAs shifts the temperature optima of the Invader reaction.

Distinguishing precursors from mature forms of miRNAs is important for accurate detection and quantitation of miRNAs. To test the capability of the Invader assay for specific detection of mature miRNA, the let-7 miRNA Invader assay was used as a model. Figure 18.3 shows the signal generation for the let-7a Invader assay in the presence of either the mature or precursor let-7a miRNA. The data show that in the presence of very high levels of the precursor let-7a (> 6.75 million copies), the Invader let-7a assay generates signal corresponding to less

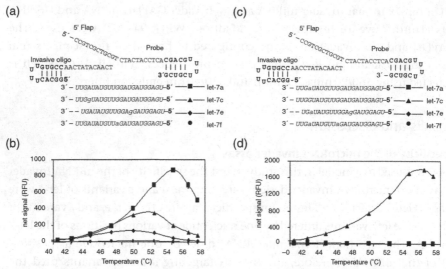

Figure 18.2. Discrimination capability of the microRNA Invader assay. (a) Invasive and probe oligonucleotides designed to detect let-7a RNA, and nucleotide sequences of let-7a RNA and closely related let-7 variants (shown in italics); nucleotide differences from let-7a RNA are in lower case or a dash. The 2′-O-methyl containing nucleotides of the regions forming hairpin structures are in bold. (b) Temperature dependence of the fluorescence signals generated in the Invader miRNA assay using as targets the let-7 variants indicated in (a). (c) Invasive and probe oligonucleotides designed to detect let-7c RNA, and nucleotide sequences of let-7 variants highlighted (in lower case or by a dash) where the sequence differs from let-7c. (d) The let-7a, let-7e and let-7f variants generated very low net signal and their temperature dependence curves are superimposed in the figure.

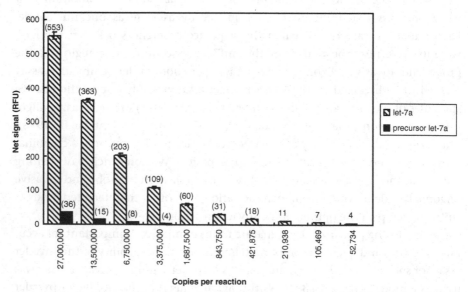

Figure 18.3. Signal generation of the mature (dashed bar) and precursor (black bar) forms of the let-7a miRNA using the let-7a microRNA Invader assay. Results shown are the average net signals obtained from three measurements and error bars are the standard deviations of these three measurements. Numbers in parentheses above the bars are average net signal values obtained.

than 5% of the observed signal for mature let-7a. This is most likely caused by the melting temperatures and stability of the precursor miRNA hairpins which are much higher than the temperatures at which the Invader miRNA assay takes place; typically between 50 °C and 60 °C. At these temperatures the precursors are mostly in the hairpin form and are unavailable for base pairing with the probe and invasive oligonucleotides of the Invader assay.

Sensitivity and quantitative capabilities of the microRNA Invader assay

The sensitivity of six miRNA Invader assays were tested by diluting their corresponding synthetic miRNA targets and performing limit of detection Invader assay experiments. The miRNA assays tested were let-7a, miR-1, miR-15, miR-16, miR-135, and miR-125b. Results show limits of detection as low as 30 zeptomoles (~20000 copies). Extending the primary and secondary reaction times of the Invader assay and increasing the reaction volume and sample volume input per reaction can further increase the detection limit of the assay (data not shown). Figure 18.4a shows an example plot of net signal versus moles of miR-15. The generated standard curves from this plot can be used to back calculate miRNA levels of unknown samples by fitting the linear portion of the curve to an exponential curve and back calculating the miRNA levels based on the obtained net signal. This shows that the microRNA Invader assay can be used to accurately detect and quantitate miRNA levels in unknown samples.

MicroRNA Invader assay using total RNA

Four Invader miRNA assays for let-7a, miR15, miR-16, and miR-125b were developed to validate the generality of the detection method and to perform their assay using total RNA isolated from different tissues. Figures 18.4b shows that these Invader miRNA assays were capable of detecting the targeted miRNA in a tissue-specific and dose dependent manner. Using standard curves generated from synthetic RNA signals for each of the miRNAs tested (not shown), we were capable of back-calculating the levels of the four miRNAs in the total RNA samples (Figure 18.4b).

MicroRNA Invader assay using cell-lysate and total RNA

Total RNA isolation methods can result in significant loss of short miRNAs. Therefore, the ability to quantify miRNAs in cell lysates without the need for total RNA isolation offers an advantage over traditional sample preparation methods which can lead to variable results. Invader miRNA assay could quantify miRNAs in cell lysates, thereby avoiding the need to isolate RNA. To assess the Invader assay's ability to detect miRNAs in the cell-lysate method, *Drosophila melanogaster* miR-1 RNA was added to HeLa or Hs578T cells and then either cell lysates prepared by treatment with NP-40 at 80 °C, or isolated total RNA prepared by phenol–chloroform extraction. Signals corresponding to ~70% of the added miR-1 were detected when assayed in either a total RNA sample or cell lysate (data not shown). This shows that the Invader miRNA assay can be reliably performed in NP-40 lysates of cells, without RNA isolation and purification.

(a)

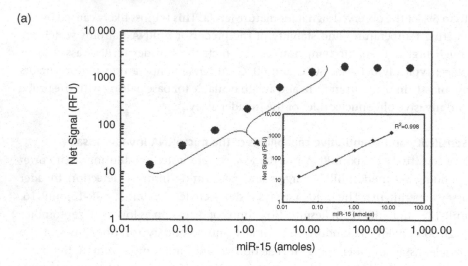

(b)

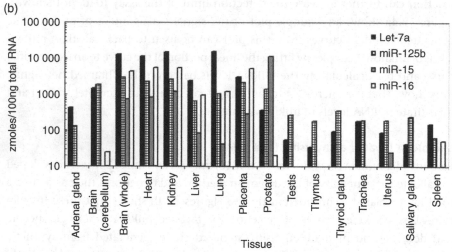

Figure 18.4. (a) Log–log plot of net signal (RFU) versus miR-15 levels (amoles) (dose response curve) obtained using the miR-15 Invader assay. An exponential fit of the linear portion of the curve generates a standard curve that can be used to calculate levels of miR-15 in unknown samples. (b) Expression profiles for let-7a (black bar), miR-125b (crossed bar), miR-15 (gray bar), and miR-16 (open bar) 100 ng of total RNA from a variety of human tissues. Exponential fits of linear portions of standard curves generated for each miRNA were used to back calculate miRNA levels in the 100 ng total RNA from each tissue.

Biplex Invader assay

Direct comparison of levels of miRNAs from different preparations can be peformed using an RNA internal standard present in both samples. This internal standard, commonly referred to as a "housekeeping gene," enables the normalization of miRNA measurements from different samples that might have different amounts of total RNA or cells or change relative to each other during development or in response to a physiological perturbation. In the biplex miRNA Invader assay, the use of two different probe oligonucleotides, each with a unique 5′-flap, allows for simultaneous detection and quantitation of two different RNAs in the same

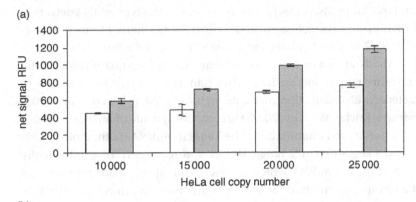

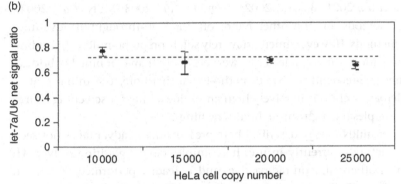

Figure 18.5. Uniform detection of let-7a RNA in HeLa cell lysates, biplexed with a U6 RNA Invader assay for normalization. (a) Net signals for let-7a RNA (open bars) and U6 RNA (gray bars) were measured by a biplex Invader assay of NP-40 performed on lysates of different numbers of HeLa cells. The signals detected for the miRNA and the U6 RNA were from spectrally distinct FAM and Redmond Red™ fluorophores, respectively. (b) The let-7a net signal values were divided by the corresponding U6 RNA signal values to obtain a normalized signal. The average of the four normalized let-7a values is shown by the dashed line.

sample (Eis *et al.*, 2001). A normalization factor for the levels of miRNA from different sample preparations can be deduced from the levels of housekeeping gene in different samples. An example of such normalization is demonstrated in Figure 18.5. Varying levels of HeLa cells in four samples were lysed and the lysates were analyzed with a biplex Invader assay designed to detect the let-7a specific signal (FAM fluorescence) and a U6 RNA specific signal (Redmond Red fluorescence). The levels of both signals were proportional to the amounts of HeLa cells analyzed (Figure 18.5a), so the normalized amount of let-7a signal remained constant across all samples regardless of the amount of HeLa cells used (Figure 18.5b). This shows that the signal of a housekeeping RNA, U6 RNA in this example, can serve for normalization of the let-7a RNA-specific signals in two preparations when the absolute number of cells is unknown.

Concluding remarks

Recent implication of miRNAs in diseases and cancer development and research focusing on using miRNA as biomarkers, have warranted the development of sensitive and specific miRNA detection assays. Some methods of miRNA detection and quantitation rely on gel-based assays such as northern blotting (Lagos-Quintana et al., 2001) and primer extension (Zeng and Cullen, 2003). These gel-based methods are labor intensive, require microgram quantities of total RNA, are low throughput and are insensitive (Lim et al., 2003). Modification to well-characterized RNA detection methods such as TaqMan (Chen et al., 2005) and microarrays (Krichevsky et al., 2003; Liu et al., 2004) made them more suitable for miRNA detection and quantitation. The TaqMan miRNA method offers low-sensitivity and a larger dynamic range. However, it requires testing individually labeled probes for each miRNA which represents a major cost for running this assay and a less specific method for small perturbations in miRNA composition (Burczynski et al., 2001; Busten, 2002; Wagner et al., 2003; Mills et al., 2004). Microarray methods, on the other hand, offer a high-throughput advantage over other methods. However, microarrays rely solely on probe-miRNA hybridization discrimination which can lead to lower specificity than combined hybridization and enzyme-recognition based methods. Furthermore, the initial cost of making microarray chips is relatively high and is ideal only for screening a large number of samples (i.e high-throughput screening).

The Invader miRNA assay described here presents many advantages not seen with other methods currently in use. It is simple, rapid (requiring only 2–3 h incubation), isothermal, non-radioactive, and is readily performed directly in cell lysates. Accordingly, it lends itself well to applications in high-throughput screening. Costs are reduced by purifying probe oligonucleotides in-house by gel electrophoresis, and by the use of a standard secondary reaction so that the same FRET pair oligonucleotide is used in all reactions. Furthermore, the Invader miRNA assay is quantitative, highly specific, and sensitive (detecting as few as 20 000 miRNA molecules).

ACKNOWLEDGMENTS

The authors wish to thank James Dahlberg and Elsebet Lund for fruitful discussions and suggestions and Dan Knauss for critical reading of the manuscript.

REFERENCES

Allawi, H. T. and SantaLucia, J., Jr. (1997). Thermodynamics and NMR of internal G.T mismatches in DNA. *Biochemistry*, **36**, 10581–10594.

Allawi, H. T. and SantaLucia, J., Jr. (1998). Nearest-neighbor thermodynamics of internal A.C mismatches in DNA: sequence dependence and pH effects. *Biochemistry*, **37**, 9435–9444.

Allawi, H. T., Dahlberg, J. E., Olson, S. et al. (2004). Quantitation of microRNAs using a modified Invader assay. *RNA*, **10**, 1153–1161.

Alvarez-Garcia, I. and Miska, E. A. (2005). MicroRNA functions in animal development and human disease. *Development*, **132**, 4653–4662.

Ambros, V. (2003). MicroRNA pathways in flies and worms: growth, death, fat, stress, and timing. *Cell*, **113**, 673–676.

Bao, N., Lye, K. W. and Barton, M. K. (2004). MicroRNA binding sites in *Arabidopsis* class III HD-ZIP mRNAs are required for methylation of the template chromosome. *Developmental Cell*, **7**, 653–662.

Bartel, D. P. (2004). MicroRNAs: genomics, biogenesis, mechanism, and function. *Cell*, **116**, 281–297.

Brennecke, J., Hipfner, D. R., Stark, A., Russell, R. B. and Cohen, S. M. (2003). bantam encodes a developmentally regulated microRNA that controls cell proliferation and regulates the proapoptotic gene hid in Drosophila. *Cell*, **113**, 25–36.

Burczynski, M. E., McMillian, M., Parker, J. B. *et al.* (2001). Cytochrome P450 induction in rat hepatocytes assessed by quantitative real-time reverse-transcription polymerase chain reaction and the RNA invasive cleavage assay. *Drug Metabolism Disposition*, **29**, 1243–1250.

Busten, S. A. (2002). Quantification of mRNA using real-time reverse transcription PCR (RT-PCR): trends and problems. *Journal of Molecular Endocrinology*, **29**, 23–29.

Calin, G. A., Sevignani, C., Dumitru, C. D. *et al.* (2004). Human microRNA genes are frequently located at fragile sites and genomic regions involved in cancers. *Proceedings of the National Academy of Sciences USA*, **101**, 2999–3004.

Cantor, C. R., Warshaw, M. M. and Shapiro, H. (1970). *Biopolymers*, **9**, 1059–1077.

Chen, C., Ridzon, D. A., Broomer, A. J. *et al.* (2005). Real-time quantification of microRNAs by stem-loop RT-PCR. *Nucleic Acids Research*, **33**, e179.

Croce, C. M. and Calin, G. A. (2005). miRNAs, cancer, and stem cell division. *Cell*, **122**, 6–7.

de Arruda, M., Lyamichev, V. I., Eis, P. S. *et al.* (2002). Invader technology for DNA and RNA analysis: principles and applications. *Expert Reviews in Molecular Diagnostics*, **2**, 487–496.

Dunn, W., Trang, P., Zhong, Q. *et al.* (2005). Human cytomegalovirus expresses novel microRNAs during productive viral infection. *Cell Microbiology*, **7**, 1684–1695.

Eis, P. S., Olson, M. C., Takova, T. *et al.* (2001). An invasive cleavage assay for direct quantitation of specific RNAs. *Nature Biotechnology*, **19**, 673–676.

Eis, P. S., Tam, W., Sun, L. *et al.* (2005). Accumulation of miR-155 and BIC RNA in human B cell lymphomas. *Proceedings of the National Academy of Sciences USA*, **102**, 3627–3632.

Hall, J. G., Eis, P. S., Law, S. M. *et al.* (2000). Sensitive detection of DNA polymorphisms by the serial invasive signal amplification reaction. *Proceedings of the National Academy of Sciences USA* **97**, 8272–8277.

Hunter, C. and Poethig, R. S. (2003). miSSING LINKS: miRNAs and plant development. *Current Opinion in Genetics & Development*, **13**, 372–378.

Johnston, R. J. and Hobert, O. (2003). A microRNA controlling left/right neuronal asymmetry in *Caenorhabditis elegans*. *Nature*, **426**, 845–849.

Kaiser, M. W., Lyamicheva, N., Ma, W. *et al.* (1999). A comparison of eubacterial and archaeal structure-specific 5′-exonucleases. *Journal of Biological Chemistry*, **274**, 21 387–21 394.

Krichevsky, A. M., King, K. S., Donahue, C. P., Khrapko, K. and Kosik, K. S. (2003). A microRNA array reveals extensive regulation of microRNAs during brain development. *RNA*, **9**, 1274–1281.

Lagos-Quintana, M., Rauhut, R., Lendeckel, W. and Tuschl, T. (2001). Identification of novel genes coding for small expressed RNAs. *Science*, **294**, 853–858.

Lagos-Quintana, M., Rauhut, R., Yalcin, A. *et al.* (2002). Identification of tissue-specific microRNAs from mouse. *Current Biology*, **12**, 735–739.

Lagos-Quintana, M., Rauhut, R., Meyer, J., Borkhardt, A. and Tuschl, T. (2003). New microRNAs from mouse and human. *RNA*, **9**, 175–179.

Lim, L. P., Glasner, M. E., Yekta, S., Burge, C. B. and Bartel, D. P. (2003). Vertebrate microRNA genes. *Science*, **299**, 1540.

Liu, C. G., Calin, G. A., Meloon, B. *et al.* (2004). An oligonucleotide microchip for genome-wide microRNA profiling in human and mouse tissues. *Proceedings of the National Academy of Sciences USA*, **101**, 9740–9744.

Lyamichev, V., Mast, A. L., Hall, J. G. *et al.* (1999). Polymorphism identification and quantitative detection of genomic DNA by invasive cleavage of oligonucleotide probes. *Nature Biotechnology*, **17**, 292–296.

Lyamichev, V. I., Kaiser, M. W., Lyamicheva, N. E. *et al.* (2000). Experimental and theoretical analysis of the invasive signal amplification reaction. *Biochemistry*, **39**, 9523–9532.

Michael, M. Z., O'Connor, S. M., van Holst Pellekaan, N. G., Young, G. P. and James, R. J. (2003). Reduced accumulation of specific microRNAs in colorectal neoplasia. *Molecular Cancer Research*, **1**, 882–891.

Mills, J. B., Rose, K. A., Sadagopan, N., Sahi, J. and de Morais, S. M. (2004). Induction of drug metabolism enzymes and MDR1 using a novel human hepatocyte cell line. *Journal of Pharmacology and Experimental Therapy*, **309**, 303–309.

Miska, E. A. (2005). How microRNAs control cell division, differentiation and death. *Current Opinion in Genetics & Development*, **15**, 563–568.

Pasquinelli, A. E. and Ruvkun, G. (2002). Control of developmental timing by microRNAs and their targets. *Annual Reviews of Cell Developmental Biology*, **18**, 495–513.

Pasquinelli, A. E., Hunter, S. and Bracht, J. (2005). MicroRNAs: a developing story. *Current Opinion in Genetics & Development*, **15**, 200–205.

Pfeffer, S., Zavolan, M., Grasser, F. A. *et al.* (2004). Identification of virus-encoded microRNAs. *Science*, **304**, 734–736.

Reinhart, B. J., Slack, F. J., Basson, M. *et al.* (2000). The 21-nucleotide let-7 RNA regulates developmental timing in *Caenorhabditis elegans*. *Nature*, **403**, 901–906.

Reinhart, B. J., Weinstein, E. G., Rhoades, M. W., Bartel, B. and Bartel, D. P. (2002). MicroRNAs in plants. *Genes & Development*, **16**, 1616–1626.

Wagner, E. J., Curtis, M. L., Robson, N. D. *et al.* (2003). Quantification of alternatively spliced FGFR2 RNAs using the RNA invasive cleavage assay. *RNA*, **9**, 1552–1561.

Walter, A. E. and Turner, D. H. (1994). Sequence dependence of stability for coaxial stacking of RNA helices with Watson-Crick base paired interfaces. *Biochemistry*, **33**, 12715–12719.

Walter, A. E., Turner, D. H., Kim, J. *et al.* (1994). Coaxial stacking of helixes enhances binding of oligoribonucleotides and improves predictions of RNA folding. *Proceedings of the National Academy of Sciences USA*, **91**, 9218–9222.

Wienholds, E. and Plasterk, R. H. (2005). MicroRNA function in animal development, *Federation of the European Biochemical Society Letters*, **579**, 5911–5922.

Wienholds, E., Kloosterman, W. P., Miska, E. *et al.* (2005). MicroRNA expression in zebrafish embryonic development. *Science*, **309**, 310–311.

Xu, P., Vernooy, S. Y., Guo, M. and Hay, B. A. (2003). The Drosophila microRNA Mir-14 suppresses cell death and is required for normal fat metabolism. *Current Biology*, **13**, 790–795.

Zeng, Y. and Cullen, B. R. (2003). Sequence requirements for micro RNA processing and function in human cells. *RNA*, **9**, 112–123.

19 A single molecule method to quantify miRNA gene expression

Sonal Patel, Joanne Garver, Michael Gallo, Maria Hackett, Stephen McLaughlin, Steven R. Gullans, Mark Nadel, John Harris, Duncan Whitney and Lori A. Neely*

Introduction

Deemed the "breakthrough of the year" by *Science* magazine in 2002, research into the biology of small RNA regulation has grown exponentially in recent years; however, the field is relatively nascent in terms of identifying and characterizing the universe of miRNAs and their expression in various biological states. According to the miRNA registry (release 6.0, www.sanger.ac.uk/Software/Rfam/mirna/index.shtml); of the 319 predicted human miRNAs the expressions of 234 have been experimentally verified by Northern blot, cloning, or microarray. Further, the total number of miRNAs within a genome is unknown. Thus, sensitive, specific, quantitative, and rapid methods for measuring the expression levels of miRNAs would significantly advance the field.

The short 21 nucleotide nature of these molecules makes them difficult to study via conventional techniques. They are not easily amplified which makes miRNA microarrays and quantitative PCR technically challenging. Despite these challenges, several groups have undertaken miRNA microarray studies to quantify miRNA gene expression. Their approaches are similar in requiring up-front enrichment for small RNAs, reverse transcription, PCR amplification, labeling, and clean-up steps. While the arrays are superior at large scale screening they lack the ability to finely discriminate expression levels and are at best semi-quantitative. Theoretically the most sensitive technique to quantify miRNAs is reverse transcription RT-PCR (real time RT-PCR). However, this method is difficult in both assay (probe) design and execution. Tissue samples must be devoid of enzyme inhibitors to enable efficient reverse transcription and amplification steps (Tichopad *et al.*, 2004). At the early stages in the reactions subtle variations in the thermal cycling conditions, reaction composition, as well as non-specific priming can have a dramatic effect on the amount of amplified target. In instances where relative quantitation is desired an internal control is needed and great care must be taken to ensure the amplification efficiencies of both the internal control

---

* Author to whom correspondence should be addressed.

MicroRNAs: From Basic Science to Disease Biology, ed. Krishnarao Appasani. Published by Cambridge University Press. © Cambridge University Press 2008.

and the target are the same. The "gold standard" for identifying and quantifying miRNA gene expression is Northern blotting which is a cumbersome and a relatively insensitive means of measuring expression levels. In addition, 10–30 µg of total RNA is typically required per well and the throughput is limited.

Several advances in single molecule detection (SMD), laser induced fluorescence (LIF) and fluorescence correlation spectroscopy (FCS) have provided highly sensitive approaches to study individual macromolecules under physiological conditions (Chirico et al., 2001; Hesse et al., 2002; Jett et al., 1989; Goodwin et al., 1996; Yanagida et al., 2000; Schwille et al., 1997; Ambrose et al., 1998). These single fluorophore, single molecule techniques have been routinely employed to quantitatively measure properties of molecules in dynamic systems such as protein folding, DNA transcription, DNA binding proteins and molecules in flowing fluid systems (Schwille et al., 1997; Ambrose et al., 1998; Haab and Mathies, 1995; Anazawa et al., 2002; Dundr et al., 2002; Bennink et al., 1999; Yamasaki et al., 1999). These methods distinguish amongst different molecules in solution based on their unique spectral properties (Schwille et al., 1997; Soper et al., 1992). Typically, individual molecules are tagged with a single fluorophore and analyzed using a confocal microscope with laser illumination and photon-burst detection to enhance signal-to-noise (Chirico et al., 2001; Haab and Mathies, 1995; Novikov et al., 2001; Schwille and Kettling, 2001; Soini et al., 2000). In dual-color FCS experiments, individual molecules diffuse into an interrogation volume and a time-dependent cross correlation provides single molecule analysis with sensitivities of 10–100 nM (Schwille et al., 1997). More recently, two-color coincidence fluorescence detection was found capable of ultra-sensitive quantitation (50 fM) when measuring individual synthetic 40 bp DNA molecules in a non-flowing system (Li et al., 2003). FCS, however, depends on diffusion and requires a complex experimental platform. Overall, these studies demonstrate the ability of single molecule techniques to detect and quantify the physico-chemical properties of individual biomolecules. To date, however, no one has used this strategy to rapidly quantify miRNA levels in complex biological samples.

We sought to develop a single molecule method for detecting and quantifying miRNAs with great sensitivity and specificity. To achieve this aim, we built a microfluidic, multi-color laser system capable of counting individual molecules as they flow at high velocity through the system. In addition, we developed a rapid solution-based hybridization assay with femtomolar sensitivity for quantitation of miRNAs using fluorescently labeled locked nucleic acid (LNA) DNA chimeric probes. This assay is homogeneous in nature requiring no target capture, enrichment, or clean-up steps and does not require any reverse transcription or amplification. Within this chapter we outline our single molecule method for miRNA quantification and demonstrate the method's sensitivity and specificity. We then apply the assay to quantify human miRNA gene expression in normal versus cancer tissues.

Methodology

Single molecule detection platform

We used a Trilogy™ 2020 confocal laser-induced fluorescence detector from US Genomics to perform the single molecule counting experiments. The instrument enables three-color fluorescent detection in a microfluidic flow stream as previously described (Chan *et al.*, 2004), with engineering modifications to automate sample handling and delivery. Its schematic is shown in Figure 19.1a. Optically this system is similar to systems described by others (Widengren and Kask, 1993; Eigen and Rigler, 1994; Nie *et al.*, 1995; Widengren, 1995). The system contains four lasers (one 488 nm and one 532 nm diode-pumped, solid-state lasers, and two 638 nm solid-state, diode lasers). In this particular study we utilized the 532 nm laser to excite the Oyster 556 fluorophore and the 638 nm lasers to excite the Oyster 656 fluorophore. The laser light is shaped into a high-eccentricity elliptical beam, reflected off a polychroic mirror and focused through a high numerical aperture oil-immersion objective (1.25 N. A. Achromat by Nikon) into a custom fused silica microfluidic capillary. At the beam waist, the focused stripes have a length of approximately 10 microns (normal to direction of fluid flow) and a width of approximately 1 micron (parallel to direction of fluid flow). Figure 19.1b shows the location of the individual laser stripes arrayed along the capillary. These highly focused stripes permit fluorescent excitation of a very small sample volume (~10 fl) within the capillary. This reduces the noise produced by Raman and Rayleigh light scattering and background fluorescence from the biological samples. Light emitted by the fluorescently tagged molecules passes through the polychroic mirror and is focused through apertures to reduce the stray light reaching the detector. It is then split by dichroic mirrors into wavelengths specific for each fluorophore, and further band-pass filtered to reject reflected laser light and any Rayleigh or Raman scattered light. The filtered light in each channel is then focused onto the active surface of an electron multiplying CCD camera. The CCD itself consists of a 128 × 128 array of pixels, each 24 microns square. The lens and objective pair provide an approximately 50 × magnification and the image of each stripe is aligned with adjustable mirrors onto a distinct band of pixels, 20 wide and 2 tall. High-speed data acquisition electronics within the camera is used to count the photons collected by each pixel with an integration time of 110 μs (9 kHz sampling rate). The camera electronics and firmware also provides automatic binning of each 2 × 2 square of pixels into a "super-pixel." It should be noted that this system has single fluorophore detection sensitivity for each of three spectrally distinct fluorophores.

We modified our experimental platform to include an automated 96-well plate compatible sample delivery system. The sample is pulled by vacuum through a 15 cm long custom fused silica microfluidic capillary which we sheath at one end with a needle. The needle is required to pierce the heat seals covering the 96-well plates.

(a)

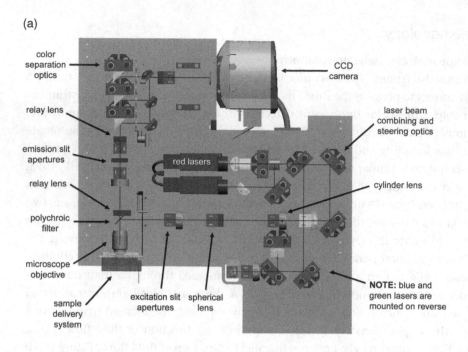

(b)

NOTE: only illumination of second
red stripe shown for clarity

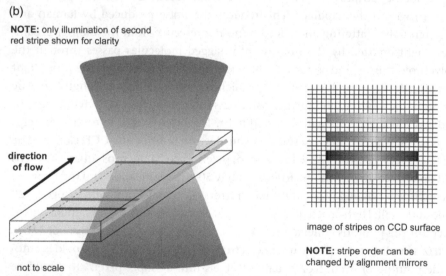

image of stripes on CCD surface

NOTE: stripe order can be
changed by alignment mirrors

not to scale

Figure 19.1. (a) An optical schematic of our four laser single molecule detection platform. (b) Laser focal volumes are focused into "stripes" and arrayed as shown within a microfluidic channel. For clarity we show only the illumination of the second red stripe. The 532 nm laser interrogation volume is focused 2 microns downstream from the first 633 nm interrogation volume (Red 1) while the second 633 nm focal volume (Red 2) is focused 8 microns downstream from the first red focal volume. The data analysis software uses a standard cross-correlation algorithm to measure the flow velocity between the two red laser interrogation volumes. This algorithm also accounts for the two micron off-set between the Red 1 and Green laser interrogation volumes and synchronizes the data traces such that a single dual color fluorescently tagged molecule will appear as a coincident peak in fluorescence emission.

The automated delivery system allows the sample injection and capillary wash routine to be pre-programmed. We flushed the capillary with conditioning solution (US Genomics (USG), Woburn, MA) for 10 min and then 1X running buffer (USG) for 20 minutes. Following this wash step, the samples are queued for interrogation. Data was collected for 36 s for all samples. Upon completion of each sample draw, the tube was automatically flushed for 10 s with 1X running buffer. The instrument also contains a rack that can hold eight 200 μl tubes. Three tubes containing 100 μl of conditioning solution were loaded into this rack and the capillary was subjected to a 10 s rinse with conditioning buffer followed by 10 s of running buffer every third sample.

Assay design

Locked nucleic acids are synthetic nucleotide analogs composed of a ribose sugar with an ethylene bridge between the 2′ oxygen and the 4′ carbon. This bridge pre-locks the sugar into the C3 endo puckered conformation that it would naturally adopt upon hybridization to a complementary nucleic acid strand. The inclusion of locked bases in DNA probes increases the thermal stability of LNA/DNA duplexes (Christensen *et al.*, 2001). In our assay we hybridize two spectrally distinguishable fluorescent LNA/DNA 10-mer probes to the small RNA of interest (Figure 19.2). Our inclusion of LNA bases within the probe increases the thermal stability of the LNA/RNA hybrids enabling us to hybridize at higher temperatures (55 °C) thereby minimizing non-specific hybridization. The use of these short probes also maximizes specificity as a single base mismatch can have a profound affect on the stability of the duplex. In our previous studies we have used tetramethylrhodamine, Alexa 546, or Oyster 556 on one of the LNA probes and Alexa 633, Cy5, or Oyster 656 on the other LNA probe. Following hybridization, the temperature is decreased to 40 °C and synthetic DNA oligonucleotides end-labeled with quencher molecules are hybridized to the remaining non-hybridized LNA probes. This quenching reaction reduces the background fluorescence thereby increasing the sensitivity of the assay. Quenched reactions are then diluted two to ten-fold (depending upon LNA probe concentration) and subjected to single molecule interrogation.

Dual-color coincident event counting

To detect and quantify molecules of interest we employ a dual-color coincident event detection strategy. As shown schematically in Figure 19.2, target molecules bearing two fluorescent probes, one labeled with Oyster 556, referred to as the "green probe," and the other labeled with Oyster 656, referred to as the "red probe," pass through the laser excitation/detection volumes (Red1, Green, and Red2) and emit photons. We record the photon bursts emanating from the laser interrogation volume in one millisecond time intervals. The laser focal volumes are arrayed within the microfluidic channel as shown in Figure 19.1b. Using a method previously described by others (Brinkmeier *et al.*, 1999) we use cross-correlation between the two red channels to measure the flow velocity of the fluorescently labeled molecules. We focus the 633 nm laser two micrometers upstream of the 532 nm laser. The off-set of the laser focal volumes eliminates spectral cross-talk. Our data analysis software accounts for the offset and

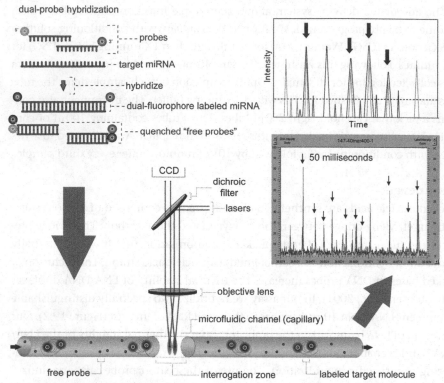

Figure 19.2. A single molecule two-color coincident detection strategy is employed to quantify miRNAs. miRNAs are detected and counted as coincident bursts in photon emission emanating from two different fluorescent dyes. The steps of the assay are as follows: miRNAs are hybridized in solution to two spectrally distinguishable fluorescent probes (in this case Oyster 556 and Oyster 656) and flowed by vacuum pressure through a capillary containing a series of femtoliter laser focal volumes. Fluorescence emission is recorded as spikes in signal intensity over time. Arrows highlight the coincident peaks in photon emission. Complementary DNA probes bearing fluorescence quencher molecules are then hybridized to the remaining unbound fluorescent probes to minimize coincident events that could be created by free probes simultaneously entering the laser interrogation spots.

synchronizes the data allowing a dual fluorescently labeled molecule to be counted as a coincident event (Figure 19.2). The data analysis method counts the number of coincident events above an established threshold. More specifically, we count the number of time periods during which the signal from each of two fluorescence channels exceeds some fixed value for that channel. The thresholds are used to eliminate the background signal from ambient fluorescence. Coincidence in two (or more) colors is used to mitigate the effect of free probes and the residual background fluorescence. A detection threshold, representing the minimum level of photon burst counted as signal from a fluorescently labeled probe, was set for each color using the average signal intensity plus five times the standard deviation of the signal observed from mock hybridization reactions containing one microgram of total RNA diluted 2- to 10-fold in 1X running buffer.

A novel feature of our data analysis method is an estimation of the number of random coincidences expected on the basis of the raw data. This estimate is subtracted from the raw coincidence count to give an estimate of the number of coincidences caused by dual-tagged molecules, and by inference, of the concentration of the analyte. A detailed description of this correction method is presented in D'Antoni *et al.* (2006).

Synthetic oligonucleotide RNAs, DNAs, and LNAs

We designed our chimeric DNA/LNA probes complementary to publicly available miRNA target sequences using in-house design tools; however, publicly available LNA design tools may be found at www.exiqon.com. We iteratively adjusted the number and placement of the locked nucleotides to achieve LNA probes with similar melting temperatures, minimal LNA homodimer and heterodimer formation, and minimal hybridization to background RNAs. The fluorescently modified chimeric DNA/LNA probes (Proligo, Boulder, CO or Exiqon, Vedbaek, Denmark), DNA quencher probes containing A-quenchers (Exiqon), and synthetic miRNA templates (Integrated DNA Technologies, Coralville, IA) were synthesized and purified. Concentrations and stoichiometry of dye-label to oligo were determined by measuring the oligonucleotide's absorbance from 650 nm to 230 nm on a Cary Varian UV/vis spectrophotometer (Palo Alto, CA). Dye extinction coefficients and correction factors were provided by Molecular Probes (Eugene, OR) or Denovo (Munster, Germany). Quencher probe labeling stoichiometry was determined by reverse phase high performance liquid chromatography.

Validating hybridization of miRNA probe pairs

To independently verify that our LNA probes hybridize efficiently to their miRNA targets and to determine the minimum concentration of probe required to drive the hybridization reactions to completion, we conducted an electrophoretic mobility shift assay using 1 pM of a radiolabeled synthetic miRNA target to various concentrations of LNA probes (100 pM to 5 nM). The miRNA targets were radiolabeled with ^{32}P using T4 PNK as previously described (Sambrook *et al.*, 1989). Hybridization reactions were incubated in 1X USG hybridization buffer (USG) in a thermocycler (MJ Research) at 80 °C for 5 min followed by a one hour incubation at 55 °C. In some cases quencher probes were then added to determine if their addition decreased the amount of hybridized product by competing with the miRNA target for binding of the LNA probes. The reactions were cooled to 4 °C and electrophoresed at 100 V through a 20% native PAGE gel (Invitrogen, Carlsbad, CA) for three hours. The gels were vacuum-dried on Whatman blotting paper (Florham Park, NJ) and exposed to a phosphorscreen. The percent hybridized miRNA target was quantified using the ImageQuant program (GE Healthcare, Piscataway, NJ).

Absolute miRNA quantitation

To quantify miRNA gene expression we employ standard curves containing a range of synthetic miRNA concentrations spiked into a complex RNA background. For our human expression profile studies we typically prepare two to three independent standard curves consisting of 0.25, 0.5, 1, 2, 5, 10, 25, and

50 pM of synthetic miRNA spiked into either human tissue total RNA depleted for small molecular weight RNAs, *E. coli* total RNA, or human universal RNA (Ambion, Austin, TX). When screening for expression in a few tissues, it is best to match the background RNA to the tissue being studied. We recommend including 6–20 no-target controls (probes and RNA background) for each miRNA. Tissue total RNA hybridization reactions consist of 50 ng to 100 ng of tissue total RNA, 200–500 pM LNA probes, and 0.5 µl of RNAse Inhibitor (US Biologicals, Swampscott, MA) in 1X USG hybridization buffer. We seal our 96-well plates with Microseal lids (MJ Research, Waltham, MA) and incubate at 80 °C for 5 min followed by incubation at 55 °C for 1 h. The plate is chilled and DNA quencher probes (2X with regard to LNA probe concentration) are added. The quenching reactions are incubated at 40 °C for 30 min. We dilute the samples to a final LNA probe concentration of 100 pM prior to single molecule interrogation. Then 95% confidence intervals based on a one-tailed student's T test assuming unequal variances are calculated for the negative controls and for each data point in the calibration curve. The lower limit of quantitation is defined as the first data point on the calibration curve that lies above the upper confidence limit of the 95% confidence interval for the no-target controls and has a coefficient of variation (standard deviation/mean of $n = 3$) of 20% or less. The data is fit with standard linear regression analyses using ordinary least squares estimations.

If the coincident event numbers counted in the tissue total RNA from three independent experiments had a coefficient of variation (CV) less than 20% and a lower confidence limit of the 95% confidence interval above the upper limit for the 95% confidence interval of the zero target control, we deem the miRNA (or a closely related family member) as expressed and calculate the concentration of the miRNA within that tissue. To avoid instances where the total RNA yielded a mean coincident event number with a CV of 20% or less but not above the upper 95% confidence limit of our zero target control, we recommend testing different amounts of total RNA for miRNA expression. We typically use 50 and 100 ng of tissue total RNA to quantify the expression of a single miRNA.

Results and discussion

Sensitivity and dynamic range of the assay

The assay can detect as little as 100 fM (~1.4 fg or 0.2 amol (attomoles) of miRNA). The lower limit of quantification for the assay is around 250 fM run concentration (corresponding to 500 fM in the hybridization reaction). Figure 19.3 shows a standard curve with a synthetic miRNA at concentrations ranging from 250 fM to 200 pM spiked into 250 ng of background RNA. All samples were diluted 2-fold prior to single molecule detection. Each data point represents the mean coincident events measured per second for three independent experiments +/− the 95% confidence interval. The linear dynamic range of the assay spans 3-logs (~200 fM to 200 pM).

To further characterize the sensitivity of the assay we tested our ability to quantify miRNA expression in 50 and 100 ng of tissue total RNA. Mir-16, mir-9,

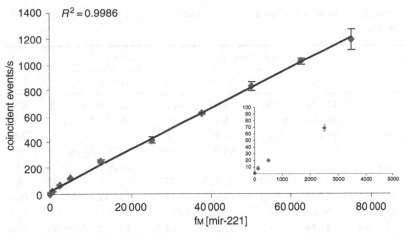

Figure 19.3. The Direct miRNA assay is sensitive to 500 fM miRNA and has a linear dynamic range spanning 3-logs (250 fM–200 pM). A synthetic RNA oligonucleotide identical in sequence to mir-187 was serially diluted from 200 pM to 250 fM and hybridized in the presence of a complex RNA background (250 ng of *E. coli* total RNA) to its complementary Oyster 556 and Oyster 656 LNA probes. Following a 30 min quenching reaction, the hybridization reactions were diluted 2-fold and analyzed on our single molecule detection platform. The scattergram is a plot of the coincident events counted per second at each miRNA run concentration. The data were fit using an ordinary least squares fit of a simple linear regression model.

mir-22, mir-126, mir-143, mir-145, mir-191, and mir-205 were selected for this test to represent a range of expression levels, and we quantified their expression in tissues where our work and others' have indicated their presence at low, moderate, and high levels. Hybridization reactions were conducted with 50, 100, and 500 ng of tissue RNA and, for ease of data comparisons, the mean number of femtograms of miRNA per microgram total RNA ($n = 3$) were reported (Table 19.1). We observe good agreement between the amount of each target miRNA measured with 50, 100, and 500 ng of total RNA with CVs less than 20% for 12 out of 14 experiments and less than 22% for all experiments. This data suggests that for most miRNAs 50 ng of tissue total RNA is sufficient for accurate and precise quantitation.

Assay specificity

To determine if our miRNA assay is capable of discrimination amongst miRNA family members, we hybridized let-7a specific probes to let-7a, let-7b, let-7c, and let-7d miRNA targets. These reactions consisted of 5, 10, 50, and 100 pM of the different let-7 family members hybridized to LNA probes designed to have higher affinity for let-7a. These reactions were diluted 5-fold prior to single molecule detection. The data shown in Figure 19.4 suggests these probes are capable of discrimination amongst the four let-7 family members. The data is plotted as coincident events detected per second for each let-7 family member with the concentrations corresponding to the run concentrations. We observe a greater than 10-fold reduction in coincident events between let-7a and let-7b and let-7d at all of our test concentrations. We observe a reduction in coincident events of 2.5-fold between let-7a and let-7c.

Table 19.1. The Direct miRNA assay can accurately quantify miRNAs in as little as 50 ng of tissue total RNA

miRNA	tissue	average fg/μg +/− standard deviation			CV
		50 ng	100 ng	500 ng	
mir 9	brain	13 556 +/− 536	13 053 +/− 267	12 216 +/− 17.6	5%
mir-16	thymus	5884 +/− 831	7558 +/− 648	5007 +/− 157	21%
mir-22	brain	1649 +/− 102	2095 +/− 622	2578 +/− 193	22%
	prostate	6129 +/− 526	6518 +/− 510	5698 +/− 84	6%
mir-126	lung	7927 +/− 655	8666 +/− 1765	6906 +/− 182	11%
mir 143	cervix	53 682 +/− 5308	52 912 +/− 3791	53 166 +/− 4043	1%
mir-145	colon	13 056 +/− 969	14 811 +/− 1175	13 742 +/− 545	6%
	liver	1050 +/− 304	991 +/− 233	1306 +/− 87.3	19%
	placenta	13 418 +/− 697	13 912 +/− 68	13 282 +/− 666	2%
	brain	1563 +/− 24	1812 +/− 125	1957 +/− 82	11%
	cervix	77 157 +/− 2974	81 084 +/− 2522	94 676 +/− 1888	11%
mir-191	spleen	816 +/− 36	ND	845 +/− 127	3%
mir 205	prostate	5556 +/− 41	4500 +/− 186	6305 +/− 980	17%
	cervix	1600 +/− 42	1433 +/− 9	1703 +/− 348	9%

Mir-9, mir-16, mir-22, mir-126, mir-143, mir-145, mir-191, and mir-205 levels were quantified in 50, 100, and 500 ng of tissue total RNA. All data were normalized to femtograms of target miRNA per microgram of tissue total RNA and presented as the mean of three independent experiments ± the standard deviation. ND: not detected.

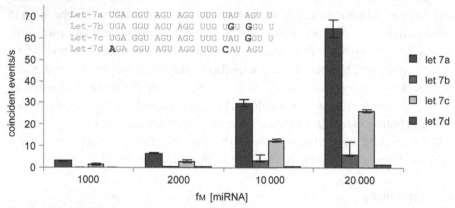

Figure 19.4. The Direct miRNA assay specifically detects the target miRNA. LNA/DNA probes designed to have high affinity for let-7a were hybridized to four members of the let-7 family (let-7a, let-7b, let-7c, and let-7d) at concentrations of 5, 10, 50, and 100 pM. Following hybridization and quenching reactions, all samples were diluted 5-fold for single molecule analysis. The sequences of the individual family members are indicated. The bar graph indicates the mean number of coincident events detected per second for three independent experiments ± the standard deviation. The miRNA concentrations indicated on the bar graph represent the run concentrations of the samples.

Quantitation of miRNA expression in cancer tissues

In a previously published study, we utilized this assay to quantify the expression of 45 different human miRNAs in 16 different tissues (Neely *et al.*, 2006). We have extended this study to measure the expression of a small subset of human miRNAs in tumor versus normal tissues. The expression of eleven human miRNAs was

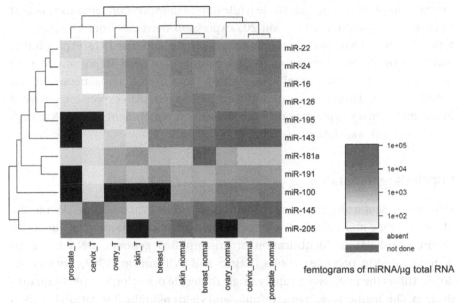

Figure 19.5. A human miRNA expression profile which quantifies aberrant miRNA gene expression found in tumor tissue. The tissue-specific expression of 11 different miRNAs within skin, breast, colon, cervix and prostate normal tissue was compared to their expression within tumor tissues. The coefficient of variation of the raw coincident event data was less than 20%. The results are the mean of three independent experiments expressed in average fg/μg tissue total RNA. Expression data was floored at the detection threshold (15 fg) and log-transformed. Hierarchical clustering using Ward's algorithm and the Euclidean distance metric was performed and a heat map was created in the R statistical environment (R Development Core Team, 2005). (See color plate 11)

quantified in breast, ovary, and prostate normal tissues and adenocarcinomas, cervix normal tissue and squamous cell carcinoma, and skin normal tissue and melanoma. Tissue total RNAs were procured from Ambion Inc. and were isolated from a single tumor with no pooling of normal or tumor samples. For each miRNA target we prepared three independent standard curves. Because of space limitations we do not show the individual standard curves but all showed strong correlation between the number of detected coincident events and the miRNA concentration with the coefficients of determination for 10 of the 11 miRNAs at 0.99 and one at 0.97. The goodness of fit of the linear data was measured not only by calculating the coefficient of determination but also by calculating the confidence intervals for the regression coefficients. Figure 19.5 is a heat map showing the miRNA expression quantified in femtograms (fg) miRNA per microgram (μg) tissue total RNA. Corroborating published results (Michael *et al.*, 2003; Iorio *et al.*, 2005) we observe significant down-regulation (greater than 2-fold) of mir-143 and mir-145 in all of our cancerous tissues. We also observe a greater than 2-fold decrease in mir-126 expression in breast tumor compared to normal breast tissue. We observe the down-regulation of mir-16, mir-22, mir-24, mir-100, mir-126, mir-191, and mir-195 in cervical tumor versus normal tissue. We observe no down-regulation of mir-181a within the cervical tumor. Mir-16, mir-22, mir-24, mir-100, mir-126, and mir-195 are down-regulated by greater than 3-fold in ovarian tumors versus normal tissue while mir-181a expression remains constant in normal versus tumor tissue. Mir-205 expression decreases from 13 pg/μg skin total RNA to undetectable amounts in melanoma and mir-181a expression is increased by 3-fold in melanoma versus normal skin. It is important to note however that the normal total RNAs were isolated from heterogeneous cellular sample (tissues). For example, the 3-fold increase in mir-181a expression observed in the melanoma sample may merely represent an enrichment of mir-181a melanocyte-specific expression.

Concluding remarks

Here we present a single molecule method for miRNA quantitation. While the utility of single molecule dual-color coincident event detection for the quantification of mRNAs or identification of un-amplified genomic DNA has been demonstrated in previous reports (Castro and Williamson, 1997; Korn *et al.*, 2003), this is the first assay to apply single molecule detection to miRNA quantification. Our method is extremely simple and yields reproducible data in less than 2 h. No preparative enrichment, ligation, or target/signal amplification steps are required, and there is no sample clean-up step. This lack of sample manipulation limits the points at which variability could be introduced into the assay thereby hindering the accuracy of quantitation. The assay is 96-well and 384-well plate compatible enabling us to quantify the expression of hundreds of miRNAs per day.

The quantitative nature of the data is unique yielding an accurate and precise amount of miRNA present in tissue total RNA. This amount is highly reproducible for a given miRNA in a particular tissue as we are able precisely and reproducibly to quantify miRNA gene expression over multiple experiments spanning several months. The accuracy of the method is evidenced by strong correlation between measured miRNA levels and Northern blot data (Neely *et al.*, 2006). The ability to measure the precise amount of a particular miRNA opens up the possibility of detecting subtle changes in miRNA expression levels that may not be observable by other techniques which measure bulk signal, such as Northern blots, microarrays, and RT-PCR.

REFERENCES

Ambrose, W. P., Semin, D. J., Robbins, D. L. *et al.* (1998). Detection system for reaction-rate analysis in a low-volume proteinase-inhibition assay. *Anal. Biochem.*, **263**, 150–157.

Anazawa, T., Matsunaga, H. and Young, E. S. (2002). Electrophoretic quantitation of nucleic acids without amplification by single-molecule imaging. *Anal. Chem.*, **74**, 5033–5038.

Bennink, M. L., Schaerer, O. D., Kanaar, R. and Sakata-Sogawa K. (1999). Single-molecule manipulation of double-stranded DNA using optical tweezers: interaction studies of DNA with RecA and YOYO-1. *Cytometry*, **36**, 200–208.

Brinkmeier, M., Dorre, K., Stephan, J. and Eigen, M. (1999). Two beam cross correlation: A method to characterize transport phenomena in micrometer-sized structures. *Analytical Chemistry*, **71**, 609–616.

Castro, A., and Williamson, J. G. K. (1997). Single molecule detection of specific nucleic acid sequences in unamplified genomic DNA. *Anal. Chem.*, **69**, 3915–3920.

Chan E. Y., Goncalves, N., Haeusler, R. A. *et al.* (2004). DNA mapping using microfluidic stretching and single-molecule detection of fluorescent site-specific tags. *Genome Research*, **6**, 1137–1146.

Chirico, G., Cannone, F., Beretta, S., Baldini, G. and Diaspro, A. (2001). Single molecule studies by means of the two-photon fluorescence distribution. *Microsc. Res. Tech.*, **55**, 359–364.

Christensen, U., Jacobsen, N., Rajwanshi, V. K., Wengel, J. and Koch, T. (2001). Stopped-flow kinetics of locked nucleic acid (LNA)-oligonucleotide duplex formation: studies of LNA-DNA and DNA-DNA interactions. *Biochem. J.*, **354**, 481–484.

D'Antoni, C. M., Fuchs, M., Harris, J. L. *et al.* (2006). Rapid quantitative analysis using a single molecule counting method. *Anal. Biochem.* (In press.)

Dundr, M., McNally, J. G., Cohen, J. and Misteli, T. (2002). Quantitation of GFP-fusion proteins in single living cells. *J. Struct. Biol.*, **140**, 92–99.

Eigen, M. and Rigler, R. (1994). Sorting single molecules: applications to diagnostics and evolutionary biotechnology. *Proc. Natl. Acad. Sci. USA*, **91**, 5740–5747.

Goodwin, P. M., Ambrose, W. P. and Keller, R. A. (1996). Single molecule detection in liquids by laser induced fluorescence. *Acc. Chem. Res.*, **29**, 603–613.

Haab, B. B. and Mathies, R. A. (1995). Single molecule fluorescence burst detection of DNA fragments separated by capillary electrophoresis. *Anal. Chem.*, **67**, 3253–3260.

Hesse, J., Wechselberger, C., Sonnleitner, M., Schindler, H., and Schutz, G. J. (2002). Single-molecule reader for proteomics and genomics. *J. Chromatogr. B. Analyt. Technol. Biomed. Life. Sci.*, **782**, 127–135.

Iorio, M. V., Ferracin, M., Liu, C. G. *et al.* (2005). MicroRNA gene expression deregulation in human breast cancer. *Cancer Research*, **65**(16), 7065–7070.

Jett, J. H., Keller, R. A., Martin, J. C. *et al.* (1989). High-speed DNA sequencing: an approach based upon fluorescence detection of single molecules. *J. Biomol. Struct. Dyn.*, **7**, 301–309.

Korn, K., Gardellin, P., Liao, B. *et al.* (2003). Gene expression analysis using single molecule detection. *Nucleic Acids Res.*, **31**(16), 1–8.

Li, H., Ying, L., Green, J. J., Balasubramanian, S. and Klenerman, D. (2003). Ultrasensitive coincidence fluorescence detection of single DNA molecules. *Anal. Chem.*, **75**, 1664–1670.

Michael, M. Z., O'Connor, S. M., van Holst Pellekaan, N. G., Young, G. P., and James, R. J. (2003). Reduced accumulation of specific microRNAs in colorectal neoplasia. *Mol. Cancer. Res.*, **12**, 882–891.

Neely, L. A., Patel, S., Garver, J. *et al.* (2006). A single-molecule method for the quantitation of microRNA gene expression. *Nature Methods*, **3**, 41–46.

Nie, S., Chiu, D. T. and Zare, R. N. (1995). Real time detection of single molecules in solution by confocal fluorescence microscpy. *Anal. Chem.*, **67**, 2849–2857.

Novikov, E., Hofkens, J., Cotlet, M. *et al.* (2001). A new analysis method of single molecule fluorescence using series of photon arrival times: theory and experiment. *Spectrochim. Acta A. Mol. Biomol. Spectrosc.*, **57**, 2109–2133.

R Development Core Team (2005). R: A language and environment for statistical computing. Vienna, Austria: R Foundation for Computing.

Sambrook, J., Fritsch, E. and Maniatis, T. (1989). *Molecular Cloning: A Laboratory Manual.* Cold Spring Harbor, NY: Cold Spring Harbor Laboratory Press.

Schwille, P. and Kettling, U. (2001). Analyzing single protein molecules using optical methods. *Curr. Opin. Biotechnol.*, **12**, 382–386.

Schwille, P., Bieschke, J. and Oehlenschlager, F. (1997). Kinetic investigations by fluorescence correlation spectroscopy: the analytical and diagnostic potential of diffusion studies. *Biophys. Chem.*, **66**, 211–228.

Soini, E., Meltola, N. J., Soini, A. E. *et al.* (2000). Two-photon fluorescence excitation in detection of biomolecules. *Biochem. Soc. Trans.*, **28**, 70–74.

Soper, S. A., Davis, L. M. and Shera, E. B. (1992). Similtaneous detection of two colors by two collinear laser beams at different wavelengths. *J. Opt. Soc. Am. B.*, **9**, 1761–1769.

Tichopad, A., Didier, A. and Pfaffl, M. W. (2004). Inhibition of real-time RT-PCR quantification due to tissue-specific contaminants. *Mol. Cell Probes*, **18**, 45–50.

Widengren, J. (1995). Fluorescence correlation spectroscopy of triplet states in solution: a theoretical and experimental study. *J. Phys. Chem.*, **99**, 13 368–13 379.

Widengren, J., and Kask, P. (1993). Fluorescence correlation spectroscopy with high count rate and low background: analysis of translational diffusion. *Eur. Biophysics* **22**, 169–175.

Yamasaki, R., Hoshino, M., Wazawa, T. *et al.* (1999). Single molecular observation of the interaction of GroEL with substrate proteins. *J. Mol. Biol.*, **292**, 965–972.

Yanagida, T., Kitamura, K., Tanaka, H., Hikikoshi Iwane, A. and Esaki, S. (2000). Single molecule analysis of the actomyosin motor. *Curr. Opin. Cell. Biol.*, **12**, 20–25.

20 Real-time quantification of microRNAs by TaqMan® assays

Yu Liang, Linda Wong, Ruoying Tan, and Caifu Chen*

Introduction

MicroRNAs are a new class of small, non-coding RNAs that control gene expression at the post-transcriptional level (Ambros, 2004; Bartel, 2004). Primary miRNA transcripts, also called pri-miRNA precursors are processed sequentially by two RNase III enzymes, Drosha and Dicer, to yield intermediate ~80-nt pre-miRNA precursors and final ~21-nt mature miRNAs. MicroRNAs are incorporated into RNA-induced silencing complex (RISC) where they identify and silence target messenger RNAs (mRNAs) through translational repression or direct cleavage (Wightman *et al.*, 1993; Olsen and Ambros, 1999; Hutvagner and Zamore, 2002; Doench and Sharp, 2004; Zhang *et al.*, 2004).

Cloning efforts and computational predictions have indicated that there are ~800 miRNA genes in human, which together regulate more than 5300 genes involved in processes including cell proliferation and metabolism, developmental timing, cell death, hematopoiesis, neuron development, human tumorigenesis, and even DNA methylation and chromatin modification (Ambros, 2003; Baehrecke, 2003; Michael *et al.*, 2003; Bao *et al.*, 2004; Bartel, 2004; Chen *et al.*, 2004; Chen and Lodish, 2005; Johnston *et al.*, 2005). Certain miRNAs are expressed ubiquitously, whereas others are expressed in a highly tissue-specific manner. Their expression levels vary greatly among species and tissues, ranging from less than 10 to more than 50,000 copies per cell (Kim *et al.*, 2004). Less abundant miRNAs routinely escape detection with technologies such as cloning, Northern hybridization, and microarray analysis (Krichevsky *et al.*, 2003; Lim *et al.*, 2003; Liu *et al.*, 2004). In this chapter, we describe a real-time quantification method, TaqMan® miRNA assays, useful for accurate and sensitive detection of miRNAs.

---

* Author to whom correspondence should be addresssed.

MicroRNAs: From Basic Science to Disease Biology, ed. Krishnarao Appasani. Published by Cambridge University Press. © Cambridge University Press 2008.

TaqMan® microRNA assays for detecting and quantifying microRNAs

Basics of the TaqMan® real-time PCR

Real-time quantitative PCR enables researchers to detect and measure small amounts of RNA by taking advantage of a polymerase chain reaction (PCR) where the amount of PCR product at the end of each PCR cycle is directly proportionate to the amount of template at the start of the PCR process. Development of novel chemistries and instrumentation platforms that enable reliable detection of PCR products on a real-time basis has led to widespread adoption of real-time RT-PCR as the method of choice for quantifying changes in gene expression (Higuchi *et al.*, 1993).

The TaqMan® PCR reaction exploits the 5′ nuclease activity of AmpliTaq Gold® DNA polymerase to cleave a TaqMan® probe during PCR (Livak *et al.*, 1995). The TaqMan® probe contains a reporter dye at the 5′ end of the probe and a quencher dye at the 3′ end of the probe. Accumulation of PCR products is detected directly by monitoring the increase in fluorescence of the reporter dye. When the probe is intact, the proximity of the reporter dye to the quencher dye suppresses the reporter fluorescence primarily by fluorescence resonance energy transfer. During PCR, if the target of interest is present, the probe specifically anneals between the forward and reverse primer sites. The 5′ to 3′ nucleolytic activity of the AmpliTaq Gold® enzyme cleaves the probe between the reporter and the quencher, freeing the reporter from the effect of the quencher thereby increasing the fluorescent signal from the reporter.

The probe fragments are then displaced from the target, and polymerization of the strands continues. The 3′ end of the probe is blocked to prevent extension of the probe during PCR. This process occurs in every cycle and does not interfere with the exponential accumulation of the products. Non-specific amplification is not detected because the increase in fluorescent signal is recorded only if the target sequence is complementary to the probe and is amplified during PCR.

Chemistry of TaqMan® microRNA assays

A forward primer, reverse primer, and a dye-labeled probe are required for conventional TaqMan®-based real-time PCR. The short length of mature miRNA molecules makes their quantification a technical challenge. Here we introduce a new miRNA quantitation method called the TaqMan® miRNA assay that overcomes these challenges (Chen *et al.*, 2005). Like the conventional TaqMan® assay, this method also includes two simple steps: reverse transcription (RT) and real-time PCR. However, the major difference is the use of a novel stem-loop primer for RT (Figure 20.1). Each stem-loop RT primer contains a 3′ overhang and a stem-loop. Its 3′ overhang spans 5–8 nucleotides that are complementary to the 3′ miRNA sequence. The RT primer extends specifically from the miRNA template in the presence of reverse transcriptase. The RT product contains not only the miRNA sequence but also sequences that come from the stem

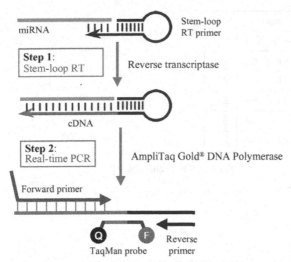

Figure 20.1. General scheme of TaqMan® microRNA assays. (See color plate 12)

and loop that are used to design a conventional real-time PCR assay with miRNA-specific forward primer and TaqMan® probe and a universal reverse primer.

There are several advantages to using a stem-loop RT primer. First, the short RT priming site improves the RT specificity for discriminating between similar miRNAs at the 3′ end. Second, it extends the length of the cDNA that can be used to design a conventional TaqMan® assay, making it more feasible to design assays for short RNAs. Third, TaqMan® miRNA assays inherit the high specificity, sensitivity, and large dynamic range of a conventional TaqMan® assay. Fourth, we observed that the stem-loop RT primers provided better specificity and sensitivity than conventional linear primers, most likely due to the base stacking of the stem-loop structure (Figure 20.2). For example, base stacking could improve the thermal stability and extend the effective footprint of RT primer/RNA duplex that are believed to be required for effective reverse transcription from relatively shorter RT primers. Fifth, unique attributes of the stem-loop RT primers may also provide better efficiency and specificity for multiplex RT reactions as well as for small RNA cloning. Finally, the spatial constraint of the stem-loop structure may prevent it from binding double-strand genomic DNA molecules and, therefore, eliminate the need for prior purification of RNA when using TaqMan® miRNA assays.

Performance of TaqMan® microRNA assays

Use of TaqMan® miRNA assays has the following advantages.

(1) **Accurate**. Increased accuracy because real-time PCR collects data in the exponential growth phase, whereas traditional PCR only measures at the end-point (plateau).

(2) **Highly specific**. Only mature miRNAs, not inactive precursors or genomic DNA are quantitated.

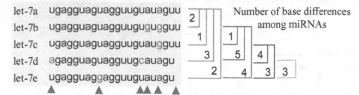

		Synthetic miRNA oligo					
		let-7a	let-7b	let-7c	let-7d	let-7e	
	let-7a	0	0.3	3.7	0.0	0.0	
	let-7b	0.0	0	0.3	0.0	0.0	
miRNA assay	let-7c	0.0	2.5	0	0.1	0.0	Non-specific detection (%)*
	let-7d	0.1	0.0	0.0	0	0.0	
	let-7e	0.0	0.0	0.0	0.0	0	

let-7a ugagguaguagguuguauaguu
let-7b ugagguaguagguugugugguu
let-7c ugagguaguagguuguaugguu
let-7d agagguaguagguugcauagu
let-7e ugagguaggagguuguauagu

Number of base differences among miRNAs

Figure 20.2. Discrimination of TaqMan® microRNA assays. The percentage of non-specific detection (indicated by the *) was calculated based on CT difference between perfectly matched and mismatched assays.

(3) **Fast, simple, and scalable**. Two-step assay provides high-quality results in less than three hours. By taking advantage of gold-standard TaqMan® reagent-based technology with universal thermal cycling conditions, TaqMan® miRNA assays are familiar, fast, easy to set up, and do not require post PCR processing since it is a closed system (i.e. does not require elec- trophoretical separation of amplified DNA).

(4) **Sensitive**. Conserves limited samples by requiring only 1–10 ng of total RNA or equivalent.

(5) **Reproducible**. Assays yield highly reproducible results reflecting the high accuracy with which miRNAs can be measured.

(6) **Wide dynamic range**. Like real-time TaqMan® PCR assays, TaqMan® miRNA assays allow measurement across a linear dynamic range of up to seven logs which range from a few to millions of copies for a synthetic miRNA or from a few picograms to tens of nanograms of total RNA.

Development of endogenous controls for TaqMan® microRNA assays

There are different factors, such as the quantity and quality of the RNA, that contribute to sample variation in gene expression studies, but the expression data can be normalized using endogenous control genes to correct for this varia- tion. Selection of an endogenous control is critical because an inadequate control will falsely affect experimental results. A reasonable endogenous control should show relative abundant and constant gene expression, but generally the expres- sion of a given control varies with types of samples, assays, and experimental conditions (Suzuki *et al.*, 2000). Therefore, it is essential to validate selected endogenous controls in our TaqMan® miRNA assays.

Housekeeping mRNA or ribosomal RNA (rRNA) genes such as β-actin, GAPDH, β2-microglobulin, and 18S rRNA are widely used to normalize gene expression

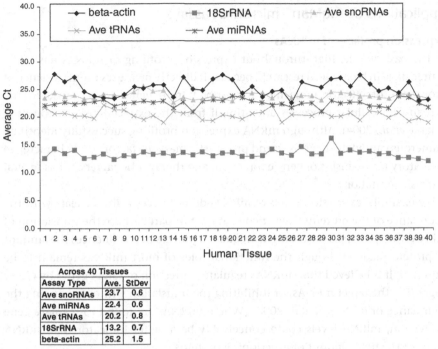

Figure 20.3. Expression pattern comparison for different types of endogenous controls in 40 human tissues.

data, but significant variation is commonly reported with these genes (Schmittgen and Zakrajsek, 2000; Suzuki *et al.*, 2000; Gorzelniak *et al.*, 2001; de Kok *et al.*, 2005). In addition, the conventional TaqMan® assays used to quantitate these housekeeping genes have a different assay design from the TaqMan® miRNA assays, so it is not ideal to use them to normalize the TaqMan® miRNA data. An alternative is to use miRNAs that show the least variable expression among a large panel of specimens, but an endogenous control for the TaqMan® miRNA assays should be less likely involved in miRNA pathways. On the contrary, other small RNAs that are similarly ubiquitous and abundant, and share size and RNA preparation method with miRNA genes (Finnegan and Matzke, 2003) would have the advantage over the housekeeping genes and miRNA controls, so we used the TaqMan® miRNA assay chemistry (Chen *et al.*, 2005) to design a number of assays for small nucleolar RNAs (snoRNAs) and transfer RNAs (tRNAs).

We examined 15 snoRNAs and 8 tRNAs across 40 normal human tissues and 59 NCI-60 cancer cell lines, and showed that these genes were both constantly and abundantly expressed. The expression variations of these snoRNAs and tRNAs in normal human tissues (Figure 20.3) and NCI-60 cancer cell lines (data not shown) were comparable to that of the selected miRNA controls (the least variable miRNAs among the specimens examined), 18S rRNA, but more stable than β-actin. Our data suggest that the small RNA endogenous controls we developed are better than the housekeeping genes and miRNA controls for the TaqMan® miRNA assay.

Applications of TaqMan® microRNA assays

Expression profiling of miRNAs

In the past decade, high-throughput expression profiling of mRNAs has taken center stage in biomedical research, because it directly measures the abundance of thousands of effector molecules transcribed from genes in response to the demand for cellular processes and stimuli from the extracellular environment (Clarke et al., 2004). Although mRNA expression profiling successfully identified numerous candidate diagnostic and prognostic markers, factors that characterize regulatory mechanisms of gene expression, and therapeutic targets, it has several intrinsic limitations.

For example, expression levels of mRNAs do not necessarily correlate with the abundance of the proteins they encode at least in part because the translation of mRNAs into proteins is also a highly regulated process that influences the amount of protein made. Although the functional roles of most miRNAs remain to be explored, it is believed that miRNAs regulate expression of target genes by either degrading the target mRNAs, or inhibiting the translation without changing the abundance of mRNAs (Pillai, 2005). Where miRNAs are found to regulate gene expression, miRNA levels could conceivably be a better indicator than mRNA levels of the final amount of functioning proteins.

Another challenge that mRNA expression profiling has experienced is difficulty in diagnosing the tissue origins of certain types of normal tissues and poorly differentiated neoplasms (Ramaswamy et al., 2001). When expression of over 16 000 mRNAs was profiled in epithelial samples using hierarchical clustering, the coherence of the samples derived from the gastrointestinal system was not observed (Lu et al., 2005). On the contrary, all the gut-derived tissues were clustered together using the miRNA expression profiles. It was hypothesized that high levels of mRNA complexity in a cell could give rise to noise that interferes with the clustering analysis. However, it is unexpected that miRNAs identify the origin of tissues better than mRNA insofar as the number of known miRNAs encoded by the genome is far less than the number of mRNAs.

MicroRNAs participate in determination of cell fate, pattern formation in embryonic development, and in control of cell proliferation, cell differentiation, and cell death (Alvarez-Garcia and Miska, 2005). Therefore, it is reasonable to speculate that miRNAs are also involved in human diseases such as cancers (Croce and Calin, 2005). Several groups of miRNAs have been identified to regulate the expression of tumor-associated genes (Chen, 2005), while others seem to hold prognostic value in predicting the survival of cancer patients (Calin et al., 2005).

Expression profile analyses have shown that the chromosomal location of a miRNA gene is an important determinant of its expression from at least two perspectives. First, most miRNA genes within 50 kb of each other have highly correlated expression patterns, consistent with the idea that they are processed from polycistronic primary transcripts (Baskerville and Bartel, 2005). Second, approximately one-third of miRNA genes are within the introns of annotated mRNAs, and their expression is frequently correlated with the expression profiles

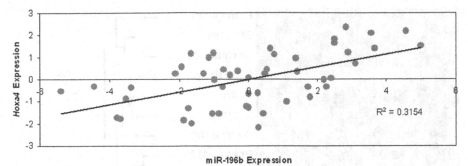

Figure 20.4. Scattered plot of the expression of *Hoxa4* mRNA and miR-196b in NCI-60 cancer cell lines ($p = 0.000\,02$ by Pearson).

of their host genes (Baskerville and Bartel, 2005). The best example may be the 4 miRNA genes (miR-196b, miR-10a, miR-196a-2, and miR-10b) that are embedded in the Hox gene clusters (Hox A, Hox B, Hox C, and Hox D, respectively). By histochemical staining and *in situ* hybridization, expression patterns of miR-10a and *Hoxb4* mRNA are very similar, suggesting that they share regulatory control of transcription (Mansfield *et al.*, 2004). We also found coherent results in our expression profiling of miRNAs in NCI-60 cancer cell lines, in which the expression of miR-196b and *Hoxa4* is significantly correlated (Figure 20.4).

The importance of miRNA gene chromosomal location also emerges as new tumor miRNA markers were identified. Given the possibility that expression of miRNA genes could be co-regulated with their neighboring genes, de-regulation of tumor-associated genes would perturb the levels of their co-expressing miRNAs; this includes changes at the genomic level such as gene amplification or deletion. Expression of miRNA genes within the regions afflicted by chromosomal aberration, a hallmark characteristic of neoplastic cells, could also be directly affected. Interestingly, miRNA genes are frequently located at fragile sites, as well as in regions of loss of heterozygosity, regions of amplification, or common breakpoint regions (Calin *et al.*, 2004), suggesting that alterations at the genomic level might be one of the pathways miRNAs use to associate with tumorigenesis. For example, miR-15a and miR-16-1 are located at a frequently deleted site in most of the B cell chronic lymphocytic leukemia patients (Calin *et al.*, 2002), and induce apoptosis in a leukemia cell line model (Cimmino *et al.*, 2005).

Because it is believed that neoplastic cells frequently recapitulate the characteristics of their normal counterparts during development, we reasoned that not only should fetal brain-specific miRNAs play roles in normal brain development, they might also participate in pathogenesis of embryonal brain tumors, such as medulloblastoma. We profiled the expression of miRNAs in human fetal brain and different regions of normal adult brain to seek fetal brain-specific miRNAs. Medulloblastoma is the most common malignant brain tumor in children and accounts for 25% of all pediatric brain tumors, and it is still not clear how aberrant signaling pathways important for CNS development lead to formation of this tumor type (Marino, 2005). Therefore, we chose medulloblastoma as our test case.

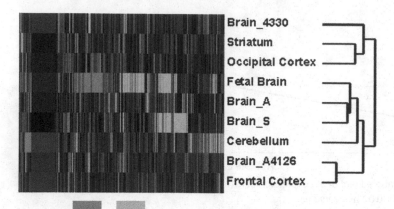

Brain_4330
Striatum
Occipital Cortex
Fetal Brain
Brain_A
Brain_S
Cerebellum
Brain_A4126
Frontal Cortex

Figure 20.5. Expression of 236 microRNAs in fetal brain and different regions of adult brain was compared using hierarchical clustering analysis, and the expression of those up-regulated (red bar) and down-regulated (green bar) in fetal brain as compared to the rest of brain specimens were selected. In order to corroborate whether expression patterns of these two sets of microRNAs are truly unique for fetal brain, we further examined their expression in specimens derived from other tissue types. (See color plate 13)

We used a two-step supervised clustering analysis to identify, at the first step, miRNAs that were differentially expressed between fetal brain and different regions of adult brain (Figure 20.5), and, at the second step, to distill to a smaller list of miRNAs that were differentially expressed between fetal brain and different normal human tissues (Figure 20.6). Chromosomal locations of this list of miRNAs showed that there are three miRNA subgroups in regions where genes known to be important for normal brain development and formation of medulloblastoma are located.

These three groups include miR-23b/24/27b, miR-29a/29b, and miR-127/299/319 that are localized at chromosomes 9q22.32, 7q32.2, and 14q32.31, respectively, where the *ptch*, *smo*, and *dlk1* genes are located. All three genes are known to be important for cerebellar development and medulloblastoma formation (Marino, 2005). These three genes are 350 kb, ~750–1750 kb, and ~200–300 kb away from miR-23b/24/27b, miR-29a/29b, and miR-127/299/319, respectively, so the expression of the three miRNA groups could be associated with that of these three genes via two mechanisms: (1) co-regulated expression at the transcription level, or (2) chromosomal abnormalities of the tumors, particularly for the first 2 miRNA groups, because deletion of chromosome 9q and gain of chromosome 7 are frequent genomic aberrations found in medulloblastoma (Russo *et al.*, 1999).

The TaqMan® miRNA assays allow us to perform expression profiling of miRNAs in multiple samples with throughputs similar to those of microarray-based technologies, but with far greater specificity and sensitivity. By using different channels of information, such as mRNA expression profiles of the samples of interest, the chromosomal locations of miRNAs, and other genomic and epigenetic changes of the samples, miRNA expression profiles would become a powerful tool to identify novel markers valuable for both basic and clinical research.

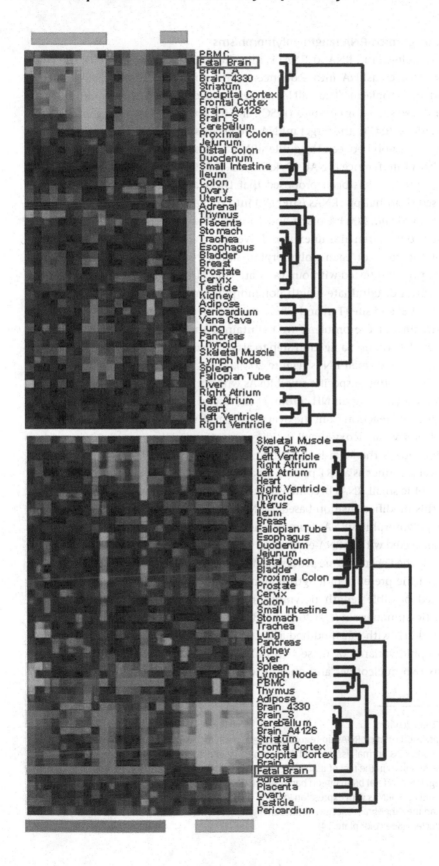

Detection of microRNA length polymorphisms

When double-strand RNA (dsRNA) is introduced into eukaryotic cells, the Dicer enzyme cleaves dsRNA into sequence-specific short interfering dsRNA species (siRNA) in cytoplasm. The siRNA duplex is incorporated into a multiprotein RNA-induced silencing complex (RISC), and the antisense strand of the unwound siRNA guides RISC to the target mRNA and cleaves at a single site in the center of the duplex region between the guide siRNA and the target mRNA. This process is called RNA interference (RNAi) (Dykxhoorn *et al.*, 2003; Bartel, 2004; Dorsett and Tuschl, 2004). It has been proposed that Dicer ambiguously processes vector-expressed short hairpin RNAs (shRNAs) into siRNAs with length polymorphisms (Dykxhoorn *et al.*, 2003; Dorsett and Tuschl, 2004), but their presence and contribution to RNAi remains unexplored. We reasoned that successful characterization of the roles of each polymorphic siRNA requires specific and sensitive quantitation and tested with our TaqMan® miRNA assay, but we found that the assay cannot discriminate length polymorphic small RNAs (siRNAs or miRNAs) differing at 3' end site (Tan and Chen, unpublished data).

To circumvent the problem, we maintained the two-step RT-PCR but added before the RT reaction a ligation step to attach a stem-loop linker to the 3' end of the polymorphic small RNAs (Figure 20.7). The 3' overhang of the linker determines the binding to specific small RNAs without discriminating different polymorphic variants, but an NH_2 group at the 3' end of the stem-loop linker blocks subsequent RT reaction from the overhang. The 3' side of the duplex region of the linker has several deoxy-uracil residues, so after ligation uracil N-glycosidase digestion opens the stem-loop structure and allows the RT reaction using the PCR reverse primer as an RT primer. As a result, cDNAs generated from individual polymorphic small RNAs become unique templates for subsequent PCR reaction. Using this modified ligation-based TaqMan® miRNA assay, we can easily differentiate polymorphic siRNAs from synthetic templates or total RNA purified from cells transfected with shRNA-expressing vector (Tan and Chen, unpublished data).

Because the Dicer enzyme processes miRNAs as it does shRNAs, it is conceivable to predict the presence of length polymorphisms for miRNAs *in vivo*. Therefore, we tested possible length polymorphisms of miR-100 and miR-137 in non-neoplastic human brain (A4126 and 4330) and malignant brain tumor specimens (38 N and 2H) with our modified assay. We designed assays specific for putative polymorphic variants of these two miRNAs, and found that the polymorphic variants two nucleotides and one nucleotide shorter than the mature miR-137

Figure 20.6. (cont.)

Top: expression of microRNAs up-regulated in fetal brain (red bar in Figure 20.5) was compared to that in another 38 normal tissues, and the microRNAs whose expression was higher in fetal brain (blue bracket) than in the rest of tissues were selected (red bars). Bottom: expression of microRNAs down-regulated in fetal brain (green bar in Figure 20.5) was compared to that in the 38 normal tissues, and microRNAs with expression lower in fetal brain than in the rest of tissues (green bar), or lower than the other brain specimens but still higher than most of the non-brain tissues (red bar), were selected. (See color plate 14)

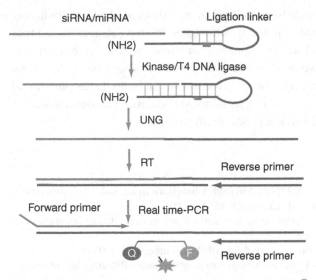

Figure 20.7. Schematic description of a ligation-based TaqMan® siRNA assay. This assay consists of four steps: (1) a stem-loop primer is phosphorylated by T4 polynucleotide kinase at the 5′ sites, then the phosphorylated linker is ligated to the 3′-end of RNA molecules; (2) uracil N-glycosidase digestion is performed to degrade the dU residues (▬) in the stem region of the linker; (3) reverse transcription (RT) is performed; (4) target-specific real-time PCR is carried out to determine the copy number of RT product.

and miR-100 listed in the Sanger database, respectively, are the predominant forms in these specimens. Our ligation-based TaqMan® miRNA assays not only provide a powerful tool to accurately measure the abundance of specific polymorphic variants of miRNAs, they will also underlie future endeavors to investigate the biology of polymorphic miRNAs.

Conclusions

MicroRNAs are a new class of small, non-coding RNAs that control gene expression at the post-transcriptional level; their short length and similarity in sequences has posed challenges for identification and quantitation of them. We developed a new method called the TaqMan® miRNA assay that overcomes these challenges, and this real-time quantitative PCR-based assay enables researchers to detect and measure small amounts of miRNA. We also modified the TaqMan® miRNA assays with a ligation reaction prior to the real-time RT-PCR to specifically identify and quantitate polymorphic variants of small RNAs.

As with any quantitative assay, the selection of an endogenous control for our TaqMan® miRNA assays is critical, but traditional housekeeping genes are not ideal for miRNA assays because of the differences in their size from miRNAs and in the chemistry of their TaqMan® assays from the miRNA assays. The miRNAs that are similarly expressed among a large panel of samples might not be appropriate controls due to their possible involvement in miRNA pathways. We have shown that snoRNAs and tRNAs are reliable endogenous controls because of their similar size, constant and abundant expression, and assay design to miRNAs.

MicroRNAs show great promise compared to mRNAs as markers for predicting cellular protein levels and identifying tissue origins. We have demonstrated that real-time TaqMan® miRNA assays are useful for these exciting miRNA applications. In addition, we have used our TaqMan® miRNA assays to identify fetal brain-specific miRNAs that might be involved in normal brain development and tumorigenesis of brain tumors. Finally, we adopted these assays to identify miRNA polymorphisms in normal and neoplastic brain tissues.

REFERENCES

Alvarez-Garcia, I. and Miska, E. A. (2005). MicroRNA functions in animal development and human disease. *Development*, **132**, 4653–4662.

Ambros, V. (2003). MicroRNA pathways in flies and worms: growth, death, fat, stress, and timing. *Cell*, **113**, 673–676.

Ambros, V. (2004). The functions of animal microRNAs. *Nature*, **431**, 350–355.

Baehrecke, E. H. (2003). miRNAs: micro managers of programmed cell death. *Current Biology*, **13**, R473–475.

Bao, N., Lye, K. W. and Barton, M. K. (2004). MicroRNA binding sites in Arabidopsis class III HD-ZIP mRNAs are required for methylation of the template chromosome. *Developmental Cell*, **7**, 653–662.

Bartel, D. P. (2004). MicroRNAs: genomics, biogenesis, mechanism, and function. *Cell*, **116**, 281–297.

Baskerville, S. and Bartel, D. P. (2005). Microarray profiling of microRNAs reveals frequent coexpression with neighboring miRNAs and host genes. *RNA*, **11**, 241–247.

Calin, G. A., Dumitru, C. D., Shimizu, M. *et al.* (2002). Frequent deletions and down-regulation of micro-RNA genes miR15 and miR16 at 13q14 in chronic lymphocytic leukemia. *Proceedings of the National Academy of Sciences USA*, **99**, 15 524–15 529.

Calin, G. A., Sevignani, C., Dumitru, C. D. *et al.* (2004). Human microRNA genes are frequently located at fragile sites and genomic regions involved in cancers. *Proceedings of the National Academy of Sciences USA*, **101**, 2999–3004.

Calin, G. A., Ferracin, M., Cimmino, A. *et al.* (2005). A microRNA signature associated with prognosis and progression in chronic lymphocytic leukemia. *New England Journal of Medicine*, **353**, 1793–1801.

Chen, C. Z. (2005). MicroRNAs as oncogenes and tumor suppressors. *New England Journal of Medicine*, **353**, 1768–1771.

Chen, C. Z. and Lodish, H. F. (2005). MicroRNAs as regulators of mammalian hematopoiesis. *Seminars in Immunology*, **17**, 155–165.

Chen, C. Z., Li, L., Lodish, H. F. and Bartel, D. P. (2004). MicroRNAs modulate hematopoietic lineage differentiation. *Science*, **303**, 83–86.

Chen, C., Ridzon, D. A., Broomer, A. J. *et al.* (2005). Real-time quantification of microRNAs by stem-loop RT-PCR. *Nucleic Acids Research*, **33**, e179.

Cimmino, A., Calin, G. A., Fabbri, M. *et al.* (2005). miR-15 and miR-16 induce apoptosis by targeting BCL2. *Proceedings of the National Academy of Sciences USA*, **102**, 13 944–13 949.

Clarke, P. A., te Poele, R. and Workman, P. (2004). Gene expression microarray technologies in the development of new therapeutic agents. *European Journal of Cancer*, **40**, 2560–2591.

Croce, C. M. and Calin, G. A. (2005). miRNAs, cancer, and stem cell division. *Cell*, **122**, 6–7.

de Kok, J. B., Roelofs, R. W., Giesendorf, B. A. *et al.* (2005). Normalization of gene expression measurements in tumor tissues: comparison of 13 endogenous control genes. *Laboratory Investigation*, **85**, 154–159.

Doench, J. G. and Sharp, P. A. (2004). Specificity of microRNA target selection in translational repression. *Genes & Development*, **18**, 504–511.

Dorsett, Y. and Tuschl, T. (2004). siRNAs: applications in functional genomics and potential as therapeutics. *Nature Reviews Drug Discovery*, **3**, 318–329.

Dykxhoorn, D. M., Novina, C. D. and Sharp, P. A. (2003). Killing the messenger: short RNAs that silence gene expression. *Nature Reviews Molecular Cell Biology*, **4**, 457–467.

Finnegan, E. J. and Matzke, M. A. (2003). The small RNA world. *Journal of Cell Science*, **116**, 4689–4693.

Gorzelniak, K., Janke, J., Engeli, S. and Sharma, A. M. (2001). Validation of endogenous controls for gene expression studies in human adipocytes and preadipocytes. *Hormone and Metabolic Research*, **33**, 625–627.

Higuchi, R., Fockler, C., Dollinger, G. and Watson, R. (1993). Kinetic PCR analysis: real-time monitoring of DNA amplification reactions. *Biotechnology*, **11**, 1026–1030.

Hutvagner, G. and Zamore, P. D. (2002). A microRNA in a multiple-turnover RNAi enzyme complex. *Science*, **297**, 2056–2060.

Johnston, R. J., Jr., Chang, S., Etchberger, J. F., Ortiz, C. O. and Hobert, O. (2005). MicroRNAs acting in a double-negative feedback loop to control a neuronal cell fate decision. *Proceedings of the National Academy of Sciences USA*, **102**, 12 449–12 454.

Kim, J., Krichevsky, A., Grad, Y. *et al.* (2004). Identification of many microRNAs that copurify with polyribosomes in mammalian neurons. *Proceedings of the National Academy of Sciences USA*, **101**, 360–365.

Krichevsky, A. M., King, K. S., Donahue, C. P., Khrapko, K. and Kosik, K. S. (2003). A microRNA array reveals extensive regulation of microRNAs during brain development. *RNA*, **9**, 1274–1281.

Lim, L. P., Glasner, M. E., Yekta, S., Burge, C. B. and Bartel, D. P. (2003). Vertebrate microRNA genes. *Science*, **299**, 1540.

Liu, C. G., Calin, G. A., Meloon, B. *et al.* (2004). An oligonucleotide microchip for genome-wide microRNA profiling in human and mouse tissues. *Proceedings of the National Academy of Sciences USA*, **101**, 9740–9744.

Livak, K. J., Flood, S. J., Marmaro, J., Giusti, W. and Deetz, K. (1995). Oligonucleotides with fluorescent dyes at opposite ends provide a quenched probe system useful for detecting PCR product and nucleic acid hybridization. *PCR Methods & Application*, **4**, 357–362.

Lu, J., Getz, G., Miska, E. A. *et al.* (2005). MicroRNA expression profiles classify human cancers. *Nature*, **435**, 834–838.

Mansfield, J. H., Harfe, B. D., Nissen, R. *et al.* (2004). MicroRNA-responsive 'sensor' transgenes uncover Hox-like and other developmentally regulated patterns of vertebrate microRNA expression. *Nature Genetics*, **36**, 1079–1083.

Marino, S. (2005). Medulloblastoma: developmental mechanisms out of control. *Trends in Molecular Medicine*, **11**, 17–22.

Michael, M. Z., O'Connor, S. M., van Holst Pellekaan, N. G., Young, G. P. and James, R. J. (2003). Reduced accumulation of specific microRNAs in colorectal neoplasia. *Molecular Cancer Research*, **1**, 882–891.

Olsen, P. H. and Ambros, V. (1999). The lin-4 regulatory RNA controls developmental timing in *Caenorhabditis elegans* by blocking LIN-14 protein synthesis after the initiation of translation. *Developmental Biology*, **216**, 671–680.

Pillai, R. S. (2005). MicroRNA function: multiple mechanisms for a tiny RNA? *RNA*, **11**, 1753–1761.

Ramaswamy, S., Tamayo, P., Rifkin, R. *et al.* (2001). Multiclass cancer diagnosis using tumor gene expression signatures. *Proceedings of the National Academy of Sciences USA*, **98**, 15 149–15 154.

Russo, C., Pellarin, M., Tingby, O. *et al.* (1999). Comparative genomic hybridization in patients with supratentorial and infratentorial primitive neuroectodermal tumors. *Cancer*, **86**, 331–339.

Schmittgen, T. D. and Zakrajsek, B. A. (2000). Effect of experimental treatment on housekeeping gene expression: validation by real-time, quantitative RT-PCR. *Journal of Biochemical & Biophysical Methods*, **46**, 69–81.

Suzuki, T., Higgins, P. J. and Crawford, D. R. (2000). Control selection for RNA quantitation. *Biotechniques*, **29**, 332–337.

Wightman, B., Ha, I. and Ruvkun, G. (1993). Posttranscriptional regulation of the heterochronic gene lin-14 by lin-4 mediates temporal pattern formation in *C. elegans*. *Cell*, **75**, 855–862.

Zhang, H., Kolb, F. A., Jaskiewicz, L., Westhof, E. and Filipowicz, W. (2004). Single processing center models for human Dicer and bacterial RNase III. *Cell*, **118**, 57–68.

21 Real-time quantification of miRNAs and mRNAs employing universal reverse transcription

Gregory J. Hurteau, Simon D. Spivack and Graham J. Brock*

Introduction

MicroRNAs have been identified as a new level of eukaryotic gene regulation. These endogenous ~22 nucleotide (nt) sequences are variably expressed in a manner that is specific to cell types, developmental stages and diseases (such as cancer). Because of their small size and variable expression levels, detection and quantification is technically challenging; various techniques have been developed to address these issues. We have modified a technique originally designed to specifically amplify mRNA and exclude pseudogene co-amplification to the real-time quantification of miRNAs. The modified technique involves the enzymatic addition of a polyA tail and the sequential hybridization of a universal RT-primer followed by reverse transcription. Transcript-specific forward primers can then be used to amplify a number of miRNAs as well as mRNA, from the same sample. We have used this technique to quantify relative miRNA expression using GAPDH mRNA as an internal standard. Relative quantification against a ubiquitously expressed miRNA, or multiplexing, is also possible, as all miRNAs in a sample are reverse transcribed. In addition, the relative quantification of both a miRNA and its predicted mRNA target can be assessed in the same sample. This allows identification of miRNA/mRNA interaction that results in cleavage or degradation of the target. The method is straightforward, needs no RNA size fractionation and involves routine enzymatic manipulations.

The identification of microRNAs and their mechanism of action was originally made in *C. elegans* (Lee *et al.*, 1993) and owing to their highly conserved nature they were subsequently described in many other species including vertebrates. The interpretation of miRNA biogenesis and mode of action has since led to the recognition of a new level of eukaryotic gene regulation (Ambros, 2001; Lagos-Quintana *et al.*, 2001). The mechanism involves endogenous ~22 nt miRNAs of which there are thought to be several hundred expressed at varying levels and in a

* Author to whom correspondence should be addressed.

MicroRNAs: From Basic Science to Disease Biology, ed. Krishnarao Appasani. Published by Cambridge University Press. © Cambridge University Press 2008.

tissue specific manner. In eukaryotes, initial expression is in the form of nascent pri-microRNA transcripts, generated by PolII and capped and polyadenylated (Cai et al., 2004). These are then processed into ~70 nt pre-miRNAs by the Drosha RNAseIII endonuclease (Lee et al., 2003) and transported from the nucleus by Ran-GTP/Exportin-5 (Yi et al., 2003). The pre-miRNAs are next processed by Dicer into the final ~22 nt noncoding single stranded mature miRNAs that regulate gene expression (Lee et al., 2003). The mature miRNAs are then incorporated into the RNA induced silencing complex (RISC) and targets recognized based on perfect (or almost perfect) complementarity to the mRNA, the result is either cleavage or transcriptional repression (Doench and Sharp, 2004; Hammond et al., 2000; Olsen and Ambros, 1999). The miRNA registry currently contains information on 3424 miRNAs, of which 326 are predicted to be human (Griffiths-Jones, 2004). Additional estimates of eukaryotic miRNAs suggest there may be 500–1000 miRNAs in the genome (Berezikov and Plasterk, 2005).

Identification of microRNAs has involved either cDNA cloning, usually following size fractionation (Lagos-Quintana et al., 2003) or computational algorithms that identify the distinct structural features of the pri-miRNAs (Berezikov et al., 2005; Ohler et al., 2004) or their targets (Xie et al., 2005). However, the function of the majority of miRNAs is currently unknown, with identification of (mRNA) targets also being predicted through computational means (Lewis et al., 2003). Such algorithms have to account for the majority of animal miRNAs (unlike plant miRNAs) lacking a perfect pairing to their targets, and predictions may alter following experimental investigation (Kiriakidou et al., 2004; Lewis et al., 2003).

Experimental validation of predicted miRNAs and their expression patterns has employed both PCR and array based techniques. These studies suggest that miRNAs have vastly different expression levels ranging from thousands of copies to only a few. Examination of the sequences identified to date has also revealed that mature miRNAs frequently have very similar sequences with "families" that differ by only one or two base pairs. This presents additional technical challenges and considerations to polymerase based amplification techniques (Chen et al., 2005; Raymond et al., 2005) and in miRNA array hybridization conditions (Liu et al., 2004).

Methodology

In this chapter, we describe the adaptation of an existing methodology to allow for the quantitation of miRNA expression levels either relative to an mRNA "housekeeping" gene or to a ubiquitously expressed miRNA. The method can also be used to measure both a miRNA and its predicted (degradation) targets in the same sample.

The technique, as outlined in Figure 21.1, is an adaptation of a method designed to specifically amplify from mRNA (Hurteau and Spivack, 2002). A conceptually similar technique was recently used for quantitative measurement of miRNA in plants (Shi and Chiang, 2005).

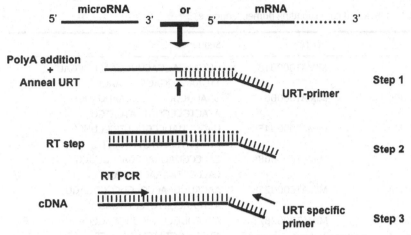

Figure 21.1. Schematic of the URT method. Mature miRNA are first modified through addition of a polyA tail as shown. The universal RT-primer, with variable 1 bp anchor (arrow), polyT and unique universal RT primer sequence, then anneals followed by reverse transcription. The resultant cDNA can then be amplified using specific primers for the miRNA and URT sequence.

Cell lysis and extraction of total RNA

The cell lines described in this chapter, MCF7 (Breast), HT1080 (Fibrosarcoma), A549 (Lung), MALME (Melanoma) and NHBE (Normal Human Bronchial Epithelium), were obtained from the ATCC (except NHBE) and cultured according to recommended conditions. Cells were pelleted at 1000 g for 5 min and washed twice in PBS. Total RNA was extracted with Trizol (Invitrogen) according to the manufacturer's recommended procedures. Briefly, cell pellets were lysed in 1 ml of Trizol reagent and incubated for 5 min at room temperature. Chloroform (200 μl) was added, mixed and incubated for 2–3 min at room temperature before centrifugation at 12 000 g for 15 min at 4 °C. The aqueous layer was transferred to a fresh tube, mixed with 500 μl of isopropanol and incubated for 10 min at room temperature. Samples were spun at 12 000 g for 10 min at 4 °C and the supernatant removed. The pellet was then washed with 1 ml of 75% EtOH, spun at 7500 g for 5 min and the supernatant removed before drying at room temperature for 10 min. After resuspending in DNase/RNase-free water, RNA quantity and quality was determined at 260/280 nm using a spectrophotometer (Nanodrop Technologies).

Enzymatic addition of polyA tails

Mature miRNAs were modified by the addition of a polyA tail [(additional polyAs are also added to nascent mRNA) (Figure 21.1, Step 1)] as follows. Total RNA (1 μg) prepared as described above in 12.2 μl of DNase/RNase-free water was added to 4 μl of 5 × E-PAP buffer, 1.5 μl of 25 nM MnCl and 1.5 μl of 10 mM ATP solution. Then 0.8 μl of *E. coli* Poly (A) Polymerase I (E-PAP) (Ambion Inc. Austin, TX) was added and the sample thoroughly mixed and incubated at 37 °C for 1 h.

Reverse transcription

Reverse transcription was performed with 10 μl (500 ng) of the E-PAP treated total RNA using Superscript III reverse transcriptase (Invitrogen Carlsbad, CA) as

Table 21.1. miRNA and corresponding primer sequences

Name	Acc No.	Sequence 5'–3'
ath-miR163	MIMAT0000184	UUGAAGAGGACUUGGAACUUCGAU
mir-163		TTGAAGAGGACTTGGAACTTC
hsa-miR-200c	MIMAT0000617	UAAUACUGCCGGGUAAUGAUGG
mir-200c		ATACTGCCGGGTAATGATGG
hsa-miR-154	MIMAT0000452	UAGGUUAUCCGUGUUGCCUUCG
mir-154		AGGTTATCCGTGTTGCCTTC
hsa-miR-99b	MIMAT0000689	CACCCGUAGAACCGACCUUGCG
mir-99b		CACCCGTAGAACCGACCTT
hsa-miR-181a	MIMAT0000256	AACAUUCAACGCUGUCGGUGAGU
mir-181a		ATTCAACGCTGTCGGTGAG
hsa-miR-502	MIMAT0002873	AUCCUUGCUAUCUGGGUGCUA
mir-502		ATCCTTGCTATCTGGGTGCT
hsa-miR-409-3p	MIMAT0001639	CGAAUGUUGCUCGGUGAACCCCU
mir-409-3p		CGAATGTTGCTCGGTGAAC
hsa-miR-518b	MIMAT0002844	CAAAGCGCUCCCCUUUAGAGGU
mir-518b		AAAGCGCTCCCCTTTAGA
hsa-miR-193b	MIMAT0002819	AACUGGCCCUCAAAGUCCCGCUUU
mir-193b		AACTGGCCCTCAAAGTCC
hsa-miR-205	MIMAT0000266	UCCUUCAUUCCACCGGAGUCUG
mir-205		CCTTCATTCCACCGGAGT
URT-primer	N/A	AACGAGACGACGACAGACTTTTTTTTTTTTTTTTV
Universal PCR primer	N/A	AACGAGACGACGACAGACTTT

follows. RNA template was added to a master mix containing 1 μl of 100 μM Universal RT primer, 1 μl of dNTP mix (each base 10 μM) and 1 μl of DNase/ RNase-free water. Total volume was adjusted to 13 μl with DNase/RNase-free water. The solution was incubated at 65 °C for 5 min and then cooled to 4 °C. A master mix containing 4 μl of 5 × first-strand buffer, 1 μl of 0.1 mM DTT, 1 μl DNase/RNase-free water and 1 μl SuperScript III per RT sample was prepared and added to each sample. The samples were incubated at 25 °C for 5 min, 50 °C for 60 min, followed by 70 °C for 15 min (Figure 21.1, Step 3). Typically, the RT reaction was diluted 1:20 and 2 μl used in the amplification of miRNAs with the transcript specific forward PCR primers (Table 21.1) and a Universal reverse primer identical to the 18 bp tag added during the RT step.

MicroRNA and mRNA primer design

The microRNA sequences used were taken from the miRNA database maintained by the Sanger Institute, http://microrna.sanger.ac.uk/sequences/index.shtml (accession numbers shown in Table 21.1), and primers purchased from Integrated DNA Technologies (IDT Skokie, IL). Primers for miRNAs and mRNAs where designed with the primer3 program from MIT, http://frodo.wi.mit.edu/ cgi-bin/primer3/primer3_www.cgi.

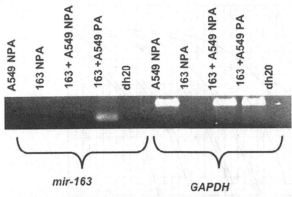

Figure 21.2. Specificity of *mir-163* primer. No mir-163 product is generated from total RNA alone (A549 NPA), oligo alone (163 NPA), or A549 RNA with mir-163 oligo template in the absence of polyA addition (163 + A549NPA), Lanes 1–3. Product is only generated when mir-163 oligo is added to RNA and polyA (PA) tail is added (163 + A549 PA), Lane 4. GAPDH transcript has an endogenous polyA tail and is therefore positive for all A549 cell line RNA extracts, Lanes 6, 8, 9, with the exception of the negative control, 163 oligo template alone (163 NPA), Lane 7.

Method sensitivity

The sensitivity of the method was assessed by using a 24 nt synthetic RNA oligonucleotide based on the sequence of the *Arabidopsis thaliana* miRNA *miR163* with an associated primer (Table 21.1). The synthetic oligonucleotide was mixed with total human RNA followed by either the enzymatic addition of a polyA tail or no treatment. Following PCR, the product of the *mir-163* primer reaction was visualized on an agarose gel (Figure 21.2). No amplification results from non-polyA treated (NPA) material with the *mir-163* specific primer. However, the addition of a polyA tail does allow the generation of a product (Figure 21.2, Lane 4). Amplification is evident with the GAPDH primer due to the endogenous polyA tail, with the exception of the synthetic oligomer alone. The dynamic range of the *mir-163* primer was then assessed using the serial dilutions of the synthetic oligonucleotide as shown Figure 21.3a. A standard curve was calculated and as shown in Figure 21.3b has a linear range when 1.5×10^{-3} to 1.5×10^{3} fM (between ~10 and ~10^{6} copies) were added.

Cloning and sequencing of miRNA amplification products

The RT-PCR products generated using a primer specific for *mir-163* were cloned using a TA-cloning kit (Invitrogen) according to the manufacturer's instructions. Subsequent ligations were transformed into chemically competent DH5α; blue/white selection identified cloned inserts for sequencing. Sequencing was performed on an ABI 3700 DNA analyzer (ABI, Foster City, CA); a representative trace is shown in Figure 21.4a. This revealed that the correct sequence was being amplified following the addition of the polyA tail. Examination of the sequence shows that the cloned insert consists of the entire synthetic oligomer (primer underlined) with the added polyA tail and URT-primer.

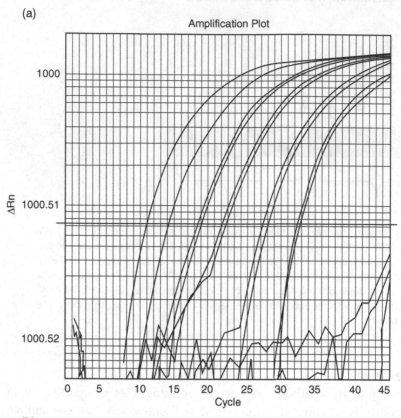

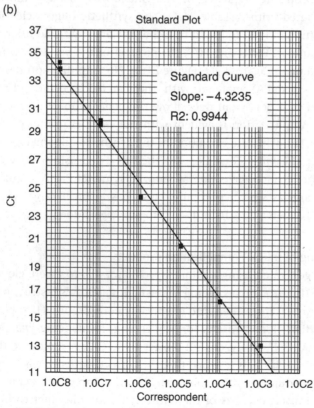

Figure 21.3. Dynamic range of synthetic oligonucleotide. (a) Dynamic range of the URT miRNA amplification assay using a synthetic oligonucleotide for *miR-163*. Serial dilutions ranged from 10^3 to 10^{-3} fм equivalent to between ~10 and 10^6 copies. (b) Showing standard curve for amplification from the *miR163* synthetic oligonucleotide.

(a)

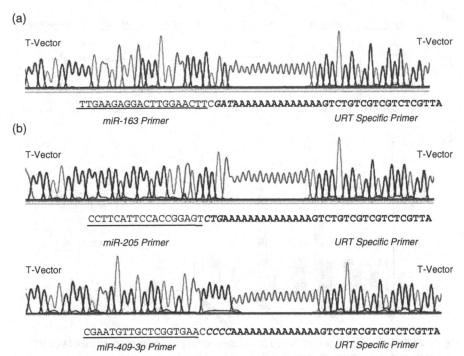

TTGAAGAGGACTTGGAACTTC*GA*TAAAAAAAAAAAAAAAGTCTGTCGTCGTCTCGTTA

miR-163 Primer   URT Specific Primer

(b)

CCTTCATTCCACCGGAGT*CTG*AAAAAAAAAAAAAAAGTCTGTCGTCGTCTCGTTA

miR-205 Primer   URT Specific Primer

CGAATGTTGCTCGGTGAAC*CCCC*AAAAAAAAAAAAAAAGTCTGTCGTCGTCTCGTTA

miR-409-3p Primer   URT Specific Primer

Figure 21.4. Sequence of cloned inserts. (a) The amplification product (Figure 21.2, Lane 4) was cloned and sequenced; a representative trace of a cloned insert (5′ to 3′) is shown. The flanking regions of the pCRII T-vector are indicated above each window. Below is the sequence of *mir-163* primer (underlined), the polyA region and sequence of the URT specific primer. (b) Sequence amplified with *mir-205* and *mir409-3p* primers. Again, base pairs not in either primer are shown in italics; these match the endogenous pri-miRNAs sequences, Accession Numbers MI0000285 and MIMAT0001639 on Chr.1 and Chr.14, respectively.

Amplification of endogenous miRNAs

As the RT step converts both miRNA and mRNA to cDNA, GAPDH or a similar ubiquitously expressed gene can be used to determine the relative expression levels in cell lines. Primers were designed for selected endogenous miRNAs, using the publicly available sequences in the Sanger Centre database. In some cases, the primer does not encompass the entire miRNA (Table 21.1) and the URT-primer has only a single variable base pair anchor. As a result, amplified products should contain one or two base pairs of unique sequence, i.e. in neither primer nor adaptor. In addition, the URT system adds a polyA tail to the "real" 3′ end, as opposed to a computationally predicted sequence. Sequencing of these products should demonstrate that they originated from the mature miRNA and not as the result of a PCR artifact. Figure 21.4b shows representative sequence chromatographs, following the cloning of the products generated using primers for *miRNA-205*, and *409-3p*. As with miR-163, the trace again shows that the amplified product contains sequence from both the miRNA (underlined) and URT specific primers. The additional base pairs shown (in italics) are not represented in either primer and therefore originated from the miRNA.

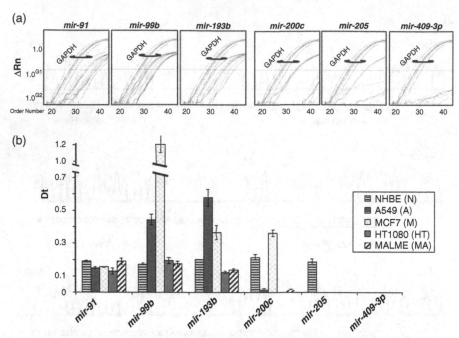

Figure 21.5. Amplification of miRNAs. (a) Duplicate amplification plots of six microRNAs, and GAPDH (circled), using RNA extracted from five cell lines. Cycle numbers are shown below, horizontal line indicates (Ct) threshold and ΔRn is the measure of relative fluorescence. (b) Graphs showing the inverse of the average ΔCt of five micro RNAs relative to GAPDH in RNA extracted from five cell lines. Lane titles correspond to cell line names in box.

Relative quantification of miRNA and mRNA

The expression levels of six microRNAs relative to GAPDH in total RNA extracted from five cell lines were then assessed (Figure 21.5a). Duplicate amplification curves are shown for each miRNA and for GAPDH with (cross threshold) Ct shown. These amplification plots are represented graphically in Figure 21.5b with the average ΔCt (inverted) plotted for each miRNA versus GAPDH in the five lines. The microRNAs *mir91, 99b* and *193b* could be detected in each sample with *mir-99b* apparently expressed at a significantly higher level in MCF7. In contrast, *mir-200c* and *205* could only be detected in NHBE and MCF7 or NHBE alone and *mir-409-3p* was undetectable in all five lines (NHBE, A549, MCF, HT1080 and MALME). However, sequencing of the amplification product generated when RNA was extracted from normal blood demonstrates that the *mir-409-3p* primer amplifies the correct sequence (Figure 21.3). All quantitative RT-PCR analysis was performed using an ABI 7900 real-time detection system and SYBER Green Dye (ABI, Foster City, CA).

Discussion

This method converts all miRNA and mRNAs into cDNA and it is possible to quantify miRNA expression using an endogenous housekeeping gene transcript as reference. It should also be possible to use an endogenous miRNA as an internal reference standard providing one that is ubiquitously expressed can be identified. This result demonstrates that this method can detect variations in expression levels of these microRNAs relative to GAPDH. A more extensive analysis of multiplexed miRNAs, samples, or cell lines would be required to establish any significance in the variations in expression shown in this limited analysis. The primers shown in Table 21.1 will amplify with specificity and sensitivity from their respective miRNAs. However, one potential limitation of the technique lies in the high level of similarity found between some families of microRNAs. As a result, a primer could amplify from highly similar families of microRNAs. The use of a Locked Nucleic Acid (LNA) forward primer as detailed by Raymond et al. (2005) is reported to improve the specificity of PCR reactions and should allow for specific amplification from highly similar miRNAs.

We have presented here a relatively simple and highly sensitive method of miRNA amplification, which allows for the detection of femtomolar (fM) levels of miRNA. However, the most significant advantage of this method is the capability of quantifying both a miRNA and its predicted mRNA target(s) relative to an internal control, from the same sample (the universal primer will lead to the conversion of both into cDNA). Currently prediction programs report several hundred and in some cases over a thousand, mRNA targets for each miRNA. In mammals, a large number of these targets may be regulated at the translational level. Nevertheless, it is possible using Affymetrix expression data to determine levels of predicted mRNA targets in various cell lines and this can be correlated with expression of miRNAs. Providing the predicted target(s) are degraded by the miRNA, this technique will enable the validation of expression levels of both, and causal inferences regarding this level of gene regulation.

ACKNOWLEDGEMENTS

We thank George Kampo for expert technical advice, and Julia Brosnan, Errin Lagow and Don Porter for useful discussions and critical reading of the manuscript.

REFERENCES

Ambros, V. (2001). MicroRNAs: tiny regulators with great potential. *Cell*, **107**, 823–826.

Berezikov, E. and Plasterk, R. H. (2005). Camels and zebrafish, viruses and cancer: a microRNA update. *Human Molecular Genetics*, **14** (Suppl. 2), R183–190.

Berezikov, E., Guryev, V., van de Belt, J. *et al.* (2005). Phylogenetic shadowing and computational identification of human microRNA genes. *Cell*, **120**, 21–24.

Cai, X., Hagedorn, C. H. and Cullen, B. R. (2004). Human microRNAs are processed from capped, polyadenylated transcripts that can also function as mRNAs. *RNA*, **10**, 1957–1966.

Chen, C., Ridzon, D. A., Broomer, A. J. *et al.* (2005). Real-time quantification of microRNAs by stem-loop RT-PCR. *Nucleic Acids Research*, **33**, e179.

Doench, J. G. and Sharp, P. A. (2004). Specificity of microRNA target selection in translational repression. *Genes & Development*, **18**, 504–511.

Griffiths-Jones, S. (2004). The microRNA Registry. *Nucleic Acids Research*, **32** (Database issue), D109–111.

Hammond, S. M., Bernstein, E., Beach, D. and Hannon, G. J. (2000). An RNA-directed nuclease mediates post-transcriptional gene silencing in Drosophila cells. *Nature*, **404**, 293–296.

Hurteau, G. J. and Spivack, S. D. (2002). mRNA-specific reverse transcription-polymerase chain reaction from human tissue extracts. *Analytical Biochemistry*, **307**, 304–315.

Kiriakidou, M., Nelson, P. T., Kouranov, A. *et al.* (2004). A combined computational-experimental approach predicts human microRNA targets. *Genes & Development*, **18**, 1165–1178.

Lagos-Quintana, M., Rauhut, R. *et al.* (2001). Identification of novel genes coding for small expressed RNAs. *Science*, **294**, 853–858.

Lagos-Quintana, M., Rauhut, R., Lendeckel, W. and Tuschl, T. (2003). New microRNAs from mouse and human. *RNA*, **9**, 175–179.

Lee, R. C., Feinbaum, R. L. and Ambros, V. (1993). The *C. elegans* heterochronic gene lin-4 encodes small RNAs with antisense complementarity to lin-14. *Cell*, **75**, 843–854.

Lee, Y., Ahn, C., Han, J. *et al.* (2003). The nuclear RNase III Drosha initiates microRNA processing. *Nature*, **425**, 415–419.

Lewis, B. P., Shih, I. H., Jones-Rhoades, M. W., Bartel, D. P. and Burge, C. B. (2003). Prediction of mammalian microRNA targets. *Cell*, **115**, 787–798.

Liu, C. G., Calin, G. A., Meloon, B. *et al.* (2004). An oligonucleotide microchip for genome-wide microRNA profiling in human and mouse tissues. *Proceedings of the National Academy of Sciences USA*, **101**, 9740–9744.

Ohler, U., Yekta, S., Lim, L. P., Bartel, D. P. and Burge, C. B. (2004). Patterns of flanking sequence conservation and a characteristic upstream motif for microRNA gene identification. *RNA*, **10**, 1309–1322.

Olsen, P. H. and Ambros, V. (1999). The lin-4 regulatory RNA controls developmental timing in *Caenorhabditis elegans* by blocking LIN-14 protein synthesis after the initiation of translation. *Developmental Biology*, **216**, 671–680.

Raymond, C. K., Roberts, B. S., Garrett-Engele, P., Lim, L. P. and Johnson, J. M. (2005). Simple, quantitative primer-extension PCR assay for direct monitoring of microRNAs and short-interfering RNAs. *RNA*, **11**, 1737–1744.

Shi, R. and Chiang, V. L. (2005). Facile means for quantifying microRNA expression by real-time PCR. *Biotechniques*, **39**, 519–525.

Xie, X., Lu, J., Kulbokas, E. J. *et al.* (2005). Systematic discovery of regulatory motifs in human promoters and 3′ UTRs by comparison of several mammals. *Nature*, **434**, 338–345.

Yi, R., Qin, Y., Macara, I. G. and Cullen, B. R. (2003). Exportin-5 mediates the nuclear export of pre-microRNAs and short hairpin RNAs. *Genes & Development*, **17**, 3011–3016.

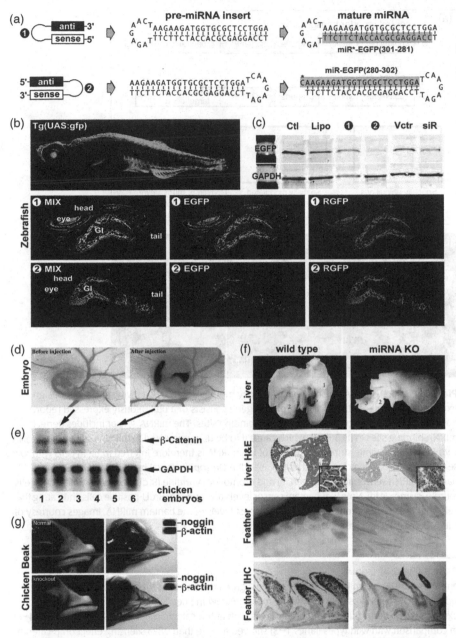

Plate 1 (Fig. 2.4) Intronic miRNA-mediated gene silencing effects *in vivo*. (a)–(c) Different preferences of RISC assembly were observed by transfection of 5′-miRNA*-stemloop-miRNA-3′ (❶) and 5′-miRNA–stemloop–miRNA*–3′ (❷) pre-miRNA structures in zebrafish, respectively. (a) One mature miRNA, namely miR-EGFP(280/302), was detected in the ❷-transfected zebrafishes, whereas the ❶-transfection produced another kind of miRNA, miR*–EGFP(301–281), which was partially complementary to the miR-EGFP(280/302). (b) The RNAi effect was only observed in the transfection of the ❷ pre-miRNA, showing less EGFP (green) in ❷ than ❶, while the miRNA indicator RGFP (red) was evenly present in all vector transfections. (c) Western blot analysis of the EGFP protein levels confirmed the specific silencing result of (b). No detectable gene silencing was observed in fishes without (Ctl) and with liposome only (Lipo) treatments. The transfection of either a U6-driven siRNA vector (siR) or an empty vector (Vctr) without the designed pre-miRNA insert resulted in no gene silencing significance. (d)–(g) Silencing of endogenous β-catenin and noggin genes in chicken

Plates 1–23 are available for download in colour from www.cambridge.org/9780521118522

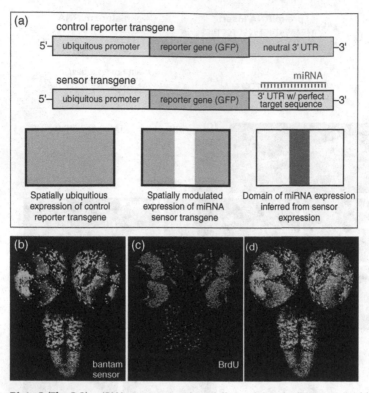

Plate 2 (Fig. 3.3) miRNA sensors reveal spatially modulated miRNA activity. (a) Sensor design and experimental interpretation. The control transgene consists of a ubiquitously expressed reporter transgene (depicted in green) lacking miRNA binding sites. The miRNA sensor includes perfect miRNA binding sites in its 3'-UTR, which cause it to be downregulated in miRNA-expressing cells in vivo (depicted as white stripe). Expression of the miRNA is therefore inferred to be in cells with low sensor expression (depicted as red stripe). (b)–(d) A *Drosophila* larval brain doubly labeled for bantam miRNA sensor activity (green, b) and bromodeoxyuridine (BrdU) to label proliferating cells (purple, c); merge (d). Note that bantam sensor levels are lowest in BrdU-positive cells, indicating that cells that are actively proliferating exhibit highest levels of the bantam miRNA. Images courtesy of Julius Brennecke and Stephen Cohen.

Plate 1 (Fig. 2.4) (cont.)

embryos. (d) The pre-miRNA construct and fast green dye mixtures were injected into the ventral side of chicken embryos near the liver primordia below the heart. (e) Northern blot analysis of extracted RNAs from chicken embryonic livers with anti-β-catenin miRNA transfections (lanes 4–6) in comparison with wild types (lanes 1–3) showed a more than 98% silencing effect on β-catenin mRNa expression, while the house-keeping gene, GAPDH, was not affected. (f) Liver formation of the β-catenin knockouts was significantly hindered (upper right two panels). Microscopic examination revealed a loose structure of hepatocytes, indicating the loss of cell-cell adhesion due to breaks in adherins junctions fromed between β-catenin and cell membrane E-cadherin in early liver development. In severely affected regions, feather growth in the skin close to the injection area was also inhibited (lower right two panels). Immunohistochemistry staining of β-catenin protein expression (brown) showed a significant decrease in the feather follicle sheaths. (g) The lower beak development was increased by the mandible injection of the anti-noggin pre-miRNA construct (down panel) in comparison to the wild type (up panel). Right panels showed bone (alizarin red) and cartilage (alcian blue) staining to demonstrate the outgrowth of bone tissues in the lower beak of the noggin knockout. Northern blot analysis (small windows) confirmed a ~60% decrease of noggin mRNA expression in the lower beak area.

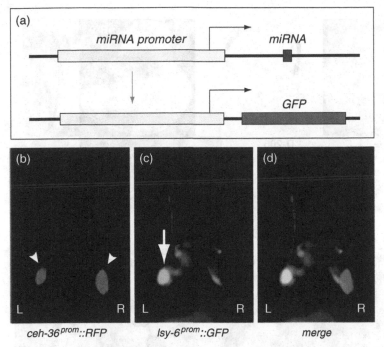

Plate 3 (Fig. 3.4) Enhancer/promoter-reporter fusions can reveal celltype-specific miRNA expression. (a) General strategy for constructing the reporter fusion. (b)–(d) A reporter fusion to regulatory sequences of the *lsy-6* miRNA reveals asymmetric expression of this miRNA in the ASE left/right chemosensory neuron pair. A *ceh-36* promoter-RFP fusion marks both left and right ASE neurons (b), while the *lsy-6* miRNA promoter-GFP transgene is active only in the left ASE neuron (c); merge is shown in (d). Images used with the permission of Oliver Hobert.

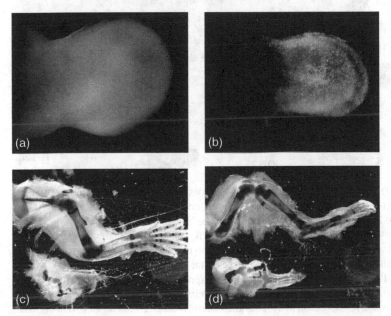

Plate 4 (Fig. 4.6) Limbs are morphologically abnormal in mice lacking functional *Dicer*. (a) and (b) Staining for cells undergoing cell death in E11.5 limbs using acridine orange shows a significant increase in cell death in mutant (b) compared to wild type (a) forelimbs. (c) and (d) Skeletal preparations of stage E20 wild type (top) and mutant (bottom) forelimbs (c) and hindlimbs (d). While *Dicer* does not appear to be required for patterning, there is a significant growth deficiency in mutant limbs (reproduced from Harfe *et al.* (2005); copyright of National Academy of Sciences USA).

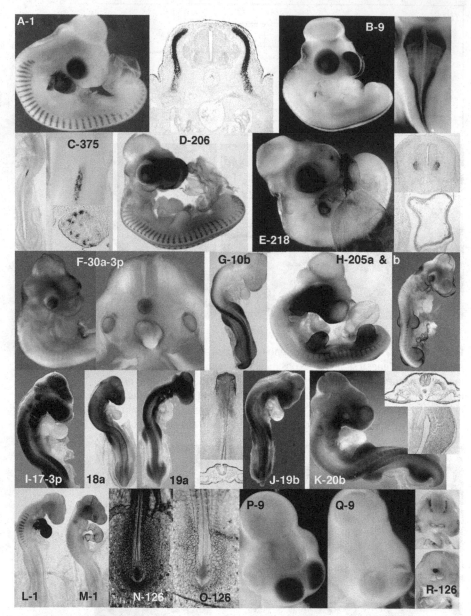

Plate 5 (Fig. 7.1) Expression of micro RNA in chick detected by whole mount *in situ* hybridization. (A–C) Expression in single tissue types. (A) miR-1 expression in the somitic myotome and cardiac myotome, stage 23 lateral view. Transverse section at the level of the spinal cord showing somitic myotome. (B) miR-9 expression in the telencephalic vesicles, midbrain, hindbrain and spinal cord, stage 24 lateral view, and dorsal view showing expression in the hindbrain with rhombomeric variability. (C) miR-375 expression in endodermal cells that will give rise to the pancreas in whole mount at stage 17, close up and in cross section. (D–H) Expression in multiple tissue types. (D) miR-206 expression in somitic myotome (mesoderm) and forebrain vesicles (ectoderm) at stage 24 lateral view. (E) miR-218 expression in the atria (mesoderm) at stage 25 lateral view, and in cross section in the spinal cord motor horns (ectoderm; top), and atria (bottom). (F) miR-30a-3p shown in whole mount at stage 19, and transverse cut showing notochord, mesonephric ducts and dorsal aorta (axial and intermediate mesoderm). (G) miR-10b expression in the entire length of the spinal cord (ectoderm) and limb buds (mesoderm). (H) miR-205a and b expression in the surface ectoderm shown at stages 23 and 21. (I–K) Nearly ubiquitous expressers, polycistronic, showing considerable

(a)

Plate 6 (Fig. 8.3) Expression of miR-134 in dendrites and its role in dendritic spine morphogenesis. (a) miR-134 (green; *in situ* hybridization signal) is present near post-synaptic sites (red; PSD-95 immunostaining) in dendrites of rat hippocampal neurons. Arrows denote sites of miR-134/PSD-95 co-localization.

Plate 5 (Fig. 7.1) (cont.)
expression at older stages in the body but much reduced in the heart. (I) miR-17-3p, -18a and -19a expression at stage 18/19, lateral view. (J) miR-19b expression at younger stages is restricted to the ectoderm and endoderm (whole mount and transverse section, stage 10), but later expands to near ubiquity, excepting heart. (K) A similar pattern seen for miR-20b, which is also not expressed in the apical ectodermal ridge of the limb (lower transverse section). (L) Stage 17 embryo strongly labeled with miR-1 double-labeled probe. (M) Stage 17 embryo weakly labeled with miR-1, 3′ single-labeled probe (same time in color reaction as (L)). (N) Stage 15 embryo tail, strongly labeled with miR-126 double-labeled probe. (O) Stage 15 embryo tail weakly labeled with miR-126 3′ single-labeled probe (same time in color reaction as (N)). (P) Stage 20 embryo strongly labeled with miR-9 double-labeled probe. (Q) Stage 20 embryo weakly labeled with miR-9 3′single-labeled probe (same time in color reaction as (P)). (R) Stage 19 embryo transected across the lumbar trunk labeled with miR-126 3′ single-labeled probe (top, no notochord label detected) vs. miR-126 tailed (bottom, strong notochord label detected).

Plate 7 (Fig. 10.3) Effects of miRNA manipulation on plant development. (a) Wt, miR319 overexpressor and transgenic plants expressing low levels of microRNA resistant TCP4. (b, c) Overexpression of miR164, (b) Cotyledon fusions in seedlings (right); (c) stem–leaf and leaf–leaf fusions (right). (d, e) Overexpression of miR156, (d) faster generation of leaves in the overexpressors (right); (e) delay in flowering time in miR156 overexpressors leading to a bushy plant (right). (f, g) Overexpression of miR172, (f) early flowering time in the transgenics (right); (g) patterning deffects on flowers overexpressing miR172 (right).

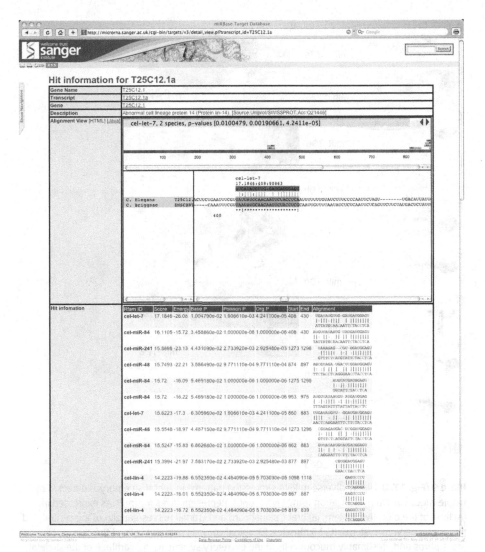

Plate 8 (Fig. 11.5) The *miRbase Targets* detailed transcript view of the worm lin-14 transcript (T25C12.1a). Predicted binding sites along the 3′ UTR for multiple miRNAs are shown (top) across a 3′ UTR schematic, which is coloured according to sequence conservation. The next pane (middle) shows an alignment (if available) between this transcript and its orthologs. In this case a predicted target site for cel-let-7 is highlighted and its conservation between *C. elegans* and *C. briggsae* is shown in orange. Finally, details for each individual site are shown (bottom) which include a full description of P-values obtained, relative coordinates of the site position in the 3′ UTR and a predicted duplex arrangement for the site.

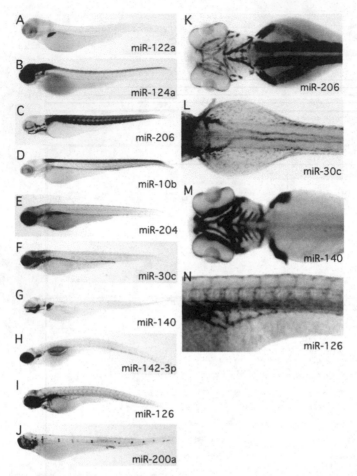

Plate 9 (Fig. 17.4) Highly diverse miRNA expression patterns in zebrafish embryos detected by whole-mount *in situ* hybridization using LNA probes. Representatives for miRNAs expressed in: (A) liver of the digestive system, (B) brain, spinal cord and cranial nerves/ganglia of the central and peripheral nervous systems, (C, M) muscles, (D) spinal cord, (E) pigment cells of the skin, (F, L) pronephros and presumably mucous cells of the excretory system, (G, M) cartilage of the skeletal system, (H) thymus, (I, N) blood vessels of the circulatory system, (J) lateral line system of the sensory organs. Embryos in (K, L, M, N) are higher magnifications of the embryos in (C, D, G, I), respectively. (A–J, N) are lateral views; (K–M) are dorsal views. All embryos are 72 h post-fertilization, except for (H), which is a 5-day old larva. From Wienholds *et al.* (2005), reprinted by permission of the American Association for the Advancement of Science.

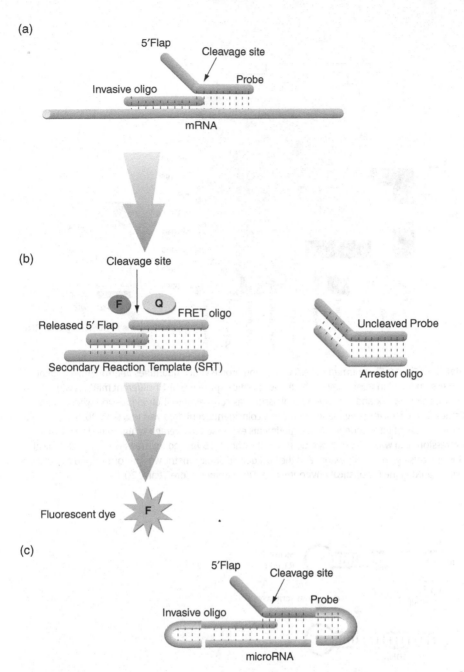

(a)

5′Flap

Cleavage site

Probe

Invasive oligo

mRNA

(b)

Cleavage site

F   Q   FRET oligo

Released 5′ Flap

Secondary Reaction Template (SRT)

Uncleaved Probe

Arrestor oligo

Fluorescent dye   F

(c)

5′Flap

Cleavage site

Probe

Invasive oligo

microRNA

Plate 10 (Fig. 18.1) Schematic representation of the RNA invasive cleavage assay. (a) mRNA Invader® assay primary reaction: an overlap–flap substrate for the structure-specific 5′ nuclease, Cleavase® *substrate* is formed upon annealing of the invasive and probe oligonucleotides to the mRNA target. The mRNA target is shown in blue and the target-specific oligonucleotides are shown in green. The non-complementary 5′-flap of the probe oligonucleotide is in purple. The site of cleavage is indicated by an arrow. (b) Secondary reaction: a secondary overlap–flap structure is formed when a Secondary Reaction Template (SRT, purple) binds the 5′-flap cleaved in the primary reaction plus a FRET oligonucleotide (purple). The FRET oligonucleotide is labeled with a fluorophore (F) and a quencher (Q). Cleavage between the fluorophore and a quencher generates a fluorescence signal. A 2′-O-methyl arrestor oligonucleotide (red) complementary to the probe is added to the secondary reaction, to sequester the uncleaved probes and prevent their binding to the SRT. (c) miRNA Invader assay primary reaction: the overall structure of the substrate resembles that shown in (a), except that the short size of the miRNA target requires the inclusion of extra stem-loop structures to the 5′ and 3′ ends of the invasive and probe oligonucleotides, respectively. The miRNA target is shown in blue, target-specific probe and invasive oligonucleotide sequences in green and 2′-O-methylated stem-loop hairpin regions are shown in red. The non-complementary 5′-flap is in purple and the arrow indicates the site of cleavage, which releases the 5′-flap.

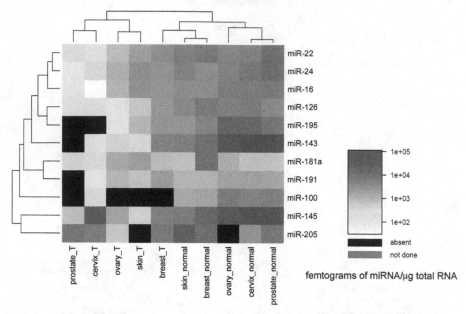

Plate 11 (Fig. 19.5) A human miRNA expression profile which quantifies aberrant miRNA gene expression found in tumor tissue. The tissue-specific expression of 11 different miRNAs within skin, breast, colon, cervix and prostate normal tissue was compared to their expression within tumor tissues. The coefficient of variation of the raw coincident event data was less than 20%. The results are the mean of three independent experiments expressed in average fg/μg tissue total RNA. Expression data was floored at the detection threshold (15 fg) and log-transformed. Hierarchical clustering using Ward's algorithm and the Euclidean distance metric was performed and a heat map was created in the R statistical environment (R Development Core Team, 2005).

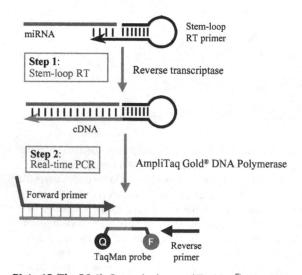

Plate 12 (Fig. 20.1) General scheme of TaqMan® microRNA assays.

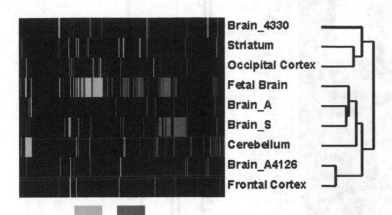

Plate 13 (Fig. 20.5) Expression of 236 microRNAs in fetal brain and different regions of adult brain was compared using hierarchical clustering analysis, and the expression of those up-regulated (red bar) and down-regulated (green bar) in fetal brain as compared to the rest of brain specimens were selected. In order to corroborate whether expression patterns of these two sets of microRNAs are truly unique for fetal brain, we further examined their expression in specimens derived from other tissue types.

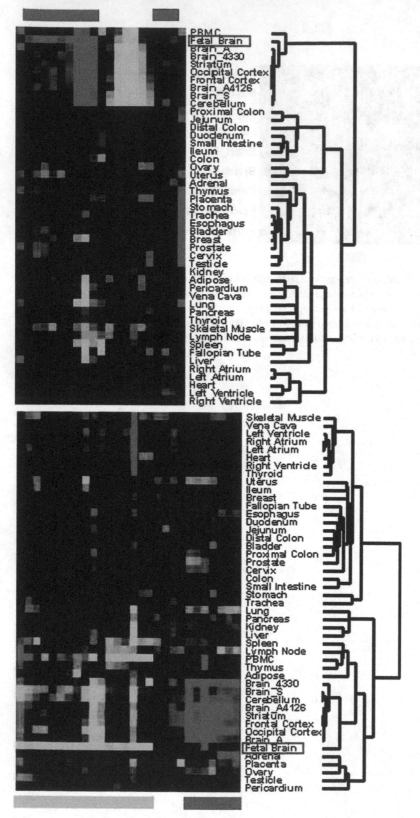

Plate 14 (Fig. 20.6) Top: expression of microRNAs up-regulated in fetal brain (red bar in Figure 20.5) was compared to that in another 38 normal tissues, and the microRNAs whose expression was higher in fetal brain (blue bracket) than in the rest of tissues were selected (red bars). Bottom: expression of microRNAs down-regulated in fetal brain (green bar in Figure 20.5) was compared to that in the 38 normal tissues, and microRNAs with expression lower in fetal brain than in the rest of tissues (green bar), or lower than the other brain specimens but still higher than most of the non-brain tissues (red bar), were selected.

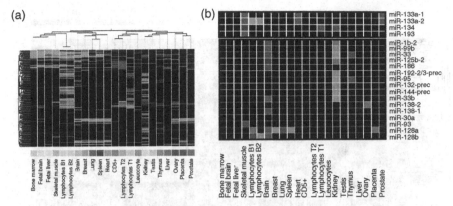

Plate 15 (Fig. 23.1) MiRNAs are differentially expressed in normal tissues. (a) Distinct patterns of miRNA expression in human adult and fetal tissues. (b) Specific overexpression of different miRNAs in skeletal muscle, heart, prostate, and brain. From Liu *et al.* (2004).

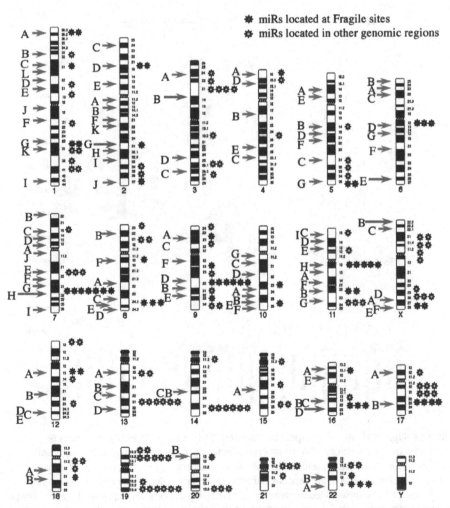

Plate 16 (Fig. 23.2) MiRNAs are located in different chromosomes and are frequently associated with CAGRs. From Calin *et al.* (2004).

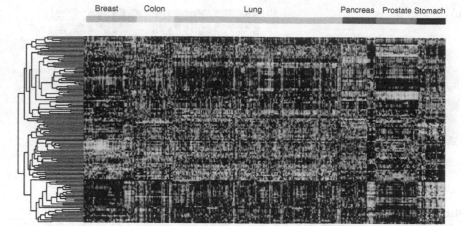

Plate 17 (Fig. 23.3) Clustering analysis of 540 samples representing six solid cancers and the respective normal tissues. The miRNome is differentially expressed in a tumor-specific manner. From Volinia *et al.* (2005).

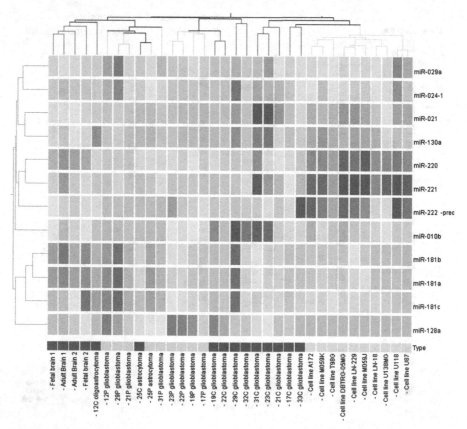

Plate 18 (Fig. 26.1) MicroRNA expression data from 11 brain tumor patients' specimens (9 glioblastomas, 1 oligoastrocytoma, 1 astrocytoma grade III), 4 normal brain (2 adult and 2 fetal) controls, and 10 human glioblastoma cell lines. Among patients' specimens, the dark blue ones represent the central area of the tumors (C samples), whereas the light blue ones are the external, control samples (P samples). Median values of 12 microRNAs differentially expressed between normal and tumoral samples were hierarchically clustered and plotted in a heat map. Red denotes the highest expression levels, while green denotes the lowest. Dendrograms indicate the correlation between groups of tissues or cells.

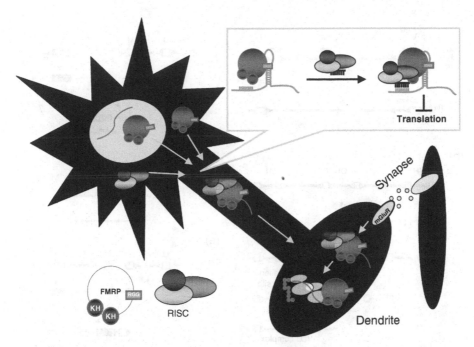

Plate 19 (Fig. 27.1) A model for FMRP-mediated translational suppression of mRNA targets via the miRNA pathway in neuron. FMRP localized to the nucleus associates with an mRNA target, possibly containing a G-quartet, through its RGG box. The mRNP complex is then exported to the cytoplasm where it is able to interact with RISC complex. Suppression of a specific mRNA target is then maintained by stabilization of miRNA:mRNA duplex by FMRP-KH2. Stabilization of this duplex leads to polyribosomal stalling and suppresses translation of the mRNA target. The RISC-containing mRNP is then transported into the dendrites where the mRNA target remains translationally suppressed. Glutamate signaling at the synapse then activates mGluR and a downstream de-phosphorylation signal for FMRP allowing for the relief of translational suppression.

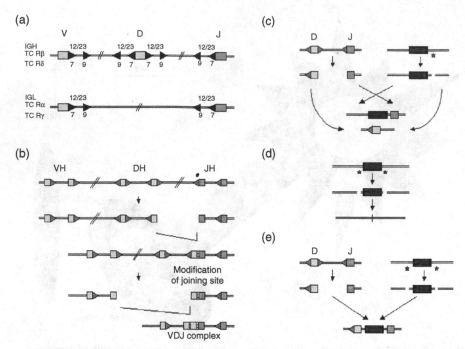

Plate 20 (Fig. 28.2) (a) Schematic structure of immunoglobulin (*IG*) and T-cell receptor (*TCR*) genes encoding *V*-region. There are diverse (*D*) segments in immunoglobulin heavy chain (*IGH*), *TCRβ*, and *TCRδ* genes but not in immunoglobulin light chain (*IGL*), *TCRα* and *TCRγ* genes. Each segment is flanked by conserved recombination signal sequences (RSSs). The RSSs consist of a heptamer and nonamer sequence separated by 12 bp or 23 bp of "spacer" sequence. Boxes and triangles indicate coding and heptamer/nonamer sequences, respectively. (b) Overview of *V(D)J* recombination events in *IGH* gene. The rearrangement of *D* to *J* takes place followed by a *V* to *D-J* complex. The joining sites are often modified by addition of nucleotides or deletion. RSSs are shown as triangles. (c) Chromosome translocation involving *IG/TCR* and/or non-*IG/TCR* loci mediated by errors of *V(D)J* recombination. Cryptic RSSs (shown by asterisks) might be found at positions close to the double-stranded DNA breakage on the non-*IG/TCR* loci. (d) Deletion can be mediated by *V(D)J* recombination machinery recognizing cryptic RSS on non-*IG/TCR* loci. (e) Insertion of a non-*IG/TCR* gene generated by *V(D)J* recombination into an *IGH* locus has been previously reported. Asterisks represent cryptic RSSs on non-*IG/TCR* loci.

(a)

DAPI   miR-181   eMHC

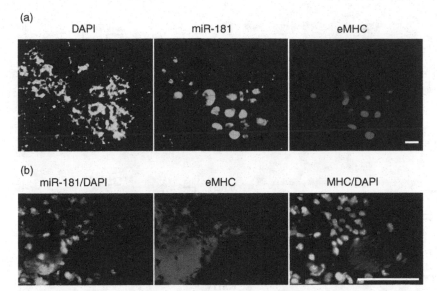

(b)

miR-181/DAPI   eMHC   MHC/DAPI

Plate 21 (Fig. 30.3) Cells expressing miR-181 are differentiating muscle cells. (a) Low magnification (x200). Cross-sections of *tibialis anterior* muscles at day 5 of cardiotoxin-induced regeneration were submitted to immuno-FISH using antibodies against embryonic MHC (eMHC) with TRITC-labeled secondary antibodies, and a DIG-labeled oligonucleotide probe complementary to miR-181 with FITC-labeled secondary antibodies; sections were also counterstained with DAPI. (b) Higher magnification ($\times$ 630). Labeling as in (a). In addition, resting fibers were stained using antibodies against sarcomeric MHC with Cy5-labeled secondary antibodies. Scale bars correspond to 25 μm.

HSV-1

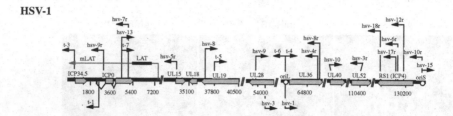

HCMV

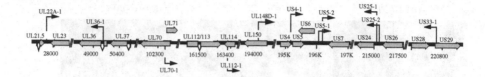

EBV   **KSHV**

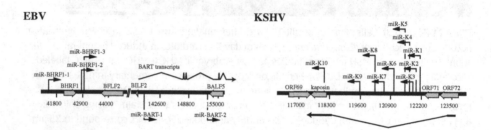

MHV-68

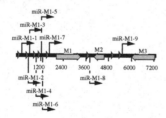

Plate 22 (Fig. 31.1) Schematic representations of herpesvirus miRNAs (black arrows) in relation to the surrounding ORFs (green arrows). Minor LAT of HSV-1 is shown in red. Position of MHV-68 "tRNAs" are shown as red arrows. Nucleotide coordinates refere to Genbank accession numbers; HSV-1 – NC_001806; HCMV – X17403; EBV – AJ507799; KSHV – AF14880. Coordinates for UL148D-1 of HCMV correspond to genbank accession number AC146906.

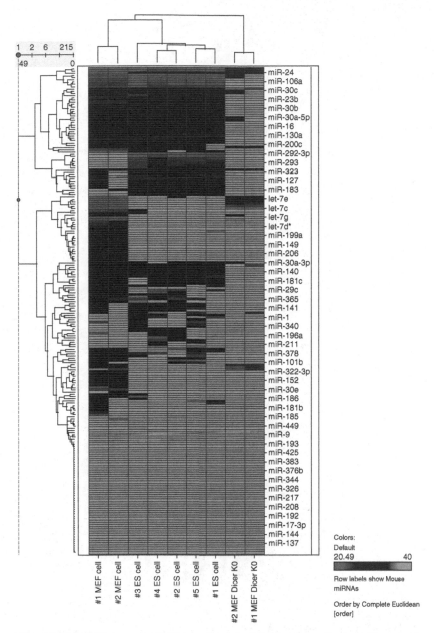

Plate 23 (Fig. 36.3) Unsupervised hierarchical clustering heatmap of microRNA expression profile for single embryonic stem cells (ES cells), single wildtype mouse embryonic fibroblasts (MEFs), and single Dicer knockout MEFs. The cluster heatmap was produced using expression levels (Ct value) of 214 miRNAs. Correlation coefficient was used as similarity measure and complete linkage method was used as clustering method. Note that higher Ct value means lower expression level.

V

MicroRNAs in
disease biology

22 Dysregulation of microRNAs in human malignancy

Kathryn A. O'Donnell and Joshua T. Mendell*

Introduction

Over the past twenty years, cancer geneticists have uncovered many of the genes responsible for the initiation and progression of the multi-step process of tumorigenesis (Vogelstein and Kinzler, 2004). Much of this research has focused on traditional protein-coding genes including oncogenes, tumor suppressors, and genes that maintain genome stability. Within the past five years, a new class of small non-coding RNAs called microRNAs (miRNAs or miRs) has been identified and recent evidence suggests that dysregulation of miRNAs is linked to the development of cancer.

In 1993, the Ambros and Ruvkun laboratories discovered that a 21-nucleotide RNA molecule called *lin-4* regulated the translation of a target message, *lin-14*, by base-pairing to its 3' untranslated region (Lee *et al.*, 1993; Wightman *et al.*, 1993). Subsequent work in this direction prompted the construction and sequencing of libraries of cloned small RNAs by several groups of investigators. Coupled with bioinformatic analyses of genomic sequence, these efforts led to the identification of several hundred miRNAs in *Drosophila*, *C. elegans*, and mammals (Lagos-Quintana *et al.*, 2001; Lau *et al.*, 2001; Lee and Ambros, 2001). More than 450 miRNAs have been identified in humans and recent estimates suggest there may be as many as 1000 (Bentwich *et al.*, 2005: Berezikov *et al.*, 2005). The biogenesis and function of microRNAs are detailed in various chapters in Part II of this book. In general, the analysis of many more miRNA–target interactions is required to better understand the mechanisms through which miRNAs elicit various effects. Delineating miRNA function requires rigorous validation of the target mRNAs that they regulate. Base-pairing at the 5' end of the miRNA is thought to be most important for mRNA target recognition. Nucleotides 2–8 of the miRNA have been termed the "miRNA seed." Many target prediction algorithms search primarily for complementarity to this sequence in the 3' UTR of target transcripts (John *et al.*, 2004; Lewis *et al.*, 2003). One potential weakness of this approach is that some

---

* Author to whom correspondence should be addressed.

MicroRNAs: From Basic Science to Disease Biology, ed. Krishnarao Appasani. Published by Cambridge University Press. © Cambridge University Press 2008.

bona fide miRNA targets lack perfect complementarity to the seed. Thus, some mammalian mRNA targets may be overlooked by solely using the seed match criterion. For example, Brennecke *et al.* demonstrated that a perfect match between the miRNA seed and its mRNA target is not required as long as there is strong compensatory pairing at the 3′ end of the miRNA (Brennecke *et al.*, 2005). Vella *et al.* also showed that the 3′ end of a miRNA influences selection of mRNA targets (Vella *et al.*, 2004). These observations will need to be incorporated into prediction algorithms to allow comprehensive identification of miRNA targets.

While great advances have been made in the identification of miRNAs, we are at the very earliest stages of understanding their functions. One of the most exciting challenges ahead will be to delineate the roles of miRNA in physiologic and pathophysiologic states. Recent studies have demonstrated that dysregulated miRNA expression is a frequent feature of human malignancies and that miRNAs directly regulate pathways that influence the development of cancer. In this chapter, we will discuss the role of microRNAs in cancer and how oncogenes regulate miRNA expression.

MicroRNAs regulate physiologic processes

MicroRNAs exhibit strict developmental and tissue-specific expression patterns. The small number of known biologic functions of microRNAs suggests a critical role for these molecules in the regulation of cellular differentiation, proliferation and apoptosis. For example, the *lin-4* and *let-7* miRNAs were identified by their mutant phenotypes in *C. elegans*. Worms with these mutations undergo abnormal differentiation of cell lineages (Lee *et al.*, 1993; Reinhart *et al.*, 2000; Wightman *et al.*, 1993). *Bantam* encodes a microRNA in *Drosophila* that controls cell proliferation and inhibits the pro-apoptotic gene *hid* (Brennecke *et al.*, 2003; Hipfner *et al.*, 2002). miR-14 and the miR-2/6/11/13/308 family have also been shown to regulate apoptosis in *Drosophila* (Leaman *et al.*, 2005; Xu *et al.*, 2003).

Other characterized microRNAs have critical functions in organ and immune system development. These include miR-1, which is involved in mammalian heart development (Zhao *et al.*, 2005), miR-375 which regulates pancreatic insulin secretion (Poy *et al.*, 2004), miR-181 which influences the differentiation of hematopoietic cells toward the B-cell lineage (Chen *et al.*, 2004), and miR-430 which is required for zebrafish brain development (Giraldez *et al.*, 2005). These studies highlight the participation of miRNAs in diverse cellular processes. It is therefore not surprising that dysregulation of miRNA function can lead to human diseases such as diabetes, neurological disorders, and cancer.

Abnormal microRNA expression and function in cancer

Three different types of genes have traditionally been implicated in cancer: oncogenes, tumor suppressors, and genome-stability genes. Until recently, these gene classes were only known to be made up of protein-coding genes. Significant evidence has now accumulated demonstrating that miRNAs also play a critical role in the development of cancer and can act as cancer-promoting and cancer-suppressing genes. The first documentation of a miRNA abnormality in cancer stemmed from studies of human chromosome 13q14. Chromosomal loss at this region is known to occur in greater than 50% of chronic lymphocytic leukemia (CLL) cases, a common form of adult leukemia (Dohner *et al.*, 2000). Deletions of 13q14 have also been documented in mantle cell lymphoma, prostate cancer, and multiple myeloma (Dong *et al.*, 2001). Although it was suspected that a tumor suppressor in this region was involved in the pathogenesis of these tumors, numerous studies failed to find the involvement of any protein-coding genes. Interestingly, Calin *et al.* identified two microRNAs, miR-15a and miR-16-1, that are located in this region and are deleted or down-regulated in 68% of CLL cases (Calin *et al.*, 2002). Cimmino and colleagues subsequently demonstrated that these miRNAs induce apoptosis by suppressing the anti-apoptotic gene BCL2. This suggested a mechanism whereby deletion of these miRNAs results in increased BCL2 expression levels, thus promoting cell survival and tumorigenesis (Cimmino *et al.*, 2005). Recently, the Croce laboratory found that miR-15a and miR-16-1 are among a set of 13 microRNAs that constitute a unique expression signature which correlates with the prognosis and progression of CLL (Calin *et al.*, 2005).

Several laboratories have investigated alterations in miRNA expression levels that correlate with the cellular transformation process and disease progression. For example, Iorio *et al.* performed miRNA expression profiling to identify miRNAs significantly dysregulated in breast carcinoma. Analysis of 10 normal and 76 breast cancer tissues led to the identification of 15 miRNAs that could accurately distinguish normal from cancerous tissue. Among these, miR-21 and miR-155 were consistently up-regulated in breast cancer and miR-10b, miR-145, and miR-125b were down-regulated (Iorio *et al.*, 2005). Another group found that the steady-state levels of miR-143 and miR-145 were reduced in pre-cancerous adenomatous polyps as well as in the advanced stages of colorectal neoplasia (Michael *et al.*, 2003).

A recent study from Lu *et al.* demonstrated the utility of miRNA expression profiling in cancer diagnosis (Lu *et al.*, 2005). A novel bead-based flow cytometric expression profiling method was developed, which allowed for profiling of mammalian miRNAs across a panel of 334 tissue and tumor samples. Using this platform, the authors found that expression profiles of approximately 200 miRNAs were surprisingly informative and could be used to precisely classify human cancers. In contrast, expression profiles of 16 000 mRNAs in these same tumors did not perform as well in classifying tumor origin or state of differentiation. Hierarchical clustering of the tumor samples using miRNA expression profiles reflected the developmental origins of the tissues from which the tumors originated. The authors concluded that miRNA expression patterns reflect the developmental history of human malignancies. Interestingly, greater than 50% of miRNAs (129 out of 217) were expressed at lower levels in tumors compared to normal tissue independent of cell type. The authors hypothesized that global miRNA expression may reflect the state of differentiation. In support of this, Lu *et al.* observed that induction of miRNAs coincided with differentiation in multiple cell systems. These observations are consistent with the notion that a major physiologic function of miRNAs is to drive terminal differentiation. The authors concluded that a small number of miRNAs provide a vast amount of diagnostic information.

Several studies have focused on specific miRNAs and the pathways they regulate in tumors. For example, Kosik and colleagues recently demonstrated that miR-21 is dramatically up-regulated in human glioblastoma tissues and cell lines. Inhibition of miR-21 expression using 2′ O-methyl antisense (AS) oligonucleotides resulted in activation of caspase activity (Chan *et al.*, 2005). These results suggest that miR-21 may contribute to malignant glioblastoma by suppressing key apoptotic factors and thereby allowing cells to evade cell death.

miR-125b-1, the human ortholog of *lin-4*, has been implicated in the pathogenesis of leukemia especially at an early step in leukemogenesis (Sonoki *et al.*, 2005). miR-125b-1 is also located in a fragile genomic region which is deleted in patients with ovary, cervical, lung, and breast cancers (Calin *et al.*, 2004). Given the role of *lin-4* in role in developmental timing and its recent association with lifespan in *C. elegans* (Boehm and Slack, 2005), it will be of great interest to identify *bona fide* targets of its mammalian ortholog.

Perhaps the first example of a miRNA demonstrated to function as an oncogene is miR-155, which is processed from the non-coding BIC (B-cell integration cluster) RNA. BIC was first identified as a common integration site in avian leukosis virus (ALV)-induced lymphomagenesis (Clurman and Hayward, 1989; McManus, 2003; Tam *et al.*, 1997). Before the discovery of miRNAs, Tam *et al.* hypothesized that BIC may function through its RNA because it appeared to lack an extensive open reading frame (Tam *et al.*, 1997). Although a highly ordered RNA structure was detected, the significance of this structure was not understood at the time (Tam, 2001). Based on its frequency as a common insertion site and its association with c-Myc activation, the BIC locus was thought to harbor a gene that could collaborate with c-Myc in lymphomagenesis. Co-expression of a conserved region in the last exon of the BIC RNA and c-Myc led to an increase in the incidence of leukemia in a virally-induced model of chicken lymphoma (Tam *et al.*, 2002). These results demonstrated that BIC cooperated with c-Myc in lymphomagenesis and provided evidence that untranslated RNAs could influence oncogenesis. It wasn't until several years later that Metzler *et al.* discovered that miR-155 is processed from the last exon of the BIC mRNA (Metzler *et al.*, 2004). Recent studies have shown that miR-155 expression is elevated in Hodgkin lymphoma samples and cell lines (Eis *et al.*, 2005; Kluiver *et al.*, 2005; van den Berg *et al.*, 2003), diffuse large B-cell lymphoma (Eis *et al.*, 2005) and in childhood Burkitt's lymphoma (Metzler *et al.*, 2004). There are conflicting reports in the literature, however. In contrast to the Metzler study, Kluiver *et al.* reported a lack of BIC and miR-155 expression in primary Burkitt's lymphoma samples (Kluiver *et al.*, 2006; van den Berg *et al.*, 2003). Nevertheless, these authors did detect elevated levels of BIC and miR-155 specifically in latency type III EBV positive Burkitt's lymphoma cell lines and in post-transplantation lymphoproliferative disorder (PTLD) cases. The authors suggest that EBV latency type III-specific proteins may influence the induction of BIC expression. Future studies are necessary to determine the role of miR-155 in the pathogenesis of Burkitt's lymphoma and other malignancies.

The tumor suppressor functions of the let-7 family of miRNAs have been well studied. *let-7* normally functions to suppress proliferation and promote differentiation in adult worms. It was previously observed that *let-7* mutants fail to exit the cell cycle and thus undergo division instead of terminal differentiation (Reinhart *et al.*, 2000). These studies lead to the hypothesis that loss of let-7 expression may lead to an increase in cellular proliferation and/or a decrease in differentiation in mammalian cells. Several years later, two groups published observations in support of this hypothesis. Let-7 genes mapped to genomic regions that are deleted in lung cancer samples (Calin *et al.*, 2004). Furthermore, Takamizawa and colleagues demonstrated that let-7 overexpression inhibited growth of a lung cancer cell line *in vitro* (Takamizawa *et al.*, 2004). These studies hinted that let-7 miRNAs might act as tumor suppressors. In an elegant series of experiments, the Slack laboratory demonstrated that the *C. elegans let-7* family of miRNAs negatively regulates *let-60*, the *C. elegans* ortholog of the RAS family of proteins (HRAS, KRAS, and NRAS), which are frequently mutated in cancer. They found that the 3' UTR of human RAS genes contain multiple binding sites for let-7 family members. A subset of these binding sites were conserved in several mammals suggesting functional relevance. Overexpression of let-7 in HepG2 cells was demonstrated to reduce RAS protein abundance. Conversely, inhibition of let-7 using antisense oligonucleotides in HeLa cells led to the accumulation of RAS proteins (Johnson *et al.*, 2005). The authors went on to show that let-7 expression was decreased in lung tumors, which inversely correlated with the levels of RAS protein (but not mRNA). This study proposes a possible mechanism leading to the overexpression of RAS in lung cancer and demonstrates that miRNAs may possess tumor suppressor activity.

Recently, a set of five microRNAs has been found to be transcriptionally up-regulated in papillary thyroid carcinoma (PTC) (H. He *et al.*, 2005). Tumors that exhibited up-regulation of these miRNAs showed loss of *KIT* mRNA and protein, a well-known tyrosine kinase receptor that regulates cell growth. In five of ten tumors, two single nucleotide polymorphisms were found in regions of *KIT* corresponding to the miRNA binding sites for miR-221, miR-222, and miR-146. The authors suggest that dysregulation of these miRNAs and the mRNA targets that they regulate may be important factors in the initiation and development of papillary thyroid carcinoma. Future studies are necessary to determine the significance of the overexpression of these miRNAs in PTC patients.

Table 22.1 summarizes what is known about each of these miRNAs implicated in human malignancy. Many important questions remain: do miRNAs cause cancer or does their altered expression simply correlate with disease progression? Analysis of one set of six microRNAs on human chromosome 13, termed the *mir-17* cluster, has shed some light on this question (discussed in the next section). Many more gain-of-function and loss-of-function studies are necessary to gain a more complete understanding of the role of miRNAs in cancer.

Table 22.1. MicroRNAs dysregulated in human malignancy

MicroRNA	Expression pattern	References
miR-15a, miR-16-1	Deleted or downregulated in chronic lymphocytic leukemia patients	Calin et al., 2002
miR-26a, miR-99a	Downregulated in lung cancer cell lines	Calin et al., 2004
miR-143, miR-145	Reduced expression in colorectal cancer	Michael et al., 2003
	Downregulated in breast, prostate, cervical, and lymphoid cancer cell lines	Iorio et al., 2005
miR-155	Upregulated in Burkitt's lymphoma	Metzler et al., 2004
	Upregulated in Hodgkin lymphoma, primary mediastinal lymphoma and diffuse large B-cell lymphoma	van den Berg et al., 2003; Eis et al., 2005; Kluiver et al., 2005
	Upregulated in breast carcinoma	Iorio et al., 2005
let-7	Maps to fragile sites in breast, lung, urothelial, and cervical cancers	Calin et al., 2004
	Downregulated in human lung cancer	Takamizawa et al., 2004
	Regulates expression of the RAS oncogene	Johnson et al., 2005
miR-17-5p, miR-18a,	LOH of the genomic region containing the miR-17 cluster in hepatocellular carcinoma	Lin et al., 1999
miR-19a, miR-20a,	Amplification of the chromosomal locus harboring the miR-17 cluster	Ota et al., 2004
miR-19b-1, and miR 92-1	Upregulation of miR-19a and miR 92-1 in cells from B-CLL patients	Calin et al., 2004
	Upregulation of the miR-17 cluster pri-miRNA in 65% of B-cell lymphoma samples	L. He et al., 2005
	Cooperates with c-Myc in a mouse model of B-cell lymphoma	L. He et al., 2005
	Directly regulated by the c-Myc oncogene	O'Donnell et al., 2005
	Overexpressed in lung cancer	Hayashita et al., 2005
miR-21	Upregulated in breast cancer tissue	Iorio et al., 2005
	Overexpressed in glioblastoma tumor tissues and cell lines	Chan et al., 2005
miR-125b-1 (lin 4 homolog)	Deleted in a subset of patients with breast, ovary, lung, and cervical cancer	Calin et al., 2004
	Insertion into a rearraged Ig heavy chain gene locus in a precursor B-cell acute lymphocytic leukemia (ALL) patient	Sonoki et al., 2005
miR-221, miR-222, miR-146	Transcriptionally upregulated in papillary thyroid carcinoma tumors	H. He et al., 2005

c-Myc regulated microRNAs

It is widely accepted that miRNAs must be expressed at specific stages and in specific tissues during development in order to function properly (Ambros, 2004). The principle mechanisms underlying the regulation of miRNA gene transcription are just beginning to be defined. In particular, very few mammalian transcription factors that regulate miRNA expression have been identified. To begin to explore the mechanisms of miRNA transcriptional regulation, our laboratory has focused on the identification of miRNAs regulated by the *c-MYC* proto-oncogene. *c-MYC* encodes a basic helix–loop–helix transcription factor that regulates a diverse set of target genes, including those that regulate cell proliferation, growth, and apoptosis (Dang, 1999). Dysregulated expression of c-Myc is one of the most common abnormalities in human malignancy (Cole and McMahon, 1999). The emerging view holds that c-Myc regulates 10%–15% of genes in the human and *Drosophila* genomes (Fernandez *et al.*, 2003; Li *et al.*, 2003; Orian *et al.*, 2003). Because miRNAs are also known to control proliferation and apoptosis, we hypothesized that c-Myc may also directly regulate miRNA expression.

In order to identify c-Myc-regulated miRNAs, we utilized a spotted array developed in our laboratory capable of measuring the expression levels of 235 human, mouse, or rat miRNAs. We analyzed miRNA expression patterns in a human B-cell line (P493–6) that expresses a tetracycline-regulated *c-MYC* transgene (Pajic *et al.*, 2000). Six miRNAs (miR-17-5p, miR-18, miR-19a, miR-19b, miR-20, and miR-92) located in a cluster on human chromosome 13 were highly up-regulated in the high c-Myc state. Collectively, these miRNAs are referred to as the *mir-17* cluster. Consistent with our data from P493–6 cells, we also observed that expression of these miRNAs was reduced by 50% in a previously described rat fibroblast cell line containing a homozygous deletion of c-Myc (Mateyak *et al.*, 1997). These data provided evidence that c-Myc expression is necessary for full physiologic expression of these miRNAs and further demonstrate that regulation of these miRNAs by c-Myc is not specific to the B-cell lineage. To determine whether regulation of the *mir-17* cluster by c-Myc is direct, we performed chromatin immunoprecipitation (ChIP) experiments in P493–6 cells and in primary human fibroblasts. ChIP analysis revealed that c-Myc binds directly to the genomic locus encoding the *mir-17* cluster providing compelling evidence that these miRNAs are indeed direct targets of this transcription factor (O'Donnell *et al.*, 2005). These findings suggest that miRNA genes are controlled by the same regulatory mechanisms that govern the expression of protein-coding genes.

The direct regulation of the *mir-17* cluster by c-Myc suggests that activation of these miRNAs may contribute to Myc-mediated tumorigenesis. Using bioinformatic prediction algorithms, several groups have identified putative targets of the *mir-17* cluster (John *et al.*, 2004; Lewis *et al.*, 2003). Interestingly, numerous mRNAs that promote growth, such as N-Myc and E2F1, as well as mRNAs that suppress growth, including Rb2/p130 and the PTEN tumor suppressor are predicted to be regulated by these miRNAs. We hypothesize that the *mir-17* cluster can participate in tumorigenesis in two ways. Amplification or overexpression of

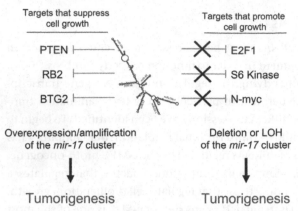

Figure 22.1. Hypothesized mechanism through which gain- or loss-of-function of the *mir-17* cluster may lead to malignancy. Left panel: overexpression or amplification of the *mir-17* cluster may result in tumorigenesis if this results in decreased expression of targets that inhibit cell growth. Right panel: deletion or loss-of-heterozygosity of the *mir-17* cluster may cause tumorigenesis if this relieves repression of targets that promote cell growth.

the *mir-17* cluster may lead to malignancy via subsequent downregulation of a tumor suppressor protein (Figure 22.1, left panel). Alternatively, deletion or loss-of-heterozygosity of the *mir-17* cluster could also promote malignant growth through the resulting elevated expression of a protein that drives tumorigenesis (Figure 22.1, right panel).

In fact, there is accumulating evidence that the *mir-17* cluster contributes to tumor formation using both mechanisms. Amplification of 13q31-q32, which contains the *mir-17* cluster, has been documented in solid tumors and in hematologic malignancies (Gordon *et al.*, 2000; Schmidt *et al.*, 2005). Using comparative genomic hybridization, Ota *et al.* determined that C13orf25, the host transcript of the *mir-17* cluster, is the target of amplification in several types of lymphoma cell lines and patient samples (Ota *et al.*, 2004). Because predicted open reading frames within C13orf25 encode small peptides that are not conserved, this transcript is unlikely to encode a protein. Instead, C13orf25 most likely solely functions as the precursor to the six miRNAs within the *mir-17* cluster. The Hammond and Hannon laboratories recently directly demonstrated that the *mir-17* cluster has oncogenic activity (L. He *et al.*, 2005). First, they confirmed that these miRNAs were highly upregulated in B cell lymphoma. These authors then directly tested whether the *mir-17* cluster contributes to tumor development utilizing the Eμ-Myc transgenic mouse which is a model of Myc-activated B cell lymphoma. Hematopoietic stem cells (HSCs) derived from fetal livers of these mice produce lymphomas late in life (4–6 months) when transplanted into lethally irradiated wild-type recipients. Mice reconstituted with Eμ-Myc HSCs overexpressing a truncated form of the *mir-17* cluster (containing five of the six miRNAs) developed lymphoma with decreased latency (less than 2 months), demonstrating that the *mir-17* cluster cooperates with c-Myc in lymphomagenesis. Furthermore, lymphomas arising from overexpression of the *mir-17* cluster primarily showed decreased apoptosis. Interestingly, animals reconstituted with Eμ-Myc HCSs expressing subsets of 96 individual miRNAs did not show accelerated tumor formation demonstrating the specificity of these findings. Taken together, these data show that the *mir-17* cluster possesses oncogenic activity. Hayashita *et al.* also recently demonstrated that overexpression of the *mir-17* cluster accelerated cell proliferation in a lung cancer cell line (Hayashita *et al.*, 2005). These findings are consistent with the scenario depicted in the left panel of Figure 22.1 whereby overexpression of the *mir-17* cluster results in decreased expression of a gene that inhibits cell growth.

There are other studies that report the LOH of 13q31-q32 in several tumor types, including breast cancer, bladder cancer, hepatocellular carcinoma, nasopharyngeal carcinoma, and pituitary adenoma (Eiriksdottir *et al.*, 1998; Honda *et al.*, 2003; Koo *et al.*, 2003; Richter *et al.*, 1999; Tsang *et al.*, 1999). Although this might indicate the presence of a tumor suppressor gene located near the *mir-17* cluster, this would also be consistent with the scenario depicted in the right panel of Figure 22.1. According to this hypothesis, deletion or decreased expression of the *mir-17* cluster may result in upregulation of a gene that promotes cell growth. It is important to stress that the function of a given miRNA will depend on the specific target mRNAs that are expressed in a particular tissue or stage of development. Moreover, the target mRNAs driving proliferation in a given tumor-type are likely to dictate whether a given miRNA behaves as a tumor suppressor or as an oncogene. It is thus important to examine the consequences of expression of the *mir-17* cluster in diverse tumor models. Additionally, it will be essential to verify the target mRNAs that these miRNAs regulate.

One predicted target of two miRNAs in the *mir-17* cluster, miR-17-5p and miR-20, is E2F1 (John *et al.*, 2004; Lewis *et al.*, 2005; Lewis *et al.*, 2003). In mammalian cells, the E2F family of transcription factors controls the G1 to S phase transition by directly stimulating expression of genes that regulate DNA replication, cell proliferation, and apoptosis (Trimarchi and Lees, 2002). Several laboratories have demonstrated that overexpression of E2F1 promotes proliferation or induces cell death under a variety of experimental conditions (Hunt *et al.*, 1997; Kowalik *et al.*, 1995; Qin *et al.*, 1994). Transcription of E2F1 is known to be directly activated by c-Myc (Fernandez *et al.*, 2003; Leone *et al.*, 1997). Additionally, expression of c-Myc is induced by E2F1, setting up a putative positive feedback loop. Nonetheless, rampant reciprocal activation of these gene products has not been observed, thus raising the question of how cells maintain control of c-Myc and E2F1 expression. We hypothesized that induction of the *mir-17* cluster by c-Myc may dampen reciprocal activation of these factors. To determine whether E2F1 is a *bona fide* target of the *mir-17* cluster, expression of miR-17-5p and miR-20 was inhibited by using 2′ O-methyl antisense oligonucleotides in Hela cells, which naturally express these miRNAs. This resulted in increased levels of E2F1 protein levels without affecting E2F1 mRNA abundance (O'Donnell *et al.*, 2005). Conversely, overexpression of the mir-17 cluster resulted in a decrease in E2F1 protein levels but did not alter mRNA levels. Reporter assays demonstrated that these miRNAs bind to two sites in the E2F1 3′ UTR. Taken together, these results demonstrated that c-Myc controls E2F1 activity by directly stimulating its transcription and by

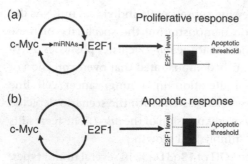

(a)

c-Myc →miRNAs┤ E2F1 ⟶   Proliferative response

(b)

c-Myc   E2F1 ⟶   Apoptotic response

Figure 22.2. (a) The *mir-17* cluster is hypothesized to dampen reciprocal activation of c-Myc and E2F1. This may serve to maintain E2F1 activity below a critical apoptotic threshold level, which would allow for a proliferative response in the absence of apoptosis. (b) In the absence of *mir-17* cluster expression, c-Myc and E2F1 co-activate their expression in an uncontrolled manner. This may lead to excessive E2F1 activity above the critical threshold level, resulting in apoptosis.

simultaneously negatively regulating its translation through activation of the *mir-17* cluster. These data support a model whereby miR-17-5p and miR-20 limit c-Myc mediated activation of E2F1 expression (Figure 22.2). This would prevent uncontrolled reciprocal activation of c-Myc and E2F1. Additionally, this regulatory circuit is predicted to maintain E2F1 levels below a critical threshold level of expression such that cell proliferation is promoted and apoptosis is avoided. Further experiments are necessary to determine whether this regulatory circuit contributes to the oncogenic activity of the *mir-17* cluster, although it should be noted that the predominant effect of expressing these miRNAs in Eμ-Myc transgenic mice was decreased apoptosis, as predicted by this model. Additionally, it will be of great interest to determine whether E2F1 abundance is decreased in lymphomas co-expressing c-Myc and the *mir-17* cluster.

Continued efforts to validate targets of these miRNAs will contribute to a more complete understanding of the role these miRNAs play in oncogenesis. In addition to E2F1, several other c-Myc target genes are predicted targets of the *mir-17* cluster, including *RPS6KA5*, *BCL11B*, *PTEN*, and *HCFC2*. It is likely that many mammalian transcription factors fine-tune expression of their target genes by activating their transcription and dampening their translation through the activity of miRNAs.

Concluding remarks

Clearly the functions of miRNAs are important in normal cellular physiology and dysregulated expression of miRNAs participates in human disease. Nevertheless, mechanistic studies of miRNAs are just beginning to be carried out. In this chapter, we have discussed specific examples of miRNAs that may function as tumor suppressors and oncogenes. Many more gain-of-function and loss-of-function studies are necessary to expand our understanding of the role miRNAs play in tumorigenesis. To partially address this issue, a miRNA consortium at the Unviersity of California, San Francisco, has recently been created. Michael McManus and colleagues have plans to systematically knock out each of the 350 known miRNAs in mice. Undoubtedly, these types of analyses will be useful for the entire miRNA and cancer research communities. Many important questions regarding the functions of miRNAs remain. For example, in addition to functioning as tumor suppressors and oncogenes, is it possible that miRNAs also act as genome stability genes which regulate cellular processes such as DNA repair? Is it possible that a given miRNA can act as both a tumor suppressor and an oncogene depending on the context of expression? Exactly how many mRNA targets do miRNAs regulate? Addressing these and other questions will be essential as we elucidate the role of miRNAs in cancer and other diseases.

ACKNOWLEDGEMENTS

The authors thank Dr. Krishnarao Appasani for helpful editing of this chapter. J.T.M. is a Rita Allen Foundation Scholar and K.A.O. is a Damon Runyon Postdoctoral Fellow.

REFERENCES

Ambros, V. (2004). The functions of animal microRNAs. *Nature*, **431**, 350–355.

Bentwich, I., Avniel, A., Karov, Y. *et al.* (2005). Identification of hundreds of conserved and nonconserved human miRNAs. *Nature Genetics*, **37**, 766–770.

Berezikov, E., Guryev, V., Belt, J. *et al.* (2005). Phylogenetic shadowing and computational identification of human microRNA genes. *Cell*, **120**, 21–24.

Boehm, M. and Slack, F. (2005). A developmental timing microRNA and its target regulate life span in *C. elegans*. *Science*, **310**, 1954–1957.

Brennecke, J., Hipfner, D. R., Stark, A., Russell, R. B. and Cohen, S. M. (2003). bantam encodes a developmentally regulated microRNA that controls cell proliferation and regulates the proapoptotic gene hid in *Drosophila*. *Cell*, **113**, 25–36.

Brennecke, J., Stark, A., Russell, R. B. and Cohen, S. M. (2005). Principles of microRNA–target recognition. *Public Library of Science Biology*, **3**, e85.

Calin, G. A., Dumitru, C. D., Shimizu, M. *et al.* (2002). Frequent deletions and down-regulation of micro-RNA genes miR15 and miR16 at 13q14 in chronic lymphocytic leukemia. *Proceedings of the National Academy of Sciences USA*, **99**, 15 524–15 529.

Calin, G. A., Sevignani, C., Dumitru, C. D. *et al.* (2004). Human microRNA genes are frequently located at fragile sites and genomic regions involved in cancers. *Proceedings of the National Academy of Sciences USA*, **101**, 2999–3004.

Calin, G. A., Ferracin, M., Cimmino, A. *et al.* (2005). A microRNA signature associated with prognosis and progression in chronic lymphocytic leukemia. *The New England Journal of Medicine*, **353**, 1793–1801.

Chan, J. A., Krichevsky, A. M. and Kosik, K. S. (2005). MicroRNA-21 is an antiapoptotic factor in human glioblastoma cells. *Cancer Research*, **65**, 6029–6033.

Chen, C. Z., Li, L., Lodish, H. F. and Bartel, D. P. (2004). MicroRNAs modulate hematopoietic lineage differentiation. *Science*, **303**, 83–86.

Cimmino, A., Calin, G. A., Fabbri, M. *et al.* (2005). miR-15 and miR-16 induce apoptosis by targeting BCL2. *Proceedings of the National Academy of Sciences USA*, **102**, 13 944–13 949.

Clurman, B. E. and Hayward, W. S. (1989). Multiple proto-oncogene activations in avian leukosis virus-induced lymphomas: evidence for stage-specific events. *Molecular and Cellular Biology*, **9**, 2657–2664.

Cole, M. D. and McMahon, S. B. (1999). The Myc oncoprotein: a critical evaluation of transactivation and target gene regulation. *Oncogene*, **18**, 2916–2924.

Dang, C. V. (1999). c-Myc target genes involved in cell growth, apoptosis, and metabolism. *Molecular and Cellular Biology*, **19**, 1–11.

Dohner, H., Stilgenbauer, S., Benner, A. *et al*. (2000). Genomic aberrations and survival in chronic lymphocytic leukemia. *The New England Journal of Medicine*, **343**, 1910–1916.

Dong, J. T., Boyd, J. C. and Frierson, H. F., Jr. (2001). Loss of heterozygosity at 13q14 and 13q21 in high grade, high stage prostate cancer. *Prostate*, **49**, 166–171.

Eiriksdottir, G., Johannesdottir, G., Ingvarsson, S. *et al*. (1998). Mapping loss of heterozygosity at chromosome 13q: loss at 13q12-q13 is associated with breast tumour progression and poor prognosis. *The European Journal of Cancer*, **34**, 2076–2081.

Eis, P. S., Tam, W., Sun, L. *et al*. (2005). Accumulation of miR-155 and BIC RNA in human B cell lymphomas. *Proceedings of the National Academy of Sciences USA*, **102**, 3627–3632.

Fernandez, P. C., Frank, S. R., Wang, L. *et al*. (2003). Genomic targets of the human c-Myc protein. *Genes and Development*, **17**, 1115–1129.

Giraldez, A. J., Cinalli, R. M., Glasner, M. E. *et al*. (2005). MicroRNAs regulate brain morphogenesis in zebrafish. *Science*, **308**, 833–838.

Gordon, A. T., Brinkschmidt, C., Anderson, J. *et al*. (2000). A novel and consistent amplicon at 13q31 associated with alveolar rhabdomyosarcoma. *Genes, Chromosomes, and Cancer*, **28**, 220–226.

Hayashita, Y., Osada, H., Tatematsu, Y. *et al*. (2005). A polycistronic microRNA cluster, miR-17-92, is overexpressed in human lung cancers and enhances cell proliferation. *Cancer Research*, **65**, 9628–9632.

He, H., Jazdzewski, K., Li, W. *et al*. (2005). The role of microRNA genes in papillary thyroid carcinoma. *Proceedings of the National Academy of Sciences USA*, **102**, 19 075–19 080.

He, L., Thomson, J. M., Hemann, M. T. *et al*. (2005). A microRNA polycistron as a potential human oncogene. *Nature*, **435**, 828–833.

Hipfner, D. R., Weigmann, K. and Cohen, S. M. (2002). The bantam gene regulates *Drosophila* growth. *Genetics*, **161**, 1527–1537.

Honda, S., Tanaka-Kosugi, C., Yamada, S. *et al*. (2003). Human pituitary adenomas infrequently contain inactivation of retinoblastoma 1 gene and activation of cyclin dependent kinase 4 gene. *Endocrine Journal*, **50**, 309–318.

Hunt, K. K., Deng, J., Liu, T. J. *et al*. (1997). Adenovirus-mediated overexpression of the transcription factor E2F-1 induces apoptosis in human breast and ovarian carcinoma cell lines and does not require p53. *Cancer Research*, **57**, 4722–4726.

Iorio, M. V., Ferracin, M., Liu, C. G. *et al*. (2005). MicroRNA gene expression deregulation in human breast cancer. *Cancer Research*, **65**, 7065–7070.

John, B., Enright, A. J., Aravin, A. *et al*. (2004). Human microRNA targets. *Public Library of Science Biology*, **2**, e363.

Johnson, S. M., Grosshans, H., Shingara, J. *et al*. (2005). RAS is regulated by the let-7 microRNA family. *Cell*, **120**, 635–647.

Kluiver, J., Poppema, S., de Jong, D. *et al*. (2005). BIC and miR-155 are highly expressed in Hodgkin, primary mediastinal and diffuse large B cell lymphomas. *Journal of Pathology*, **207**, 243–249.

Kluiver, J., Haralambieva, E., de Jong, D. *et al*. (2006). Lack of BIC and microRNA miR-155 expression in primary cases of Burkitt lymphoma. *Genes, Chromosomes, and Cancer*, **45**, 147–153.

Koo, S. H., Ihm, C. H., Kwon, K. C. *et al*. (2003). Microsatellite alterations in hepatocellular carcinoma and intrahepatic cholangiocarcinoma. *Cancer Genetics and Cytogenetics*, **146**, 139–144.

Kowalik, T. F., DeGregori, J., Schwarz, J. K. and Nevins, J. R. (1995). E2F1 overexpression in quiescent fibroblasts leads to induction of cellular DNA synthesis and apoptosis. *Journal of Virology*, **69**, 2491–2500.

Lagos-Quintana, M., Rauhut, R., Lendeckel, W. and Tuschl, T. (2001). Identification of novel genes coding for small expressed RNAs. *Science*, **294**, 853–858.

Lau, N. C., Lim, L. P., Weinstein, E. G. and Bartel, D. P. (2001). An abundant class of tiny RNAs with probable regulatory roles in *Caenorhabditis elegans*. *Science*, **294**, 858–862.

Leaman, D., Chen, P. Y., Fak, J. *et al*. (2005). Antisense-mediated depletion reveals essential and specific functions of microRNAs in *Drosophila* development. *Cell*, **121**, 1097–1108.

Lee, R. C. and Ambros, V. (2001). An extensive class of small RNAs in *Caenorhabditis elegans*. *Science*, **294**, 862–864.

Lee, R. C., Feinbaum, R. L. and Ambros, V. (1993). The *C. elegans* heterochronic gene lin-4 encodes small RNAs with antisense complementarity to lin-14. *Cell*, **75**, 843–854.

Leone, G., DeGregori, J., Sears, R., Jakoi, L. and Nevins, J. R. (1997). Myc and Ras collaborate in inducing accumulation of active cyclin E/Cdk2 and E2F. *Nature*, **387**, 422–426.

Lewis, B. P., Shih, I. H., Jones-Rhoades, M. W., Bartel, D. P. and Burge, C. B. (2003). Prediction of mammalian microRNA targets. *Cell*, **115**, 787–798.

Lewis, B. P., Burge, C. B. and Bartel, D. P. (2005). Conserved seed pairing, often flanked by adenosines, indicates that thousands of human genes are microRNA targets. *Cell*, **120**, 15–20.

Li, Z., Van Calcar, S., Qu, C. *et al.* (2003). A global transcriptional regulatory role for c-Myc in Burkitt's lymphoma cells. *Proceedings of the National Academy of Sciences USA*, **100**, 8164–8169.

Lin, Y. W., Sheu, J. C., Liu, L. Y. *et al.* (1999). Loss of heterozygosity at chromosome 13q in hepatocellular carcinoma: identification of three independent regions. *European Journal of Cancer*, **35** (12), 1730–1734.

Lu, J., Getz, G., Miska, E. A. *et al.* (2005). MicroRNA expression profiles classify human cancers. *Nature*, **435**, 834–838.

Mateyak, M. K., Obaya, A. J., Adachi, S. and Sedivy, J. M. (1997). Phenotypes of c-Myc-deficient rat fibroblasts isolated by targeted homologous recombination. *Cell Growth and Differentiation*, **8**, 1039–1048.

McManus, M. T. (2003). MicroRNAs and cancer. *Seminars in Cancer Biology*, **13**, 253–258.

Metzler, M., Wilda, M., Busch, K., Viehmann, S. and Borkhardt, A. (2004). High expression of precursor microRNA-155/BIC RNA in children with Burkitt lymphoma. *Genes, Chromosomes, and Cancer*, **39**, 167–169.

Michael, M. Z., O'Connor, S. M., van Holst Pellekaan, N. G., Young, G. P. and James, R. J. (2003). Reduced accumulation of specific microRNAs in colorectal neoplasia. *Molecular Cancer Research*, **1**, 882–891.

O'Donnell, K. A., Wentzel, E. A., Zeller, K. I., Dang, C. V. and Mendell, J. T. (2005). c-Myc-regulated microRNAs modulate E2F1 expression. *Nature*, **435**, 839–843.

Orian, A., van Steensel, B., Delrow, J. *et al.* (2003). Genomic binding by the Drosophila Myc, Max, Mad/Mnt transcription factor network. *Genes & Development*, **17**, 1101–1114.

Ota, A., Tagawa, H., Karnan, S. *et al.* (2004). Identification and characterization of a novel gene, C13orf25, as a target for 13q31-q32 amplification in malignant lymphoma. *Cancer Research*, **64**, 3087–3095.

Pajic, A., Spitkovsky, D., Christoph, B. *et al.* (2000). Cell cycle activation by c-myc in a burkitt lymphoma model cell line. *The International Journal of Cancer*, **87**, 787–793.

Poy, M. N., Eliasson, L., Krutzfeldt, J. *et al.* (2004). A pancreatic islet-specific microRNA regulates insulin secretion. *Nature*, **432**, 226–230.

Qin, X. Q., Livingston, D. M., Kaelin, W. G., Jr. and Adams, P. D. (1994). Deregulated transcription factor E2F-1 expression leads to S-phase entry and p53-mediated apoptosis. *Proceedings of the National Academy of Sciences USA*, **91**, 10 918–10 922.

Reinhart, B. J., Slack, F. J., Basson, M. *et al.* (2000). The 21-nucleotide let-7 RNA regulates developmental timing in *Caenorhabditis elegans*. *Nature*, **403**, 901–906.

Richter, J., Wagner, U., Schraml, P. *et al.* (1999). Chromosomal imbalances are associated with a high risk of progression in early invasive (pT1) urinary bladder cancer. *Cancer Research*, **59**, 5687–5691.

Schmidt, H., Bartel, F., Kappler, M. *et al.* (2005). Gains of 13q are correlated with a poor prognosis in liposarcoma. *Modern Pathology*, **18**, 638–644.

Sonoki, T., Iwanaga, E., Mitsuya, H. and Asou, N. (2005). Insertion of microRNA-125b-1, a human homologue of lin-4, into a rearranged immunoglobulin heavy chain gene locus in a patient with precursor B-cell acute lymphoblastic leukemia. *Leukemia*, **19**, 2009–2010.

Takamizawa, J., Konishi, H., Yanagisawa, K. *et al.* (2004). Reduced expression of the let-7 microRNAs in human lung cancers in association with shortened postoperative survival. *Cancer Research*, **64**, 3753–3756.

Tam, W. (2001). Identification and characterization of human BIC, a gene on chromosome 21 that encodes a noncoding RNA. *Gene*, **274**, 157–167.

Tam, W., Ben-Yehuda, D. and Hayward, W. S. (1997). bic, a novel gene activated by proviral insertions in avian leukosis virus-induced lymphomas, is likely to function through its noncoding RNA. *Molecular and Cellular Biology*, **17**, 1490–1502.

Tam, W., Hughes, S. H., Hayward, W. S. and Besmer, P. (2002). Avian bic, a gene isolated from a common retroviral site in avian leukosis virus-induced lymphomas that encodes a noncoding RNA, cooperates with c-myc in lymphomagenesis and erythro-leukemogenesis. *Journal of Virology*, **76**, 4275–4286.

Trimarchi, J. M. and Lees, J. A. (2002). Sibling rivalry in the E2F family. *Nature Reviews Molecular and Cell Biology*, **3**, 11–20.

Tsang, Y. S., Lo, K. W., Leung, S. F. *et al.* (1999). Two distinct regions of deletion on chromosome 13q in primary nasopharyngeal carcinoma. *The International Journal of Cancer*, **83**, 305–308.

van den Berg, A., Kroesen, B. J., Kooistra, K. *et al.* (2003). High expression of B-cell receptor inducible gene BIC in all subtypes of Hodgkin lymphoma. *Genes, Chromosomes, and Cancer*, **37**, 20–28.

Vella, M. C., Reinert, K. and Slack, F. J. (2004). Architecture of a validated microRNA: target interaction. *Chemical Biology*, **11**, 1619–1623.

Vogelstein, B. and Kinzler, K. W. (2004). Cancer genes and the pathways they control. *Nature Medicine*, **10**, 789–799.

Wightman, B., Ha, I. and Ruvkun, G. (1993). Posttranscriptional regulation of the heterochronic gene lin-14 by lin-4 mediates temporal pattern formation in *C. elegans*. *Cell*, **75**, 855–862.

Xu, P., Vernooy, S. Y., Guo, M. and Hay, B. A. (2003). The *Drosophila* microRNA miR-14 suppresses cell death and is required for normal fat metabolism. *Current Biology*, **13**, 790–795.

Zhao, Y., Samal, E. and Srivastava, D. (2005). Serum response factor regulates a muscle-specific microRNA that targets Hand2 during cardiogenesis. *Nature*, **436**, 214–220.

23 High throughput microRNAs profiling in cancers

Muller Fabbri, Ramiro Garzon, Amelia Cimmino, George Adrian Calin
and Carlo Maria Croce*

Introduction

MicroRNAs are small noncoding RNAgenes (18–24 nucleotides in length) that
have been identified in different organisms from the nematode *C. elegans* to
humans (for reviews see Bartel, 2004; He and Hannon, 2004). Recently it has
become more and more evident that microRNAs play important roles in regulating the translation and degradation of mRNAs through base pairing to perfectly
(in plants) or partially (in mammals) complementary sites, mainly but not
exclusively in the untranslated region (UTR) of the target mRNA (Lagos-
Quintana *et al.*, 2001; Lau *et al.*, 2001; Lee and Ambros, 2001). MicroRNAs are
initially transcribed by RNA polymerase II (pol II) as long primary transcripts
called primary-miRNAs (pri-miRNAs). A double-stranded RNA-specific ribonuclease
called Drosha is responsible for the processing of pri-miRNAs into hairpin RNAs of
70–100bp known as pre-miRNAs, which contain a two nucleotide 3' overhang
characteristic of RNase III cleavage products (Cullen, 2004). Pre-miRNAs are
transported to the cytoplasm by the nuclear export factor exportin 5. Once
in the cytoplasm pre-miRNAs are processed by a second, double-stranded specific
ribonuclease III called Dicer in a 18–24 nucleotide duplex. The product of
Dicer's cleavage is incorporated into a large protein complex called RISC
(RNA-induced silencing complex), which includes as core components the
Argonaute proteins (Ago1–4 in humans). One strand of the miRNA duplex
remains stably associated with RISC and becomes the mature miRNA. The opposite strand, called passenger strand or miRNA*, is discarded through two different
mechanisms. When the miRNA duplex is loaded into a RISC complex containing
Ago2, the passenger strand may be cleaved. Alternatively, RISC containing
any other Ago protein may remove the passenger strand through a bypass
mechanism which does not involve cleavage but presumably is based on duplex

---

* Author to whom correspondence should be addressed.

MicroRNAs: From Basic Science to Disease Biology, ed. Krishnarao Appasani. Published by
Cambridge University Press. © Cambridge University Press 2008.

unwinding (Gregory *et al.*, 2005; Matranga *et al.*, 2005; Rand *et al.*, 2005). The mature miRNA guides the RISC complex to target mRNAs which are subsequently cleaved or translationally silenced. When miRNA and mRNA present perfect base pair complementarity, the target mRNA is cleaved (this mechanism has been described as predominant in plants). Imperfect complementarity between miRNA and target mRNA (predominant in *C. elegans, D. melanogaster*, and mammals) leads to translational silencing of the target, although also in case of imperfect base pairing a reduction of the target mRNA has been described (He and Hannon, 2004). As a result miRNAs negatively regulate the expression of target mRNAs. Physiologically they regulate basic cellular functions such as differentiation, proliferation, development and apoptosis. MiRNA deregulation in human diseases (cancer, diabetes, immuno or neurodegenerative disorders) is being extensively investigated by many researchers (for review see Sevignani *et al.*, 2006).

Several groups, including our own, have studied how the miRNome (defined as the full complement of miRNAs in a genome) is differentially expressed in normal tissue differentiation and in the neoplastic transformation. Consequently miRNA "signatures" (in both hematological malignancies and solid tumors) which allow us to distinguish between tumoral and normal cells, and that in some instances are associated with the prognosis and the progression of the neoplasia, were identified.

MicroRNAs are expressed in tissue-specific pattern

The development of tools able to detect the expression of miRNAs has clearly demonstrated the existence of distinct and specific miRNA signatures in normal tissues. Oligonucleotide microRNA microarray chips, containing hundreds of human precursor and mature microRNA probes, identified different and tissue-specific patterns of miRNA expression in human and mouse tissues (Liu *et al.*, 2004; Esquela-Kerscher and Slack, 2004). The microarrays have several advantages. First, they allow the identification of the global expression of several hundred genes in the same sample at one time point. Second, they detect both mature and precursor miRNA molecules (through a careful design of the oligonucleotide probes). Third, they require less RNA than a normal Northern blotting analysis with no need for radioactive isotopes. Figure 23.1 shows the different expression of a panel of RNAs from 7 hematopoietic and 11 solid tissues and from 2 samples of fetal origin (fetal liver and brain). The microarray data confirm that different tissues have distinctive and specific patterns of miRNome expression, and that the specificity of the pattern is maintained among different individuals. Interestingly the hematopoietic tissues presented two different clusters, one including CD5+ cells, T lymphocytes, and leukocytes, and the second containing bone marrow, fetal liver and B lymphocytes. A difference was observed in the miRNA pattern of expression between fetal or adult origin of the same tissue (e.g. brain). This approach identified miRNA highly expressed in only one or few tissues: *miR-1b-2, miR-99b, miR-125*, and *miR 128* (brain); *miR-133a* and

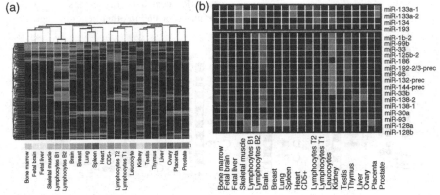

Figure 23.1. MiRNAs are differentially expressed in normal tissues. (a) Distinct patterns of miRNA expression in human adult and fetal tissues. (b) Specific overexpression of different miRNAs in skeletal muscle, heart, prostate, and brain. From Liu *et al.* (2004). (See color plate 15)

miR-133b (skeletal muscle, heart, and prostate), *miR-223* (spleen), and *miR-16-1* (in CD5+ cells). In mice it has been observed that a tissue-specific microRNA is predominant in the whole group of expressed miRNA, suggesting its important role in tissue differentiation. In spleen *miR-143* is the most abundant, in colon *miR-142*, whereas in mouse liver *miR-122a* and *miR-122b* represent 72% of all cloned miRNAs (Lagos-Quintana *et al.*, 2002; Seitz *et al.*, 2004). Another recently developed tool to detect miRNA levels is based on a new, bead-based flow cytometric miRNA method (Lu *et al.*, 2005). The expression analysis of 217 mammalian miRNAs from 334 samples (including multiple human cancers) has confirmed the existence of a tissue-specific miRNA signature. The results of both the microarray and the flow cytometric assay have been confirmed by various techniques including Northern blotting and quantitative RT-PCR.

MicroRNAs are frequently located in cancer-associated genomic regions (CAGRs)

In order to study a possible involvement of miRNAs in cancer, our group performed a genome-wide systematic search for correlations between the genomic localizations of miRNAs and CAGRs (Calin *et al.*, 2004). After having mapped 186 miRNAs and compared their location to the location of previously reported non-random genetic alterations, it was observed that more than half of miRNAs are in CAGRs (Figure 23.2). Overall, 19% of miRNAs are located inside or close to fragile sites (FRA), including FRA in which no known tumor suppressor genes map (e.g. FRA7 H and *miR-29a* and *miR-29b-1*). About half of miRNAs are in LOH regions or in regions of amplification. *MiR-142* is located at 50 nucleotides from the t(8;17) breakpoint region, which involves chromosome 17 and *MYC*. The translocation juxtaposes the *MYC* gene close to the *miR-142* promoter inducing an abnormal *MYC* overexpression. *MiR-180* maps at 1 kb from the *MN1* gene, involved in the t(4;22) in meningioma. The translocation inactivates *MN1* and the miRNA gene

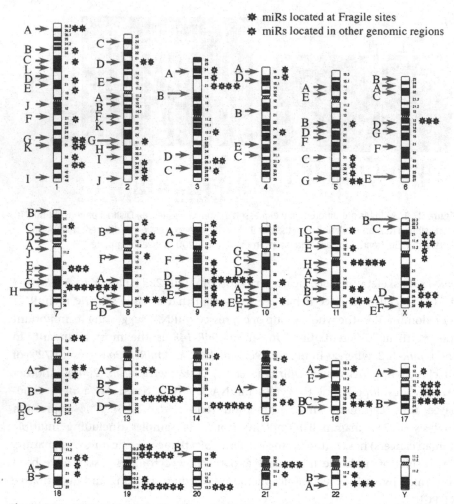

Figure 23.2. MiRNAs are located in different chromosomes and are frequently associated with CAGRs. From Calin *et al.* (2004). (See color plate 16)

located in the same position. *MiR-122a* maps in the minimal amplicon around the *MALT1* gene in aggressive marginal zone lymphoma, and about 160 kb from the breakpoint region of t(11;18) in mucosa-associated lymphoid tissue lymphoma (Sanchez-Izquierdo *et al.*, 2003). Other miRNAs are localized in target regions for viral integration (e.g. *miR-142*, at fragile site FRA17B, a target for HPV16 integration in cervical tumors). A strong correlation was also observed between miRNA localization and the *HOX* genes, which are a family of encoding transcription factors important in normal development and in cancerogenesis (Calin *et al.*, 2004). This association underlines the role of miRNAs in differentiation and (together with the CAGRs mapping) in cancer.

MicroRNA expression signature in hematological malignancies

A correlation between miRNAs and cancer was identified for the first time in chronic lymphocytic leukemia (CLL), the most common form of adult leukemia in the Western world (Calin *et al.*, 2002). Since hemizygous and/or homozygous deletions at 13q14 occur in more than half of cases and represent the most frequent chromosomal alteration in CLL, an extensive research of one or more tumor suppressor genes (TSGs) at 13q14 began. Deletions at 13q14 also occur in about 50% of mantle cell lymphomas (Stilgenbauer *et al.*, 1998), in 30%–40% of multiple myeloma (Elnenaei *et al.*, 2003) and in about 70% of prostate carcinomas (Dong *et al.*, 2001), suggesting a wider role of 13q14 TSGs in human cancers. Despite the exstensive study of a region spanning more than 1Mb and the identification of eight genes located in this genomic area (*LEU1, LEU2, LEU5, CLLD6, KPNA3, CLLD7, LOC51131,* and *CLLD8*), a detailed genetic analysis has failed to demonstrate the consistent involvement of any of the genes located in the deleted region. Comparing 60 B-CLL patients and 30 human cancer cell lines to a panel of normal tissues, including CD5 + B cells isolated from tonsils of normal individuals (representing normal cells corresponding to CLL malignant counterpart), we identified a cluster of two miRNAs (*miR-15a* and *miR-16-1*) within the minimal region of deletion (about 30-kb), which are deleted or downregulated in 68% of B-CLL patients. Albeit another similar cluster, named *miR-15b* and *miR-16-2* is located on chromosome 3q25-26.1 (Lagos-Quintana *et al.*, 2002), its expression is very low in lymphoid cells, therefore suggesting a key role of *miR-15a* and *miR-16-1* deregulation in the pathogenesis of CLL (for review on CLL and miRNAs, see Calin and Croce, 2006b). One of the most consistent biological landmarks of CLL is the overexpression of the oncogenic protein Bcl2 (B-cell lymphoma 2). Its abnormal upregulation has been reported in many types of human cancers, both solid and hematologic tumors (Sanchez-Beato *et al.*, 2003). In follicular lymphomas and in a fraction of diffuse B-cell lymphomas, the translocation t(14;18) (q32;q21) places the *BCL2* gene under the control of immunoglobulin heavy-chain enhancers, resulting in a hyperexpression of the gene (Tsujimoto *et al.*, 1984). We demonstrated that Bcl2 is a direct target of *miR-15a* and *miR-16-1* and that the restoration of *miR-15a* and/or *miR-16-1* in MEG-01 (a leukemia-derived cell line with the deletion of one allele and alteration of the other *miR-15a/16-1* locus and no expression of *miR-15a* and *miR-16-1* genes) induced apoptosis and tumor growth arrest in nude mice, enlightening an interaction of considerable therapeutic implications for the treatment of tumors overexpressing Bcl2 (Cimmino *et al.*, 2005). Several factors that can predict the clinical courses of CLL have been identified, mainly the levels of 70-kDzeta-associated protein (ZAP-70), and the mutational status of the rearranged immunoglobulin heavy-chain variable-region (IgV$_H$). In a group of 94 CLL samples, we identified a unique miRNA expression signature, composed of 13 miRNAs (out of 190 analyzed), which was associated with prognosis and disease progression (Calin *et al.*, 2005). This signature distinguishes cases of CLL with indolent behavior and better prognosis (low levels of ZAP-70, and mutated IgV$_H$) from cases in which the disease has a more aggressive course with a shorter time interval from the diagnosis to the beginning of the therapy (high levels of ZAP-70, and unmutated IgV$_H$). We also described germ-line or somatic mutations in 5/42 sequenced miRNAs in 11/75 patients with CLL, with respect to no mutations found in 160 normal individulas ($P < 0.001$). Interestingly, a germ-line point mutation in the *miR-16-1/15a* primary precursor was responsible for a reduction of miRNAs expression both in vitro and in vivo, and that was associated with a deletion of the normal allele (Calin *et al.*, 2005). This, in addition to the wide downregulation in CLL cases, represents a strong proof for the roles of *miR-15a*, and *miR-16-1* as tumor suppressors.

Deregulation of miRNA expression could lead both to a reduction of miRNA expression levels (as in the case of *miR-15a* and *miR-16-1* in CLL), through different mechanisms (e.g. deletion, mutation, methylation) or to increased miRNA levels, mainly through amplifications and translocations. In hematologic malignancies an abnormally high expression of *miR-155* and of its host gene *BIC* has been described in several types of B-cell lymphomas. These include diffuse large B cell lymphoma (DLBCL), expecially in the activated B cell phenotype (poor prognosis), with respect to the germinal center phenotype (better prognosis) (Eis *et al.*, 2005), primary mediastinal B cell lymphoma (PMBL), and Hodgkin's lymphomas (HL) (Kluiver *et al.*, 2005). To confirm the role of *miR-155* as an oncogene, we have recently developed an Eμ miR155 transgenic mouse (able to express the *miR-155* selectively in B lymphocytes) and we observed the development of a preleukemic pre B cell proliferation evident in spleen and bone marrow, followed by frank B cell malignancy (Costinean *et al.*, 2006). In pediatric Burkitt's lymphoma (BL) it has been reported a 100-fold upregulation of the hairpin precursor *miR-155* (Metzler *et al.*, 2004), whereas more recently another group has observed the lack of *BIC* and *miR-155* expression in primary cases of BL (Kluiver *et al.*, 2006), showing a difference which could be related to the age of onset of BL. The *MYC* oncogene plays a key role in the development of BL.

Chromosomal translocations are mainly responsible for an hyperexpression of the *MYC* oncogene, through its juxtaposition next to an immunoglobulin enhancer (Joos *et al.*, 1992). The *MYC* oncogene binds directly to a cluster of six miRNAs (*miR-17-92*) on chromosome 13 and activates the expression of these miRNAs (O'Donnell *et al.*, 2005). The *miR-17-92* cluster negatively regulates the E2F1 transcription factor. As E2F1 is a target of *MYC* itself, these results indicate a complex regulatory network in which miRNAs play a role as fine tuners of cell proliferation. Another group has also demonstrated that the *miR-17-92* cluster cooperates with *MYC* to promote the formation of B cell lymphomas in the mouse (L. He *et al.*, 2005). Tumors characterized by upregulation of Myc and of certain members of the cluster (*miR-17-5p* and *miR-20*) showed absence of apoptosis, a phenomenon commonly observed in Myc-induced lymphomas (L. He *et al.*, 2005).

MiRNAs are involved also in erythro and megakaryocytopoiesis. A downregulation of *miR-221* and *miR-222* has been described in human erythropoietic cultures of CD34 + cord blood progenitor cells (Felli *et al.*, 2005). These two miRNAs target the oncogene *c-KIT*. Therefore it was observed that their expression inhibits normal erythropoiesis and cell growth of the Kit positive TF-1 erythroleukemic cell line, via Kit receptor down-modulation. Recently our group was able to identify the miRNA signature involved in human megakaryocytopoiesis (Garzon *et al.*, 2006). MicroRNA expression profiling of *in vitro* differentiated megakaryocytes derived from CD34 + hematopoietic progenitors, revealed the down-regulation of six miRNAs (*miR-10a, miR-126, miR-106, miR-10b, miR-17,* and *miR20*), which can be considered a signature of megakaryocytic differentiation. Interestingly *miR-130a* (down regulate itself) targets the transcription factor MAFB, which is involved in the activation of the GPIIB promoter, an important protein for platelet physiology. Upregulation of *miR-101, miR-126, miR-99a, miR-135,* and *miR-20* has been observed in megakaryoblastic leukemic cell lines, with respect to the *in vitro* differentiated megakaryocytes, revealing a regulatory role of miRNAs in megakaryocytopoiesis and the importance of their deregulation in megakaryocytic malignancies.

Recently a regulatory function of miRNAs in granulocytopoiesis has been described (Fazi *et al.*, 2005). These authors have demonstrated that the human granulocytic differentiation is controlled by a regulatory circuitry involving *miR-223* and the transcription factors NFI-A and C/EBPα. The NFI-A factor keeps *miR-223* at low levels, whereas its replacement by C/EBPα, following retinoic acid-induced differentiation, upregulates *miR-223,* expression. On the other hand, *miR-223* represses NFI-A translation, creating a minicircuitry in which *miR-223* plays a central role as regulator of this important molecular pathway of granulopoiesis.

MicroRNA expression signature in solid tumors

The first correlation between miRNAs and solid tumors was seen in colorectal cancer (Michael *et al.*, 2003). These authors investigated possible changes in miRNA levels during colorectal carcinogenesis and found a consistent down-regulation of mature *miR-143* and *miR-145* (but not of the unprocessed hairpin precursors) at the adenomatous and cancer stages of the colorectal neoplasia, when compared to normal tissues. These observations are consistent with a role of these miRNAs in the early stages of colorectal cancerogenesis.

A reduction of expression > 80% by Northern blotting was observed for *let-7* in 60% of lung cancer cell lines and in 44% (7 out of 16) of lung tumors, when compared with normal lung tissues (Takamizawa *et al.*, 2004). This study (which included 143 lung cancer cases followed for more than five years after potentially curative resection) showed for the first time a statistically significant correlation between low levels of *let-7* and a shorter survival after surgery. The prognostic impact of *let-7* levels was independent of the stage disease. Moreover it was demonstrated that overexpression of *let-7* in A549 lung cancer cell lines inhibited tumor cell growth *in vitro*, suggesting its role as a tumor suppressor gene in lung cancer. Several *let-7* family members, including *let-7-a-2, let-7c,* and *let-7g*, are located to minimally deleted regions in lung cancers (Calin *et al.*, 2004). Recently it has been demonstrated that the *let-7* family negatively regulates the

RAS gene both in *C. elegans* and in human cells, and that the downregulation of *let-7* could induce upregulation of Ras, resulting in human lung oncogenesis (Johnson *et al.*, 2005). A miRNA microarray analysis conducted on 104 pairs of primary lung cancers and corresponding non tumoral tissues, identified unique miRNA profiles that could discriminate tumoral from normal lung tissues as well as molecular signatures that differ in tumor histology (Yanaihara *et al.*, 2006). In this study, 43 miRNAs resulted differentially expressed in lung cancer versus normal lung tissues, including miRNAs located in fragile sites (*miR-21* at FRA17B, *miR-27b* at FRA9D, and *miR-32* at FRA9E), in well documented regions of amplification (*miR-21 at 17q23.2, miR-205 at 1q32.2*) or deletion (*miR-126*, miR-126 at 9q34.3*) in lung cancer. These 43 differentially expressed miRNAs allow to distinguish among lung cancer and normal lung. This study identified also 17 miRNAs differentially expressed in lung adenocarcinoma versus normal lung and 16 miRNAs deregulated in squamous cell carcinoma versus non tumoral tissues. The existence of a tumor- and histological-specific miRNA signature has important clinical implications and suggests a role of miRNAs in the differential diagnosis of cancers of unknown origin. A panel of six miRNAs altered in both tumor histotypes versus normal tissues was indentified (*miR-21, miR-191, miR-155, miR-210, miR-126**, and *miR-224*), whereas six miRNAs were differentially expressed between the two analyzed lung cancer histotypes (*miR-205, miR-99b, miR-203, miR-202, miR-102*, and *miR-204prec.*). Interestingly in this study a statistically significant correlation was observed between high levels of *miR-155 precursor* as well as low levels of *let-7-a-2 precursor* and poor survival in 32 adenocarcinoma patients. These data confirm a role of *miR-155* as an oncogene and of the *let-7* family members as tumor suppressor genes. An overexpression of the *miR-17-92* cluster (Hayashita *et al.*, 2005) has been reported in lung cancers, expecially with the small cell histotype, defining a consistent role of this cluster as tumorigenic both in hematologic and solid malignancies.

In breast cancer, our group was able to identify the existence of a specific miRNA signature (Iorio *et al.*, 2005). The analysis of 76 breast cancer samples and 10 normal breast tissues showed 29 miRNAs whose expression is significantly deregulated, and a set of 15 miRNAs that correctly predicted the nature of the analyzed samples (tumor versus normal), with 100% accuracy. The most deregulated miRNAs (of the 29 members of the breast cancer-specific signature) were *miR-125b, miR-145, miR-21*, and *miR-155*. MiRNAs were also found differentially expressed in various biopathological features distinctive of human breast cancer. *MiR-30a, miR-30b, miR-30c, miR30d*, and *miR-30e* were all downregulated in both estrogen and progesterone-negative tumors. As in lung cancer, in breast cancer a reduced expression of *let-7* was correlated to a bad prognosis (lymph node metastases and high proliferative index). *MiR-9-3* was downregulated in breast cancers with either high vascular invasion or presence of lymph node metastases, suggesting that its downregulation is triggered during progression, and in particular, during the acquisition of cancer metastatic potential.

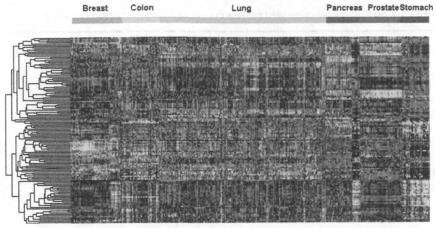

Figure 23.3. Clustering analysis of 540 samples representing six solid cancers and the respective normal tissues. The miRNome is differentially expressed in a tumor-specific manner. From Volinia *et al.* (2005). (See color plate 17)

In papillary thyroid carcinoma (PTC) a set of five miRNAs (including *miR-221, miR-222,* and *miR-146b,*) are upregulated and distinguish between PTC and normal thyroid tissue (H. He *et al.*, 2005). *MiR-221* is upregulated also in peritumoral, normal tissues, revealing that its deregulation is an early event in PTC tumorigenesis. In this neoplasia it was observed an upregulation of *miR-21*, and *miR-155* as well. The signature of miRNAs deregulated in PTC includes 23 miRNAs (17 highly expressed, and 6 with low expression), but only 5 upregulated miRNAs (*miR-221, miR-222, miR-146b, miR-21,* and *miR-181a*) are sufficient to successfully predict the malignant status. In this study polymorphisms in the 3′ UTR region of the *KIT* gene, in sites of interaction with the targeting miRNAs, were described.

A role of *miR-21* in human glioblastoma, most likely by influencing expression of genes involved in apoptosis, has been recently reported (Chan *et al.*, 2005). An extensive analysis of 245 miRNAs in primary glioblastomas has identified 9 miRNAs strongly upregulated in tumor versus normal peritumoral tissues, and 4 miRNAs downregulated (Ciafrè *et al.*, 2005). In particular, *miR-221* has been identified as a candidate for the role of tumor-specific marker in human glioblastoma.

Our group has recently conducted a large-scale miRNome analysis on 540 samples including lung, breast, stomach, prostate, colon, and pancreatic tumors (Volinia *et al.*, 2005). The clustering analysis of the 540 samples representing six solid cancers and the respective normal tissues clearly shows a tumor-specific miRNA signature (Figure 23.3). In a comparison between tumor specimens versus normal tissue, a total of 43 miRNAs result deregulated (26 upregulated and 17 downregulated). Table 23.1 shows the list of the miRNAs whose deregulation is shared by the signatures of the six solid cancers included in the study. In synthesis, *miR-21* is strongly upregulated in all six types of solid cancers considered, and globally we observed a large portion of overexpressed miRNAs, some of them with a well-characterized cancer association (*miR-17-5p, miR-20a, miR-21, miR-92, miR-106a,* and *miR-155*).

Table 23.1. Most commonly upregulated miRNAs in three or more types of solid tumors

MiR	Chromosomal location	Breast	Colon	Lung	Pancreas	Prostate	Stomach
21	17q23.2	X	X	X	X	X	X
17-5p	13q31.3	X	X	X	X	X	
191	3p21.31		X	X	X	X	X
29b-2	1q32.2	X	X		X	X	
223	Xq12		X		X	X	X
128b	3p22.3		X	X	X		
199a-1	19p13.2			X	X	X	
24-1	9q22.32		X		X		X
24-2	19p13.12		X		X		X
146	5q33.3	X			X	X	
155	21q21.3	X	X	X			
181b-1	1q31.3	X			X	X	
20a	13q31.3		X		X	X	
107	10q23.31		X		X		X
32	9q31.3		X		X	X	
92-2	Xq26.2				X	X	X
214	1q24.3				X	X	X
30c	1p34.2		X		X	X	
25	7q22.1				X	X	X
221	Xp11.3		X		X		X
106a	Xq26.2		X		X	X	

Conclusions

Overall, the studies conducted by our and other groups clearly indicate the existence of miRNA expression signatures which are tissue- and tumor-specific. Combining the available data from both hematological and solid tumors, it appears clear that some miRNAs have mainly a tumor-suppressor function and are almost constantly downregulated in cancers (e.g. *miR-15a, miR-16-1, let-7, miR-145*), whereas others behave as oncogenic miRNAs (e.g. *miR-155, miR-21, cluster 17–92*) (for reviews see Calin and Croce, 2006a; Croce and Calin, 2006). The identification of the mRNA targeted by these and other abnormally expressed miRNAs will clarify to what extent each of the single miRNA deregulations affects the acquisition of the malignant phenotype and at which level of the neoplastic transformation the aberrant miRNA expression is needed in cancerogenesis. The definition of patterns of deregulated miRNAs could be of great clinical and therapeutic value, helping to identify the histotype of tumors of unknown or unclear origin, and allowing a better therapy for patients whose treatment could otherwise be misled by an incorrect diagnosis. In some cases (CLL, lung cancer) it has been demonstrated that the altered expression of some miRNAs has a prognostic value, and correlates with the progression of the disease. The identification of important targets for some miRNAs (e.g. *BCL2* for *miR-15a/16-1*, *KIT* for *miR-221/222*, *RAS* for *let-7*, *MAFB* for *miR-130a*, *and E2F1 for the 17/92 cluster*) shows the possibility of using miRNAs as therapeutic tools able to restore the physiology of biological pathways, whose deregulation has a key role in human cancerogenesis.

ACKNOWLEDGEMENTS

Dr. Croce is supported by Program Project Grants from the National Cancer Institute, and Dr. Calin is supported by a Kimmel Foundation Scholar award and by the CLL Global Research Foundation. We apologize to many colleagues whose work was not cited owing to space limitations.

REFERENCES

Bartel, D. P. (2004). MicroRNAs: genomics, biogenesis, mechanism, and function. *Cell*, **23**, 281–297.

Calin, G. A. and Croce, C. M. (2006a). MicroRNA – cancer connection: the beginning of a new tale. *Cancer Research*. (In press.)

Calin, G. A. and Croce, C. M. (2006b). Genomics of CLL: MicroRNAs as new players with clinical significance. *Seminars in Oncology*, **33**, 167–173.

Calin, G. A., Dumitru, C. D., Shimizu, M. *et al.* (2002). Frequent deletions and down-regulation of micro-RNA genes miR15 and miR16 at 13q14 in chronic lymphocytic leukemia. *Proceedings of the National Academy of Sciences USA*, **99**, 15 524–15 529.

Calin, G. A., Sevignani, C., Dumitru, C. D. *et al.* (2004). Human microRNA genes are frequently located at fragile sites and genomic regions involved in cancers. *Proceedings of the National Academy of Sciences USA*, **101**, 2999–3004.

Calin, G. A., Ferracin, M., Cimmino, A. *et al.* (2005). A microRNA signature associated with prognosis and progression in chronic lymphocytic leukemia. *New England Journal of Medicine*, **353**, 1793–1801.

Chan, J. A., Krichevsky, A. M. and Kosik, K. S. (2005). MicroRNA-21 is an antiapoptotic factor in human glioblastoma cells. *Cancer Research*, **65**, 6029–6033.

Ciafrè, S. A., Galardi, S., Mangiola, A. *et al.* (2005). Extensive modulation of a set of microRNAs in primary glioblastoma. *Biochemical Biophysical Research Communications*, **334**, 1351–1358.

Cimmino, A., Calin, G. A., Fabbri, M. *et al.* (2005). MiR-15 and miR-16 induce apoptosis by targeting BCL2. *Proceedings of the National Academy of Sciences USA*, **102**, 13 944–13 949.

Costinean, S., Zanesi, N., Pekarsky, Y. *et al.* (2006). Pre B cell proliferation and lymphoblastic leucemia/high grade lymphoma in Eμ miR155 transgenic mice. *Proceedings of the National Academy of Sciences USA*, 2006 Apr 25 [Epub ahead of print].

Croce, C. M. and Calin, G. A. (2006). miRNAs, cancer, and stem cell division. *Cell*, **15**, 122, 6–7.

Cullen, B. R. (2004). Transcription and processing of human microRNA precursors. *Molecular Cell*, **16**, 861–865.

Dong, J. T., Boyd, J. C. and Frierson, H. F., Jr. (2001). Loss of heterozygosity at 13q14 and 13q21 in high grade, high stage prostate cancer. *Prostate*, **49**, 166–171.

Eis, P. S., Tam, W., Sun, L. *et al.* (2005). Accumulation of miR-155 and BIC RNA in human B cell lymphomas. *Proceedings of the National Academy of Sciences USA*, **102**, 3627–3632.

Elnenaei, M. O., Hamoudi, R. A., Swansbury, J. *et al.* (2003). Delineation of the minimal region of loss at 13q14 in multiple myeloma. *Genes Chromosomes Cancer*, **36**, 99–106.

Esquela-Kerscher, A. and Slack, F. J. (2004). The age of high-throughput microRNA profiling. *Nature Methods*, **1**, 106–107.

Fazi, F., Rosa, A., Fatica, A. *et al.* (2005). A minicircuitry comprised of microRNA-223 and transcription factors NFI-A and C/EBPalpha regulates human granulopoiesis. *Cell*, **123**, 819–831.

Felli, N., Fontana, L., Pelosi, E. *et al.* (2005). MicroRNAs 221 and 222 inhibit normal erythropoiesis and erythroleukemic cell growth via kit receptor down-modulation. *Proceedings of the National Academy of Sciences USA*, **102**, 18 081–18 086.

Garzon, R., Pichiorri, F., Palumbo, T. *et al.* (2006). MicroRNA fingerprints during human megakaryocytopoiesis. *Proceedings of the National Academy of Sciences USA*, **103**, 5078–5083.

Gregory, R. I., Chendrimada, T. P., Cooch, N. and Shiekhattar, R. (2005). Human RISC couples microRNA biogenesis and posttranscriptional gene silencing. *Cell*, **123**, 631–640.

Hayashita, Y., Osada, H., Tatematsu, Y. *et al.* (2005). A polycistronic microRNA cluster, miR-17-92, is overexpressed in human lung cancers and enhances cell proliferation. *Cancer Research*, **65**, 9628–9632.

He, H., Jazdzewski, K., Li, W. *et al.* (2005). The role of microRNA genes in papillary thyroid carcinoma. *Proceedings of the National Academy of Sciences USA*, **102**, 19075–19080.

He, L. and Hannon, G. J. (2004). MicroRNAs: small RNAs with a big role in gene regulation. *Nature Reviews Genetics*, **5**, 522–531.

He, L., Thomson, J. M., Hemann, M. T. *et al.* (2005). A microRNA polycistron as a potential human oncogene. *Nature*, **435**, 828–833.

Iorio, M. V., Ferracin, M., Liu, C. G. *et al.* (2005). MicroRNA gene expression deregulation in human breast cancer. *Cancer Research*, **65**, 7065–7070.

Johnson, S. M., Grosshans, H., Shingara, J. *et al.* (2005). RAS is regulated by the let-7 microRNA family. *Cell*, **120**, 635–647.

Joos, S., Haluska, F. G., Falk, M. H. *et al.* (1992). Mapping chromosomal breakpoints of Burkitt's t(8;14) translocations far upstream of c-myc. *Cancer Res*, **52**, 6547–6552.

Kluiver, J., Poppema, S., de Jong, D. *et al.* (2005). BIC and miR-155 are highly expressed in Hodgkin, primary mediastinal and diffuse large B cell lymphomas. *Journal of Pathology*, **207**, 243–249.

Kluiver, J., Haralambieva, E., de Jong, D. *et al.* (2006). Lack of BIC and microRNA miR-155 expression in primary cases of Burkitt lymphoma. *Genes Chromosomes Cancer*, **45**, 147–153.

Lagos-Quintana, M., Rauhut, R., Lendeckel, W. and Tuschl, T. (2001). Identification of novel genes coding for small expressed RNAs. *Science*, **294**, 853–858.

Lagos-Quintana, M., Rauhut, R., Yalcin, A. *et al.* (2002). Identification of tissue-specific microRNAs from mouse. *Current Biology*, **12**, 735–739.

Lau, N. C., Lim, L. P., Weinstein, E. G. and Bartel, D. P. (2001). An abundant class of tiny RNAs with probable regulatory roles in *Caenorhabditis elegans*. *Science*, **294**, 858–862.

Lee, R. C. and Ambros, V. (2001). An extensive class of small RNAs in *Caenorhabditis elegans*. *Science*, **294**, 862–864.

Liu, C. G., Calin, G. A., Meloon, B. *et al.* (2004). An oligonucleotide microchip for genome-wide microRNA profiling in human and mouse tissues. *Proceedings of the National Academy of Sciences USA*, **101**, 9740–9744.

Lu, J., Getz, G., Miska, E. A. *et al.* (2005). MicroRNA expression profiles classify human cancers. *Nature*, **435**, 834–838.

Matranga, C., Tomari, Y., Shin, C., Bartel, D. P. and Zamore, P. D. (2005). Passenger-strand cleavage facilitates assembly of siRNA into Ago2-containing RNAi enzyme complexes. *Cell*, **123**, 607–620.

Metzler, M., Wilda, M., Busch, K., Viehmann, S. and Borkhardt, A. (2004). High expression of precursor microRNA-155/BIC RNA in children with Burkitt lymphoma. *Genes Chromosomes Cancer*, **39**, 167–169.

Michael, M. Z., O' Connor, S. M., van Holst Pellekaan, N. G., Young, G. P. and James, R. J. (2003). Reduced accumulation of specific microRNAs in colorectal neoplasia. *Molecular Cancer Research*, **1**, 882–891.

O'Donnell, K. A., Wentzel, E. A., Zeller, K. I., Dang, C. V. and Mendell, J. T. (2005). c-Myc-regulated microRNAs modulate E2F1 expression. *Nature*, **435**, 839–843.

Rand, T. A., Petersen, S., Du, F. and Wang, X. (2005). Argonaute2 cleaves the anti-guide strand of siRNA during RISC activation. *Cell*, **123**, 621–629.

Sanchez-Beato, M., Sanchez-Aguilera, A. and Piris, M. A. (2003). Cell cycle deregulation in B-cell lymphomas. *Blood*, **101**, 1220–1235.

Sanchez-Izquierdo, D., Buchonnet, G., Siebert, R. *et al.* (2003). MALT1 is deregulated by both chromosomal translocation and amplification in B-cell non-Hodgkin lymphoma. *Blood*, **101**, 4539–4546.

Seitz, H., Royo, H., Bortolin, M. L. *et al.* (2004). A large imprinted microRNA gene cluster at the mouse Dlk1-Gtl2 domain. *Genome Research*, **14**, 1741–1748.

Sevignani, C., Calin, G. A., Siracusa, L. and Croce, C. M. (2006). Mammalian microRNAs: a small world for fine-tuning gene expression. *Mammalian Genome*, **17**, 189–202.

Stilgenbauer, S., Nickolenko, J., Wilhelm, J. *et al.* (1998). Expressed sequences as candidates for a novel tumor suppressor gene at band 13q14 in B-cell chronic lymphocytic leukemia and mantle cell lymphoma. *Oncogene*, **16**, 1891–1897.

Takamizawa, J., Konishi, H., Yanagisawa, K. *et al.* (2004). Reduced expression of the let-7 microRNAs in human lung cancers in association with shortened postoperative survival. *Cancer Research*, **64**, 3753–3756.

Tsujimoto, Y., Finger, L. R., Yunis, J., Nowell, P. C. and Croce, C. M. (1984). Cloning of the chromosome breakpoint of neoplastic B cells with the t(14;18) chromosome translocation. *Science*, **226**, 1097–1099.

Volinia, S., Calin, G. A., Liu, C. G. *et al.* (2005). A microRNA expression signature of human solid tumors defines cancer gene targets. *Proceedings of the National Academy of Sciences USA*, **103**, 2257–2261.

Yanaihara, N., Caplen, N., Bowman, E. *et al.* (2006). Unique microRNA molecular profiles in lung cancer diagnosis and prognosis. *Cancer Cell*, **9**, 189–198.

24 Roles of microRNAs in cancer and development

Andrea Ventura, Madhu S. Kumar and Tyler Jacks*

Introduction

The recent discovery of microRNAs (miRNAs) and the factors that control their biogenesis and function has revealed novel and unexpected mechanisms for gene regulation (Ambros, 2004; Bartel, 2004; Bagga et al., 2005). The realization that microRNAs are involved in biological processes including cell proliferation, apoptosis and differentiation has led many to speculate that their abnormal function might contribute to the pathogenesis of human cancer (McManus, 2003; Chen, 2005; Couzin, 2005; Croce and Calin, 2005; Eder and Scherr, 2005; Gregory and Shiekhattar, 2005; Hammond, 2006). More recently, these speculations have been supported by experimental evidence including miRNA profiling of human and murine cancers, mutation analysis of human tumors and characterization of mice and cells carrying conditional and constitutive loss of function alleles of genes involved in microRNA biogenesis. Although considerable additional research is required, these initial results suggest that both a global change in miRNA expression and more specific alteration of individual miRNAs might contribute to tumorigenesis.

Here we will first review the evidence pointing to a role for global changes in miRNA expression in tumorigenesis and the methodologies used to detect these changes. We will further discuss the phenotypic consequences of loss of function mutations of Dicer1, an enzyme essential for miRNA biogenesis. Finally, we will explore individual miRNAs that have been implicated in the regulation of genes involved in oncogenesis.

Cloning and profiling miRNAs

Recognizing the importance of microRNAs (miRNAs) in the regulation of cell fate decisions during invertebrate development, especially cell differentiation, proliferation and apoptosis, several groups have cloned small RNAs, including miRNAs,

---

* Author to whom correspondence should be addressed.

MicroRNAs: From Basic Science to Disease Biology, ed. Krishnarao Appasani. Published by Cambridge University Press. © Cambridge University Press 2008.

from various mammalian cell types and developmental stages (Houbaviy *et al.*, 2003; Lagos-Quintana *et al.*, 2002; Poy *et al.*, 2004). These efforts have cataloged over 300 unique miRNAs from mouse and human tissues (for an updated registry visit: http://microrna.sanger.ac.uk/sequences/index.shtml). Because the cloning frequency of a given miRNA strongly correlates with its abundance, these cloning methods provide a rough estimate of the miRNA profile of a given cell type, developmental stage, or disease state. There are, however, several disadvantages to this approach for the determination of miRNA content. First, large-scale cloning efforts are both experimentally and computationally laborious. Moreover, there are no standardized workflows for the cloning of small RNAs, requiring a large amount of empirical determination that consequently reduces throughput. Additionally, rare cell populations, such as adult progenitor/stem cells and human tissue specimens, are much less amenable to common cloning strategies, which generally use whole tissues or well-characterized cell lines. Finally, while cloning strategies do serve a discovery function, in that they allow the identification of novel miRNAs, they inevitably lead to the isolation of various other RNAs, such as endogenous siRNAs, ribosomal/transfer RNA products, and other uncharacterized small RNAs. This increase in complexity reduces the likelihood of detecting known miRNAs of lower abundance in a given tissue type.

In order to determine the expression levels of a large number of known miRNAs in a given tissue or cell type in a rapid and cost-efficient manner, several laboratories have generated systems for profiling miRNAs. Earlier methods used glass-slide microarrays composed of printed oligonucleotides corresponding to the known collection of miRNAs, with labeling performed by either biotintylated cDNA synthesis or fluorophore-conjugated RNA ligation (Liu *et al.*, 2004; Thomson *et al.*, 2004). These initial studies showed that miRNA microarrays could be used to detect specific changes in miRNA expression levels with good sensitivity and specificity. They also demonstrated that many miRNAs have highly restricted tissue expression patterns, a finding that has recently received elegant confirmation by *in situ* RNA hybridization studies on zebrafish and mouse embryos performed by using locked nucleic acids (LNA) (Kloosterman *et al.*, 2006; Wienholds *et al.*, 2005).

More recently, the Golub group has developed a highly sensitive bead-based hybridization method to simultaneously determine the relative expression levels of hundreds of microRNAs (Lu *et al.*, 2005). In this method, oligonucleotide-capture probes complementary to human and mouse miRNAs are coupled to polystyrene beads impregnated with various mixtures of fluorescent dyes such that each individual miRNA is represented by a distinct color. Following adaptor ligation, reverse-transcribed miRNAs are amplified using a common primer, hybridized to the capture beads and stained. Finally, the beads are analyzed on a high-speed flow cytometer capable of identifying bead color and intensity. These data can be used to determine the abundance of individual miRNA species. The bead method was capable of efficiently discriminating between single base-pair mismatches in a spiking experiment involving 11 closely related sequences. Its main limitation is that currently only about 100 unique fluorophores are available, limiting the number of miRNAs that can be simultaneously analyzed in a single experiment.

This method was applied to profile the expression of 217 miRNAs in human tumors compared to normal tissues, leading to the striking observation that tumor cells display a global reduction in miRNA expression levels (Lu et al., 2005). An analogous difference was observed when normal lungs and lung tumors from K-Ras mutant mice were compared. Although this finding does not prove that microRNAs play a functional role in tumorigenesis, the observation that more differentiated cells express higher levels of miRNAs than undifferentiated or poorly differentiated cells suggests that miRNA might be involved in determining and maintaining the differentiated state. It remains to be seen whether the global pattern of miRNA expression in cancer reflects a down-regulation of these genes in the course of tumorigenesis and/or the over-representation of a particular cell lineage in the tumor compared to the corresponding normal tissue. This is of particular interest given recent work linking stem cells and cancer, especially since at least some miRNAs are thought to be regulators of stem cell fates (Bartel, 2004; Houbaviy et al., 2003; Kuwabara et al., 2004; Suh et al., 2004). Additionally, the miRNA profile was shown to be an accurate predictor of both the differentiation state and developmental origin of the tumor. Moreover, classifiers based on these miRNA profiles were found to be more informative than mRNA profiles in classifying poorly differentiated tumors. In total, these reports demonstrate that miRNA profiling is a powerful method for cataloging the miRNA content in a given cell type, developmental stage, or disease state.

MicroRNAs in mouse development

Though essential in determining where and when different miRNAs are expressed, profiling studies provide relatively little information regarding the role of a particular miRNA, or even of miRNAs in general, within a given biological process. One way to determine such biological functions is the generation and characterization of mutant alleles in model organisms. In invertebrates, such as the nematode *C. elegans* and the fruit fly *D. melanogaster*, loss-of-function and gain-of-function alleles of several miRNAs have been characterized, providing crucial evidence for the role of individual miRNAs in processes including, but not limited to, developmental timing (Reinhart et al., 2000; Sulston and Horvitz, 1981), apoptosis (Brennecke et al., 2003), neuronal differentiation (Chang et al., 2004; Johnston and Hobert, 2003), muscle differentiation (Sokol and Ambros, 2005) and fat metabolism (Xu et al., 2003).

Although the characterization of miRNA knockout alleles in mammals has not yet been described, *Dicer1* knockout mice have been characterized (Bernstein et al., 2003). Since *Dicer1* is essential for miRNA biogenesis, these studies have provided evidence that miRNAs play an essential role in mouse embryonic development. *Dicer1*$^{-/-}$ mice are not viable with lethality occurring at about embryonic day 7.5 (Bernstein et al., 2003). When early *Dicer1*$^{-/-}$ embryos were examined, they were found to have reduced expression of Oct-4, a proposed master regulator of embryonic stem cell pluripotency, suggesting that Dicer1 deficiency impairs stem cell maintenance in early embryos. Consistent with a central role for Dicer in stem cell maintenance, the authors were unable to derive viable embryonic stem (ES) cells from early $Dicer1^{-/-}$ embryos. Dicer-null ES cells were later successfully generated by using $Dicer1$ conditional knockout alleles (Kanellopoulou et al., 2005; Murchison et al., 2005). In both studies, the ES cells exhibited normal morphology and, in contrast to the Dicer-deficient embryos, maintained Oct-4 expression. However, both Dicer-deficient ES cells had reduced proliferation, failed to form embryoid bodies and teratomas, and did not contribute to chimeras, thus confirming the essential role of $Dicer1$ during early development. Both groups also reported increased levels of transcripts from the major and minor satellite repeats, which are normally silenced during the establishment of heterochromatin. Kanellopoulou and colleagues found this increase in satellite transcription was coupled to decreased DNA and histone methylation at sites of heterochromatin, which was not observed by Murchison and colleagues. The reason for this discrepancy is not entirely clear, although several possible explanations are possible, including different targeting strategies and genetic backgrounds. Yet both reports show that loss of Dicer has a fundamental impact on ES cells, including but not limited to changes in heterochromatin structure.

The use of conditional alleles of $Dicer1$ has extended beyond the early embryo to the evaluation of Dicer's role in advanced stages of development. One example is the study of loss of Dicer function in the developing limb, where miR-196a was known to regulate genes of the homeobox (Hox) cluster (Hornstein et al., 2005; Yekta et al., 2004). When Cre was expressed from the limb mesoderm in $Dicer1$ conditional homozygotes, a coordinate decrease in limb size along with a developmental delay was observed (Harfe et al., 2005; see also Chapter 4 in this book). This decrease in limb size was not the result of a failure to proliferate but instead of an increase in programmed cell death. Interestingly, while there was increased apoptosis, all the differentiated cell types of the limb mesoderm were present and there were no obvious patterning defects or changes in expression of known patterning genes. One caveat is that the authors found that Dicer-deficient murine embryo fibroblasts (MEFs) exhibit chromosome segregation defects, triggering the formation of anaphase bridges. Thus, it remains unclear whether the increased apoptosis was caused by checkpoint activation due to chromosome segregation defects or some uncharacterized change in Dicer-null limbs. Further studies will be needed to clarify the role of Dicer in the contribution to the limb development phenotype.

The consequences of loss of Dicer function on murine hematopoiesis have also been examined, particularly in T-cell development (Cobb et al., 2005; Muljo et al., 2005). Cobb et al. achieved T-cell specific deletion of $Dicer1$ with the Lck-Cre transgene, whose expression is limited to early T-cell development, while Muljo et al. used CD4-Cre transgenic mice to specifically delete $Dicer1$ at a later stage in T-cell development. The effects were relatively mild, with both groups observing increased apoptosis along with distinct changes in lineage

Table 24.1. Mammalian microRNAs with known physiological roles

miRNA	Cell/tissue type	Reference
miR-181	hematopoietic	(Chen *et al.*, 2004)
miR-223	hematopoietic	(Fazi *et al.*, 2005; Chen *et al.*, 2006)
miR-1/miR-133	skeletal muscle	(Zhao *et al.*, 2005)
miR-375	pancreatic islets	(Poy *et al.*, 2004)
miR-196	limb development	(Hornstein *et al.*, 2005)
miR-134	dendrites	(Schratt *et al.*, 2006)
miR-124	neural progenitors	(Conaco *et al.*, 2006)

selection in Dicer-deficient progenitors. In contrast to ES cells, no change in major or minor satellite repeat transcripts in Dicer-deficient T cells was observed, implying that such changes in heterochromatin structure might be cell-context specific.

Ablation of Dicer activity in the developing murine lung has also been characterized (Harris *et al.*, 2006). Through the use of a Shh-Cre transgene, Harris and colleagues deleted *Dicer1* in the developing lung epithelium and observed an immediate arrest in lung branching morphogenesis. It should be noted this impaired branching occurred before the expected increase in pro-grammed cell death of the lung epithelium, suggesting that the function of Dicer in lung epithelial branching is independent of its role in cell viability. Intriguingly, Dicer-deficient lung epithelium appears to affect the expression of *Fgf10* in the lung mesenchyme, suggesting a non-cell autonomous function for Dicer in branching morphogenesis. Taken together, these studies underscore the role of Dicer in cell lineage decisions and the necessity of conditional alleles to specifically probe biological function throughout development and in the adult.

Though studies of *Dicer1* loss-of-function alleles provide the opportunity to examine the general role of miRNAs in mammalian development, they fail to address the specific contribution of individual miRNAs and miRNA families. Thus, several groups have initiated studies into the role of mammalian miRNAs exclusively (or highly expressed) within a given cell type or tissue (Table 24.1). In these studies, particular miRNAs have been shown to be involved in hema-topoietic lineage choice (miR-181 and miR-223) (Chen *et al.*, 2004; Fazi *et al.*, 2005), skeletal muscle differentiation (miR-1 and miR-133) (Chen *et al.*, 2006; Yang *et al.*, 2005), insulin secretion by pancreatic islets (miR-375) (Poy *et al.*, 2004), limb development (miR-196) (Harfe *et al.*, 2005), and neuronal differentiation and activity (miR-124 and miR-134) (Conaco *et al.*, 2006; Schratt *et al.*, 2006). Although compelling, all of these studies were based on either the over-expression of particular miRNAs or on their inhibition in cell-based assays using synthetic antagonists. Clearly, *in vivo* loss of function studies through the generation of constitutive and conditional null alleles are required and will likely greatly contribute to our understanding of mammalian miRNA biology.

Oncogenic and anti-oncogenic miRNAs

miRNAs may play direct causal roles in the pathogenesis of cancer by negatively regulating the expression of tumor suppressor genes (TSG) or positively affecting the expression of oncogenes. In this context, it is straightforward to imagine how mutations in microRNAs or their cognate sites might have such an effect (Table 24.2). For example, amplification/overexpression of a miRNA targeting a tumor suppressor gene would lead to its down-regulation, while deletion/silencing of a miRNA targeting an oncogene would have the opposite effect. In both cases the net result would be to promote cellular transformation. Indeed recurrent amplification or overexpression and deletion or downregulation of individual miRNAs or miRNA clusters has been recently reported (see below). At least in theory more subtle molecular changes involving miRNAs and their cognate binding sites are also possible. A point mutation in a miRNA might increase its affinity for a TSG (and even lead to mRNA cleavage rather than translation repression) or decrease its affinity for an oncogene. In principle, such mutation might also lead to an off-target activity and downregulate a tumor suppressor gene that is not a physiological target of that particular miRNA. Finally, the mutation could directly involve the oncogene or the tumor suppressor gene and lead, respectively, to loss or creation of a binding site for a co-expressed miRNA. Although to date only deletion or amplification of miRNAs have been identified in human cancers, it is possible that point mutations and mutations affecting the 3′ UTR of oncogenes and tumor suppressor genes have so far escaped detection because they have not been thoroughly examined.

Whatever the molecular nature of the mutation, the idea that the oncogenic activity of a miRNA is the result of its action on a single target is probably an oversimplification. Computational and experimental approaches suggest that miRNAs are part of intricate regulatory networks where individual miRNAs

Table 24.2. Possible direct and indirect mechanisms through which miRNA can lead to transformation

Mutation/change in expression affecting:	Nature of the mutation/change in expression	Consequence
Oncogenic miRNA	Amplification, overexpression or increased stability	Downregulation of tumor suppressor gene
Oncogenic miRNA	Point-mutation changing specificity or increasing affinity	Downregulation of tumor suppressor gene
Tumor suppressive miRNA	Deletion, downregulation or decreased stability	Upregulation of oncogene
Tumor suppressive miRNA	Point-mutation decreasing affinity	Upregulation of oncogene
Oncogene	Deletion of miRNA binding site(s)	Upregulation of oncogene
Tumor suppressor gene	Acquisition of a novel miRNA binding site or mutation leading to increased affinity of an existing miRNA binding site	Downregulation of tumor suppressor gene

simultaneously regulate the expression of hundreds of different genes (Lewis *et al.*, 2003; John *et al.*, 2004; Bentwich, 2005; Farh *et al.*, 2005; Grun *et al.*, 2005; Krek *et al.*, 2005; Lim *et al.*, 2005). The complexity is further increased by the fact that many 3′ UTRs have binding sites for multiple miRNAs (Stark *et al.*, 2005) resulting in the combinatorial effect of different miRNAs converging to regulate a single mRNA (Bartel and Chen, 2004). Finally, many miRNAs belong to families of small RNAs with highly related sequence (in some cases identical miRNAs being encoded by multiple loci) suggesting a high degree of compensation and functional overlap. As a result, the net effect of loss or ectopic expression of a given miRNA will largely depend on the particular cellular context, the microRNA milieu in which the mutation happens to originate and the potential targets co-expressed and therefore available for modulation.

The miR-15∼16 cluster and the leukemia connection

Calin and colleagues noted in 2004 (Calin *et al.*, 2004b) that although miRNAs compose only about 1% of the human genome, over 50% of them are located in cancer-associated genomic regions, including fragile sites, frequently deleted or amplified regions and break-points for chromosomal translocations. Among such microRNAs, the miR-15∼16 cluster is of particular interest because it is located on Ch 13q14, a region deleted in more than 65% of cases of B-cell chronic lymphocytic leukemia (B-CLL). Despite several years of intense study, no bona fide tumor suppressor genes have been identified in this region (reviewed in Pekarsky *et al.* (2005)). The connection between miR-15∼16 and B-CLL is strengthened by the demonstration that hemizygous deletion of this region is associated with downregulation of miR-15∼16 and that the antiapoptotic gene Bcl-2 is a direct miR-15∼16 target (Figure 24.1) (Calin *et al.*, 2002; Cimmino *et al.*, 2005). This suggests a possible scenario where loss of miR-15∼16 or reduced miR-15∼16 expression leads to over-expression of Bcl-2 and suppression of apoptosis, thus contributing to the pathogenesis of B-CLL. Croce's group (Calin *et al.*, 2004a) identified a microRNA signature that correlated with prognosis and progression in a panel of patients with B-CLL.

Among the 13 miRNAs that constitute this signature, lower levels of miR-15 and miR-16 were shown to associate with worse prognosis and a shorter interval between diagnosis and disease progression, a result that is consistent with the previous observation that a more favorable prognosis is associated with deletion at 13q14 (and hence with low levels of miR-15∼16). In addition, mutational analysis revealed a germ-line single base-change in the pri-miR-15∼16 in two B-CLL patients that led to reduced miR-15∼16 expression *in vitro* and *in vivo*. Interestingly, one patient had a familiar history of CLL and breast cancer and the same change was not observed in 160 control individuals, suggesting that it is either a germ-line mutation or a very rare polymorphism.

In the same study, single base changes were observed in 7 additional pri- or pre-miRNAs (out of 42 analyzed) in 11 (out of 75) patients. However, with the

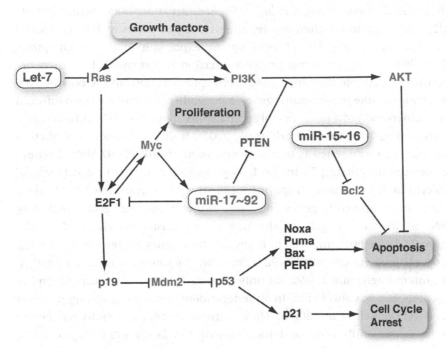

Figure 24.1. A schematic representation of oncogenic and tumor suppressive pathways affected by mammalian microRNAs.

exception of the miR-15~16 change, none of the alterations was shown to directly affect miRNA expression levels. Although no changes were in the mature miRNA sequence, these results point to an important role of miR-15~16 and other miRNAs in the pathogenesis of B-CLL and clearly warrant further study.

miR-17~92

Perhaps the most convincing evidence for a direct causal link between microRNAs and human cancer has come from the study of the miR-17~92 cluster. Following the first report of its frequent amplification in B-cell lymphomas (Ota *et al.*, 2004), overexpression or increased copy number of this cluster has been reported in colon and small cell lung cancers (Hammond, 2006; L. He *et al.*, 2005). A strong indication of a direct causal link came from the work of He and colleagues that showed that forced overexpression of a truncated version of this cluster (miR-17~19b) cooperates with c-Myc overexpression in a mouse model of B-cell lymphomas (L. He *et al.*, 2005). An essential, and still unanswered, question is how does miR-17~92 affect tumorigenesis? What is the critical set of genes whose modulation is responsible for its oncogenic properties?

The model used in the study by He *et al.* – a Eμ-Myc transgenic mouse – might suggest where to look for an answer. Because c-Myc can stimulate both cell proliferation and apoptosis, a cooperating mutation leading to suppression of cell death seems to be required to develop a B-cell lymphoma in these mice

(Schmitt *et al.*, 1999; Soengas *et al.*, 1999; Schmitt *et al.*, 2000; Schmitt *et al.*, 2002). These mutations often involve the p19Arf-p53 pathway, but accelerated tumorigenesis can also be achieved by overexpression of the antiapoptotic protein Bcl-2 or of a dominant-negative version of caspase-9. It is therefore tempting to speculate that miR-17~92 overexpression leads to the downregulation of one or more pro-apoptotic genes, a possibility that seems to be reinforced by the observation of a lower rate of spontaneous apoptosis in the miR-17~19b-overexpressing lymphomas (L. He *et al.*, 2005). If this is the case, a prediction that could be easily tested is that overexpression of miR-17~92 should relieve the pressure to mutate p53 in the Eμ-Myc model. A search in a database of predicted miRNA targets (http://microrna.sanger.ac.uk/targets/v2/) reveals a number of proapoptotic genes and other tumor suppressor genes, including PTEN, as potential targets of the miR-17~92 cluster (Lewis *et al.*, 2005) (Figure 24.1). Determining which, if any, of these genes are responsible for the oncogenic properties of miR-17~92 is currently the subject of intense investigation. Interestingly, miR-17~92 not only cooperates with c-Myc during lymphomagenesis, but was also found, in an independent study, to be among its direct targets (O'Donnell *et al.*, 2005). These authors described a molecular circuit linking c-Myc, miR-17~92 and the transcription factor E2F1 (Figure 24.1). According to this model, c-Myc can directly induce transcription of the miR-17~92 cluster and miR-17/miR-20 can in turn repress the translation of E2F1 (via binding to conserved sites in its 3' UTR). What is the significance of these findings in relation to the oncogenic activity of miR-17~92? E2F1 is a well-studied transcription factor that can, similarly to c-Myc, induce cell proliferation and apoptosis. The apoptotic function seems to be due largely to its ability to activate the p19-Arf/p53 axis, and it is, therefore, tempting to speculate that its downregulation by miR-17~92 might explain the reduced apoptosis observed by He *et al.* Although targeted inactivation of E2F1 does not accelerate Eμ-Myc-induced lymphomagenesis (it actually suppresses it) (Baudino *et al.*, 2003), this does not exclude that a more subtle modulation of E2F1 levels by miR-17~92 might preferentially affect apoptosis versus proliferation. In this regard it would be informative to study the consequences of the miR-17~92 overexpression on c-Myc-induced apoptosis especially in cell-based models.

As with most informative experiments, the work of He and O'Donnell raises more questions than it answers. For example, does miR-17~92 cooperate with other oncogenes and tumor suppressor genes? Is it capable of acting as an oncogene itself? Can individual members of the cluster, or particular combinations, recapitulate the oncogenic properties?

Finally, it must be noted that the biology of miR-17~92 is complicated by the existence of two paralogs, one on chromosome X and one on chromosome 5 (Figure 24.2). It will be important to determine whether these clusters are also subject to amplification/overexpression in human cancer and whether they share the oncogenic properties of miR-17~92. It is likely that *in vivo* gain-of-function and loss-of-function studies will provide answers to many of these questions.

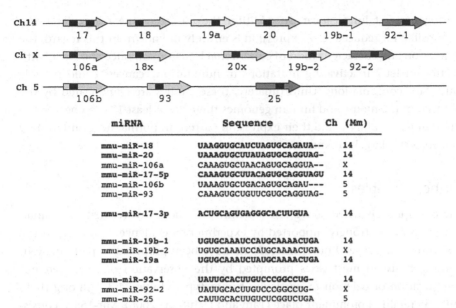

miRNA	Sequence	Ch (Mm)
mmu-miR-18	UAAGGUGCAUCUAGUGCAGAUA--	14
mmu-miR-20	UAAAGUGCUUAUAGUGCAGGUAG-	14
mmu-miR-106a	CAAAGUGCUAACAGUGCAGGUA--	X
mmu-miR-17-5p	CAAAGUGCUUACAGUGCAGGUAGU	14
mmu-miR-106b	UAAAGUGCUGACAGUGCAGAU---	5
mmu-miR-93	CAAAGUGCUGUUCGUGCAGGUAG-	5
mmu-miR-17-3p	ACUGCAGUGAGGGCACUUGUA	14
mmu-miR-19b-1	UGUGCAAAUCCAUGCAAAACUGA	14
mmu-miR-19b-2	UGUGCAAAUCCAUGCAAAACUGA	X
mmu-miR-19a	UGUGCAAAUCUAUGCAAAACUGA	14
mmu-miR-92-1	UAUUGCACUUGUCCCGGCCUG-	14
mmu-miR-92-2	UAUUGCACUUGUCCCGGCCUG-	X
mmu-miR-25	CAUUGCACUUGUCUCGGUCUGA	5

Figure 24.2. Genomic organization and sequence alignment of the murine miR-17~92 cluster and its paralogs.

let-7 and the Ras connection

Among the first miRNAs to be discovered in *C. elegans*, and still among the best characterized, members of the lethal-7 (let-7) family of microRNAs have recently been the subject of intense investigation as potential miRNAs with tumor suppressor activities. In *C. elegans*, let-7 expression is induced in a particular cell type (seam cells) at the time when these cells have to undergo terminal differentiation and exit cell cycle. Inactivating mutations in let-7 prevent this transition and the cells instead continue to divide, a behavior reminiscent of cancer cells (Pasquinelli *et al.*, 2000; Reinhart *et al.*, 2000). Croce and colleagues (Calin *et al.*, 2004b) have reported recurrent deletions in loci encoding let-7 miRNAs in human lung cancer and other tumor types. In addition, work from Takamizawa *et al.* (2004) has shown a correlation between reduced let-7 expression and poor prognosis in human lung cancers, a finding further substantiated by the observation that ectopic expression of let-7 can inhibit the growth of a human cancer cell line.

A possible mechanistic explanation for the tumor suppressor activity of let-7 came from the Slack group (Johnson *et al.*, 2005) who demonstrated that, in *C. elegans*, members of the let-7 family of miRNAs negatively regulate the expression of *let-60*, a gene involved in vulval development and the worm ortholog of the human Ras family of oncogenes. Significantly, human let-7 miRNAs were also shown to bind and repress the expression of human Ras proteins and a striking inverse correlation between let-7 and Ras protein levels was observed in human lung cancers (Johnson *et al.*, 2005). As for the previous examples, many questions remain to be addressed. Despite the suggestive evidence that Ras proteins are indeed targets of let-7 in mammalian cells, none of the studies reported so far demonstrates a direct causal link between loss of let-7 expression and the pathogenesis of lung cancer. In addition the nature of the molecular event responsible for reduced let-7 expression is entirely unknown. In this regard, the generation and characterization of conditional let-7 mutant mice and the identification of let-7 inactivating mutations in human lung cancers could provide important confirmations. Unfortunately, these studies are complicated by the fact that in the mouse and human genomes there are at least 12 members of the let-7 family of miRNAs and their expression pattern in normal lung and in lung cancers is still largely unknown.

Further examples

The examples discussed above refer to miRNAs whose involvement in human cancers is more strongly supported by experimental evidence. However, in the past three years several other microRNAs have been proposed to participate in tumorigenesis, in most cases prompted by the observation of their frequent amplification or deletion in particular tumor types (Table 24.3). Among these miR-155 retains a prominent place (Tam and Dahlberg, 2006). Mir-155 is consistently found amplified/overexpressed in large cell lymphomas and pediatric Burkitt's lymphomas (Metzler *et al.*, 2004; Eis *et al.*, 2005; Kluiver *et al.*, 2005). Indeed, the miR-155 primary transcript had been originally identified with the name of *BIC* as a common proviral DNA insertion site in lymphomas induced by the avian leucosis virus (Clurman and Hayward, 1989; Tam *et al.*, 1997). Although the authors noted that the *BIC* transcript could form extensive double-stranded RNA structures, the absence of an obvious open reading frame remained a puzzling feature of this potential oncogene, even after it was shown that BIC could cooperate with Myc in chicken lymphomagenesis and erythroleukemogenesis assay (Tam *et al.*, 2002).

Table 24.3. MicroRNAs and microRNA cluster with known association to human cancers

MiRNA or miRNA cluster	Tumor types	Change	Reference
Let-7	Lung cancer	Reduced expression	(Johnson *et al.*, 2005; Takamizawa *et al.*, 2004)
MiR-15~16	B-CLL	Deleted or downregulated	(Calin *et al.*, 2002)
miR-17~92	B-cell lymphomas, colon cancers	Increased expression	(L. He *et al.*, 2005; Ota *et al.*, 2004)
miR-155/BIC	Burkitt's, Hodgkin's and large B-cell lyphomas	Increased expression	(Metzler *et al.*, 2004; Tam *et al.*, 1997; Tam and Dahlberg, 2006)
miR-143, miR-145	Colorectal cancer	Reduced expression	(Iorio *et al.*, 2005)
miR-21	Glioblastoma and breast cancer	Increased expression	(Ciafrè *et al.*, 2005; Iorio *et al.*, 2005)

Although it is formally possible that BIC exerts its oncogenic properties independently of miR-155, the fact that the only phylogenetically conserved region of BIC RNA corresponds to the one encoding this miRNA strongly suggests that this is unlikely. It will be of great interest to know whether, analogously to miR-17~92, miR-155 can cooperate with Eμ-Myc in murine lymphomagenesis models.

Conclusion

In 1993, with the discovery of the first miRNA in *C. elegans* (Lee *et al.*, 1993) few would have guessed the tremendous impact that this new class of small non-coding RNAs was going to have on the study of development and cancer. Even now, after over a decade of discoveries and thousands of miRNAs identified in organisms ranging from the nematode to *Homo sapiens*, it is difficult to predict what lies ahead. With regard to the study of human cancers, will miRNAs emerge as central players or will we find that with few exceptions, they play only a marginal role? If individual miRNAs will indeed be confirmed to play a causative role in tumorigenesis, it is likely that they will also become potentially excellent drug targets. The technology to systemically knockdown miRNA expression *in vivo* (Krutzfeldt *et al.*, 2005) is already emerging and promises to become a powerful therapeutic tool. At the same time, it is easy to imagine that synthetic miRNAs might be delivered *in vivo* to replace non-functional ones. Perhaps a new generation of rationally designed, gene-specific and highly effective drugs will soon see the light of day. Even if these initial hopes will not be confirmed, miRNAs are already demonstrating their power as prognostic and diagnostic tools (Calin *et al.*, 2004a; H. He *et al.*, 2005; Lu *et al.*, 2005; Volinia *et al.*, 2006), often proving superior to conventional gene-expression profiles. Perhaps more important, the advent of these small RNAs has changed the way we think about gene regulation and is attracting and exciting a new generation of young researchers. It is easy to predict that they will be busy for a long time to come.

ACKNOWLEDGEMENTS

T. J. is an Investigator of the Howard Hughes Medical Institute. This work was supported by United States Public Health Service MERIT Award R37-GM34277 from the National Institute of Health (to T. J.). A. V. is the recipient of an American Italian Cancer Foundation postdoctoral fellowship and M. K is the recipient of a Graduate Research Fellowship from the National Science Foundation.

REFERENCES

Ambros, V. (2004). The functions of animal microRNAs. *Nature*, **431**, 350–355.
Bagga, S., Bracht, J., Hunter, S. *et al.* (2005). Regulation by let-7 and lin-4 miRNAs results in target mRNA degradation. *Cell*, **122**, 553–563.
Bartel, D. P. (2004). MicroRNAs: genomics, biogenesis, mechanism, and function. *Cell*, **116**, 281–297.
Bartel, D. P. and Chen, C. Z. (2004). Micromanagers of gene expression: the potentially widespread influence of metazoan microRNAs. *Nature Reviews Genetics*, **5**, 396–400.

Baudino, T. A., Maclean, K. H., Brennan, J. *et al.* (2003). Myc-mediated proliferation and lymphomagenesis, but not apoptosis, are compromised by E2f1 loss. *Molecular Cell*, **11**, 905–914.

Bentwich, I. (2005). Prediction and validation of microRNAs and their targets. *Federation of European Biochemical Society Letters*, **579**, 5904–5910.

Bernstein, E., Kim, S. Y., Carmell, M. A. *et al.* (2003). Dicer is essential for mouse development. *Nature Genetics*, **35**, 215–217.

Brennecke, J., Hipfner, D. R., Stark, A., Russell, R. B. and Cohen, S. M. (2003). bantam encodes a developmentally regulated microRNA that controls cell proliferation and regulates the proapoptotic gene hid in *Drosophila*. *Cell*, **113**, 25–36.

Calin, G. A., Dumitru, C. D., Shimizu, M. *et al.* (2002). Frequent deletions and down-regulation of micro-RNA genes miR15 and miR16 at 13q14 in chronic lymphocytic leukemia. *Proceedings of the National Academy of Sciences USA*, **99**, 15 524–15 529.

Calin, G. A., Liu, C. G., Sevignani, C. *et al.* (2004a). MicroRNA profiling reveals distinct signatures in B cell chronic lymphocytic leukemias. *Proceedings of the National Academy of Sciences USA*, **101**, 11 755–11 760.

Calin, G. A., Sevignani, C., Dumitru, C. D. *et al.* (2004b). Human microRNA genes are frequently located at fragile sites and genomic regions involved in cancers. *Proceedings of the National Academy of Sciences USA*, **101**, 2999–3004.

Chang, S., Johnston, R. J., Jr., Frokjaer-Jensen, C., Lockery, S. and Hobert, O. (2004). MicroRNAs act sequentially and asymmetrically to control chemosensory laterality in the nematode. *Nature*, **430**, 785–789.

Chen, C. Z. (2005). MicroRNAs as oncogenes and tumor suppressors. *New England Journal of Medicine*, **353**, 1768–1771.

Chen, C. Z., Li, L., Lodish, H. F. and Bartel, D. P. (2004). MicroRNAs modulate hematopoietic lineage differentiation. *Science*, **303**, 83–86.

Chen, J. F., Mandel, E. M., Thomson, J. M. *et al.* (2006). The role of microRNA-1 and microRNA-133 in skeletal muscle proliferation and differentiation. *Nature Genetics*, **38**, 228–233.

Ciafrè, S. A., Galardi, S., Mangiola, A. *et al.* (2005). Extensive modulation of a set of microRNAs in primary glioblastoma. *Biochemical Biophysical Research Communications*, **334**, 1351–1358.

Cimmino, A., Calin, G. A., Fabbri, M. *et al.* (2005). miR-15 and miR-16 induce apoptosis by targeting BCL2. *Proceedings of the National Academy of Sciences USA*, **102**, 13 944–13 949.

Clurman, B. E. and Hayward, W. S. (1989). Multiple proto-oncogene activations in avian leukosis virus-induced lymphomas: evidence for stage-specific events. *Molecular Cell Biology*, **9**, 2657–2664.

Cobb, B. S., Nesterova, T. B., Thompson, E. *et al.* (2005). T cell lineage choice and differentiation in the absence of the RNase III enzyme Dicer. *Journal of Experimental Medicine*, **201**, 1367–1373.

Conaco, C., Otto, S., Han, J. J. and Mandel, G. (2006). Reciprocal actions of REST and a microRNA promote neuronal identity. *Proceedings of the National Academy of Sciences USA*, **103**, 2422–2427.

Couzin, J. (2005). Cancer biology. A new cancer player takes the stage. *Science*, **310**, 766–767.

Croce, C. M. and Calin, G. A. (2005). miRNAs, cancer, and stem cell division. *Cell*, **122**, 6–7.

Eder, M. and Scherr, M. (2005). MicroRNA and lung cancer. *New England Journal of Medicine*, **352**, 2446–2448.

Eis, P. S., Tam, W., Sun, L. *et al.* (2005). Accumulation of miR-155 and BIC RNA in human B cell lymphomas. *Proceedings of the National Academy of Sciences USA*, **102**, 3627–3632.

Farh, K. K., Grimson, A., Jan, C. *et al.* (2005). The widespread impact of mammalian microRNAs on mRNA repression and evolution. *Science*, **310**, 1817–1821.

Fazi, F., Rosa, A., Fatica, A. *et al.* (2005). A minicircuitry comprised of microRNA-223 and transcription factors NFI-A and C/EBPalpha regulates human granulopoiesis. *Cell*, **123**, 819–831.

Gregory, R. I. and Shiekhattar, R. (2005). MicroRNA biogenesis and cancer. *Cancer Research*, **65**, 3509–3512.

Grun, D., Wang, Y. L., Langenberger, D., Gunsalus, K. C. and Rajewsky, N. (2005). MicroRNA target predictions across seven *Drosophila* species and comparison to mammalian targets. *Public Library of Science Computational Biology*, **1**, e13.

Hammond, S. M. (2006). MicroRNAs as oncogenes. *Current Opinions in Genetics and Development*, **16**, 4–9.

Harfe, B. D., McManus, M. T., Mansfield, J. H., Hornstein, E. and Tabin, C. J. (2005). The RNaseIII enzyme Dicer is required for morphogenesis but not patterning of the vertebrate limb. *Proceedings of the National Academy of Sciences USA*, **102**, 10 898–10 903.

Harris, K. S., Zhang, Z., McManus, M. T., Harfe, B. D. and Sun, X. (2006). Dicer function is essential for lung epithelium morphogenesis. *Proceedings of the National Academy of Sciences USA*, **103**, 2208–2213.

He, H., Jazdzewski, K., Li, W. *et al.* (2005). The role of microRNA genes in papillary thyroid carcinoma. *Proceedings of the National Academy of Sciences USA*, **102**, 19 075–19 080.

He, L., Thomson, J. M., Hemann, M. T. *et al.* (2005). A microRNA polycistron as a potential human oncogene. *Nature*, **432**, 828–833.

Hornstein, E., Mansfield, J. H., Yekta, S. *et al.* (2005). The microRNA miR-196 acts upstream of Hoxb8 and Shh in limb development. *Nature*, **438**, 671–674.

Houbaviy, H. B., Murray, M. F. and Sharp, P. A. (2003). Embryonic stem cell-specific microRNAs. *Developmental Cell*, **5**, 351–358.

Iorio, M. V., Ferracin, M., Liu, C. G. *et al.* (2005). MicroRNA gene expression deregulation in human breast cancer. *Cancer Research*, **65**, 7065–7070.

John, B., Enright, A. J., Aravin, A. *et al.* (2004). Human microRNA targets. *Public Library of Science Biology*, **2**, e363.

Johnson, S. M., Grosshans, H., Shingara, J. *et al.* (2005). RAS is regulated by the let-7 microRNA family. *Cell*, **120**, 635–647.

Johnston, R. J. and Hobert, O. (2003). A microRNA controlling left/right neuronal asymmetry in *Caenorhabditis elegans*. *Nature*, **426**, 845–849.

Kanellopoulou, C., Muljo, S. A., Kung, A. L. *et al.* (2005). Dicer-deficient mouse embryonic stem cells are defective in differentiation and centromeric silencing. *Genes & Development*, **19**, 489–501.

Kloosterman, W. P., Wienholds, E., de Bruijn, E., Kauppinen, S. and Plasterk, R. H. (2006). In situ detection of miRNAs in animal embryos using LNA-modified oligonucleotide probes. *Nature Methods*, **3**, 27–29.

Kluiver, J., Poppema, S., de Jong, D. *et al.* (2005). BIC and miR-155 are highly expressed in Hodgkin, primary mediastinal and diffuse large B cell lymphomas. *Journal of Pathology*, **207**, 243–249.

Krek, A., Grun, D., Poy, M. N. *et al.* (2005). Combinatorial microRNA target predictions. *Nature Genetics*, **37**, 495–500.

Krutzfeldt, J., Rajewsky, N., Braich, R. *et al.* (2005). Silencing of microRNAs *in vivo* with "antagomirs". *Nature*, **438**, 685–689.

Kuwabara, T., Hsieh, J., Nakashima, K., Taira, K. and Gage, F. H. (2004). A small modulatory dsRNA specifies the fate of adult neural stem cells. *Cell*, **116**, 779–793.

Lagos-Quintana, M., Rauhut, R., Yalcin, A. *et al.* (2002). Identification of tissue-specific microRNAs from mouse. *Current Biology*, **12**, 735–739.

Lee, R. C., Feinbaum, R. L. and Ambros, V. (1993). The *C. elegans* heterochronic gene lin-4 encodes small RNAs with antisense complementarity to *lin-14*. *Cell*, **75**, 843–854.

Lewis, B. P., Shih, I. H., Jones-Rhoades, M. W., Bartel, D. P. and Burge, C. B. (2003). Prediction of mammalian microRNA targets. *Cell*, **115**, 787–798.

Lewis, B. P., Burge, C. B. and Bartel, D. P. (2005). Conserved seed pairing, often flanked by adenosines, indicates that thousands of human genes are microRNA targets. *Cell*, **120**, 15–20.

Lim, L. P., Lau, N. C., Garrett-Engele, P. *et al*. (2005). Microarray analysis shows that some microRNAs downregulate large numbers of target mRNAs. *Nature*, **433**, 769–773.

Liu, C. G., Calin, G. A., Meloon, B. *et al*. (2004). An oligonucleotide microchip for genome-wide microRNA profiling in human and mouse tissues. *Proceedings of the National Academy of Sciences USA*, **101**, 9740–9744.

Lu, J., Getz, G., Miska, E. A. *et al*. (2005). MicroRNA expression profiles classify human cancers. *Nature*, **435**, 834–838.

McManus, M. T. (2003). MicroRNAs and cancer. *Seminars in Cancer Biology*, **13**, 253–258.

Metzler, M., Wilda, M., Busch, K., Viehmann, S. and Borkhardt, A. (2004). High expression of precursor microRNA-155/BIC RNA in children with Burkitt lymphoma. *Genes Chromosomes Cancer*, **39**, 167–169.

Muljo, S. A., Ansel, K. M., Kanellopoulou, C. *et al*. (2005). Aberrant T cell differentiation in the absence of Dicer. *Journal of Experimental Medicine*, **202**, 261–269.

Murchison, E. P., Partridge, J. F., Tam, O. H., Cheloufi, S. and Hannon, G. J. (2005). Characterization of Dicer-deficient murine embryonic stem cells. *Proceedings of the National Academy of Sciences USA*, **102**, 12 135–12 140.

O'Donnell, K. A., Wentzel, E. A., Zeller, K. I., Dang, C. V. and Mendell, J. T. (2005). c-Myc-regulated microRNAs modulate E2F1 expression. *Nature*, **435**, 839–843.

Ota, A., Tagawa, H., Karnan, S. *et al*. (2004). Identification and characterization of a novel gene, C13orf25, as a target for 13q31-q32 amplification in malignant lymphoma. *Cancer Research*, **64**, 3087–3095.

Pasquinelli, A. E., Reinhart, B. J., Slack, F. *et al*. (2000). Conservation of the sequence and temporal expression of let-7 heterochronic regulatory RNA. *Native*, **408**, 86–89.

Pekarsky, Y., Calin, G. A. and Aqeilan, R. (2005). Chronic lymphocytic leukemia: molecular genetics and animal models. *Current Topics in Microbiology and Immunology*, **294**, 51–70.

Poy, M. N., Eliasson, L., Krutzfeldt, J. *et al*. (2004). A pancreatic islet-specific microRNA regulates insulin secretion. *Nature*, **432**, 226–230.

Reinhart, B. J., Slack, F. J., Basson, M. *et al*. (2000). The 21-nucleotide let-7 RNA regulates developmental timing in *Caenorhabditis elegans*. *Nature*, **403**, 901–906.

Schmitt, C. A., McCurrach, M. E., de Stanchina, E., Wallace-Brodeur, R. R. and Lowe, S. W. (1999). INK4a/ARF mutations accelerate lymphomagenesis and promote chemoresistance by disabling p53. *Genes & Development*, **13**, 2670–2677.

Schmitt, C. A., Rosenthal, C. T. and Lowe, S. W. (2000). Genetic analysis of chemoresistance in primary murine lymphomas. *Nature Medicine*, **6**, 1029–1035.

Schmitt, C. A., Fridman, J. S., Yang, M. *et al*. (2002). Dissecting p53 tumor suppressor functions *in vivo*. *Cancer Cell*, **1**, 289–298.

Schratt, G. M., Tuebing, F., Nigh, E. A. *et al*. (2006). A brain-specific microRNA regulates dendritic spine development. *Nature*, **439**, 283–289.

Soengas, M. S., Alarcon, R. M., Yoshida, H. *et al*. (1999). Apaf-1 and caspase-9 in p53-dependent apoptosis and tumor inhibition. *Science*, **284**, 156–159.

Sokol, N. S. and Ambros, V. (2005). Mesodermally expressed *Drosophila* microRNA-1 is regulated by Twist and is required in muscles during larval growth. *Genes & Development*, **19**, 2343–2354.

Stark, A., Brennecke, J., Bushati, N., Russell, R. B. and Cohen, S. M. (2005). Animal microRNAs confer robustness to gene expression and have a significant impact on 3′ UTR evolution. *Cell*, **123**, 1133–1146.

Suh, M. R., Lee, Y., Kim, J. Y. *et al*. (2004). Human embryonic stem cells express a unique set of microRNAs. *Developmental Biology*, **270**, 488–498.

Sulston, J. E. and Horvitz, H. R. (1981). Abnormal cell lineages in mutants of the nematode *Caenorhabditis elegans*. *Developmental Biology*, **82**, 41–55.

Takamizawa, J., Konishi, H., Yanagisawa, K. *et al*. (2004). Reduced expression of the let-7 microRNAs in human lung cancers in association with shortened postoperative survival. *Cancer Research*, **64**, 3753–3756.

Tam, W. and Dahlberg, J. E. (2006). miR-155/BIC as an oncogenic microRNA. *Genes Chromosomes Cancer*, **45**, 211–212.

Tam, W., Ben-Yehuda, D. and Hayward, W. S. (1997). bic, a novel gene activated by proviral insertions in avian leukosis virus-induced lymphomas, is likely to function through its noncoding RNA. *Molecular Cell Biology*, **17**, 1490–1502.

Tam, W., Hughes, S. H., Hayward, W. S. and Besmer, P. (2002). Avian bic, a gene isolated from a common retroviral site in avian leukosis virus-induced lymphomas that encodes a noncoding RNA, cooperates with c-myc in lymphomagenesis and erythro-leukemogenesis. *Journal of Virology*, **76**, 4275–4286.

Thomson, J. M., Parker, J., Perou, C. M. and Hammond, S. M. (2004). A custom microarray platform for analysis of microRNA gene expression. *Nature Methods*, **1**, 47–53.

Volinia, S., Calin, G. A., Liu, C. G. *et al.* (2006). A microRNA expression signature of human solid tumors defines cancer gene targets. *Proceedings of the National Academy of Sciences USA*, **103**, 2257–2261.

Wienholds, E., Kloosterman, W. P., Miska, E. *et al.* (2005). MicroRNA expression in zebrafish embryonic development. *Science*, **309**, 310–311.

Xu, P., Vernooy, S. Y., Guo, M. and Hay, B. A. (2003). The *Drosophila* microRNA Mir-14 suppresses cell death and is required for normal fat metabolism. *Current Biology*, **13**, 790–795.

Yang, W. J., Yang, D. D., Na, S. *et al.* (2005). Dicer is required for embryonic angiogenesis during mouse development. *Journal of Biological Chemistry*, **280**, 9330–9335.

Yekta, S., Shih, I. H. and Bartel, D. P. (2004). MicroRNA-directed cleavage of HOXB8 mRNA. *Science*, **304**, 594–596.

Zhao, Y., Samal, E. and Srivastava, D. (2005). Serum response factor regulates a muscle-specific microRNA that targets Hand2 during cardiogenesis. *Nature*, **436**, 214–220.

25 miR-122 in mammalian liver

Jinhong Chang and John M. Taylor*

Introduction

MicroRNAs (miRNAs) are small non-coding RNA molecules about 22 nucleotides (nt) in length. They are derived by cleavage from larger precursor RNAs, most of which seem to be polyadenylated polymerase II transcripts (Lee et al., 2002). The processing of miRNA depends on two consecutive actions. The first is by the endonuclease drosha, which acts in the nucleus and cleaves the primary miRNA transcripts (pri-miRNAs) into ~70–80 nt hairpin-like miRNA precursors (pre-miRNAs)(Lee et al., 2003). This miRNA precursor is then transported to the cytoplasm (Bohnsack et al., 2004; Lund et al., 2004) where it is further cleaved by the nuclease dicer, to release the mature miRNA species (Ketting et al., 2001; Lee et al., 2003). These miRNAs can then regulate the expression of target genes with complementary sequence by either cleavage of the target mRNA or inhibition of the translation (Bartel, 2004).

In mammals, hundreds of miRNAs have been identified, some of which are expressed in tissue-specific (Lagos-Quintana et al., 2002) and developmental stage-specific manner (Krichevsky et al., 2003). Altered expression of specific miRNA genes has been associated with the development of cancers (McManus, 2003; see also Chapter 23 in this book). miR-122 is a liver-specific miRNA and in a recent study we examined the expression of miR-122 during mouse liver development and in liver tumors (Chang et al., 2004). We examined the pre-miRNA from which miR-122 is derived and we provided evidence that a target in the liver for miR-122 might be the mRNA for a high affinity cationic amino acid transporter referred to as CAT-1. The objective of the present review article is not only to provide a reprise for that earlier study, but also to include previously unpublished data, and reassess the understanding of miR-122 in the light of subsequent publications by others.

---

* Author to whom correspondence should be addressed.

MicroRNAs: From Basic Science to Disease Biology, ed. Krishnarao Appasani. Published by Cambridge University Press. © Cambridge University Press 2008.

Biogenesis of miR-122

In 2001, Tuschl and colleagues conducted an extensive cloning and sequencing project to identify small RNAs present in various adult mouse tissues (Lagos-Quintana *et al.*, 2002). miR-122 was found to be present only in liver tissue; within the liver, it was the most frequent miRNA isolate. At that time, when the sequence of miR-122 was examined against the available expressed sequence tag (EST) database, the only match was the transcript of a woodchuck gene called *hcr*.

Woodchucks are a widely used animal model for studying hepadnavirus replication and the association with liver tumors. Woodchuck hepatitis virus, like human hepatitis B virus, frequently causes chronic infections that can progress to hepatocellular carcinoma (Moroy *et al.*, 1986; Etiemble *et al.*, 1989). In 1989, Buendia and colleagues studied many such woodchuck liver tumors in order to understand changes that might have occurred at the level of the chromosomal DNA. In one particular tumor they detected a translocation of part of the *hcr* gene to the locus for c-*myc*. This translocation led to overexpression of an altered form of c-*myc* (Moroy *et al.*, 1989), as we will see in a later section.

In their original study, Buendia and co-workers characterized the mRNA transcript for *hcr* gene present in normal woodchuck liver. They reported that 5' of the transcription start site there were elements resembling a classical eukaryotic promoter. As shown in Figure 25.1, they found that the *hcr* transcript was a 4.7 kb unspliced polyadenylated RNA with a 37 amino acid open reading frame located almost at the 5'-end. For reasons they could not understand at that time, 95% of *hcr* RNA was somewhat shorter, about 4.5 kb, and lacked poly(A). Results from a nuclease protection assay suggested that these RNAs had undergone endonucleolytic processing at one or more of several sites located about 200 nt from the site of poly(A).

The more recent studies of Lagos-Quintana *et al.* (2002) showed that mouse miR-122 was located in this 3'-end of the *hcr* transcript, and they considered the predicted folding of this region to be roughly consistent with it being the precursor for miR-122 generation. Other observations also supported *hcr* being the primary transcript for miR-122. Buendia and colleagues demonstrated that the *hcr* locus was highly expressed in normal woodchuck liver but not in other tissues examined. This was consistent with the liver-specific expression of miR-122. In addition, unlike the conventional polyadenylated RNA, the *hcr* transcripts were found mainly located in nucleus. This sub-cellular localization seemed consistent with the interpretation made for other miRNAs (Lee *et al.*, 2002), that in the nucleus drosha cleaved the primary transcript and released the pre-miRNA for further processing by dicer. As represented in Figure 25.1, one of the 3'-ends of the *hcr* was located 10 nt upstream of the 5'-end of miR-122. Another mapped exactly at the 3'-end of miR-122. We note that while these data are supportive of the *hcr* transcript being a precursor for miR-122, the available processing data are more complicated than a single simple pathway (Lee *et al.*, 2002) in which drosha cleaves out the pre-miRNA in the nucleus, which in turn is processed by dicer, to release the miRNA. It is possible that there are alternative processing pathways,

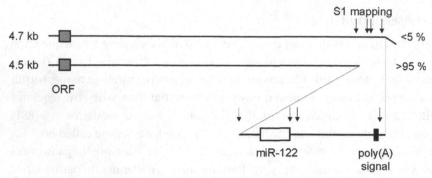

Figure 25.1. Relationship between woodchuck *hcr* transcripts and the miR-122 sequence. Details of the *hcr* transcripts are from Moroy *et al.* (1989). The predicted ORF on the *hcr* transcript is indicated by a gray box and its poly(A) signal by a black box. The miR-122 is indicated by an open box. Arrows indicate the sites obtained by S1 mapping, that correspond to the 3′-ends of transcripts that are no longer polyadenylated.

and maybe not all are productive for miRNA generation. Certainly, more detailed studies are needed for the processing of *hcr*.

Since the time of the studies of Lagos-Quintana *et al.* (2002), the complete nucleotide sequences of many animal genomes have been made available. This has allowed a genomics and evolutionary assessment of *hcr* sequences that are uniquely represented in several such complete sequences. For example, Figure 25.2a shows an alignment of 160 nt of woodchuck sequence, at and around the miR-122, in comparison with sequences for human, mouse, and fugu. From the indicated consensus sequence it can be seen that miR-122 sequence (shaded) is absolutely conserved in these four species. However, in the sequences adjacent to miR-122, there was significant but less than 100% conservation. Outside of this region, the conservation was much less.

We have also considered the predicted folding of RNAs containing this miR-122. For example, Figure 25.2b shows the predicted stem-loop structure of 66 nt at and around mouse miR-122 (http://microrna.sanger.ac.uk). This structure has been predicted to be the pre-miRNA, possibly with some flanking sequence. An asterisk is used to indicate those nucleotides not conserved among the four sequences summarized in Figure 25.2a. It can be seen that most of the variations are limited to the unpaired loop region of the stem-loop structure. That is, the secondary structure of miR-122 precursor is well conserved, between mammals and fugu.

In order to experimentally prove that *hcr* contained the precursor for miR-122, we cloned and expressed the 160 nt woodchuck *hcr* sequence in human embryonic kidney 293 cells (Figure 25.2a). As shown in lane 1 of Figure 25.2c, these cells normally have no detectable miR-122. However, after transfection, these cells processed and accumulated significant levels of miR-122 (lane 2), as assayed by a Northern analysis. The amount detected was even more than the endogenous level present in Huh7 cells, a human liver line (lane 3), although it was not as much as detected in mouse and woodchuck liver (lanes 4 and 5, respectively) or in primary human hepatocytes (lane 6). As described later, the miR-122 expressed in the 293 cells was functional.

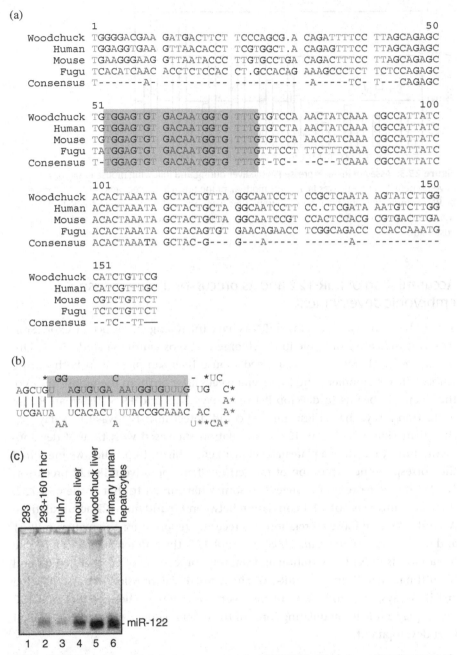

(a)

```
              1                                                    50
Woodchuck     TGGGGACGAA GATGACTTCT TCCCAGCG.A CAGATTTTCC TTAGCAGAGC
    Human     TGGAGGTGAA GTTAACACCT TCGTGGCT.A CAGAGTTTCC TTAGCAGAGC
    Mouse     TGAAGGGAAG GTTAATACCC TTGTGCCTGA CAGACTTTCC TTAGCAGAGC
     Fugu     TCACATCAAC ACCTCTCCAC CT.GCCACAG AAAGCCCTCT TCTCCAGAGC
Consensus     T-------A- ---------- ---------- -A-----TC- T---CAGAGC

              51                                                  100
Woodchuck     TGTGGAGTGT GACAATGGTG TTTGTGTCCA AACTATCAAA CGCCATTATC
    Human     TGTGGAGTGT GACAATGGTG TTTGTGTCTA AACTATCAAA CGCCATTATC
    Mouse     TGTGGAGTGT GACAATGGTG TTTGTGTCCA AACCATCAAA CGCCATTATC
     Fugu     TATGGAGTGT GACAATGGTG TTTGTTTCCT TTCTTTCAAA CGCCATTATC
Consensus     T-TGGAGTGT GACAATGGTG TTTGT-TC-- --C--TCAAA CGCCATTATC

              101                                                 150
Woodchuck     ACACTAAATA GCTACTGTTA GGCAATCCTT CCGCTCAATA AGTATCTTGG
    Human     ACACTAAATA GCTACTGCTA GGCAATCCTT CC.CTCGATA AATGTCTTGG
    Mouse     ACACTAAATA GCTACTGCTA GGCAATCCGT CCACTCCACG CGTGACTTGA
     Fugu     ACACTAAATA GCTACAGTGT GAACAGAACC TCGGCAGACC CCACCAAATG
Consensus     ACACTAAATA GCTAC-G--- G---A----- -------A-- ----------

              151
Woodchuck     CATCTGTTCG
    Human     CATCGTTTGC
    Mouse     CGTCTGTTCT
     Fugu     TCTCTGTTCT
Consensus     --TC--TT—
```

(b)

```
          *  GG            C              -  *UC
    AGCUGU    AGUGUGA  AAUGGUGUUUG  UG   C*
    ||||||    |||||||  |||||||||||  ||        A*
    UCGAUA    UCACACU  UUACCGCAAAC  AC   A*
          AA            A              U**CA*
```

(c)

miR-122

1 2 3 4 5 6

Figure 25.2. The *hcr* transcript contains precursor for miR-122. Panel (a) shows the alignment of 160 nt DNA sequences from woodchuck (X13234), human (NT_025028), mouse (NT_039674.1) and fugu (UCSC Bioinformatics Site) that flank miR-122. Sequences were selected using the BLAT analysis/alignment tool at UCSC Genomic Bioinformatics Site (http://genome.ucsc.edu). The miR-122 sequences are shaded gray. Panel (b) shows the predicted folding of stem-loop structure around the mouse miR-122 sequence (http://microrna.sanger.ac.ul). The sequences with gray shade are miR-122. Indicated by asterisks are the nucleotides not conserved among the four species. Panel (c) shows the expression of miR-122 in livers and cell lines, as assayed by Northern analysis. Lanes 1 and 2 are 293 cells without or with exogenous expression of the 160 nt woodchuck *hcr* sequence. Lane 3 is RNA from Huh 7 cells, a liver tumor cell line. Lanes 4–6 are RNAs from mouse liver, woodchuck liver and primary cultured human hepatocytes, respectively.

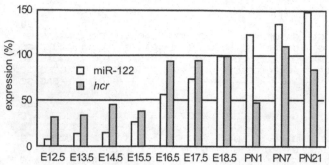

Figure 25.3. Assay of RNAs expressed in the liver during and following mouse embryonic development. Liver tissue was taken post mortem, at the times indicated. miR-122 was assayed by RNase protection and the mouse *hcr* transcripts were assayed by real-time RT-PCR. Both levels were normalized to E18.5 as 100%. E refers to days of embryonic development, and PN to days post natal.

Accumulation of miR-122 and its precursor during mouse embryonic development

Studies have shown that many miRNAs have intriguing stage-specific expression patterns, indicative of a role in develpoment (Lagos-Quintana *et al.*, 2002; Lim *et al.*, 2003). Therefore, we examined mouse liver samples taken both during embryonic development and for several weeks after birth. In the mouse embryo, the liver bud begins to develop between days E8.5 and E9.5. Using an RNase protection assay, the earliest time we could detect miR-122 was E12.5 days. As shown in Figure 25.3, miR-122 accumulation increased with time of development, almost reaching a plateau level just before birth. In parallel, we measured the corresponding expression of transcripts of the putative miR-122 precursor, *hcr*. While the pattern of changes was somewhat similar to that of the miR-122, there certainly was not a 1:1 correlation between *hcr* and miR-122 accumulation. A similar kind of loose correlation was recently reported for the mature miR-21 and its precursor (Jiang *et al.*, 2005). For miR-122, the difference may well be due to variations in *hcr* transcription and stability, processing of *hcr* to miR-122, and the ultimate stabilization of miR-122. Since miR-122 expression was first detected at E12.5 days, after formation of the liver bud, we hypothesized that miR-122 might play a role in modulating some of the diverse genes expressed during fetal liver development.

miR-122 functionality

miR-122 and liver tumor

As described above, chronic woodchuck hepatitis infections frequently progress to liver cancer. Back to 1989, Buendia and colleagues found many tumors with enhanced expression of N-*myc* and c-*myc* (Moroy *et al.*, 1986), most of which resulted from viral insertional mutagenesis. However, one of 30 tumors, designated as W64, was characterized as an unusual chromosomal translocation. At the

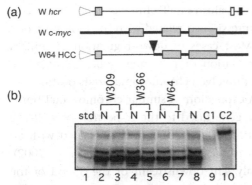

Figure 25.4. miR-122 in woodchuck liver tumors. Panel (a) shows the translocation between *hcr* gene and c-*myc* in W64 liver tumor. The *hcr* ORF and *myc* exons are shaded gray. The miR-122 sequence is indicated by an open box and the poly(A) signal by a black box. The point of translocation of c-*myc* to *hcr* is indicated by an arrow. The open arrows refer to the promoters and direction of transcription. These data are extracted from the report of Moroy *et al.* (1989). Panel (b) shows miR-122 expression in woodchuck liver tumor tissues by RNase protection assay. Lane 1 is a 2 pg standard for miR-122. Lanes 2–5 are RNAs from two pairs of non-tumor and tumor tissues. Lanes 6 and 7 are non-tumor and tumor tissues from W64. Lanes 9 and 10 are controls with or without RNase in absence of woodchuck RNA.

breakpoint, the 5′-end of the *hcr* gene was translocated to in front of the second exon of c-*myc*, leading to a 50-fold increase in c-*myc* expression, and a rapidly progressing liver tumor (Moroy *et al.*, 1989). As shown in Figure 25.4a, we expected this translocation of one chromosome should lead to less expression of normal full-length *hcr* transcript containing miR-122 sequences and this would subsequently decrease the level of processed miR-122.

We obtained the paired normal and tumor tissues of W64 and quantitated the expression of miR-122 using a RNase protection assay (Chang *et al.*, 2004). As shown in Figure 25.4b, lanes 6 and 7, we found the miR-122 level in W64 tumor tissue was about 2-fold less than the normal liver. This result was an expected consequence of the translocation that occurred on one chromosome. We next tested whether changes in the level of miR-122 might also occur in other wood-chuck liver tumors. We tested another 10 pairs of tumor and non-tumor tissue taken from woodchucks chronically infected with woodchuck hepatitis virus, as studied recently at the Fox Chase Cancer Center. Only in 2/10 cases did we observe a decrease in the miR-122 level (Figure 25.4b, lanes 2–5). None of these decreases was as much as the 2-fold reduction in miR-122 for W64. Therfore, such a translocation between *hcr* and c-*myc* is not a common event in woodchuck liver tumors.

While others have observed that altered expression of miRNAs can be associated with certain types of cancer (Calin *et al.*, 2002), we realize that even for W64 liver tumor, we have no evidence that the reduction in miR-122 level *per se* cooperated with up-regulated c-*myc* to promote development of the woodchuck tumor. Studies of miRNA expression profiles in different human tumors have generally observed a down-regulation and occasionally an up-regulation. In one case, the *myc* protein was found to bind directly to the miR-17–92 locus on chromosome

13 and activate expression of this miRNA cluster (O'Donnell *et al.*, 2005; see also Chapter 22 in this book). This overexpression correlated with the formation of B cell lymphomas (He *et al.*, 2005). Most cases of down-regulation of miRNA involved a homozygous deletion, mutation or promoter hyper-methylation. The W64 woodchuck liver tumor that involved chromosomal translocation was a special example of reduced miRNA expression. A similar chromosomal trans-location, producing a truncated *myc* gene fused to the 5'-portion of miR-142 precursor gene, led to an up-regulation of c-*myc* that was associated with an aggressive B cell leukemia (Gauwerky *et al.*, 1989; Lagos-Quintana *et al.*, 2002). In this case, the breakpoint was only 4 nt downstream of the 3'-end of the predicted miR-142 precursor. It remains unknown whether this translocation interfered with miR-142 processing.

In our study we also assayed two human liver tumor cell lines: Huh7 cells, which are derived from a primary human hepatoma, and HepG2 cells, which are derived from a hepatoblastoma. Interestingly, we found that in both cell lines, the expression of miR-122 was significantly reduced. In Huh7 it was only about 10%–20% compared with primary cultured human hepatocytes (Figure 25.2c) (Chang *et al.*, 2004), and it was undetectable in HepG2 cells (Chang *et al.*, 2004). When the putative human primary transcript for miR-122 was assayed by real-time RT-PCR, the level in Huh7 was comparable to human hepatocytes; however, in HepG2 cells, it was almost undetectable. This indicated that regulation of miR-122 was either at a transcriptional or processing level. It might be interesting to further investigate the mechanism for the down-regulation of miR-122 in human tumor cell lines. However, such down-regulation did not occur in most of the woodchuck tumors studied. Others have recently reported a similar lack of corre-lation in human liver tumors (Jiang *et al.*, 2005).

Validation of a potential mRNA target for miR-122

Currently, out of hundreds of identified miRNAs, it is only in a few cases that the biological function of the miRNA is known (Bartel, 2004). Several groups pre-dicted the potential targets for mammalian miRNA. Two groups (Lewis *et al.*, 2003; John *et al.*, 2004) predicted that a target for miR-122 would be the cationic acid transporter protein, CAT-1 (also known as solute carrier family 7 cationic amino acid transporter y+ system member 1, or SLC7A1) (MGI:88117). They also showed that the predicted pairing between miR-122 and CAT-1 was conserved not only for human, mouse and rat, but also for fugu. Using algorithms previously used to predict miRNA targets in *Drosophila* and human, six sites (indicated as a–f) were predicted within the 2 kb 3'-non-coding sequence of human CAT-1 (Chang *et al.*, 2004; John *et al.*, 2004).

To test these predictions experimentally we used a reporter system that employs two constructs which express mRNAs encoding different forms of the hepatitis delta virus protein, the 214 amino acid large delta protein (δAg-L) and the 195 amino acid small delta protein (δAg-S). These two proteins have the same N-terminal sequence and can be quantitated in immunoblot assays using the same polyclonal rabbit antibody. Into the 3'-end non-coding region of the

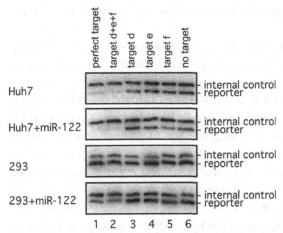

Figure 25.5. Validation of CAT-1, the predicted miR-122 target, using a protein reporter system. As indicated, two cell lines, Huh 7 and 293 were used in the absence or presence of exogenous expression of miR-122. The internal control and reporter constructs expressed two forms of hepatitis delta protein (δAg-L and δAg-S, respectively) that were assayed by immunoblot. In addition, into the 3'-untranslated region of the vector to δAg-S, were inserted potential miR-122 target sequences. Lane 1 is a positive control, which contained the insert of a single target with 100% complementarity to miR-122. Lane 2 is a region of human CAT-1 with three potential miR-122 target sites, indicated as d–f. Lanes 3–5, contain targets d, e, or f, respectively. Lane 6 is a negative control lacking any miR-122 target sequence. Three days after co-transfection of the two constructs, total cellular protein was analyzed by immunoblot using rabbit polyclonal antibody to detect and quantitate the two forms of δAg.

construct expressing δAg-S, we added one or more of the predicted miR-122 target sequences. The construct expressing δAg-L was co-transfected as an internal control. Three days after transfection we used an immunoblot to determine the ratio of δAg-S to δAg-L as a measure of the inhibition attributable to the presence in the δAg-S construct of the putative miR-122 target sequences.

As shown in Figure 25.5, in Huh7 cells, which express significant endogenous levels of miR-122, the δAg-S expression was strongly inhibited when three target sequences (d, e, and f) were added into the 3'-non-coding region (Figure 25.5, top, lane 2). On the contrary, the presence of just one of the three targets did not give sufficient inhibition (Figure 25.5, top, lanes 3–5). We also tested the consequences of expressing more miR-122 in the Huh7 cells. To achieve this we transiently transfected the cells with a vector expressing 160 nt of the woodchuck *hcr* precursor sequence. In this way the inhibition of the reporter construct was increased, but again, only when all three putative miR-122 targets were present. Such findings agree with others who suggest that multiple target sites are essential for proper miRNA regulation *in vivo* and *in vitro* (Doench and Sharp, 2004; Brennecke *et al.*, 2005). In 293 cells that do not express endogenous miR-122, there was no observable inhibition. However, when miR-122 was overexpressed exogenously, there was inhibition when three targets were linked to the reporter. Similarly, one target on the reporter RNA did not give significant inhibition. In addition, this study further confirmed that miR-122 produced from the 160 nt of woodchuck *hcr* sequence surrounding miR-122 was also functional. This also supports the interpretation that the *hcr* transcript is the precursor for miR-122.

It is generally considered that when there is less than 100% pairing between a miRNA and its target sequence that inhibition can only be at the level of translation (Bartel, 2004). Surprisingly, when we assayed the mRNA expressed from our reporter constructs, we found that there was also a comparable decrease when three target sites were present (Chang *et al.*, 2004). While there is a precedent for miRNAs directing cleavage reactions *in vivo*, this has so far been shown only for target sites with perfect or near-perfect complementarity (Yekta *et al.*, 2004). It must therefore be made clear that we did not obtain direct experimental evidence that miR-122, with its significant mismatches to each of the target sites, actually cleaved the reporter mRNA. It is relevant that recent studies have shown that a mRNA whose translation has been inhibited by miRNA or siRNA, can either be associated with non-functional polysomes or be recruited to cytoplasmic structures referred to as P-bodies. Such structures are separate from the translation machinery, and are sites of mRNA storage or degradation (Liu *et al.*, 2005a; Liu *et al.*, 2005b).

Our reporter study indicated that *in vivo*, CAT-1 could be negatively regulated by miR-122. Consistent with this result, it has been shown that expression of CAT-1 mRNA is unique in that it is detected in all adult mammalian tissues with the exception of the liver (Ito and Groudine, 1997; Vekony *et al.*, 2001). We also showed that the expression profile of miR-122 and CAT-1 during mouse liver embryonic development was in an inverse correlation (Chang *et al.*, 2004). It would be interesting if one can demonstrate miR-122 action on the full-length endogenous CAT-1 mRNA. Even then, it is still a big task to understand the biological relevance of this regulation. In mammals, many amino acids are essential in that they need to be transported into cells. CAT-1 is a high affinity transporter for Arg and Lys. Given that CAT-1 is down-regulated in liver, CAT-2, which has functional redundancy with CAT-1, is actually overexpressed.

Other potential targets of miR-122

Several effective algorithms have been developed for prediction of microRNA and their targets in mammals. For miR-122, in addition to the above-mentioned CAT-1, a dozen other targets were predicted. These included Ras related protein RAB-10, cytoplasmic polyadenylation element binding protein 2, transcription factor sox-3, and DNA polymerase zeta catalytic subunit (John *et al.*, 2004). None of these predictions has yet been evaluated in the experimental or biological setting. In a recent study conducted by Krutzfeldt and colleagues (Krutzfeldt *et al.*, 2005), chemically engineered oligonucleotides, termed "antagomirs", were used to silence miR-122 in mouse liver. Gene expression and bioinformatics analysis of mRNA from antagomir-treated animals revealed that more than 300 genes were up-regulated. Some of these genes are usually repressed in hepatocytes. In addition, among the genes with annotated 3'-UTR, ~50% had at least one miR-122 recognition motif, allowing that they might be direct targets for miR-122. These results suggest that the role of miR-122 might involve the maintenance of adult liver phenotype. Furthermore, since many of these genes had not even been previously predicted (John *et al.*, 2004; Krek *et al.*, 2005; Lewis *et al.*, 2005), the authors suggested that the total number of direct miR-122 targets might be quite high.

A recent study by Peter Sarnow and colleagues (Jopling *et al.*, 2005) showed that hepatitis C virus replicated more efficiently in Huh7 cells when miR-122 was present. Blocking of miR-122 led to a substantial reduction in viral RNA level. A functional miR-122 binding site was mapped to the 5′-region of the viral RNA. The mechanism of this unusual form of miR-122 action is not yet clear.

Conclusions

We have shown that the liver-specific microRNA, miR-122, is processed from a liver specific pol II transcript (*hcr* transcript) with limited coding ability. The sequence of miR-122 as well as the predicted stem-loop structure flanking the miR-122 sequence are highly conserved among human, mouse, woodchuck and fugu. In one unique example of a woodchuck liver tumor, the *hcr* gene was translocated to in front of the second exon of the c-*myc* gene. This led to the 50-fold overexpression of c-*myc* and decreased the level of processed miR-122 from the normal *hcr* transcript. However, it remains unclear whether this decrease of miR-122 coordinated with c-myc to facilitate this progressive form of liver tumor. We have attempted to validate CAT-1 as one of the predicted miR-122 targets. Nevertheless, there are many other potential targets for miR-122 that need to be further studied in order to understand the important function of miR-122 in liver development and liver diseases. Such studies may eventually lead to the establishment of therapeutic strategies for related liver diseases.

ACKNOWLEDGMENTS

JMT was supported by NIH grants AI-26522 and CA-06927, and by an appropriation from the Commonwealth of Pennsylvania. We acknowledge the contributions of the co-authors on our original article (Chang *et al.*, 2004). We thank Eric Moss and Richard Katz for constructive comments on the manuscript.

REFERENCES

Bartel, D. P. (2004). MicroRNAs: genomics, biogenesis, mechanism, and function. *Cell*, **116**, 281–297.

Bohnsack, M. T., Czaplinski, K. and Gorlich, D. (2004). Exportin 5 is a RanGTP-dependent dsRNA-binding protein that mediates nuclear export of pre-miRNAs. *RNA*, **10**, 185–191.

Brennecke, J., Stark, A., Russell, R. B. and Cohen, S. M. (2005). Principles of microRNA-target recognition. *Public Library of Science Biology*, **3**, e85.

Calin, G. A., Dumitru, C. D., Shimizu, M. *et al.* (2002). Frequent deletions and down-regulation of micro-RNA genes miR15 and miR16 at 13q14 in chronic lymphocytic leukemia. *Proceedings of the National Academy of Sciences USA*, **99**, 15 524–15 529.

Chang, J., Nicolas, E., Marks, D. *et al.* (2004). miR-122, a mammalian liver-specific microRNA, is processed from *hcr* mRNA and may downregulate the high affinity cationic amino acid transporter CAT-1. *RNA Biology*, **1**, 106–113.

Doench, J. G. and Sharp, P. A. (2004). Specificity of microRNA target selection in translational repression. *Genes & Development*, **18**, 504–511.

Etiemble, J., Moroy, T., Jacquemin, E., Tiollais, P. and Buendia, M. A. (1989). Fused transcripts of c-*myc* and a new cellular locus, *hcr* in a primary liver tumor. *Oncogene*, **4**, 51–57.

Gauwerky, C. E., Huebner, K., Isobe, M., Nowell, P. C. and Croce, C. M. (1989). Activation of MYC in a masked t(8;17) translocation results in an aggressive B-cell leukemia. *Proceedings of the National Academy of Sciences USA*, **86**, 8867–8871.

He, L., Thomson, J. M., Hemann, M. T. *et al.* (2005). A microRNA polycistron as a potential human oncogene. *Nature*, **435**, 828–833.

Ito, K. and Groudine, M. (1997). A new member of the cationic amino acid transporter family is preferentially expressed in adult mouse brain. *Journal of Biological Chemistry*, **272**, 26 780–26 786.

Jiang, J., Lee, E. J., Gusev, Y. and Schmittgen, T. D. (2005). Real-time expression profiling of microRNA precursors in human cancer cell lines. *Nucleic Acids Research*, **33**, 5394–5403.

John, B., Enright, A. J., Aravin, A. *et al.* (2004). Human microRNA targets. *Public Library of Science Biology*, **2**, e363.

Jopling, C. L., Yi, M., Lancaster, A. M., Lemon, S. M. and Sarnow, P. (2005). Modulation of hepatitis C virus RNA abundance by a liver-specific MicroRNA. *Science*, **309**, 1577–1581.

Ketting, R. F., Fischer, S. E. J., Bernstein, E. *et al.* (2001). Dicer functions in RNA interference and in synthesis of small RNA involved in developmental timing in *C. elegans. Genes & Development*, **15**, 2654–2659.

Krek, A., Grun, D., Poy, M. N. *et al.* (2005). Combinatorial microRNA target predictions. *Nature Genetics*, **37**, 495–500.

Krichevsky, A. M., King, K. S., Donahue, C. P., Khrapko, K. and Kosik, K. S. (2003). A microRNA array reveals extensive regulation of microRNAs during brain development. *RNA*, **9**, 1274–1281.

Krutzfeldt, J., Rajewsky, N., Braich, R. *et al.* (2005). Silencing of microRNAs in vivo with "antagomirs". *Nature*, **438**, 685–659.

Lagos-Quintana, M., Rauhut, R., Yalcin, A. *et al.* (2002). Identification of tissue-specific microRNAs from mouse. *Current Biology*, **12**, 735–739.

Lee, Y., Ahn, C., Han, J. *et al.* (2003). The nuclear RNase III Drosha initiates microRNA processing. *Nature*, **425**, 415–419.

Lee, Y. H., Jeon, K., Lee, J.-T., Kim, S. and Kim, V. N. (2002). MicroRNA maturation, stepwise processing and subcellular localization. *European Molecular Biology Organization Journal*, **21**, 4663–4670.

Lewis, B. P., Shih, I. H., Jones-Rhoades, M. W., Bartel, D. P. and Burge, C. B. (2003). Prediction of mammalian microRNA targets. *Cell*, **115**, 787–798.

Lewis, B. P., Burge, C. B. and Bartel, D. P. (2005). Conserved seed pairing, often flanked by adenosines, indicates that thousands of human genes are microRNA targets. *Cell*, **120**, 15–20.

Lim, L. P., Glasner, M. E., Yekta, Y., Burge, C. B. and Bartel, D. P. (2003). Vertebrate microRNA genes. *Science*, **299**, 1540.

Liu, J., Rivas, F. V., Wohlschlegel, J. *et al.* (2005a). A role for the P-body component GW182 in microRNA function. *Nature Cell Biology*, **7**, 1161–1166.

Liu, J., Valencia-Sanchez, M. A., Hannon, G. J. and Parker. R. (2005b). MicroRNA-dependent localization of targeted mRNAs to mammalian P-bodies. *Nature Cell Biology*, **7**, 719–723.

Lund, E., Guttinger, S., Calado, A., Dahlberg, J. E. and Kutay, U. (2004). Nuclear export of microRNA precursors. *Science*, **303**, 95–98.

McManus, M. T. (2003). MicroRNAs and cancer. *Seminars in Cancer Biology*, **13**, 253–258.

Moroy, T., Marchio, A., Etiemble, J. *et al.* (1986). Rearrangement and enhanced expression of c-myc in hepatocellular carcinoma of hepatitis virus infected woodchucks. *Nature*, **324**, 276–279.

Moroy, T., Etiemble, J., Bougueleret, L. *et al.* (1989). Structure and expression of *hcr*, a locus rearranged with c-*myc* in a woodchuck hepatocellular carcinoma. *Oncogene*, **4**, 59–65.

O'Donnell, K. A., Wentzel, E. A., Zeller, K. I., Dang, C. V. and Mendell, J. T. (2005). c-Myc-regulated microRNAs modulate E2F1 expression. *Nature*, **435**, 839–843.

Vekony, N., Wolf, S., Boissel, J. P., Gnauert, K. and Closs, E. I. (2001). Human cationic amino acid transporter hCAT-3 is preferentially expressed in peripheral tissues. *Biochemistry*, **40**, 12 387–12 394.

Yekta, S., Shih, I. H. and Bartel, D. P. (2004). MicroRNA-directed cleavage of HOXB8 mRNA. *Science*, **304**, 594–596.

26 MiRNAs in glioblastoma

Silvia Anna Ciafrè

Introduction

Glioblastoma multiforme: a lethal brain tumor still missing a thorough comprehension of molecular oncogenesis

Glioblastoma multiforme (GBM) is one of the most lethal forms of cancers and the most common brain tumor in adults, accounting for approximately 12%–15% of all intracranial neoplasms and 50%–60% of all astrocytic tumours (Zhu and Parada, 2002; Hulleman and Helin, 2005). In most European and North American countries, incidence is approximately 2–3 new cases per 100 000 people per year. Composed of poorly differentiated neoplastic astrocytes, glioblastomas are located preferentially in the subcortical white matter of the cerebral hemispheres, and either may develop from lower grade astrocytic tumors (secondary or progressive glioblastoma multiforme) or may arise very rapidly *de novo* (primary glioblastoma multiforme) (Wechsler-Reya and Scott, 2001). Despite progress in research on the molecular aspects of malignant gliomas, the prognosis of these brain tumors continues to be dismal, with a median survival time of 12 months from the time of diagnosis. Neither continuous improvements in surgery and radiation techniques, nor in chemotherapy, have been able to change glioblastoma patients' life expectancy over decades.

This highly malignant tumor is thought to arise from astrocytes or astrocyte precursors, but the heterogeneity of GBM tumor morphology and behavior (as indicated by the term "multiforme") makes conclusions about its origin extremely difficult (Wechsler-Reya and Scott, 2001). Recently, several experimental observations have led to formulate the hypothesis that this type of malignancy might arise from the transformation of adult neural stem cells, normally present in the brain just in the main areas of distribution of brain tumors, and able to trigger gliomagenesis in response to oncogenic mutations (Singh *et al.*, 2003; Singh *et al.*, 2004; Galli *et al.*, 2004; Tunici *et al.*, 2004; Yuan *et al.*, 2004; Sanai *et al.*, 2005). Currently, primary brain tumors are classified in a traditional manner that reflects their histological appearance and location (Kleihues *et al.*, 2002). The widely accepted notion that cancer is a disorder of genes, however, has opened the

MicroRNAs: From Basic Science to Disease Biology, ed. Krishnarao Appasani. Published by Cambridge University Press. © Cambridge University Press 2008.

possibility of classifying tumors according to the genetic alterations that underlie their pathogenesis and that regulate their malignant behavior. Cytogenetic and molecular studies have identified several recurrent, non-random genetic abnormalities associated with glial tumors. Both oncogene activation and tumor suppressor gene inactivation have been shown to be a part of the multi-step process of glial tumorigenesis and tumor progression (Reifenberger and Collins, 2004). However, the current knowledge of tumor genetics still does not allow identifying clinically relevant factors predictive for outcome or response to therapy, and much more information is needed to understand the fundamental biological properties which make many gliomas so refractory to therapy.

To date, many attempts have been made in search of gene expression profiles that specifically mark different classes of tumors (van den Boom *et al.*, 2003; Mischel *et al.*, 2003; Nigro *et al.*, 2005), ideally the identification of defined groups of co-regulated genes, possibly sharing common biological functions, should shed a new light on the roles of such genes in initiation and progression of cancer. In recent years, great attention has been devoted to a novel class of noncoding RNAs, named microRNAs, which were demonstrated to have critical functions across several biological processes, to such an extent that microRNA expression profiling resulted extremely informative to determine developmental lineages and differentiation states of human tissues (Pasquinelli *et al.*, 2005). In this frame, the study of tumor-specific expression of microRNAs might give important results in the understanding of tumor biology.

MicroRNAs and cancer

MicroRNAs are a wide class of small, noncoding RNAs that regulate protein expression at the post-transcriptional level, by specifically targeting the 3′ UTRs of multicellular eukaryotic mRNAs, and impairing their translation through a block of translation initiation (Kim, 2005). The expression of many microRNAs was shown to be temporally and spatially regulated, while the disruption of their physiological expression patterns was associated with several examples of human tumorigenesis, suggesting that they may play a role as a novel class of oncogenes or tumor suppressor genes (Gregory and Shiekhattar, 2005). Indeed, several micro-RNA genes are located in chromosomal regions frequently deleted, rearranged or amplified in human cancers (Calin *et al.*, 2004a); one recent example is that of miR-125b, whose coding sequence is located at chromosome 11q23–24, a region extensively deleted in breast, ovarian, and lung tumors (Iorio *et al.*, 2005). MiR-125b, demonstrated to be down-regulated in breast cancer, might be regarded too as a tumor suppressor gene involved in the tumorigenesis of those tumors. Another pair of microRNAs whose aberrant expression has been linked to human tumorigenesis is represented by miR-15 and miR-16, located in a cluster at 13q14.3, and frequently down-regulated in B cell chronic lymphocytic leukemia, as well as in pituitary adenomas and in prostate cancer (Calin *et al.*, 2002; Calin *et al.*, 2004b; Bottoni *et al.*, 2005; see also Chapter 23 in this book). Recently, the antiapoptotic protein Bcl2 has been identified as a target for these two microRNAs, providing a clue that links the widely represented Bcl2 over-expression in many forms of human cancers to the down-regulation of miR-15 and miR-16 (Cimmino *et al.*, 2005). A similar collection of evidence relates let-7 microRNA down-regulation observed in lung tumors to the over-expression of RAS, a critical oncogene, suggesting let-7 negative regulation of RAS as a mechanism for let-7 in lung oncogenesis (Johnson *et al.*, 2005; see also Chapter 22 in this book). A further notable observation is that miR-143 and miR-145 show consistently reduced expression in both adenomatous and cancerous stages of colorectal neoplasia (Michael *et al.*, 2003), while miR-143 expression has been related to adipocyte differentiation (Esau *et al.*, 2004). These and other microRNAs clearly under-expressed in cancer can be classified as a new form of tumor-suppressors, critically interfering with basic processes commonly dis-regulated in cancer, such as apoptosis or proliferation.

On the other hand, the over-expression of some microRNAs was documented in other human cancers, such as the miR-155 precursor BIC in paediatric Burkitt's and other lymphomas (Metzler *et al.*, 2004; Kluiver *et al.*, 2005), or a whole microRNA cluster, located in chromosome 13q32–33, in B cell lymphoma (He *et al.*, 2005). The microRNAs encoded in this cluster were demonstrated to co-operate with c-myc in accelerating tumor development; moreover, their oncogenic action is played also through the negative regulation of the transcription factor E2F1, previously defined as a tumor-suppressor (O'Donnell *et al.*, 2005). In this class of "oncogenic" microRNAs we can place also miR-21, strongly over-expressed in glioblastoma, and recently recognized as an antiapoptotic factor (Chan, 2005). Overall, these recent findings provide solid evidence that microRNA expression profiles can be employed to develop a molecular taxonomy of cancer.

MicroRNAs in brain

To date, several microRNAs have been identified whose expression is restricted, or even specific, to one tissue, organ, or developmental stage (Wienholds and Plasterk, 2005). Interestingly, approximately one half of specifically expressed microRNAs are brain-specific, or at least significantly enriched in that organ. However, the great complexity of both developing and mature brain accounts for the variable and variegated microRNA expression patterns found in distinct areas, cell types, and developmental stages of vertebrate central nervous system. For example, two microRNAs, miR-124 and miR-128, were found to be mostly active in neurons, while miR-23, miR-26 and miR-29 in astrocytes. Others, like miR-9 and miR-125, are more evenly distributed throughout the brain (Smirnova *et al.*, 2005). It is possible that diverse microRNA populations in astrocytic and neural cells may induce the divergence of the specific protein sets in the two lineages. Of note, the specific expression of subsets of microRNAs in the brain over time is generally conserved across species, and particularly in mammals, where temporal waves of microRNA expression were observed during development (Dostie *et al.*, 2003; Krichevsky *et al.*, 2003; Miska *et al.*, 2004; Rogelj and Giese, 2004; Sempere *et al.*, 2004), implicating microRNAs as a co-ordinated family of potent regulators of gene expression for the establishment of cell identity in the brain (Klein *et al.*, 2005).

Methodology

RNA extraction from glioblastoma samples and cell lines

Tissue samples were obtained after informed consent from adult patients diagnosed with glioblastoma *de novo*, one astrocytoma grade II, and one oligoastrocytoma grade III, freshly resected during surgery and immediately frozen in liquid nitrogen for subsequent total RNA extraction. Human cell lines used in this study included 10 glioblastoma cell lines, DBTRG-05MG, U118, U87, A172, LN18, M059J, M059K, LN229, T98G, U138MG; a neuroblastoma cell line, SHSY5Y, and HeLa carcinoma line. RNA was isolated from tissues and cell lines using Trizol reagent (Invitrogen, Carlsbad, CA) according to the manufacturer's instructions.

Microarray analysis of microRNA expression

The microarray experiments were performed as in Liu *et al.* (2004). Briefly, 5 μg of total RNA was biotin-labeled and used for hybridization on each miRNA microarray chip containing 368 probes in triplicate, corresponding to 245 human and mouse miRNA genes. All probes on these microarrays were 40-mer oligonucleotides spotted by contacting technologies and covalently attached to a polymeric matrix. Raw data were normalized and analyzed in GENESPRING software version 7.2 (Silicon Genetics, Redwood City, CA), as described in detail in Ciafrè *et al.* (2005). Statistical comparisons were performed by using the ANOVA (Analysis of Variance) tool together with the Benjamini and Hochberg correction for false positives reduction. Clustering analysis was made by using Pearson correlation as a measure of similarity.

Results and discussion

Microarray analysis of the microRNA signature of glioblastoma tissue samples

In our study, we performed a genome-wide expression profiling by using a microRNA microarray in glioblastoma patients' samples, in search of an association between the expression of specific microRNAs and the glioblastoma state. By this research, which is, by definition of the microarray technique itself, a wide range one, we meant to take the first step towards the identification of possible new molecular markers of glioblastoma, to "pave the way" to further analysis by which it will be theoretically possible to cluster glioblastoma patients on the basis of their specific molecular aberrations, and consequently propose targeted molecular therapy approaches.

An important peculiarity of the glioblastoma samples we tested in our work resides in the availability, for each tumor sample, of a specific, "internal" control: for each patient we resected a central tumor area (C), surgically and histopathologically recognized as frankly tumoral, and one or more samples from a peripheral glial area (P), at an average distance of 2 cm from the border of the enhanced tumor, which did not show any evidence of tumor presence, by macroscopical surgeon's evaluation. We believe that the possibility of collecting

Table 26.1. A selection of the most significantly modulated microRNAs in glioblastoma patients' samples

		C/P mean (min–max)	Chromosome location
Up-regulated			
miR10b	5/9	2.78 (0.86–10.36)	2q31
miR130a	5/9	1.82 (0.62–5.77)	11q12
miR221	5/9	1.50 (0.72–7.32)	Xp11.3
miR21	4/9	2.08 (0.76–10.55)	17q23.2
Down-regulated			
miR128a	5/9	0.68 (0.34–1.82)	2q21
miR181a	2/9	0.61 (0.29–2.42)	9q33.1-q34.13
miR18b	2/9	0.64 (0.31–2.63)	1q31.2-q32.1
miR181c	3/9	0.71 (0.30–2.51)	19p13.3

In the second column, the number of glioblastoma samples showing the modulation is described out of a total of nine. C/P mean represents the average ratio between tumor sample values (C: center of the tumor) and the control sample values (P: peripheral brain area from the same patient). The minimum and the maximum values are also shown in parentheses.

patient-specific "normal" controls gave us the important opportunity of critically evaluating the common individual variability, and consequently of highlighting only the truly significant variations.

We started our analysis with nine primary glioblastoma samples, from five female and four male adult patients; to allow the identification of miRNAs involved in tumor progression, the ratio between each tumor sample (C) and the corresponding normal sample (P) was calculated. We selected all miRNAs over-expressed (C/P ratio ≥ 1.5) in at least four out of nine tumor samples (C), with respect to the matching "normal" sample (P). The same analysis was performed to identify all miRNAs under-expressed (C/P ratio from 0 to 0.7) in at least four samples. We also compared all samples as two groups, all Cs versus all Ps (Welch t-test, p-value cut-off 0.05). By using these selective parameters, we were able to identify four over-represented microRNAs and only one under-expressed microRNA (Table 26.1), miR-128a. However, when we adopted less selective criteria, we were able to find three other down-regulated microRNAs, notably all members of the same "family," the miR-181 one. Many microRNAs exist, in fact, that are usually named with the same numerical name (i.e. "181"), indicating the family they belong to, but that are distinguished by a suffix (i.e. "a", "b", or "c"), indicating that they reside in distinct chromosomal locations and that they bear slight sequence differences, yet usually not affecting their target binding activity. Our observation that the expression of three members of the same miR-181 family is negatively modulated in glioblastoma patients' samples is interesting, as it suggests the existence of a possible common target mRNA, concomitantly regulated by these microRNAs, and that would be likely de-repressed in glioblastoma.

Among over-expressed miRNAs, the one whose expression reached the highest levels in tumors compared to peripheral tissues was miR-10b, which was in an excess of 0.86 to 10.36 fold in five out of nine samples. Notably, also when we compared all samples as two groups, all Cs versus all Ps, miR-10b showed an average over-expression of 2.78 fold, being the most statistically significant ($p = 0.0088$). No differences were found for any of the modulated miRNAs depending on the patients' sex.

All the microRNAs whose modulation we observed in our samples were mature ones; the microarray chip we used, in fact, is able to distinguish between the mature and precursor forms (pre-miRNA) of 76 microRNA sequences. We think that this selective accumulation or depletion of the processed, active microRNAs more likely reflects a modulation of their expression rather than more "mechanical" causes, like chromosomal rearrangements of the DNA regions where their genes reside. In favor of this hypothesis comes the observation that the genomic locations harboring the genes coding for the modulated microRNAs are not usually deleted, amplified or rearranged in glioblastoma.

We then validated and extended our microarray study by performing Northern blot analysis on three out of nine samples already analyzed (data not shown), and on ten new sample pairs. By this approach, we measured the expression of miR-10b and of miR-221, chosen among over-expressed microRNAs, and of the down-regulated miR-128a and of miR-181 microRNAs, and we confirmed the results obtained by microarray. This result is significant, especially if one takes into account the lower sensitivity of Northern blot compared to microarray, meaning that the variations previously observed are strong enough to be detected also by Northern, and are true also in a sample population different from that analyzed by microarray. Actually, in agreement with the experimental features of the techniques we employed, miR-181c expression was not detectable, as predictable from the very low level of miRNA expression revealed by our microarray. In conclusion, a few microRNAs are selectively modulated in tumor samples versus their matched normal controls.

MicroRNA expression analysis in human glioblastoma cell lines

A further confirmation of our data, showing the involvement of microRNA modulation in glioblastoma, came to us when we performed a similar set of analyses (microarray and Northern blot) on human glioblastoma cell lines, and compared them with miRNA expression profiles of two adult and two fetal normal brain samples. In this experimental group, we searched for an *in vitro* reproducible model of the glioblastoma very complex and, by definition, variable state, and we hoped to find a more homogeneous distribution of the modulated microRNAs, in comparison to patients' tumor samples. Among the expected data homogeneity, we looked for the microRNAs whose expression was clearly different in cell lines versus normal tissues, and, among those, we specifically devoted our attention to the microRNAs showing the same modulation (up- or down-) pattern as in patients' samples, in view of the definition of a glioblastoma-specific microRNA signature. MiR-221 was the most interesting over-expressed microRNA (see Table 26.2), being significantly up-regulated in 10 out of 10 cell lines (cell line/ normal brain ratio = 5.32; $p = 0.000\,656$), notably flanked in its over-expression by the frequent upregulation of miR-222 precursor form (cell line/normal brain

Table 26.2. Microarray analysis of microRNA expression in human glioblastoma cell lines

		L/B mean (min–max)	Chromosome location
Up-regulated			
miR221	10/10	5.32 (1.61–16.46)	Xp11.3
miR222-prec	10/10	2.66 (1.06–6.36)	Xp11.3
miR24-1	10/10	1.95 (1.44–3.97)	9q22.1
miR29a	10/10	1.47 (1.12–3.03)	7q32
miR21	9/10	1.55 (1.03–2.12)	17q23.2
Down-regulated			
miR181b	9/10	0.29 (0.17–0.53)	1q31.2–q32.1
miR181a	9/10	0.32 (0.16–0.60)	9q33.1–q34.13
miR128a	10/10	0.47 (0.38–0.59)	2q21
miR181c	9/10	0.55 (0.35–0.74)	19p13.3

The modulation of either active miRNA molecules or precursors (prec) is specified. In the second column, the number of glioblastoma cell line samples showing the modulation is described out of a total of 10. L/B ratio represents the average ratio between cell line samples values (L) and the control sample values (B). The minimum and the maximum values are also shown in parentheses.

ratio $= 2.66$; $p = 0.0394$), which is encoded in the same genomic cluster as miR-221. In our data we also found the confirmation of a previously published work (Chan *et al.*, 2005), as we showed that miR-21 was the only other microRNA whose over-expression was common to both patients' and cell line samples. MiR-21 was recognized as an antiapoptotic factor that is over-expressed not only in glioblastoma but also in another neoplasm of the central nervous system, neuroblastoma (Saito-Ohara *et al.*, 2003). Also under-expressed microRNAs confirmed the results already described for patients' samples, and a specific attention was drawn by miR-128a (cell line/normal brain average ratio $= 0.47$, in 10/10 cell lines), and by the miR-181 family, of which miR-181a and miR-181b represented the most statistically significant repressed molecules (cell line/normal brain ratio $= 0.32$ and 0.29, respectively) in glioblastoma cell lines. In conclusion, we observed that the modulation of a restricted group of microRNAs is consistent with that observed in patients' samples.

Expression profiling-based grouping of glioblastoma patients' and cell line samples

Microarray analysis is a very powerful and versatile research tool, as it allows us to study the "raw" numerical data stemming from the bioinformatic interpretation of the microchip hybridization and to exploit them to watch several aspects of the same experiment, and, consequently, of the same data and sample set. Thus, we employed our microarray data to cluster hierarchically all our glioblastoma samples, including both patients' and cell lines. The selected group of microRNAs that we chose to this aim included all those that were consistently modulated in both cell lines and in glioblastoma patients' samples, together with the most "notable" ones among patients' samples, such as miR-10b and miR-130a, which were the most statistically significant over-expressed in tumor specimens, but not in cell lines.

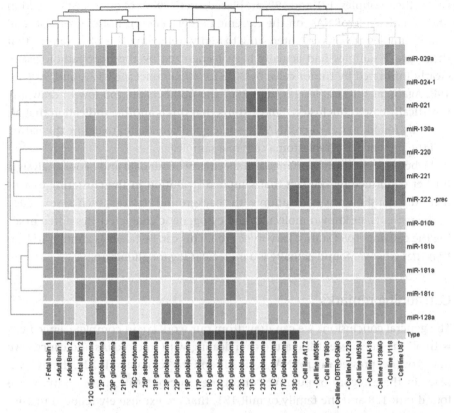

Figure 26.1. MicroRNA expression data from 11 brain tumor patients' specimens (9 glioblastomas, 1 oligoastrocytoma, 1 astrocytoma grade III), 4 normal brain (2 adult and 2 fetal) controls, and 10 human glioblastoma cell lines. Among patients' specimens, the dark blue ones represent the central area of the tumors (C samples), whereas the light blue ones are the external, control samples (P samples). Median values of 12 microRNAs differentially expressed between normal and tumoral samples were hierarchically clustered and plotted in a heat map. Red denotes the highest expression levels, while green denotes the lowest. Dendrograms indicate the correlation between groups of tissues or cells. (See color plate 18)

It was very exciting to observe the result of the clustering (Figure 26.1): it is evident that the neat separation of over-expressed from under-expressed microRNAs has as a direct consequence the non-random partitioning of the samples into at least four major groups: the cluster harboring four normal brain controls is located opposite to the branch containing all the glioblastoma cell lines, in turn clearly distinct from all other samples; immediately close to this branch were significantly grouped most of the C samples. The fourth cluster contained all P samples and two C specimens, which were the only two non-glioblastoma samples: one of these, 25C, was an astrocytoma II grade, and the other, 12C, an oligoastrocytoma III grade.

It is obviously impossible to draw any conclusion from the expression pattern of only two samples, but it would be tempting to speculate that the partly different, less aggressive nature of those tumors resulted in their different pattern of expression of the glioblastoma-specific set of miRNAs. Moreover, by looking at the central part of the sample distribution in the dendrogram, one can notice that two of the P samples, 17P and 19P, are placed, on the basis of our specific selection of distinctive microRNAs, on the extreme border between normal controls and tumor samples; it was very interesting to verify, following this observation, that those two P samples had actually been recognized by the histopathologist as clearly infiltrated by tumor cells, by 20% for what concerned 17P, and 50% for 19P. Again, as for samples 12C and 25C, the numbers are too low to draw any statistically based conclusion, but we think that this is another indication that we probably found, in our microRNA set, some good candidates to define the nature and the presence of glioblastoma cells in a sample.

We believe that the importance and specificity of our data and of our analysis is further confirmed by the fact that sample distribution in the clustering experiment did not show any correlation with "background" factors, such as sex or age of the patients, or known molecular and cytogenetic characteristics differentiating the cell lines (p53 status, karyotype etc.). This can be taken as another indication that what we found actually is a molecular tag for glioblastoma.

Concluding remarks

The general conclusions we can draw from our work are in complete agreement with all the most recent literature data dealing with microRNAs and cancer: we found over-expressed microRNAs, miR-221 and the cognate miR-222, that we will place in the category of putative "oncogene" microRNAs; on the contrary, we found miR-128 and the family of miR-181, that are extensively reduced in glioblastoma. However, regarding these under-expressed microRNAs, we do not have any solid evidence allowing us to classify them as "tumor suppressors," and we think that their reduced expression in central nervous system tumors, arising in an astroglial environment, might rather represent a marker of the loss, or lack, of differentiation already known to characterize many tumor cells, and glioblastoma in particular.

The full comprehension of the role played by our selected microRNAs in glioblastoma is necessarily linked to the identification of their target mRNAs. This ultimate goal must be pursued through sequential and strictly connected steps, involving a first theoretical, bioinformatic approach: several algorithms have now been developed for the prediction of microRNA targets, and they are becoming progressively more sophisticated as they take into consideration more and more variable parameters likely influencing the actual binding between a microRNA and its natural target. The most used and reliable prediction programs freely available on-line at the time we are writing are based on the optimization of sequence complementarity by using position-specific rules, and rely on strict requirements of interspecies conservation. They are Targetscan: http://genes.mit.edu/targetscan/; Pictar: http://pictar.bio.nyu.edu/; and MiRBase: http://microrna.sanger.ac.uk/targets/v0/. The first step that must be taken for the search of a microRNA target is the critical comparison of the lists of hypothetical mRNAs that are picked up as highly ranked by all algorithms, or at least by two of them. By "critical comparison" we mean the evaluation of many biological, rather than

```
miR222 5' AGCUACAUCUGGCUACUGGGUCUC 3'
        | | | | | | | |          | |  |
miR221 5' AGCUACAUUGUCUGCUGGGUUUC   3'
```

```
miR181c 5'AACAUUCAAC–CUGUCGGUGAGU 3'
          | | | | | | | | | | |  | | | | | | | | | | | | | |
miR181a 5'AACAUUCAACGCUGUCGGUGAGU 3'
          | | | | | | | |   | | | | | | | | | | | |  |
miR181b 5'AACAUUCAUUGCUGUCGGUGGG  3'
```

Figure 26.2. Alignment of miR-221 and miR-222 sequences (top) and of three members of the miR-181 family (bottom). The shading shows the "seed" region which is the major determinant for mRNA target recognition by microRNAs. As shown, the sequence similarity is extensive all along the miR-181 sequences, whereas it is restricted mainly to the seed nucleotides for the miR-221/222 pair.

bioinformatic, features, of the selected targets, such as the tissue specificity, the relative abundance in the different phases of the development, and their role in fundamental cellular processes, like the progression of the cell cycle, apoptosis, migration or differentiation. After these preliminary considerations, the experimental validation of the target is then performed by the usual molecular biology studies involving protein levels measurement (e.g. Western blot) and the definition of the exact sites in the mRNA target 3' UTR which are recognized by the microRNA (e.g. reporter constructs harboring the 3' UTR or parts of it downstream of a reporter gene, together with similar constructs where the candidate sites are mutagenized).

In our experimental setting, we are currently on the way to searching for miR-221 and miR-222 targets; we have started form these two microRNAs because they are the over-expressed ones, and we believe that their action could be a real "oncogenic" one. It is interesting to notice that those two microRNAs, besides being part of a cluster, and thus likely co-transcribed, are also very similar in their mature sequences (Figure 26.2). This leads us to think that their targets might be common, and that their action might be co-operative, in some cases. In fact, in the preliminary bioinformatic searches that we have already performed, the majority of the predicted targets are "hit" by both microRNAs (Table 26.3). On the opposite side, we are also looking for glioblastoma-specific repressed microRNA targets, and for three of them, the miR-181 "family," we are finding the same frame as for the miR-221/miR-222 couple: the miR-181s are even more similar one to each other than miR-221 to miR-222 (Figure 26.2), and in fact the predicted overlapping of targets is really evident, when performing the bioinformatic search (Table 26.3).

For all these cases, it will be now necessary to experimentally verify the biological meaning of the predicted targets and the consequences produced by the recognition of the same target by more than one microRNA (i.e. co-operation or competition?). In conclusion, by our work we have identified members of the microRNA group that are significantly modulated in glioblastoma, and we now want to deepen our comprehension of their specific roles by the identification of their target mRNAs.

Table 26.3. Putative targets predicted for microRNAs specifically modulated in glioblastoma

microRNA	Putative targets
Up-regulated	
miR221	Regulating synaptic membrane exocytosis 3 (RIMS3); HECT domain containing 2 (HECTD2); cyclin-dependent kinase inhibitor 1B (CDKN1B or p27 or Kip1); RNA binding motif protein 24 (RBM 24)
miR222	Regulating synaptic membrane exocytosis (RIMS3); cyclin-dependent kinase inhibitor 1B (CDKN 1B or p27 or Kip1); RNA binding motif protein 24 (RBM24)
miR21	Pleiomorphic adenoma gene 1 (PLAG1); retinitis pigmentosa 2 (RP2); activity-dependent neuroprotector (ADNP)
Down-regulated	
miR128a	Putative MAPK activating protein PM20, PM21 (DKFZp566C0424); ribosomal protein S6 kinase, 90kDa (RPS6KA5); mitochondrial glycerol 3-phosphate acyltransferase (RPAM); polo-like kinase 2 (PLK2); PHD-like zinc finger protein (PHF6)
miR181	Oxysterol-binding protein-like proten 3 (OSBPL3); hypermethylated in cancer 2 (HIC2)

Up to five putative targets are listed for each microRNA. Targets were chosen among the first 10 ranked by Pictar (http://pictar.bio.nyu.edu/), provided that they are recognized also by Targetscan (http://genes.mit.edu/targetscan/). In some cases (PLK2 as a target for miR-128a and RBM24 as a target for miR-222) the same target was identified also by a third algorithm, MiRBase (http://microrna.sanger.ac.uk/targets/v0/). Note that targets are collectively identified for miR-181a, b, and c, as all prediction programs did not significantly distinguish among them.

REFERENCES

Bottoni, A., Piccin, D., Tagliati, F. *et al.* (2005). miR-15a and miR-16-1 down-regulation in pituitary adenomas. *Journal of Cell Physiology*, **204**, 280–285.

Calin, G. A., Dumitru, C. D., Shimizu, M. *et al.* (2002). Frequent deletions and down-regulation of micro-RNA genes miR15 and miR16 at 13q14 in chronic lymphocytic leukemia. *Proceedings of the Natural Academy of Sciences USA*, **99**, 15 524–15 529.

Calin, G. A., Sevignani, C., Dumitru, C. D. *et al.* (2004a). Human microRNA genes are frequently located at fragile sites and genomic regions involved in cancers. *Proceedings of the Natural Academy of Sciences USA*, **101**, 2999–3004.

Calin, G. A., Liu, C. G., Sevignani, C. *et al.* (2004b). MicroRNA profiling reveals distinct signatures in B cell chronic lymphocytic leukemias. *Proceedings of the Natural Academy of Sciences USA*, **101**, 11 755–11 760.

Chan, J. A., Krichevsky, A. M. and Kosik, K. S. (2005). MicroRNA-21 is an antiapoptotic factor in human glioblastoma cells. *Cancer Research*, **65**, 6029–6033.

Ciafrè, S. A., Galardi, S., Mangiola, A. *et al.* (2005). Extensive modulation of a set of microRNAs in primary glioblastoma. *Biochemical and Biophysical Research Communications*, **334**, 1351–1358.

Cimmino, A., Calin, G. A., Fabbri, M. *et al.* (2005). miR-15 and miR-16 induce apoptosis by targeting BCL2. *Proceedings of the Natural Academy of Sciences USA*, **102**, 13 944–13 949.

Dostie, J., Mourelatos, Z., Yang, M., Sharma, A. and Dreyfuss, G. (2003). Numerous microRNPs in neuronal cells containing novel microRNAs. *RNA*, **9**, 180–186.

Esau, C., Kang, X., Peralta, E. *et al.* (2004). MicroRNA-143 regulates adipocyte differentiation. *Journal of Biological Chemistry*, **279**, 52 361–52 365.

Galli, R., Binda, E., Orfanelli, U. *et al.* (2004). Isolation and characterization of tumorigenic, stem-like neural precursors from human glioblastoma. *Cancer Research*, **64**, 7011–7021.

Gregory, R. I. and Shiekhattar, R. (2005). MicroRNA biogenesis and cancer. *Cancer Research*, **65**, 3509–3512.

He, L., Thomson, J. M., Hemann, M. T. *et al.* (2005). A microRNA polycistron as a potential human oncogene. *Nature*, **435**, 828–833.

Hulleman, E. and Helin, K. (2005). Molecular mechanisms in gliomagenesis. *Advances in Cancer Research*, **94**, 1–27.

Iorio, M. V., Ferracin, M., Liu, C. G. *et al.* (2005). MicroRNA gene expression deregulation in human breast cancer. *Cancer Research*, **65**, 7065–7070.

Johnson, S. M., Grosshans, H., Shingara, J. *et al.* (2005). RAS is regulated by the let-7 microRNA family. *Cell*, **120**, 635–647.

Kim, V. N. (2005). MicroRNA biogenesis: coordinated cropping and dicing. *Nature Reviews in Molecular and Cellular Biology*, **6**, 376–385.

Kleihues, P., Louis, D. N., Scheithauer, B. W. *et al.* (2002). The WHO classification of tumors of the nervous system. *Journal of Neuropathology and Experimental Neurology*, **61**, 215–225; discussion 226–229.

Klein, M. E., Impey, S. and Goodman, R. H. (2005). Role reversal: the regulation of neuronal gene expression by microRNAs. *Current Opinions in Neurobiology*, **15**, 507–513.

Kluiver, J., Poppema, S., de Jong, D. *et al.* (2005). BIC and miR-155 are highly expressed in Hodgkin, primary mediastinal and diffuse large B cell lymphomas. *Journal of Pathology*, **207**, 243–249.

Krichevsky, A. M., King, K. S., Donahue, C. P., Khrapko, K. and Kosik, K. S. (2003). A microRNA array reveals extensive regulation of microRNAs during brain development. *RNA*, **9**, 1274–1281.

Liu, C. G., Calin, G. A., Meloon, B. *et al.* (2004). An oligonucleotide microchip for genome-wide microRNA profiling in human and mouse tissues. *Proceedings of the Natural Academy of Sciences USA*, **101**, 9740–9744.

Metzler, M., Wilda, M., Busch, K., Viehmann, S. and Borkhardt, A. (2004). High expression of precursor microRNA-155/BIC RNA in children with Burkitt lymphoma. *Genes Chromosomes and Cancer*, **39**, 167–169.

Michael, M. Z., O'Connor, S. M., van Holst Pellekaan, N. G., Young, G. P. and James, R. J. (2003). Reduced accumulation of specific microRNAs in colorectal neoplasia. *Molecular Cancer Research*, **1**, 882–891.

Mischel, P. S., Shai, R., Shi, T. *et al.* (2003). Identification of molecular subtypes of glioblastoma by gene expression profiling. *Oncogene*, **22**, 2361–2373.

Miska, E. A., Alvarez-Saavedra, E., Townsend, M. *et al.* (2004). Microarray analysis of microRNA expression in the developing mammalian brain. *Genome Biology*, **5**, R68.

Nigro, J. M., Misra, A., Zhang, L. *et al.* (2005). Integrated array-comparative genomic hybridization and expression array profiles identify clinically relevant molecular subtypes of glioblastoma. *Cancer Research*, **65**, 1678–1686.

O'Donnell, K. A., Wentzel, E. A., Zeller, K. I., Dang, C. V. and Mendell, J. T. (2005). c-Myc-regulated microRNAs modulate E2F1 expression. *Nature*, **435**, 839–843.

Pasquinelli, A. E., Hunter, S. and Bracht, J. (2005). MicroRNAs: a developing story. *Current Opinions in Genetic Development*, **15**, 200–205.

Reifenberger, G. and Collins, V. P. (2004). Pathology and molecular genetics of astrocytic gliomas. *Journal of Molecular Medicine*, **82**, 656–670.

Rogelj, B. and Giese, K. P. (2004). Expression and function of brain specific small RNAs. *Reviews in Neurosciences*, **15**, 185–198.

Saito-Ohara, F., Imoto, I., Inoue, J. *et al.* (2003). PPM1D is a potential target for 17q gain in neuroblastoma. *Cancer Research*, **63**, 1876–1883.

Sanai, N., Alvarez-Buylla, A. and Berger, M. S. (2005). Neural stem cells and the origin of gliomas. *New England Journal of Medicine*, **353**, 811–822.

Sempere, L. F., Freemantle, S., Pitha-Rowe, I. *et al.* (2004). Expression profiling of mammalian microRNAs uncovers a subset of brain-expressed microRNAs with possible roles in murine and human neuronal differentiation. *Genome Biology*, **5**, R13.

Singh, S. K., Clarke, I. D., Terasaki, M. *et al.* (2003). Identification of a cancer stem cell in human brain tumors. *Cancer Research*, **63**, 5821–5828.

Singh, S. K., Clarke, I. D., Hide, T. and Dirks, P. B. (2004). Cancer stem cells in nervous system tumors. *Oncogene*, **23**, 7267–7273.

Smirnova, L., Grafe, A., Seiler, A., *et al.* (2005). Regulation of miRNA expression during neural cell specification. *European Journal of Neurosciences*, **21**, 1469–1477.

Tunici, P., Bissola, L., Lualdi, E. *et al.* (2004). Genetic alterations and *in vivo* tumorigenicity of neurospheres derived from an adult glioblastoma. *Molecular Cancer*, **3**, 25.

van den Boom, J., Wolter, M., Kuick, R. *et al.* (2003). Characterization of gene expression profiles associated with glioma progression using oligonucleotide-based microarray analysis and real-time reverse transcription-polymerase chain reaction. *American Journal of Pathology*, **163**, 1033–1043.

Wechsler-Reya, R. and Scott, M. P. (2001). The developmental biology of brain tumors. *Annual Reviews of Neurosciences*, **24**, 385–428.

Wienholds, E. and Plasterk, R. H. (2005). MicroRNA function in animal development. *Federation of the European Biochemical Sciences Letters*, **579**, 5911–5922.

Yuan, X., Curtin, J., Xiong, Y. *et al.* (2004). Isolation of cancer stem cells from adult glioblastoma multiforme. *Oncogene*, **23**, 9392–400.

Zhu, Y. and Parada, L. F. (2002). The molecular and genetic basis of neurological tumours. *Nature Reviews Cancer*, **2**, 616–626.

27 Role of microRNA pathway in Fragile X mental retardation

Keith Szulwach and Peng Jin*

Introduction

Loss of the Fragile X Mental Retardation Protein (FMRP) has been identified as the major cause of Fragile X syndrome, one of the most common forms of inherited mental retardation. FMRP's RNA binding character has implicated it in translational regulation. Recently, FMRP has also been linked to the microRNA pathway that is involved in translational suppression. Current work on Fragile X syndrome strives to determine the functional role of FMRP in translational suppression of associated mRNA targets and how components of the microRNA pathway may help to mediate this function.

Clinical phenotypes of Fragile X syndrome

Fragile X syndrome is one of the most common forms of inherited mental retardation with an estimated prevalence of about 1 in 4000 males and 1 in 8000 females. The syndrome is transmitted as an X-linked dominant trait with reduced penetrance (80% in males and 30% in females). The clinical presentations of Fragile X syndrome include mild to severe mental retardation, with IQ between 20 and 70, mildly abnormal facial features of a prominent jaw and large ears, mainly in males, and macroorchidism in post-pubescent males (Crawford *et al.*, 2001; Terracciano *et al.*, 2005). Many patients also display subtle connective tissue abnormalities. Behaviorally, affected males tend to exhibit hyperactivity, social anxiety, preservative speech and language, tactile defensiveness, and hand biting (Crawford *et al.*, 2001; Terracciano *et al.*, 2005). In addition, there seems to be an association between Fragile X syndrome and autism, as up to 25% of male patients display autistic behavior (Crawford *et al.*, 2001).

---

* Author to whom correspondence should be addressed.

MicroRNAs: From Basic Science to Disease Biology, ed. Krishnarao Appasani. Published by Cambridge University Press. © Cambridge University Press 2008.

Molecular basis of Fragile X syndrome

Fragile X syndrome is mainly caused by an unstable trinucleotide repeat in the 5' untranslated region (5'-UTR) of the FMR1 gene. The 5'-UTR of FMR1 contains a highly polymorphic CGG trinucleotide repeat that can be characterized into four different categories: common alleles (6–40 repeats), intermediate alleles (41–55 repeats), premutation alleles (56–200 repeats), and full mutation alleles (>200 repeats). Common and intermediate alleles, often punctuated by AGG interruptions, are usually transmitted from parent to offspring in a stable manner. However, premutation alleles are unstable and could expand into full mutation or different-size premutation alleles during germline transmission (Crawford et al., 2001). The risk of expansion of a premutation to a full mutation in the next generation exponentially increases with the number of repeats, a phenomenon known as "genetic anticipation" or "Sherman Paradox" (Crawford et al., 2001). Interestingly, premutation alleles expand to full mutations only when transmitted by a female carrier. This is because of the fact that full mutation males have only premutation-size alleles in their sperm. In full mutations, the CGG repeat is massively expanded and hypermethylated, which results in transcriptional repression of the FMR1 gene and leads to the loss of FMRP. Identification of other mutations in the FMR1 gene, such as deletions and point mutations among patients with usual phenotype, but without fragile site expression, firmly established the FMR1 gene as the only gene involved in the pathogenesis of Fragile X syndrome. Thus, the loss of functional FMRP is the cause of Fragile X syndrome.

Biological functions of FMRP

FMRP belongs to a small family of highly conserved proteins referred to as the Fragile X-related proteins (Siomi et al., 1996; Zhang et al., 1995). Proteins of this family are characterized by the presence of two ribonucleoprotein K homology domains (KH domains), a cluster of arginine and glycine residues (RGG box), and both nuclear localization and nuclear export signals (Figure 27.1) (Feng et al., 1997b). The KH domain and RGG box are common among RNA-binding proteins. Indeed, FMRP has been found to bind RNA homopolymers and mRNAs in vitro, indicating a potential role for FMRP in the regulation of RNA metabolism (Ashley et al., 1993). Association of FMRP with polyribosomes in an RNA-dependant manner via messenger ribonucleoprotein (mRNP) particles has in fact been demonstrated, while both in vitro and in vivo studies have established a role for FMRP in translational regulation (Ashley et al., 1993; Feng et al., 1997a; Laggerbauer et al., 2001; Li et al., 2001). As a result, much effort has been made to identify mRNA targets of FMRP so as to identify the genes translationally misregulated in the absence of FMRP, which may themselves cause the phenotypes associated with Fragile X syndrome. Notably, a portion of such mRNA targets were found in studies aimed at identifying molecular requirements for FMRP–RNA interactions. Two groups independently found that binding of FMRP

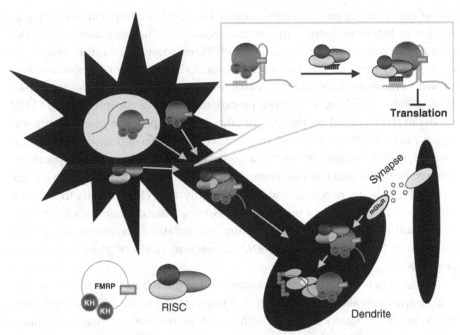

Figure 27.1. A model for FMRP-mediated translational suppression of mRNA targets via the miRNA pathway in neuron. FMRP localized to the nucleus associates with an mRNA target, possibly containing a G-quartet, through its RGG box. The mRNP complex is then exported to the cytoplasm where it is able to interact with RISC complex. Suppression of a specific mRNA target is then maintained by stabilization of miRNA:mRNA duplex by FMRP-KH2. Stabilization of this duplex leads to polyribosomal stalling and suppresses translation of the mRNA target. The RISC-containing mRNP is then transported into the dendrites where the mRNA target remains translationally suppressed. Glutamate signaling at the synapse then activates mGluR and a downstream de-phosphorylation signal for FMRP allowing for the relief of translational suppression. (See color plate 19)

to RNA requires a particular nucleic acid structure, known as a G-quartet, as well as specific sequences surrounding this structure (Figure 27.1) (Darnell *et al.*, 2001; Schaeffer *et al.*, 2001). Furthermore, mutation analysis identified the RGG box, but not KH domain, as the portion of FMRP responsible for binding of G-quartet RNA (Darnell *et al.*, 2004; Ramos, *et al.*, 2003). However, a role for RNA binding by the KH2 domain of FMRP has been noted as an important characteristic allowing for proper translational suppression of mRNA targets. Identification of a severely affected patient with a missense mutation (I304N) in the second KH domain of FMRP pointed towards the significance of this domain to FMRP function. The I304 N missense mutation was found to be located in a region of the KH2 domain important for protein-RNA interaction, and subsequent *in vitro* RNA selection experiments revealed consensus motifs that may contribute to the binding of FMRP-KH2 to mRNA targets. Structural analysis of *in vitro* selected RNA indicated the presence of a loop–loop–pseudoknot or "kissing complex." Interestingly, these structured RNAs were capable of competing FMRP off of polyribosomes entirely, while high-affinity G-quartet RNAs could not, indicating that binding of "kissing complex" RNA to the KH2 domain of FMRP allows for proper translational suppression (Darnell *et al.*, 2005).

Post-translational modification of FMRP has also been implicated in the regulation of FMRP-mediated translational regulation. Phosphorylation of FMRP on a highly conserved serine residue (Ser499) was found to result in an association of FMRP with stalled ribosomes. Phosphorylation of Ser499, however, was shown to have no effect on FMRP's ability to bind RNA homopolymer (Ceman *et al.*, 2003). Therefore, it has been proposed that while phosphorylated FMRP results in translational suppression of bound transcripts, a de-phosphorylation signal causes FMRP to dissociate from stalled ribosomes and allows the expression of mRNA targets. Recently, a second post-translational modification, methylation, has also been observed on FMRP. Arginines within the RGG box of FMRP can be methylated, and *in vitro* methylated FMRP results in reduced binding of the protein to a minimal sc1 RNA sequence containing a G-quartet. This result indicates that FMRP arginine methylation may be an additional mechanism by which the FMRP–RNA interaction could be regulated (Stetler *et al.*, 2006).

Normally, expression of FMRP is widespread, although not necessarily ubiquitous, being most abundant in brain and testis (Devys *et al.*, 1993). In neurons, FMRP has been found localized within and at the base of dendritic spines in association with polyribosomes (Feng *et al.*, 1997b). Dendritic spines in Fragile X patients and *Fmr-1* knockout mice are denser apically, elongated, thin, and tortuous (Terracciano *et al.*, 2005). This association is RNA-dependent as well as microtubule dependent indicating a role for FMRP in mRNA trafficking and dendritic development (Figure 27.1). Support for a role of FMRP in neural dysgenesis has been provided in studies identifying MAP1B, a key component of microtubule stability, as an mRNA target of FMRP. Lack of FMRP in mice was found to cause elevated levels of MAP1B protein and increased microtubule stability in neurons. Thus, loss of FMRP results in altered microtubule dynamics that affect neural development and indicates a potential role for FMRP in synaptic plasticity (Lu *et al.*, 2004; Zhang *et al.*, 2001).

A role for FMRP in synaptic plasticity, particularly long-term depression (LTD), seems logical since LTD is a protein synthesis dependent phenomenon. Indeed, such a role for FMRP has been established. Activation of the metabotropic glutamate receptor 5 (mGluR5) by DHPG agonist stimulates LTD under normal conditions. LTD activated in this manner requires new protein synthesis but does not require transcription (Huber *et al.*, 2000). This data suggests that mGluR5 stimulation allows the removal of translational repression of transcribed and localized mRNA messages necessary for LTD. Fitting with such a model is the observation that mGluR dependent LTD is exaggerated in an *Fmr1*-knockout mouse (Huber *et al.*, 2002). It has been suggested that the presence of FMRP represses translation of proteins involved in LTD, and that stimulation of mGluR5 results in the localized translation of these transcripts. Currently, the effect of mGluR5 antagonists on the balance between FMRP-mediated translational repression and mGluR5 activated protein synthesis is seen as a potential target for new therapy treating Fragile X syndrome.

Role of microRNA pathway in FMRP-mediated translational control

Initial biochemical studies in *Drosophila* to identify protein components of *dFmrp*-containing complexes and RISC led to the identification of *dFmrp* as a component of RISC. Further analysis revealed specific interactions between *dFmrp* and two functional RISC proteins, *dAGO2* and *Dicer* (Caudy *et al.*, 2002; Ishizuka *et al.*, 2002). Although *dAGO2* is generally associated with siRNA-mediated gene silencing in *Drosophila*, the loss of *dFmr1* does not seem to affect RNAi (Caudy *et al.*, 2002; Ishizuka *et al.*, 2002; Okamura *et al.*, 2004). In addition, the endogenous miRNAs have been found associated with FMRP in both flies and mammals (Caudy *et al.*, 2002; Ishizuka *et al.*, 2002; Jin *et al.*, 2004). Therefore, it has been proposed that FMRP-mediated translational suppression occurs via the microRNA pathway and involves miRNA. This view is further supported by the association of FMRP with a mammalian Argonaute, EIF2C2, which is itself a component of miRNA-containing mRNP complexes. In adult mice brains, Dicer and eIF2c2 (the mouse homologue of AGO1) have also been observed to interact with FMRP at post-synaptic densities (Jin *et al.*, 2004; Lugli *et al.*, 2005).

Importantly, the genetic interactions between *dFmr1* and the components of miRNA pathway have been demonstrated in *Drosophila*. *dAGO1* was shown to dominantly interact with *dFmr1* in *dFmr1* loss-of-function and overexpression models. Overexpression of *dFmr1* leads to a mildly rough eye phenotype that is the result of increased neuronal cell death. By introducing a recessive lethal allele of *AGO1*, containing a P-element insertion that reduces its expression level, suppression of the mild rough eye phenotype could be observed (Figure 27.2). The loss-of-function model revealed that *AGO1* was required for *dFmr1* regulation of synaptic plasticity. In the absence of *dFmr1*, pronounced synaptic overgrowth can be observed in the neuromuscular junction (NMJ) of *Drosophila* larvae, similar to the dendritic overgrowth seen on the brains of *Fmr1* knock-out mice and human patients. It was shown that while both *dFmr1* and *dAGO1* heterozygotes had normal NMJs, a transheterozygote displayed strong synaptic overgrowth and over-elaboration of synaptic terminals (Jin *et al.*, 2004). This result suggests that a limiting factor for function of *dFmr1* at synapses is the function of *AGO1*. Rather significantly, this directly implicates the potential role of miRNA pathway in human disease since *dAGO1* modulates translational suppression mediated by *dFmrp*. The genetic interaction between *dFmr1* and *dAGO2* has also been demonstrated in experiments concerning the gene *pickpocket1* (*ppk1*), which was found to control rhythmic locomotion in flies. *ppk1* was found to be an mRNA target of dfmr1 protein and the expression of PPK1 seems to be regulated by both *dFmr1* and *dAGO2* (Xu *et al.*, 2004).

Experiments concerning the role of FMRP-KH2 in binding "kissing complex" RNA for FMRP mediated translational suppression may hint at a mechanism by which FMRP cooperates with the miRNA pathway. The *in vitro* RNA selection experiments were able to identify "kissing complex" as a structure that specifically interacts with the KH2 domain of FMRP and prevents association of FMRP with polyribosomes (Darnell *et al.*, 2005). However, it is unclear whether the KH2

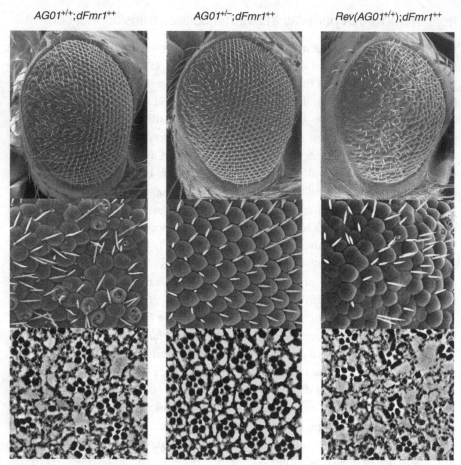

$AGO1^{+/+};dFmr1^{++}$ $AGO1^{+/-};dFmr1^{++}$ $Rev(AGO1^{+/+});dFmr1^{++}$

Figure 27.2. Genetic interaction between FMRP and miRNA pathway. Overexpression of *dfmr1* by a *Sevenless* promoter ($AGO1^{+/+}$; $dFmr1^{++}$) results in a mildly rough eye phenotype owing to neuronal apoptosis. Introduction of a recessive lethal allele of *AGO1* ($AGO1^{+/-}$; $dFmr1^{++}$), containing a P-element insertion that reduces *AGO1* expression, results in suppression of the mildly rough eye phenotype. An *AGO1* revertant ($Rev(AGO1^{+/+})$; $dFmr1^{++}$), generated by precise excision of the P-element insertion, is unable to suppress the mildly rough eye phenotype.

domain of FMRP binds to intermolecular or intramolecular "kissing complexes." Since the *in vitro* RNA selection experiments were performed with random RNA libraries it remains possible that the "kissing complex" identified represents an intermolecular interaction whereby FMRP-KH2 is stabilizing a duplex between a miRNA and an mRNA target which would, therefore, suppress translation of that target (Darnell *et al.*, 2005). Such an explanation would help account for the role of FMRP in miRNA mediated translation suppression but has yet to be validated experimentally.

Small RNAs in FMR1 gene silencing?

The role of non-coding portions of the FMR1 gene, particularly the 5′-UTR, have been proposed to be involved in RNAi-directed methylation of expanded CGG

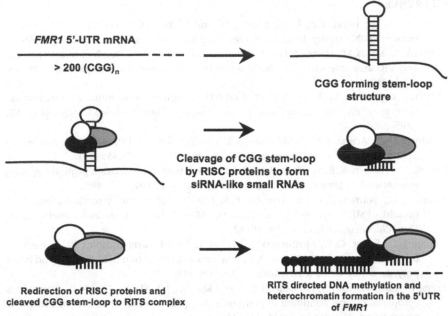

FMR1 5'-UTR mRNA

> 200 (CGG)n

CGG forming stem-loop structure

Cleavage of CGG stem-loop by RISC proteins to form siRNA-like small RNAs

Redirection of RISC proteins and cleaved CGG stem-loop to RITS complex

RITS directed DNA methylation and heterochromatin formation in the 5'UTR of FMR1

Figure 27.3. Potential mechanism for RNAi directed transcriptional silencing of *FMR1*. Transcription of the expanded CGG repeats results in the formation of stable hairpins in the *FMR1* transcript. The stable hairpins are then recognized and cleaved by Dicer to form siRNA-like small RNAs. The small RNAs are then loaded into a RITS complex and are redirected back to the 5'-UTR of *FMR1*. The RITS complex then initiates *de novo* DNA methylation of CpGs and formation of heterochromatin. The *FMR1* gene is then transcriptionally silenced and FMRP is lost.

repeats. An *in vitro* transcribed RNA mimicking the 5'-UTR of a premutation FMR1 allele was found to form stable hairpins, especially in regions of uninterrupted CGG repeats. Moreover, the structured RNA was cleavable by Dicer, producing ~20 nt RNAs (Handa *et al.*, 2003). It is possible that processing of the CGG repeat by Dicer allows for small RNAs derived from CGG repeats to be loaded into a RITS (RNA-induced initiator of transcriptional gene silencing) complex that is guided by these siRNA-like RNAs to homologous DNA sequences. The RITS complex may then initiate chromatin modification and DNA methylation of the FMR1 gene leading to transcriptional silencing (Figure 27.3).

Summary

Despite the accumulation of data on the genetic, biochemical, and molecular interactions between FMRP and the components of the miRNA pathway, the molecular mechanism by which known participants of the RNAi pathway are able to cooperate with FMRP to suppress translation of specific mRNA targets important for synaptic plasticity remains to be determined. Understanding the molecular basis of Fragile X syndrome, the first human disease linked to microRNA pathway, will provide mechanistic insight into not only this disorder but also the microRNA pathway itself as well as learning and memory in general.

REFERENCES

Ashley, C. T., Jr., Wilkinson, K. D., Reines, D. and Warren, S. T. (1993). FMR1 protein: conserved RNP family domains and selective RNA binding. *Science*, **262**, 563–566.

Caudy, A. A., Myers, M., Hannon, G. J. and Hammond, S. M. (2002). Fragile X-related protein and VIG associate with the RNA interference machinery. *Genes & Development*, **16**, 2491–2496.

Ceman, S., O'Donnell, W. T., Reed, M. *et al.* (2003). Phosphorylation influences the translation state of FMRP-associated polyribosomes. *Human Molecular Genetics*, **12**, 3295–3305.

Crawford, D. C., Acuna, J. M. & Sherman, S. L. (2001). FMR1 and the fragile X syndrome: human genome epidemiology review. *Genetics in Medicine*, **3**, 359–371.

Darnell, J. C., Jensen, K. B., Jin, P. *et al.* (2001). Fragile X mental retardation protein targets G quartet mRNAs important for neuronal function. *Cell*, **107**, 489–499.

Darnell, J. C., Warren, S. T. and Darnell, R. B. (2004). The fragile X mental retardation protein, FMRP, recognizes G-quartets. *Mental Retardation and Developmental Disabilities Research Reviews*, **10**, 49–52.

Darnell, J. C., Fraser, C. E., Mostovetsky, O. *et al.* (2005). Kissing complex RNAs mediate interaction between the Fragile-X mental retardation protein KH2 domain and brain polyribosomes. *Genes & Development*, **19**, 903–918.

Devys, D., Lutz, Y., Rouyer, N., Bellocq, J. P. and Mandel, J. L. (1993). The FMR-1 protein is cytoplasmic, most abundant in neurons and appears normal in carriers of a fragile X premutation. *Nature Genetics*, **4**, 335–340.

Feng, Y., Absher, D., Eberhart, D. E. *et al.* (1997a). FMRP associates with polyribosomes as an mRNP, and the I304N mutation of severe fragile X syndrome abolishes this association. *Molecular Cell*, **1**, 109–118.

Feng, Y., Gutekunst, C. A., Eberhart, D. E. *et al.* (1997b). Fragile X mental retardation protein: nucleocytoplasmic shuttling and association with somatodendritic ribosomes. *Journal of Neuroscience*, **17**, 1539–1547.

Handa, V., Saha, T. and Usdin, K. (2003). The fragile X syndrome repeats form RNA hairpins that do not activate the interferon-inducible protein kinase, PKR, but are cut by Dicer. *Nucleic Acids Research*, **31**, 6243–6248.

Huber, K. M., Kayser, M. S. and Bear, M. F. (2000). Role for rapid dendritic protein synthesis in hippocampal mGluR-dependent long-term depression. *Science*, **288**, 1254–1257.

Huber, K. M., Gallagher, S. M., Warren, S. T. and Bear, M. F. (2002). Altered synaptic plasticity in a mouse model of fragile X mental retardation. *Proceedings of the National Academy of Sciences USA*, **99**, 7746–7750.

Ishizuka, A., Siomi, M. C. and Siomi, H. (2002). A Drosophila fragile X protein interacts with components of RNAi and ribosomal proteins. *Genes & Development*, **16**, 2497–2508.

Jin, P., Zarnescu, D. C., Ceman, S. *et al.* (2004). Biochemical and genetic interaction between the fragile X mental retardation protein and the microRNA pathway. *Nature Neuroscience*, **7**, 113–117.

Laggerbauer, B., Ostareck, D., Keidel, E. M., Ostareck-Lederer, A. and Fischer, U. (2001). Evidence that fragile X mental retardation protein is a negative regulator of translation. *Human Molecular Genetics*, **10**, 329–338.

Li, Z., Zhang, Y., Ku, L. *et al.* (2001). The fragile X mental retardation protein inhibits translation via interacting with mRNA. *Nucleic Acids Research*, **29**, 2276–2283.

Lu, R., Wang, H., Liang, Z. *et al.* (2004). The fragile X protein controls microtubule-associated protein 1B translation and microtubule stability in brain neuron development. *Proceedings of the National Academy of Sciences USA*, **101**, 15 201–15 206.

Lugli, G., Larson, J., Martone, M. E., Jones, Y. and Smalheiser, N. R. (2005). Dicer and eIF2c are enriched at postsynaptic densities in adult mouse brain and are modified by neuronal activity in a calpain-dependent manner. *Journal of Neurochemistry*, **94**, 896–905.

Okamura, K., Ishizuka, A., Siomi, H. and Siomi, M. C. (2004). Distinct roles for Argonaute proteins in small RNA-directed RNA cleavage pathways. *Genes & Development*, **18**, 1655–1666.

Ramos, A., Hollingworth, D. and Pastore, A. (2003). G-quartet-dependent recognition between the FMRP RGG box and RNA. *RNA*, **9**, 1198–1207.

Schaeffer, C., Bardoni, B., Mandel, J. L. *et al.* (2001). The fragile X mental retardation protein binds specifically to its mRNA via a purine quartet motif. *European Molecular Biology Organization Journal*, **20**, 4803–4813.

Siomi, M. C., Zhang, Y., Siomi, H. & Dreyfuss, G. (1996). Specific sequences in the fragile X syndrome protein FMR1 and the FXR proteins mediate their binding to 60 S ribosomal subunits and the interactions among them. *Molecular and Cellular Biology*, **16**, 3825–3832.

Stetler, A., Winograd, C., Sayegh, J. *et al.* (2006). Identification and characterization of the methyl arginines in the fragile X mental retardation protein Fmrp. *Human Molecular Genetics*, **15**, 87–96.

Terracciano, A., Chiurazzi, P. & Neri, G. (2005). Fragile X syndrome. *American Journal of Medical Genetics. Part C, Seminars in Medical Genetics*, **137**, 32–37.

Xu, K., Bogert, B. A., Li, W. *et al.* (2004). The fragile X-related gene affects the crawling behavior of *Drosophila* larvae by regulating the mRNA level of the DEG/ENaC protein pickpocket1. *Current Biology*, **14**, 1025–1034.

Zhang, Y., O'Connor, J. P., Siomi, M. C. *et al.* (1995). The fragile X mental retardation syndrome protein interacts with novel homologs FXR1 and FXR2. *European Molecular Biology Organization Journal*, **14**, 5358–5366.

Zhang, Y. Q., Bailey, A. M., Matthies, H. J. *et al.* (2001). Drosophila fragile X-related gene regulates the MAP1B homolog Futsch to control synaptic structure and function. *Cell*, **107**, 591–603.

28 Insertion of *miRNA125b-1* into immunoglobulin heavy chain gene locus mediated by *V(D)J* recombination in precursor B cell acute lymphoblastic leukemia

Takashi Sonoki and Norio Asou*

Introduction

Acute lymphoblastic leukemia (ALL) is a malignant tumor of progenitor cell committed to an immunophenotypically and genetically differential pathway of lymphoid lineage. The leukemic cells lose the property of further differentiation and gain the ability of autonomic proliferation, allowing clonal expansion of the malignant cells in the bone marrow. The bone marrow, which is replaced by enormous leukemic cells, no longer produces sufficient normal blood cells, so that individuals suffering from this disease show a reduced number of red blood cells, platelets and normal white blood cells. In addition to expansion in the bone marrow, the leukemic cell can infiltrate other organs such as lymph nodes, spleen, liver, central nervous system, and gonads; and sometimes develops solid tumor called lymphoblastic lymphoma (Head, 2004; Berg *et al.*, 2005; Pui, 2006).

The lymphoid cells express cell surface antigens sequentially during the differential pathway (Figure 28.1). Since ALL cells share many of the features of normal lymphoid progenitor cells, the vast majority of ALL cells can be broadly classified by their cellular origins using cytochemical and immunophenotypic analyses into precursor B-cell ALL, B-cell ALL and precursor T-cell ALL (Figure 28.1). The most common ALL is the precursor B-ALL, the second T-ALL, and the third B-ALL, which account for ~75%, ~20%, and ~5% of all ALL patients, respectively. The precursor B-cell ALL is defined by cell surface expression of B-lineage associated antigens including CD19, HLA-DR and CD10. Expression of terminal deoxynucleotide transferase (TdT), which is a cellular enzyme that catalyzes addition of non-germline encoded nucleotide (N-region) to the site of gene segment joining, is a useful marker of precursor B-ALL. The B-cell ALL expresses surface immunoglobulin and B-lineage associated antigens lacking TdT expression. The leukemic cells of this phenotype represent a unique morphology, with deeply basophilic

---

* Author to whom correspondence should be addressed.

MicroRNAs: From Basic Science to Disease Biology, ed. Krishnarao Appasani. Published by Cambridge University Press. © Cambridge University Press 2008.

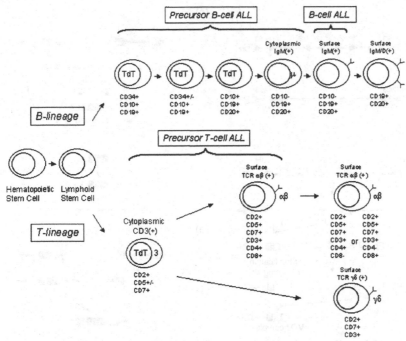

Figure 28.1. Early differential pathway of lymphoid cell and origins of ALL cells. Lymphoid cell committed to B-lineage differentiates in bone marrow, whereas T-lineage differentiates in thymus. TdT: terminal deoxynucleotidyl transferase.

cytoplasm bearing numerous vacuoles, corresponding to L3 or Burkitt type (French–American–British classification) (Bennett *et al.*, 1976). T-cell ALL is defined by expression of the T-cell associated antigens including cytoplasmic CD3, with CD7 plus CD2 or CD5, of which cytoplasmic CD3 is diagnostic.

Lymphocytes naturally undergo rearrangement of noncontiguous DNA segments of immunoglobulin (*IG*) or T-cell receptor (*TCR*) genes to obtain diversity of their antigen repertoire (Figure 28.2a, b). The primary mechanism responsible for generating diversity is an assembly of variable (*V*), diversity (*D*), and joining (*J*) segments by *V(D)J* recombination that occurs in early stages of lymphocyte development (Okada and Alt, 1995). Every single lymphocyte has a unique *IG/TCR* gene rearrangement; thus, detection of clonal *IG/TCR* gene rearrangement by Southern blot analysis is a useful genetic marker of lymphoid malignancies, especially in ALL cells of unknown cellular origin immunophenotypically (Asou *et al.*, 1991).

The *V(D)J* recombination involves the recognition of conserved recombination signal sequence (RSS), double-strand DNA cleavage, and re-ligation of the two ends of the cleaved double strand DNA (Figure 28.2b) (Okada and Alt, 1995). The consensus RSS is composed of a heptamer (CACAGTG) and a nonamer (ACAAAAACC) separated by either 12 +/− 1 or 23 +/− 1 bp called "spacer" sequences (Figure 28.2a). Prior to the re-ligation, the boundary sites are further modified, resulting in increase of diversity of the antigen repertoires. There are several different modification mechanisms including nucleotide deletion and

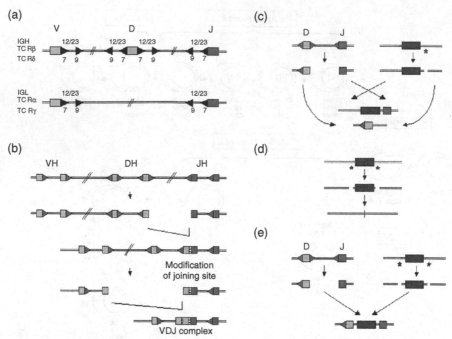

Figure 28.2. (a) Schematic structure of immunoglobulin (*IG*) and T-cell receptor (*TCR*) genes encoding *V*-region. There are diverse (*D*) segments in immunoglobulin heavy chain (*IGH*), *TCRβ*, and *TCRδ* genes but not in immunoglobulin light chain (*IGL*), *TCRα* and *TCRγ* genes. Each segment is flanked by conserved recombination signal sequences (RSSs). The RSSs consist of a heptamer and nonamer sequence separated by 12 bp or 23 bp of "spacer" sequence. Boxes and triangles indicate coding and heptamer/nonamer sequences, respectively. (b) Overview of *V(D)J* recombination events in *IGH* gene. The rearrangement of *D* to *J* takes place followed by a *V* to *D-J* complex. The joining sites are often modified by addition of nucleotides or deletion. RSSs are shown as triangles. (c) Chromosome translocation involving *IG/TCR* and/or non-*IG/TCR* loci mediated by errors of *V(D)J* recombination. Cryptic RSSs (shown by asterisks) might be found at positions close to the double-stranded DNA breakage on the non-*IG/TCR* loci. (d) Deletion can be mediated by *V(D)J* recombination machinery recognizing cryptic RSSs on non-*IG/TCR* loci. (e) Insertion of a non-*IG/TCR* gene generated by *V(D)J* recombination into an *IGH* locus has been previously reported. Asterisks represent cryptic RSSs on non-*IG/TCR* loci. (See color plate 20)

N-region addition. The N-region addition is catalyzed by TdT without template DNAs and often appears as GC rich sequences.

Errors in *V(D)J* recombination have been proposed to be implicated in genetic alterations such as chromosome translocations, deletion and insertion involving *IG* or *TCR* gene loci (Kuppers and Dalla-Favera, 2001). Chromosome translocation brings cellular oncogenes close to *IG/TCR* genes (Figure 28.2c). Cryptic RSSs are, although not in all cases, found at flanking region of the double-strand DNA breakage on the non-*IG/TCR* loci. The molecular consequence of the transloca-tion is a deregulated expression of the cellular oncogenes controlled by strong transcription enhancers within *IG/TCR* loci. Deletion of non-*IG/TCR* loci can be also mediated by *V(D)J* recombination (Figure 28.2d). In this case, cryptic RSSs are often identified near the affected genes. Finally, insertion of *BCL2* into an *IGH* locus resulting from illegitimate *V(D)J* recombination has been previously found in lymphoma patients with abundant BCL2 protein expression (Figure 28.2e) (Vaandrager *et al.*, 2000). RSS-like sequences were identified around the *BCL2* breakpoints, suggesting that the excision of *BCL2* was also mediated by *V(D)J* recombination machinery.

Here, we describe an insertion of *miRNA125b-1* into an *IGH* identified by molecular cloning of a rearranged *IGHJ* allele from precursor B-ALL cells (Sonoki *et al.*, 2005a). A detailed sequence analysis indicated that the insertion was partially mediated by *V(D)J* recombination. Our result suggests that the *miR125b-1* has a role in early B-cell development and in leukemogenesis.

Case report and results

A 35-year-old woman presented with lower abdominal pain. Seven years ago, she was diagnosed to have precursor B-cell ALL positive for CD19, CD20 and TdT expression. She had received bone marrow transplantation from her HLA-matched sister after achieving a complete remission with chemotherapy. Several imaging studies showed bilateral ovarian tumors that were implicated in her abdominal pain. The ovarian tumor was surgically resected and was histologically consistent with lymphoblastic lymphoma and immunophenotypically positive for TdT, CD10 and CD19, the same immunophenotype of the initial ALL cells seven years ago (Sonoki *et al.*, 2005b).

Southern blot analyses of the ovarian tumor cells demonstrated a single rearranged joining segment of the *IGH* gene (*IGHJ*) with a faint germ-line band (Figure 28.3a). This result indicated that one *IGHJ* allele underwent rearrangement and another deletion. The germ-line band was weakly detected, indicating that the ovarian tumor appeared to be composed of almost pure ALL cells. A 4.2 kb-rearranged band detected by Southern blot after *Pst* I digest was amplified as a 1.8 kb product by long distance inverse-PCR (LDI-PCR) and sequenced (Figure 28.3a) (Sonoki *et al.*, 2005a).

Sequence analysis using the Blast search program (http://www.ncbi.nlm.nih.gov/BLAST) revealed that the rearranged *IGH* allele was composed of a *V* (*VH4-33*) segment, a 241 bp DNA fragment derived from chromosome 11q24, a 18bp nucleotide of *D* (*DH2-21*) segment, and a *J* (*JH5*) segment, from the 5′ to 3′ direction (Figure 28.3b). Interestingly, the 11q24-derived sequence contained both precursor (88 bp) and mature (22 bp) sequences of *miR 125b-1*, in the same orientation of *IGH* sequences, indicating that the *IGH* rearrangement resulted in "in frame" insertion into the rearranged *IGH* allele. Because of the "in frame" insertion, the *miR125b-1* can be transcribed with the *IGH* segments; however, we have not yet checked the expression owing to limitation of the material.

To address whether the *V(D)J* recombination machinery is implicated in this insertion event, we compared the *VH4-33/-miR125b-1/DH2-21/JH5*, germinal *VH4-33*, germinal *DH2-21* and germinal *miR125b-1* sequences. The *VH4-33* was truncated just upstream of a conserved RSS and five GC-rich nucleotides (CCCGG), a putative N-region, were added at the *VH4-33/miR125b-1* junction (Figure 28.4a). The *DH2-21* lacked several coding sequences at both junctions,

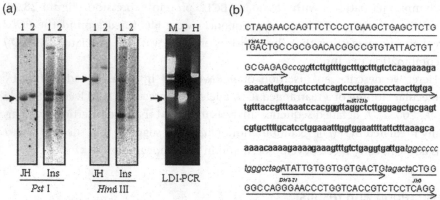

Figure 28.3. (a) Results of Southern blot analysis of *IGH* joining region (*IGHJ*), co-migration study using an 11q24 fragment and LDI-PCR. The ovarian tumor cell exhibited one rearranged band (~4.2 kb with Pst I and ~6.8 kb with Hind III digestions) of *IGHJ* and very faint germ line. LDI-PCR yielded corresponding PCR products. A 1.8 kb LDI-PCR amplified using *Pst* I digest was sequenced. The rearranged *IGHJ* contained a 241 bp of DNA fragment originated from chromosome 11 band q23. The 11q24 sequence showed the same sized rearrangement. Lane 1: patient's ovarian tumor cell, lane 2: germ-line control, M: DNA size marker (λ/Hind III), P: *Pst* I digested DNA, H: *Hind* III digested DNA. (b) Sequence alignment of the PCR product. A 262 bp DNA fragment originated from 11q24 was inserted between *VH4-33* and *DH2-21* segments of the rearranged *IGH* allele. The *VH4-33*, mature *miR-125b-1*, *DH2-21*, and *JH5* sequence are underlined by arrows. The direction of the arrows indicates 5′ to 3′. Capital and small/bold letters represent *IGH* and 11q24 sequences, respectively. Italic letters are thought to result from junctional modifications of *V(D)J* recombination.

presumably because of nucleotide deletion during *V(D)J* recombination. At the *miR125b-1/D2-21* boundary, there were 17 bp of GC-rich nucleotides suggesting that an addition of N-region occurred. In contrast, at the *D2-21/JH5* boundary, four nucleotides (GACT) were found in both *DH2-21* coding and boundary sequences. This repetition might result from other modification mechanisms than N-region addition (Figure 28.4b). Findings of sequence analysis on the derivative *IGH* allele supported the idea that *V(D)J* recombination machinery participated in the double-strand cleavage of *IGH* segments and re-ligation of *IGH/miR125b-1* joining sites. On the other hand, we found no RSS-like sequences around the germ line of *miR125b-1*, suggesting that an unknown mechanism rather than *V(D)J* recombination initiated the excision of the *miR125b-1* from its own gene locus.

Discussion

Three *miR-125*s (*miR-125a*, *miR-125b-1* and *miR-125b-2*) are thought to be homologues of *C. elegans lin-4* miRNA (Lagos-Quintana *et al.*, 2002). The *miR-125a* and *miR-125b* differ only by a central di-uridine insertion, a uridine to cytosine change, and an adenine addition. The *miR-125b-1* and *miR-125b-2* are the same sequence but originate from independent precursors located on different chromosome loci. Interestingly, previous studies have reported that the *C. elegans lin-4* regulates the timing of larval development (Ambros, 2004) and that

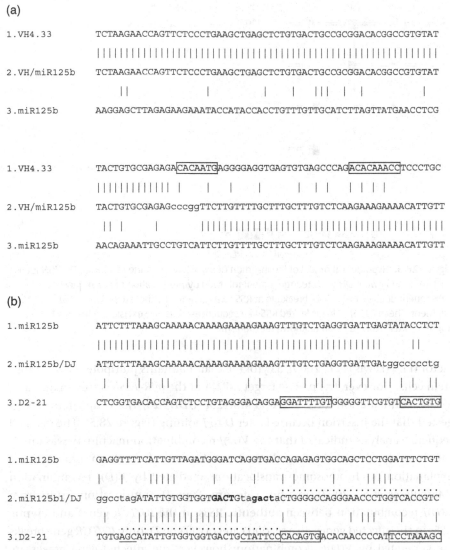

Figure 28.4. (a) Comparison of *VH4-33* (lane 1), *VH4-33/miR125b-1* (lane 2) and *miR125b-1* (lane 3) sequences at the *VH/miR125b-1* junction. The *VH4-33* was truncated just upstream of RSS. The heptamer and nonamer sequences are shown in the boxes. Five putative N-region nucleotides (cccgg) were added at the junction. The dotted letters represent coding sequence of *VH4-33*.
(b) Comparison of *DH2-21* (lane 1), *miR125b-1/DH2-21JH5* (lane 2) and *miR125b-1* (lane 3) sequences at the *miR125b-1/DH2-21JH5* junction. The heptamer and nonamer sequences are shown in the boxes. Note the RSS at the 5′ side of *DH2-21* is inverted sequence. The dotted letters represent coding sequence of *DH2-21* or *JH5* (JH5 sequence is shown in lane 2). Underlined sequences are deleted coding sequences of *DH2-21*. Duplicated sequences (gact) are shown by bold letters.

D. melanogaster miR-125b expression is detectable only in adult but not in egg (Lagos-Quintana *et al.*, 2002). These results suggest that the *miR-125* is involved in the timing of tissue development or cell differentiation. The insertion of *miR-125b-1* into the *IGH* allele might impair the expression of the small RNA in precursor B-cell leading to aberrant differentiation. Several messenger RNAs have been listed as target genes of the *miR-125b*, including leukemia inhibitory

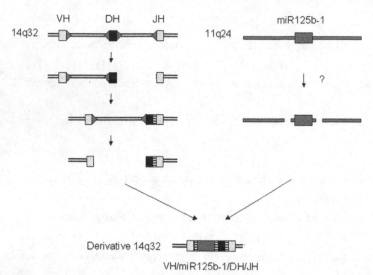

Figure 28.5. Hypothetical model of the insertion of *miR125b-1* into the *IGH* locus. The *IGH* locus on 14q32 underwent *DH* to *JH* rearrangement followed by modification of *DH/JH* junction. Subsequent double-strand DNA breakage at RSSs was invaded by the excised *miR125b-1* fragment. The *miR125b-1* locus lacked RSS-like sequences; thus, the excision of the *miR125b-1* fragment was involved by an unknown mechanism.

factor (LIF) (Lewis *et al.*, 2003), a cytokine that maintains pluripotency of various stem cells. However, the precise target mRNA of the *miR-125b* is still unknown.

The configuration of 5'-*VH4-33/miR125b-1/DH2-21/JH5*-3' apparently suggested that the insertion occurred after *D* to *J* joining (Figure 28.5). The detailed sequence analysis indicated that the *V(D)J* recombination machinery was implicated in the insertion. Two distinct mechanisms have been proposed for explanation of chromosome translocations mediated by *V(D)J* recombination (Marculescu *et al.*, 2002). In a first category (Type 1), translocation results from *V(D)J* recombination between authentic RSSs in the *IG/TCR* genes and cryptic RSSs in the affected genes. In a second category (Type 2), only *IG/TCR* gene breaks are generated by *V(D)J* recombination; however, the affected gene breaks are created by a yet unknown mechanism. A previous report demonstrated that insertion of *BCL2* into *IGH* locus was mediated by the *V(D)J* recombination between RSSs of *IGH* and cryptic RSS of *BCL2* genes, therefore resulting from the Type 1 machinery. In our case presented here, there were several characteristics of *V(D)J* recombination in the rearranged *IGH* allele but there was no cryptic RSS around *miR125b-1*, indicating that the insertion might result from the Type 2 machinery.

Previous studies have proven that almost all target genes of the *IG/TCR* alterations play significant roles in development of both normal and malignant lymphocytes (Willis and Dyer, 2000). Emerging evidence shows that some microRNAs modulate hematopoietic cell differentiation (Chen *et al.*, 2004) or are implicated in lymphomagenesis (Calin *et al.*, 2005; He *et al.*, 2005). Taken together, our results may suggest that *miR125b-1* contributes to the early stage of B-lymphocyte differentiation and that it has a pathological role when the expression is altered.

REFERENCES

Ambros, V. (2004). The functions of animal microRNAs. *Nature*, **431**, 350–355.

Asou, N., Suzushima, H., Hattori, T. *et al.* (1991). Acute unclassified leukemia originating from undifferentiated cells with the aberrant rearrangement and expression of immunoglobulin and T-cell receptor genes. *Leukemia*, **5**, 293–299.

Bennett, J. M., Catovsky, D., Daniel, M. T. *et al.* (1976). Proposals for the classification of the acute leukaemias. French–American–British (FAB) co-operative group. *British Journal of Haematology*, **33**, 451–458.

Berg, S. L., Steuber, C. P. and Poplack, D. G. (2005). Clinical manifestations of acute lymphoblastic leukemia. In *Hematology: Basic Principles and Practice*, Fourth Edition, Hoffman, R., Benz, E. J. Jr, Shattil, S. J., Furie, B., Cohen, H. J., Silberstein, L. E. and McGlave, P. (eds.). Philadelphia, PA: Elsevier, pp. 1155–1162.

Calin, G. A., Ferracin, M., Cimmino, A. *et al.* (2005). A microRNA signature associated with prognosis and progression in chronic lymphocytic leukemia. *New England Journal of Medicine*, **353**, 1793–1801.

Chen, C. Z., Li, L., Lodish, H. F. and Bartel, D. P. (2004). MicroRNAs modulate hematopoietic lineage differentiation. *Science*, **303**, 83–86.

He, L., Thomson, J. M., Hemann, M. T. *et al.* (2005). A microRNA polycistron as a potential human oncogene. *Nature*, **435**, 828–833.

Head, D. R. (2004). Classification and differentiation of the acute leukemias. In *Wintrobe's Clinical Hematology*, Eleventh Edition, Greer, J. P., Foerster, J., Lukens, J. N., Rodgers, G. M., Paraskevas, F. Glader, B. (eds.). Philadelphia, PA: Lippincott Williams & Wilkins, pp. 2063–2076.

Kuppers, R. and Dalla-Favera, R. (2001). Mechanisms of chromosomal translocations in B cell lymphomas. *Oncogene*, **20**, 5580–5594.

Lagos-Quintana, M., Rauhut, R., Yalcin, A. *et al.* (2002). Identification of tissue-specific microRNAs from mouse. *Current Biology*, **12**, 735–739.

Lewis, B. P., Shih, I. H., Jones-Rhoades, M. W., Bartel, D. P. and Burge, C. B. (2003). Prediction of mammalian microRNA targets. *Cell*, **115**, 787–798.

Marculescu, R., Le, T., Simon, P., Jaeger, U. and Nadel, B. (2002). V(D)J-mediated translocations in lymphoid neoplasms: a functional assessment of genomic instability by cryptic sites. *Journal of Experimental Medicine*, **195**, 85–98.

Okada, A. and Alt, F. W. (1995). The variable region assembly mechanism. In *Immunoglobulin Genes*, Second edition, Honjo, T. and Alt, F. W. (eds.). London: Academic Press, pp. 205–234.

Pui, C-H. (2006). Acute lymphoblastic leukemia. In *Williams Hematology*, Seventh Edition, Lichtman, M. A., Beutler, E., Kipps, T. J., Seligsohn, U., Kaushansky, K. and Prchal, J. T. (eds.). New York: McGraw-Hill Medical, pp. 1321–1342.

Sonoki, T., Iwanaga, E., Mitsuya, H. and Asou, N. (2005a). Insertion of microRNA-125b-1, a human homologue of lin-4, into a rearranged immunoglobulin heavy chain gene locus in a patient with precursor B-cell acute lymphoblastic leukemia. *Leukemia*, **19**, 2009–2010.

Sonoki, T., Iwanaga, E., Mitsuya, H. and Asou, N. (2005b). Ovarian relapse seven years after bone marrow transplantation for B-cell acute lymphoblastic leukemia: an unusual Krukenberg tumor. *American Journal of Hematology*, **80**, 75–76.

Vaandrager, J. W., Schuuring, E., Philippo, K. and Kluin, P. M. (2000). V(D)J recombinase-mediated transposition of the BCL2 gene to the IGH locus in follicular lymphoma. *Blood*, **96**, 1947–1952.

Willis, T. G. and Dyer, M. J. (2000). The role of immunoglobulin translocations in the pathogenesis of B-cell malignancies. *Blood*, **96**, 808–822.

29 miRNAs in TPA-induced differentiation of HL-60 cells

Tomoko Kozu

Introduction

miRNA is a group of small RNAs about 22 nt in length involved in the regulation of gene expression at the translational level (Grosshans and Slack, 2002; Bartel, 2004). Growing evidence indicates that miRNAs act as guides for recognizing their target mRNAs in RISC complexes by hybridizing to target sites in the 3′ UTR of mRNAs with at least a 7 nt complete match to the 5′ 2 to 8 sequence of miRNA and about a 70% complementarity in total (Brennecke et al., 2005; Lewis et al., 2005). Most miRNAs are phylogenetically highly conserved as well as in their sequences that recognize target mRNA sites mRNAs (Berezikov et al., 2005). Given the broad target recognition of miRNAs, a vast number of target sequences should exist in the genome. To date, about 330 species of miRNA have been identified in the human genome, and it is predicted that each miRNA regulates several hundred target genes and that more than 30% of the total genes are regulated by miRNAs (Bartel and Chen, 2004; John et al., 2004; Lim et al., 2005). miRNAs are encoded in genes transcribed by RNA polymerase II as an independent gene or in the intron of other genes, so that miRNA expression is regulated according to their function in a tissue- and/or stage-specific manner (Lagos-Quintana et al., 2002; Lee et al., 2004; Hsu et al., 2006). In order to clarify the exact function of each miRNA, it is necessary to determine the spatial and temporal interaction between miRNAs and their requisite targets.

Hematopoiesis is a highly regulated multi-step process that maintains hematopoietic stem cells by self-renewal and facilitates differentiation of stem cells to each blood cell lineage. Hematopoiesis is an excellent system for analyzing molecular mechanisms of cell differentiation and apoptosis. Some miRNAs are expressed specifically in hematopoietic cells, such as miR-181 in the thymus, miR-223 in myeloid cells, and miR-142 in blood cells (Chen and Lodish, 2005). The gene encoding miR-15a/-16 cluster exists at translocation breakpoints in B cell chronic lymphocytic leukemia (Pekarsky et al., 2005; Ramkissoon et al., 2005). In the t(8;17) translocation of pro-lymphoid leukemia, the miR-142 gene fuses upstream of the Myc gene and subsequently activates c-Myc expression. A miR-cluster gene encoding six species of miRNAs, miR-17, miR-18, miR-19a,

MicroRNAs: From Basic Science to Disease Biology, ed. Krishnarao Appasani. Published by Cambridge University Press. © Cambridge University Press 2008.

miR-20, miR-19b and miR-92, mapped on 11q23 are amplified and are over-expressed in B cell lymphoma (Hayashita *et al.*, 2005; L. He *et al.*, 2005). Moreover, ectopic expression of miR-181 has been shown to lead to increased B cells in mice (Chen *et al.*, 2004). These data suggest that miRNAs perform important roles in hematopoietic cell differentiation.

In this chapter, we outline the characteristics of TPA-induced differentiation of HL-60 cells, summarize alterations in expression profiles of mRNAs and miRNAs during differentiation, and finally discuss miRNA function and the problems that remain to be solved.

Differentiation of HL-60 cells

HL-60 is a myeloid leukemia cell line established from a patient with acute promyelocytic leukemia (Collins *et al.*, 1977; Collins, 1987). The *in vitro* growth of HL-60 cells requires transferrin and insulin and is stimulated in an autocrine fashion by colony stimulating factor (CSF)-like factors. HL-60 cells exhibit a 15- to 30-fold genomic amplification of *Myc* and express c-MYC protein at high levels (Collins and Groudine, 1982; Nowell *et al.*, 1983). One of the p53 loci on 17p is extensively deleted and the other chromosome 17 is lost, so that p53 protein is absent (or null) in HL-60 cells (Wolf and Rotter, 1985). Additionally, HL-60 cells contain N-RAS codon 61 mutations. In spite of these genetic aberrations including at least three major oncogenic mutations, HL-60 cells are able to differentiate terminally *in vitro* to several cell types of the myelomonocytic lineage following treatment with inducer agents. For example, retinoic acid, DMSO and Actinomycin D induce differentiation of HL-60 cells to granulocytes. A variety of natural compounds including Vitamin D_3 induce monocytic differentiation of HL-60 cells. The primary inducer of macrophage-like HL-60 cells is the phorbol ester tumor promoter 12-O-tetradecanoylphorbol-13-acetate (TPA). GM-CSF and culturing in slightly alkaline media (pH 7.6 to 7.8) induce eosinophils. TPA is the most potent inducer of the aforementioned differentiation inducers. TPA-induced HL-60 differentiation occurs in the absence of cell division with new DNA synthesis, whereas granulocytic differentiation by retinoic acid occurs through several cell divisions. TPA-induced HL-60 cells become adherent and exhibit monocyte/macrophage surface antigens, expression of nonspecific ester-ase and phagocytic activity within 48 h following exposure to TPA. Thus TPA induces rapid chromatin remodeling in HL-60 cells without cell division. Considerable evidence shows that the cellular receptor of TPA in HL-60 cells is protein kinase C (PKC)(Gilmore and Martin, 1983; Salehi *et al.*, 1988; Slosberg *et al.*, 2000). PKC is a calcium-activated phospholipid-dependent protein kinase that plays a crucial role in transducing various extracellular signals across the cell membrane. PKC is normally activated transiently by endogenous diacylglycerol, but TPA substitutes for endogenous diacylglycerol, resulting in prolonged activa-tion of PKC. For the analysis of TPA-induced differentiation, a variant of HL-60, HL-525 cells, are available (Tonetti *et al.*, 1992). These cells are resistant to TPA-induced macrophage maturation, but sensitive to inducers of monocyte and granulocyte differentiation. HL-525 cells are deficient in PKCβ and PRKX. PRKX is a myeloid-specific protein kinase, the expression of which is under control of PKCβ, and a key mediator of macrophage and granulocyte development (Semizarov *et al.*, 1998). At least 14 proteins are specifically phosphorylated in HL-60 cells within 30 min following TPA treatment. In particular, the transferrin-receptor is phosphorylated and receptor levels in the membrane decrease, contributing to a rapid loss of proliferative activity. As a direct target of PKC, topoisomerase II is phosphorylated, which functions in the regulation of DNA replication and the transcription of specific genes. Phosphorylation of topoisomerase II activates unknotting and relaxation of DNA, resulting in the alteration of gene expression (Sahyoun *et al.*, 1986). Alteration of the chromatin structure might activate Myc intron factor (MIF-1) binding to the X-box site located in intron 1 of the *MYC* gene, resulting in premature termination of *MYC* transcription and leading to the rapid down-regulation of c-MYC expression (Salehi *et al.*, 1988; Meulia *et al.*, 1992). NF-κB activation is also observed immediately following TPA exposure. Although the precise mechanism concerning TPA activation of signaling pathways remains unknown, temporal and coordinated gene changes have been shown through the use of DNA microarray analyses (Seo *et al.*, 2000; Zheng *et al.*, 2002).

Alteration in expression profiles of mRNAs and miRNAs

Extensive chromatin remodeling occurs in HL-60 cells within 48 h following TPA induction. Gene expression profiles induced by TPA have been chronologically analyzed using DNA microarrays (Zheng *et al.*, 2002). Initially, a rapid shut-down in the expression of over one hundred genes occurs, including that of *Myc*, *Fibrillarin*, *HMG-2*, *PCNA*, *RPS6K* (ribosomal protein S6 kinase), *DNAJB2*, *GEM*, *GAB1* and *TLX1*. Many of these are associated with cell proliferation at the transcriptional or post-transcriptional levels. In fact, the amount of total RNA per cell decreases to 40% at 15 min following TPA induction, and recovers gradually to initial levels at 48 h (unpublished observations). On the other hand, the early response genes are induced within 15 min following treatment with TPA and expression was undetectable by 2–3 h (Figure 29.1). The early response genes induced by TPA (0.25 h and 0.5 h) include TGF-β_2, CSF1R, SMARCB1 (INI1, subunit of SWI/SNF chromatin remodeling complex), *Rab2* and transcription factors such as *JUND*, *FOSB*, homeobox protein NKX2.1, transcription factor 3 (E2A immunoglobulin enhancer binding factors E12/E47, *TCF3*) and Kruppel-like zinc finger factor (KLF4). A wave of transcriptional activation of genes then occurs by these early response genes. The onset of induction of intermediate response genes occurs 2–3 h following TPA treatment and expression is shut off by 12–24 h. These genes include interleukin 8 and the small inducible cytokines A2, A3, A4 and A7, early growth response-1 and -2 (EGR-1 and -2) and Pac-1. The late response genes are induced or repressed between 12 h and 24 h and expression either peaks or declines by 48 h. Some of the late genes include transcriptional factors that function in hematopoietic differentiation such as *RUNX1*, *HOXB7*, *TSSC1*, *MLL*, *ETS2* and *KLF1*, and differentiation markers such as villin, vimentin

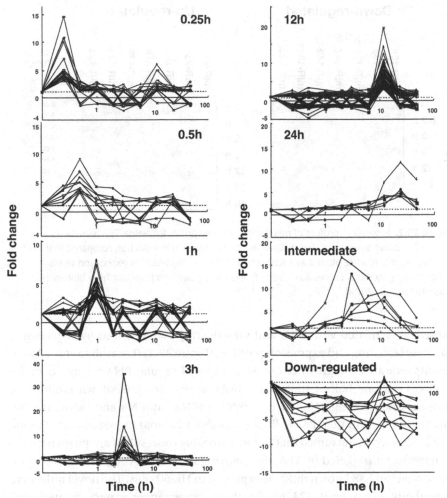

Figure 29.1. Gene expression profiles of TPA-induced differentiation of HL-60 cells. Genes were grouped by the time point of the highest expression after TPA-induction using the normalized cDNA microarray data (Zheng *et al.*, 2002).

and activin A. No further activation of gene expression appears to occur at 48 h, so that remodeling to monocyte/macrophage may be completed at 48 h following TPA induction. Protein analysis using two-dimensional gel electrophoresis showed 136 spots (19.9% of total spots) exhibited quantitative changes between HL-60 control and macrophages (Juan *et al.*, 2002). However, only four proteins (G protein, histidyl-tRNA synthetase homolog, hnRNP A2/B1, and A1) were matched between proteomic analysis and cDNA microarray data, and the rest showed a poor correlation between mRNA expression and protein abundance. This result suggests the existence of regulation of protein levels by miRNAs.

Expression profile of miRNA during TPA-induced differentiation

In order to clarify the function of miRNAs in macrophage differentiation, we isolated several hundred RNAs ranging from 18 to 26 nucleotides from HL-60 cells

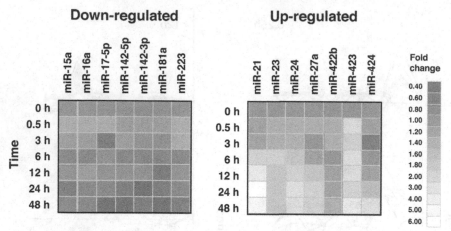

Figure 29.2. Expression profiles of miRs at various time points following TPA induction. Up-regulated and down-regulated miRNAs at 48 h following TPA-induction compared to that of the start point (0 h) were plotted as a function of time in relation to their expression levels exhibited by gray scale. Expression levels of miRs were quantified from Northern blot analysis (Kasashima *et al.*, 2004).

untreated or treated with TPA, and subsequently characterized by sequencing, database searching and expression profiling (Figure 29.2) (Kasashima *et al.*, 2004). Results indicated the presence of several kinds of cellular RNA fragments in the library. Over one half of the cloned RNAs represented breakdown products of abundant non-coding RNAs such as rRNA, snRNA, snoRNA and tRNA. The frequency of rRNAs in the TPA (+) library was about 2.5-fold higher than that in the TPA (−) library. We assumed that this was probably the result of apoptosis or traces of remodeling triggered by TPA. By removing non-miRNA sequences, we found three novel and 38 known miRNAs expressed in HL-60 cells. The novel miRs were named miR-423 and miR-424 by Rfam (http://www.sanger.ac.uk/Software/Rfam/mirna)(Griffiths-Jones *et al.*, 2006). We also found that antisense strands of miR-30e and miR-378 were expressed in HL-60 cells (Table 29.1). Four methods are now available to analyze the expression profile of miRNAs. These include direct cloning of miRNAs, microarrays, Northern blotting and quantitative RT-PCR. Novel species of miRNA can be identified by cloning as mentioned above, but vast amounts of clones need to be analyzed in order to obtain quantitative populations of all expressed miRNAs (Elbashir *et al.*, 2001). Expression profiles of known miRNAs species can be analyzed by hybridization to microarrays (Ciafrè *et al.*, 2005; Iorio *et al.*, 2005). With Northern blotting, the size and amount of mature and precursor miRNA can be analyzed. With real-time RT-PCR, mature miRNAs can be quantified using specific primers for individual miRNAs (Chen *et al.*, 2005). We analyzed the expression profile of miRNAs in HL-60 cells without and with TPA induction using a library of primers for real-time RT-PCR consisting of 150 known species of miRNA (TaqMan MicroRNA Assays, Applied Biosystems). Seventy-one species of miRNA were commonly expressed in HL-60 cells and the expression of an additional three species increased at 48 h following TPA induction (our unpublished data). Lu *et al.* analyzed the miRNA expression profile in

Table 29.1. Expression profiles of HL-60 cells analyzed by four methods

miRNAs in HL-60 cells	MicroBeads array	Cloning[a]		Northern blot[b]		Real-time PCR[c]		
		TPA−	TPA+	TPA−	TPA+	TPA−	TPA+	Ratio
let-7a	+	1	3			0.106	0.127	**1.20**
let-7b	+	2	0			0.076	0.035	**0.46**
let-7c	+							
let-7d	+	3	0			0.179	0.182	1.02
let-7e	−	0	1			0.0004	0.0021	**5.71**
let-7f	+	0	1					
let-7g	+					0.174	0.200	1.15
let-7i						0.027	0.029	1.07
miR-15a	+	5	**2**	1	**0.84**	0.021	0.022	1.08
miR-15b	+					0.090	0.058	**0.64**
miR-16	+	1	0	1	**0.78**	1.000	0.780	**0.78**
miR-17-3p	+	0	1			0.009	0.0061	**0.66**
miR-17-5p	+	5	**1**	1	**0.41**	0.842	0.638	**0.76**
miR-18	−	2	0					
miR-19a	+					1.229	0.373	**0.30**
miR-19b	+	1	0					
miR-20	+	1	0			0.733	0.253	**0.34**
miR-21	+	1	21	1	4.95	0.095	0.515	**5.44**
miR-23a	+	1	3	1	1.8	0.108	0.237	**2.20**
miR-23b	+					0.015	0.013	**0.87**
miR-24	+	0	6	1	2.34			
miR-25	+	1	1			0.302	0.138	**0.46**
miR-26a	+	2	4			0.121	0.156	**1.29**
miR-26b	+	1	2			0.064	0.082	**1.29**
miR-27a				1	2.06	0.036	0.275	**7.61**
miR-27b	+	8	10			0.005	0.007	**1.42**
miR-28						0.003	0.011	**3.86**
miR-29a	−	1	0			0.220	0.457	**2.08**
miR-29b	+					0.103	0.067	**0.65**
miR-29c	+	0	1			0.369	0.430	1.16
miR-30a-3p	−	1	0			0.0051	0.0063	1.24
miR-30a-5p	−	1	0					
miR-30b	+					0.216	0.140	**0.65**
miR-30c						0.126	0.154	**1.22**
miR-30d	+					0.037	0.032	**0.85**
miR-30e	+	0	1			0.009	0.019	**2.18**
miR-30e-3'		1	3					
miR-33	+							
miR-34a						0.00006	0.0033	**58.71**
miR-92	+	2	1			1.402	0.902	**0.64**
miR-93	+	1	0					
miR-98	+					0.019	0.022	1.16
miR-99a						0.009	0.030	**3.40**
miR-100						0.009	0.035	**4.01**
miR-101	+	1	0					
miR-103	+					0.143	0.104	**0.73**
miR-106a	+					0.501	0.302	**0.60**
miR-106b	+							
miR-107	+					0.0031	0.0023	**0.74**

Table 29.1. (cont.)

miRNAs in HL-60 cells	MicroBeads array	Cloning[a]		Northern blot[b]		Real-time PCR[c]		
		TPA−	TPA+	TPA−	TPA+	TPA−	TPA+	Ratio
miR-124a	++	0	2			0.0075	0.033	**4.41**
miR-124b						0.0019	0.0046	**2.42**
miR-125a	+					0.0086	0.062	**7.16**
miR-125b						0.0064	0.019	**2.95**
miR-128a	−	1	0			0.0016	0.0032	**1.98**
miR-130b						0.130	0.056	**0.43**
miR-139	+					0.0006	0.0008	**1.30**
miR-140*	+					0.055	0.054	0.98
miR-142-3p	++	6	**1**	1	**0.57**	2.571	1.030	**0.40**
miR-142-5p	++	13	**9**	1	**0.42**	0.113	0.075	**0.67**
miR-146						0.002	0.075	**49.80**
miR-148a						0.058	0.052	**0.89**
miR-152	+							
miR-153	+							
miR-155	−	0	1			0.000	0.002	**9.44**
miR-181a	+			1	**0.61**	0.159	0.105	**0.66**
miR-181b	+					0.257	0.073	**0.28**
miR-181c	+					0.005	0.004	**0.73**
miR-186						0.119	0.137	1.15
miR-191						0.172	0.193	1.12
miR-193	+					−		
miR-195						0.008	0.012	**1.54**
miR-197						0.020	0.033	**1.68**
miR-199a*						0.048	0.024	**0.50**
miR-199b	+					0.042	0.011	**0.26**
miR-200c						0.001	0.003	**4.03**
miR-210						0.010	0.003	**0.35**
miR-213						0.009	0.001	**0.08**
miR-221	+	1	1			0.071	0.776	**10.98**
miR-222	+	0	3			0.038	0.349	**9.30**
miR-223	+			1	**0.61**	0.516	0.321	**0.62**
miR-294	+							
miR-301						0.014	0.101	**7.24**
miR-302a	+					0.078	0.068	**0.87**
miR-320	+	4	**1**					
miR-324-5p						0.009	0.004	**0.44**
miR-328	+					0.001	0.001	**1.68**
miR-330	+					−		
miR-339	+	0	1			0.030	0.010	**0.35**
miR-342						0.058	0.108	**1.86**
miR-352	+							
miR-374						0.014	0.022	**1.59**
miR-422b		23	11	1	**1.25**			
miR-423		1	0	1	**2.7**			
miR-424		1	3	1	**2.92**			

[a] Clone number
[b] Relative amount as TPA- = 1
[c] 2 normalized with miR-16
Normal type, down-regulated; bold type, up-regulated.

HL-60 cells during differentiation to neutrophils following treatment with *all-trans* retinoic acid using beads-based microarrays (Lu *et al.*, 2005). It was found that 59 miRNAs are commonly expressed, which is comparable with our data. We have obtained many miRNA clones or positive results in RT-PCR, which were undetected or undetermined in the microarray (Table 29.1). In total, 95 species of miRNAs were detected in HL-60 cells. These miRNAs could be further classified into subsets of miRNAs that responded differently following TPA induction, being either up-regulated or down-regulated. Thirty-two miRNAs are down-regulated and 35 miRNAs are up-regulated. The down-regulated miRNAs include miR-15, miR-16, the miR-17~92 cluster, and hematopoietic-specific miRNAs such as miR-181, miR-142 and miR-223, and the expression of most of these decrease to between 40%–80%. miR-16 regulates the expression of short-lived transcription factors with AU-rich elements (Jing *et al.*, 2005). The miR-17–92 cluster is amplified in lymphoma, and has experimentally been shown to be oncogenic (L. He *et al.*, 2005). miR-181 promotes B-cell differentiation when expressed in hematopoietic stem/progenitor cells (Chen *et al.*, 2004). miR-142 is located at a translocation site found in a case of aggressive B-cell leukemia. Therefore, the down-regulated miRs during TPA-induced differentiation might be related to a cancerous state. The up-regulated miRs include miR-21, miR-24, miR-27a, miR-124a, miR-125a, miR-146, miR-155, miR-221, miR-222, miR-423 and miR-424. miR-124 is expressed specifically in brain and functions in neurogenesis (Smirnova *et al.*, 2005). miR-125, miR-155 and miR-21 are deregulated in human breast cancer (Iorio *et al.*, 2005). miR-221, miR-222 and miR-146 are up-regulated in papillary thyroid carcinoma and suppress KIT expression (H. He *et al.*, 2005). Some of these might function in halting cell proliferation. In order to clarify the role of each miRNA in the differentiation of HL-60 cells, the molecular targets of miRNAs in this context must be determined.

Co-existence of miRNAs and their target mRNA during the differentiation of HL-60 cells

In order to clarify the role of miRNAs in differentiation and to glean the meaning of miRNA up- and down-regulation during the differentiation process, genes activated by TPA were searched containing miR-target sites in the 3′ UTR using miRanda software (http://www.targetscan.org/). The predicted target sites of miRs in the transcription factors activated at each time point, which co-exist in HL-60 cells at the same time, were plotted (Figure 29.3). The miR-17–92 cluster and miR-142 are both down-regulated at 48 h following TPA-induction, but are activated once at 12 h, showing a bi-phasic expression pattern, and several target transcription factors are activated at 12 h. The predicted target sites of the let-7 family exist in transcription factors activated at 12 h, but not in early response genes. As shown in Figure 29.3, many miR-target sites are found in one gene, so that such a gene must be regulated by several kinds of miRNA at the same time. miR-target sites are also predicted in many genes other than transcription factors, such as receptors, adaptors, and protein kinases relating to cell signaling. Delineation of

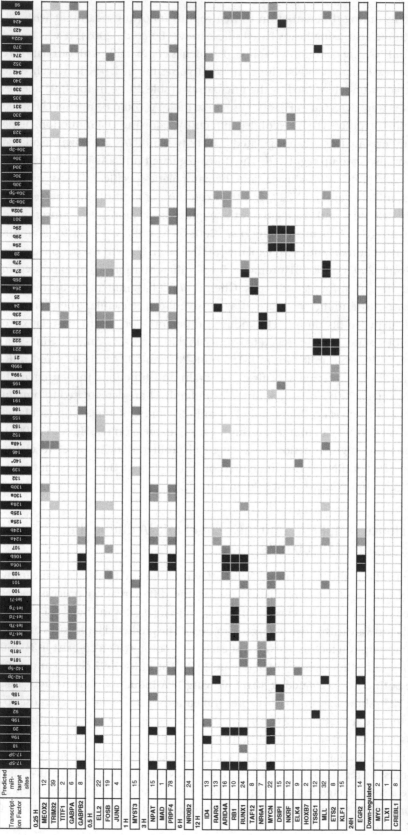

Figure 29.3. Prediction of miRNA targets in transcription factors activated by TPA. The predicted target sites of miRs expressing in HL-60 cells in transcription factors are indicated by three colors: red, up-regulated; blue, down-regulated; gray, unchanged or unknown. Numbers in the lane of predicted target sites are total number of miR-target sites predicted by Target Scan software (Release 3.1). Clustered miRs are shown in the same colors.

the critical regulation in TPA-induced macrophage differentiation represents the focus of the next subject.

Concluding remarks

Recently, miR-223 and transcription factors NFI-A and C/EBPa were reported to form a negative-feedback loop, which is important for granulocytic differentiation induced by retinoic acid (Fazi *et al.*, 2005). Also, miR-181 was shown to down-regulate the homeobox protein Hox-A11 (a repressor of the differentiation process) during myoblast differentiation, and to contribute towards establishing the muscle phenotype (Naguibneva *et al.*, 2006). Novel mechanisms of differentiation regulation using miRNAs can manifest in the macrophage differentiation system of HL-60 cells. Since the transcription factors activated following TPA-induction include several proto-oncogenes relating to leukemogenesis, investigating the regulatory mechanism of these factors by miRNAs will be important in contributing towards our understanding of leukemogenesis. Additionally, it is important to determine the mechanism of acute chromatin remodeling induced by TPA in HL-60 cells. Thus, the differentiation of HL-60 cells by TPA provides a useful *in vitro* system to analyze hematopoietic differentiation.

ACKNOWLEDGEMENTS

Own studies, cited in this review, were supported by Grants-in-Aid for Scientific Research on Priority Areas from the Ministry of Education, Culture, Sports, Science, and Technology of Japan.

REFERENCES

Bartel, D. P. (2004). MicroRNAs: genomics, biogenesis, mechanism, and function. *Cell*, **116**, 281–297.

Bartel, D. P. and Chen, C. Z. (2004). Micromanagers of gene expression: the potentially widespread influence of metazoan microRNAs. *Nature Reviews in Genetics*, **5**, 396–400.

Berezikov, E., Guryev, V., van de Belt, J. *et al.* (2005). Phylogenetic shadowing and computational identification of human microRNA genes. *Cell*, **120**, 21–24.

Brennecke, J., Stark, A., Russell, R. B. and Cohen, S. M. (2005). Principles of microRNA–target recognition. *Public Library of Science Biology*, **3**, e85.

Chen, C., Ridzon, D. A., Broomer, A. J. *et al.* (2005). Real-time quantification of microRNAs by stem-loop RT-PCR. *Nucleic Acids Research*, **33**, e179.

Chen, C. Z. and Lodish, H. F. (2005). MicroRNAs as regulators of mammalian hematopoiesis. *Seminars in Immunology*, **17**, 155–165.

Chen, C. Z., Li, L., Lodish, H. F. and Bartel, D. P. (2004). MicroRNAs modulate hematopoietic lineage differentiation. *Science*, **303**, 83–86.

Ciafrè, S. A., Galardi, S., Mangiola, A. *et al.* (2005). Extensive modulation of a set of microRNAs in primary glioblastoma. *Biochemical Biophysical Research Communications*, **334**, 1351–1358.

Collins, S. and Groudine, M. (1982). Amplification of endogenous myc-related DNA sequences in a human myeloid leukaemia cell line. *Nature*, **298**, 679–681.

Collins, S. J. (1987). The HL-60 promyelocytic leukemia cell line: proliferation, differentiation, and cellular oncogene expression. *Blood*, **70**, 1233–1244.

Collins, S. J., Gallo, R. C. and Gallagher, R. E. (1977). Continuous growth and differentiation of human myeloid leukaemic cells in suspension culture. *Nature*, **270**, 347–349.

Elbashir, S. M., Lendeckel, W. and Tuschl, T. (2001). RNA interference is mediated by 21- and 22-nucleotide RNAs. *Genes & Development*, **15**, 188–200.

Fazi, F., Rosa, A., Fatica, A. *et al.* (2005). A minicircuitry comprised of microRNA-223 and transcription factors NFI-A and C/EBPalpha regulates human granulopoiesis. *Cell*, **123**, 819–831.

Gilmore, T. and Martin, G. S. (1983). Phorbol ester and diacylglycerol induce protein phosphorylation at tyrosine. *Nature*, **306**, 487–490.

Griffiths-Jones, S., Grocock, R. J., van Dongen, S., Bateman, A. and Enright, A. J. (2006). miRBase: microRNA sequences, targets and gene nomenclature. *Nucleic Acids Research*, **34**, D140–144.

Grosshans, H. and Slack, F. J. (2002). Micro-RNAs: small is plentiful. *Journal of Cell Biology*, **156**, 17–21.

Hayashita, Y., Osada, H., Tatematsu, Y. *et al.* (2005). A polycistronic microRNA cluster, miR-17-92, is overexpressed in human lung cancers and enhances cell proliferation. *Cancer Research*, **65**, 9628–9632.

He, H., Jazdzewski, K., Li, W. *et al.* (2005). The role of microRNA genes in papillary thyroid carcinoma. *Proceedings of the National Academy of Sciences USA*, **102**, 19075–19080.

He, L., Thomson, J. M., Hemann, M. T. *et al.* (2005). A microRNA polycistron as a potential human oncogene. *Nature*, **435**, 828–833.

Hsu, P. W., Huang, H. D., Hsu, S. D. *et al.* (2006). miRNAMap: genomic maps of microRNA genes and their target genes in mammalian genomes. *Nucleic Acids Research*, **34**, D135–139.

Iorio, M. V., Ferracin, M., Liu, C. G. *et al.* (2005). MicroRNA gene expression deregulation in human breast cancer. *Cancer Research*, **65**, 7065–7070.

Jing, Q., Huang, S., Guth, S. *et al.* (2005). Involvement of microRNA in AU-rich element-mediated mRNA instability. *Cell*, **120**, 623–634.

John, B., Enright, A. J., Aravin, A. *et al.* (2004). Human microRNA targets. *Public Library of Science Biology*, **2**, e363.

Juan, H. F., Lin, J. Y., Chang, W. H. *et al.* (2002). Biomic study of human myeloid leukemia cells differentiation to macrophages using DNA array, proteomic, and bioinformatic analytical methods. *Electrophoresis*, **23**, 2490–2504.

Kasashima, K., Nakamura, Y. and Kozu, T. (2004). Altered expression profiles of microRNAs during TPA-induced differentiation of HL-60 cells. *Biochemical Biophysical Research Communications*, **322**, 403–410.

Lagos-Quintana, M., Rauhut, R., Yalcin, A. *et al.* (2002). Identification of tissue-specific microRNAs from mouse. *Current Biology*, **12**, 735–739.

Lee, Y., Kim, M., Han, J. *et al.* (2004). MicroRNA genes are transcribed by RNA polymerase II. *European Molecular Biology Organization Journal*, **23**, 4051–4060.

Lewis, B. P., Burge, C. B. and Bartel, D. P. (2005). Conserved seed pairing, often flanked by adenosines, indicates that thousands of human genes are microRNA targets. *Cell*, **120**, 15–20.

Lim, L. P., Lau, N. C., Garrett-Engele, P. *et al.* (2005). Microarray analysis shows that some microRNAs downregulate large numbers of target mRNAs. *Nature*, **433**, 769–773.

Lu, J., Getz, G., Miska, E. A. *et al.* (2005). MicroRNA expression profiles classify human cancers. *Nature*, **435**, 834–838.

Meulia, T., Krumm, A., Spencer, C. and Groudine, M. (1992). Sequences in the human c-myc P2 promoter affect the elongation and premature termination of transcripts initiated from the upstream P1 promoter. *Molecular Cell Biology*, **12**, 4590–4600.

Naguibneva, I., Ameyar-Zazoua, M., Polesskaya, A. *et al.* (2006). The microRNA miR-181 targets the homeobox protein Hox-A11 during mammalian myoblast differentiation. *Nature Cell Biology*, **8**, 278–284.

Nowell, P., Finan, J., Dalla-Favera, R. *et al.* (1983). Association of amplified oncogene c-myc with an abnormally banded chromosome 8 in a human leukaemia cell line. *Nature*, **306**, 494–497.

Pekarsky, Y., Calin, G. A. and Aqeilan, R. (2005). Chronic lymphocytic leukemia: molecular genetics and animal models. *Current Topics in Microbiology and Immunology*, **294**, 51–70.

Ramkissoon, S. H., Mainwaring, L. A., Ogasawara, Y. *et al.* (2005). Hematopoietic-specific microRNA expression in human cells. *Leukemia Research*, **30**, 643–647.

Sahyoun, N., Wolf, M., Besterman, J. *et al.* (1986). Protein kinase C phosphorylates topoisomerase II: topoisomerase activation and its possible role in phorbol ester-induced differentiation of HL-60 cells. *Proceedings of the National Academy of Sciences USA*, **83**, 1603–1607.

Salehi, Z., Taylor, J. D. and Niedel, J. E. (1988). Dioctanoylglycerol and phorbol esters regulate transcription of c-myc in human promyelocytic leukemia cells. *Journal of Biological Chemistry*, **263**, 1898–1903.

Semizarov, D., Glesne, D., Laouar, A., Schiebel, K. and Huberman, E. (1998). A lineage-specific protein kinase crucial for myeloid maturation. *Proceedings of the National Academy of Sciences USA*, **95**, 15 412–15 417.

Seo, J., Kim, M. and Kim, J. (2000). Identification of novel genes differentially expressed in PMA-induced HL-60 cells using cDNA microarrays. *Molecular Cells*, **10**, 733–739.

Slosberg, E. D., Yao, Y., Xing, F. *et al.* (2000). The protein kinase C beta-specific inhibitor LY379196 blocks TPA-induced monocytic differentiation of HL60 cells. *Molecular Carcinogenesis*, **27**, 166–176.

Smirnova, L., Grafe, A., Seiler, A. *et al.* (2005). Regulation of miRNA expression during neural cell specification. *European Journal of Neurosciences*, **21**, 1469–1477.

Tonetti, D. A., Horio, M., Collart, F. R. and Huberman, E. (1992). Protein kinase C beta gene expression is associated with susceptibility of human promyelocytic leukemia cells to phorbol ester-induced differentiation. *Cell Growth & Differentiation*, **3**, 739–745.

Wolf, D. and Rotter, V. (1985). Major deletions in the gene encoding the p53 tumor antigen cause lack of p53 expression in HL-60 cells. *Proceedings of the National Academy of Sciences USA*, **82**, 790–794.

Zheng, X., Ravatn, R., Lin, Y. *et al.* (2002). Gene expression of TPA induced differentiation in HL-60 cells by DNA microarray analysis. *Nucleic Acids Research*, **30**, 4489–4499.

30 MiRNAs in skeletal muscle differentiation

Irina Naguibneva*, Anna Polesskaya and Annick Harel-Bellan

Introduction

MicroRNAs (miRNAs) represent an important class of short natural RNAs that act as post-transcriptional regulators of gene expression. Genetic studies in *Caenorhabditis elegans and Drosophila* revealed that miRNAs are involved in fine tuning the spatial and temporal regulation of developmental events, including precursor cell proliferation, differentiation and programmed death (Ambros, 2003; Brennecke *et al.*, 2003; Sempere *et al.*, 2003, Xu *et al.*, 2003; Biemar *et al.*, 2005). MiRNAs have been found essentially in every cell type analyzed to date. A recent systematic analysis of spatial expression of miRNA in developing zebrafish embryos showed that most tissues have a unique time-dependent pattern of miRNA expression (Wienholds *et al.*, 2005). *In silico* methods predicted that the individual miRNAs have, on average, hundreds of target mRNAs, suggesting that miRNAs have enormous regulatory roles in different genetic programs (Lewis *et al.*, 2003; Brennecke *et al.*, 2005; Krek *et al.*, 2005; Xie *et al.*, 2005). However, the number of functional miRNA/target pairs experimentally characterized to date is minimal.

We have addressed the function of miRNAs in mammalian skeletal muscle. Muscle formation (Figure 30.1) involves the proliferation of myoblast precursor cells, which subsequently exit from the cell cycle and enter a terminal differentiation program that includes myoblast fusion into large multi-nucleated cells (myotubes) and expression of muscle specific markers such as myosin heavy chain (MHC) and muscle creatine kinase (MCK) (Figure 30.1). Differentiation can be recapitulated in *ex vivo* models, using either totipotent ES cells directed toward the muscle lineage (Dinsmore *et al.*, 1998), or established myoblast cell lines that by default enter the skeletal muscle differentiation pathway when they are deprived of growth factors (Bains *et al.*, 1984).

The "master switch" MyoD family of basic helix–loop–helix (bHLH) myogenic factors, as well as the MEF protein family, directly control expression of

---

* Author to whom correspondence should be addressed.

MicroRNAs: From Basic Science to Disease Biology, ed. Krishnarao Appasani. Published by Cambridge University Press. © Cambridge University Press 2008.

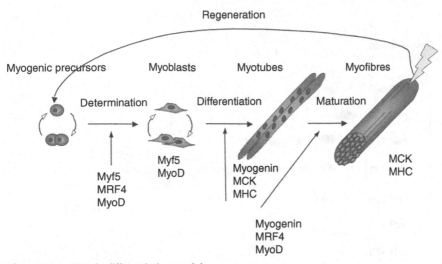

Figure 30.1. Muscle differentiation model.

muscle-specific genes (for review, see Buckingham *et al.* (2003)). Other important transcription regulators of myogenesis, such as proteins of Pax and Hox families, can either act upstream of MyoD (Pax3) (Tajbakhsh *et al.*, 1997) or directly modify MyoD expression (HoxA-11 and HoxA-13) (Yamamoto and Kuroiwa, 2003). While there has been significant progress in understanding the transcriptional regulation of myogenesis, few studies concentrate on post-transcriptional mechanisms and on regulation by non-protein factors.

Very little data about involvement of miRNAs in regulation of muscle differentiation are available to date. Pioneer studies of miRNAs began with cloning of these molecules from adult tissues and embryonic cell lines (Lagos-Quintana *et al.*, 2002; Lagos-Quintana *et al.*, 2003; Houbaviy *et al.*, 2003; Poy *et al.*, 2004). Tissue-specific miRNA expression has been observed during development and in adults (Sempere *et al.*, 2004), and in particular in the central nervous system (Krichevsky *et al.*, 2003). Skeletal muscle tissue appears to be a poor source of miRNAs. Only three miRNA species have been shown to be "skeletal muscle-enriched," miR-133, miR-1, and miR-206 (Sempere *et al.*, 2004). In mouse, the miR-1 family (miR-1-1 and miR-1-2) is specifically expressed in cardiac and skeletal muscle precursor cells. These miRNAs control proliferation and survival of cardiomyocytes, and their expression is regulated by the serum response factor (SRF). In skeletal muscle, miR-1 transcription is responsive to MyoD and MEF-2 transcription factors (Zhao *et al.*, 2005).

MiRNA expression during muscle differentiation

In order to gain knowledge about miRNA function in mouse muscle, we have used Northern blots to characterize miRNA expression patterns in differentiating totipotent ES cells, as well as in terminally differentiating myoblast cell lines (C2C12). Examples are shown in Figure 30.2a, and the results are summarized in Table 30.1. A subset of miRNAs was not detectable at any step of muscle determination and

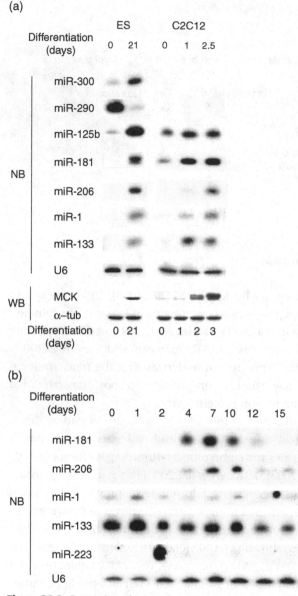

Figure 30.2. Expression of selected miRNAs during muscle differentiation. (a) Time-course of miRNAs and muscle marker expression in differentiating ES or C2C12 cells. NB: Northern blot, WB: Western blot, MCK: muscle creatine kinase, α-tub: α-tubulin (loading control). (b) Time-course of miRNAs expression during muscle regeneration.

differentiation. Detectable miRNAs fell into several classes. A small number of miRNAs, such as miR-300, were up-regulated in differentiating ES cells, but undetectable or low in C2C12 myoblasts (Table 30.1 and Figure 30.2a). In ES cells, differentiation is far from being homogeneous, and significant non-muscle differentiation is observed within the embryoid bodies. MiR-300 may thus be involved in these non-muscle differentiation pathways. Alternatively, this class of miRNAs may participate in a mechanism related to muscle formation from totipotent stem

Table 30.1. Expression of various miRNAs during muscle differentiation

MiRNA	ES	EB	MB	MT
296	+/−	+	+/−	+/−
298	+/−	+	+/−	+/−
300	+/−	++	−	−
290	++	−	−	−
291s	++	−	−	−
291as	++	−	−	−
295	++	−	−	−
298	+	−	−	−
299	+	+/−	−	−
16	−	+	+/−	+/−
21	−	++	++	++
22	+/−	+	++	++
34a	+/−	++	+/−	+
99a	−	nd	+/−	+/−
125b	+/−	++	+	++
143	+/−	++	++	++
let7d	−	nd	+	+
let7i	−	nd	+	+
let7g	−	nd	+	+
301	+/−	+/−	+/−	+/−
133	−	++	−	+
1	−	+	−	+
181	−	++	+/−	++
206	−	+	−	+
208	−	+	−	+
297	−	+	−	+
let7c	−	+	+/−	++
188	nd	nd	−	+
23a	nd	nd	+/−	++
30b	nd	nd	−	+
339	nd	nd	−	+
9	−	nd	−	−
10b	−	−	−	−
15a	−	−	−	−
15b	−	−	−	−
99b	−	−	−	−
106	−	−	−	−
124a	−	−	−	−
129	−	−	−	−
131	−	−	−	−
141	−	nd	−	−
142s	−	−	−	−
142as	−	−	−	−
213	−	−	−	−
302	−	−	−	−
223	−	−	−	−

Expression of miRNAs was analyzed by Northern blot in embryonic stem cells, either proliferating (ES) or forming muscle-oriented embryonic bodies (EB), as well as in C2C12 cells, either proliferating (myoblasts, MB) or differentiated into myotubes (MT). MiRNAs are clustered according to their expression profile. Expression levels: high: ++; medium: +; low: +/−; undetectable: −; nd: not done.

cells, but lost in established myoblast cell lines such as C2C12. A second class, exemplified by miR-290-295, was down-regulated on ES cell differentiation and remained undetectable in myoblasts under all conditions tested. An interesting hypothesis would be that these miRNAs are associated with cellular totipotency, as previously suggested (Houbaviy *et al.*, 2003).

A third class of miRNAs, exemplified by miR-125b, were expressed at low or undetectable levels in proliferating ES cells, but were up-regulated upon muscle-oriented differentiation. Analysis of differentiating myoblasts did not show further strong induction during the terminal step, suggesting that these miRNAs might be associated with determination. Finally, a group of miRNAs was induced by triggering differentiation in ES cells, as well as by inducing terminal differentiation of myoblasts. These miRNAs might thus be involved in the terminal differentiation program. Among those were miR-133, miR-1, miR-206, previously shown to be expressed in muscle (Sempere *et al.*, 2004), miR-208, miR-297, let-7c and, more unexpectedly, miR-181, known as an important regulator of the B-lymphocyte lineage (Chen *et al.*, 2004). The expression patterns of these miRNAs roughly coincided with those of muscle-specific proteins such as muscle creatine kinase (MCK) (Figure 30.2a).

To date, there are few data available as to the function and targets of miRNAs in skeletal muscle *in vivo*. MiRNAs 125a/b and let-7c, which are associated with myogenic determination and differentiation in our studies, are highly expressed in the developing mouse embryo, from day 10.5 to the newborn stage. Recent publications suggest that formation of skeletal muscle tissue in limb buds can be regulated by specific down-regulation of lin-41 protein by let-7c and/or miR125a/b (Mansfield *et al.*, 2004, Schulman *et al.*, 2005). The mechanism of action and downstream targets of Lin-41 in muscle remain to be elucidated. Another possible target for miRNA regulation in developing skeletal muscle is the RNA-binding protein Lin-28, which is highly expressed in embryonic muscle tissue (Yang and Moss, 2003; Polesskaya *et al.*, 2007). Lin-28 can be regulated by let-7c and miR-125a/b in an *in vitro* differentiation assay (Sempere *et al.*, 2004), and our current work suggests that regulation of skeletal muscle development and maintenance is dependent on the amounts of Lin-28 present in muscle cells (Polesskaya *et al.*, 2007). Indeed, it is expected that regulation of target proteins by miRNAs will not have a "black-and-white" effect (the miRNA is expressed, the target protein is not), but rather correspond to a very fine balance resulting in quite opposite developmental phenomena with only slight changes in the target protein levels.

miRNAs are differentially expressed in muscle regeneration

To characterize the function of miRNAs in skeletal muscle *in vivo*, we analyzed the expression of several miRNAs that are specifically associated with terminal muscle differentiation *in vitro*, in an *in vivo* model. Skeletal muscle regeneration can be induced by a local cardiotoxin injection into adult mouse *tibialis anterior* (TA) muscle, as described by Charge and Rudnicki (2004). Briefly, injection of cardio-toxin induces the rapid necrosis of 10%–12% of fibers within the TA muscle,

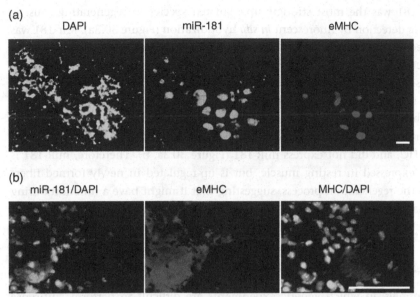

Figure 30.3. Cells expressing miR-181 are differentiating muscle cells. (a) Low magnification (x200). Cross-sections of *tibialis anterior* muscles at day 5 of cardiotoxin-induced regeneration were submitted to immuno-FISH using antibodies against embryonic MHC (eMHC) with TRITC-labeled secondary antibodies, and a DIG-labeled oligonucleotide probe complementary to miR-181 with FITC-labeled secondary antibodies; sections were also counterstained with DAPI. (b) Higher magnification (× 630). Labeling as in (a). In addition, resting fibers were stained using antibodies against sarcomeric MHC with Cy5-labeled secondary antibodies. Scale bars correspond to 25 μm. (See color plate 21)

followed by activation of muscle stem cells (satellite cells), and proliferation of the resulting myoblast population during the first 3 days following injection. Formation of new muscle fibers, expressing an embryonic form of myosin, starts on day 2–3 of regeneration. Maturation of fibers starts on day 5, and continues until day 7–8, as evidenced by the replacement of embryonic myosin by adult sarcomeric myosin. Regeneration is functionally complete in 10–16 days, and histologically complete in a month.

Interestingly, two muscle-specific miRNAs, miR-133 and miR-1, demonstrated constitutive expression throughout the regeneration (Figure 30.3a). MiR-206, as well as miR-181, were scarcely detected in resting muscle, but were strongly up-regulated upon regeneration of muscle fibers. Their expression increased concomitantly with the formation of new muscle fibers. A brief inflammation reaction, accompanied by the recruitment of non-muscle cells, is induced by necrosis between days 2–3 of regeneration. It can be followed by the expression of miR-223 (Figure 30.2b), a miRNA specific for cells of myeloid origin such as granulocytes and monocytes (Ramkissoon *et al.*, 2005; Monticelli *et al.*, 2005). However, the up-regulation of miR-206 and miR-181 occurred between days 4–7 of regeneration, and thus cannot be accounted for by local inflammation (Figure 30.1b). Moreover, we have performed histological studies of regenerating muscle, to specifically characterize the expression of miRNAs with regard to cell type and subcellular localization.

MiR-181 was the most strongly up-regulated species in regenerating muscle, allowing detection by fluorescent *in situ* hybridization (Figure 30.3a). Mir-181 was detected in regeneration areas that are characterized by the presence of high numbers of nuclei with a disorganized topology, when stained by DAPI. High expression of miR-181 was observed in multinucleated, newly forming fibers, as well as in adjacent mononucleated myoblasts, that were positively counter-stained for embryonic myosin heavy chain (eMHC), a well-characterized protein marker of regenerating fibers. Intact fibers stained positive for adult sarcomeric MHC, but not for eMHC, and did not express miR-181 (Figure 30.3a, b). Therefore, miR-181 is poorly expressed in resting muscle, but is up-regulated in newly formed fibers during the regeneration process, suggesting that it might have a function during terminal muscle differentiation. To address this question, we have developed a loss-of-function assay, based on an antisense approach (Naguibneva *et al.*, 2006a).

A loss-of-function assay for miRNA

In mammals, in which genetic experiments are difficult to perform, antisense approaches can be useful to deplete miRNAs. We used complementary oligonucleotides including locked nucleic acids (LNA), developed by Wengel and co-workers and Imanishi and co-workers (for review, see Petersen and Wengel (2003)). LNA bases are ribonucleotide analogs containing a methylene bridge between the 2′ oxygen and the 4′ carbon of the ribose ring (Figure 30.4a). The chemistry of LNAs and their use in *in situ* hybridizations were detailed earlier in this book by S. Kauppinen. The introduction of LNA bases into oligonucleotides confers exceptional stability to duplexes formed with complementary sequences. Moreover, these LNA-containing oligonucleotides prove to be easily soluble, stable under physiological conditions, and non-toxic at concentrations less than 100 nM. We designed anti-miR-181 LNA/DNA mixmers containing stretches of at least seven DNA bases (Figure 30.4b).

Although LNAs are not sensitive to RNase H, according to the rules developed by Erdmann, seven consecutive natural DNA bases are enough to allow activation of RNase H upon binding to RNA (Kurreck *et al.*, 2002). An LNA/DNA antisense oligonucleotide, complementary to miR-181a, abrogated miR-181 detection in Northern blots (Figure 30.4c). This effect was sequence specific, as it was not observed with mutated oligonucleotides. Inhibition of target miRNA lasted for at least several days in myoblast cells, and the LNA/DNA oligonucleotides persisted for the same duration, a time compatible with the functional analysis of miRNAs during *in vitro* differentiation of myoblast cell line (Naguibneva *et al.*, 2006b). Therefore, this approach is suitable to address miRNA function during myoblast terminal differentiation.

MiR-181 is required for muscle cell terminal differentiation

The anti-miR-181 LNA/DNA oligonucleotides, transfected into C2C12 cells, dramatically affected myoblast differentiation, as assessed both by myotube

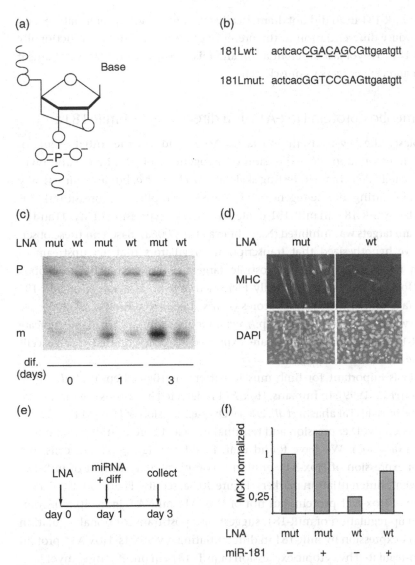

Figure 30.4. Inhibition of miR-181 affects myoblast differentiation. (a) Locked nucleic acid. (b) Sequences of anti-miR-181 LNAs; capitals are modified bases, mutations are underlined; oligonucleotide with four mutations has been used as control. (c) Northern blot. MiR-181 is undetectable after LNA transfection. (d) C2C12 cells were transfected with the anti-miR-181a antisense LNA (50 nM), either wild type or mutated; MHC expression was analyzed by immunofluorescence. (e, f) Rescue experiment. C2C12 cells were transfected with the mutant (mut) or wild type (WT) miR-181 antisense (miR-181) LNA oligonucleotide; after 24 h cells were transfected again with synthetic miR-181a (+) or a control double-stranded RNA sequence (−) and placed under differentiation conditions; at day 3 cell extracts were analyzed by Western blot. The MCK expression was normalized to α-tubulin.

formation and the expression of MHC and MCK (Figure 30.4d). Mutated sequences did not have any effect. This abnormal phenotype was rescued by transfection of synthetic double-stranded miR-181 into differentiating cells. These results demonstrate that the inhibitory effect was due to specific binding of antisense oligonucleotides to the complementary miR-181 (Figure 30.4e, f).

Synthetic miR-181 itself did not dramatically affect myoblast differentiation, and did not induce differentiation in the presence of serum. These data functionally link miR-181 to myoblast differentiation and raise the question of its mechanism of action and gene targets in muscle cells.

The homeobox protein Hox-A11 is a direct target of miR-181a

In myoblasts, the key transcription factor MyoD induces the initial events of terminal differentiation, i.e. expression of myogenin and p21. MyoD and myogenin are barely detectable in resting skeletal muscle *in vivo*, but are dramatically up-regulated during muscle regeneration, showing a profile of expression that is very similar to miR-181. In miR-181-depleted cells, the expression of MyoD and its downstream targets was inhibited (Naguibneva *et al*, 2006a). Based on these observations, we hypothesized that transcriptional regulators that act upstream of MyoD in myogenesis might be possible targets of miR-181. The homeobox protein Hox-A11 has been repeatedly identified *in silico* as a target of miR-181 isoforms (a, b and c) with various scores (Lewis *et al.*, 2003; Krek *et al.*, 2005; see http://www.microrna.org/ and http://pictar.bio.nyu.edu/). Our data indicate that miR-181a and b, but not c, are expressed in differentiated C2C12 cells (Naguibneva *et al.*, 2006a).

Hox-A11 is important for limb muscle patterning (Beauchemin *et al.*, 1994; Hashimoto *et al.*, 1999). In humans, Hox-A11 is detected in various organs including skeletal muscle (Takahashi *et al.*, 2004). Over-expression of Hox-A11 in C2C12 cells represses MyoD expression and terminal muscle differentiation (Yamamoto and Kuroiwa, 2003). We have found that, in differentiating C2C12 cells, the pattern of expression of Hox-A11 protein is complementary to that of miR-181 and myogenic differentiation markers (Figure 30.5a; see also Figure 30.2a). Down-regulation of Hox-A11 protein, but not of Hox-A11 mRNA (Figure 30.5b), coincides with up-regulation of miR-181, suggesting a post-transcriptional regulation of Hox-A11 expression by miR-181 in differentiating myoblasts. Hox-A11 protein was down-regulated by ectopic expression of miR-181a in proliferating myoblasts and, conversely, protein levels of Hox-A11 were up-regulated by miR-181 inhibition: the absolute level of Hox-A11 protein was higher in cells treated with the LNA antisense than in control cells (Naguibneva *et al.*, 2006a).

To test if the predicted miR-181 site in the 3′ untranslated region (UTR) of Hox-A11 mRNA is responsible for silencing of Hox-A11 expression by miR-181, we cloned tandem repeats of the putative target site downstream of a luciferase reporter, and co-transfected this vector with different miR-181 isoforms into miR-181-negative cells. Insertion of wild type sequences rendered the reporter sensitive to ectopic miR-181a, whereas mutation of the target sequences abolished this effect (Naguibneva *et al.*, 2006a). Interestingly, the effect was far more pronounced with miR-181a than with miR-181b; moreover, co-transfection of the 181b isoform did not increase inhibition by the 181a isoform in the luciferase assay.

We also tested the interactions between miR-181 and HoxA11 at the functional level. We reasoned that, if Hox-A11 is a direct target of miR-181a, and depletion of

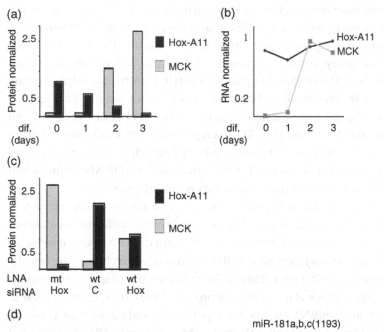

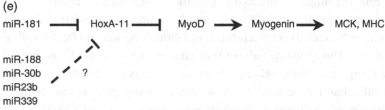

Figure 30.5. Hox-A11 is a target for miR-181. (a) Hox-A11 expression during myoblast cell differentiation *ex vivo*: extracts of differentiating C2C12 cells were analyzed by Western blot at indicated time points, using anti-Hox-A11 or anti MCK antibodies. Normalized to α-tubulin. (b) Hox-A11 and MCK expression analyzed by Q-RT-PCR. Normalized to 36b4. (c) Hox-A11 down-regulation suppresses the phenotype induced by miR-181 inhibition. C2C12 cells were transfected with a siRNA against Hox-A11 (Hox) or a control scrambled sequence (C) along with mutant (m) or wt (wt) miR-181 antisense LNAs as indicated, and placed in differentiation medium for 2 days. MCK and Hox-A11 expression was monitored by Western blot. Normalized to α-tubulin. (d) Predicted binding sites for several miRNAs in 3′ UTR of Hox-A11 (from PicTar: http://pictar.bio.nyu.edu). (e) The miR-181 pathway in terminally differentiating myoblasts.

miR-181a by the antisense LNA disturbs both Hox-A11 down-regulation and terminal myoblast differentiation; then artificial down-regulation of Hox-A11 by RNA interference might rescue the differentiation phenotype. Indeed, down-regulation of Hox-A11 with an siRNA significantly, albeit not fully, restored myoblast differentiation in cells treated with the miR-181 LNA antisense. Moreover, in these experiments, differentiation was inversely correlated with Hox-A11 expression (Figure 30.5c); Hox-A11 was absent in control cells that differentiated normally, and up-regulated in LNA-treated cells that did not differentiate. The siRNA reduced Hox-A11 protein levels in LNA-treated cells, resulting in an intermediate level of Hox-A11 protein as well as an intermediate level of differentiation. Together these data indicate that Hox-A11 is an important and direct target of miR-181. Since Hox-A11 is a repressor of myoblast differentiation *ex vivo*, our data support a model in which miR-181 participates in differentiation by alleviating repression by Hox-A11, which in turn results in MyoD induction and triggers the expression of muscle specific proteins (Figure 30.5e).

We noted, however, that inhibition of miR-181 by LNA did not completely abolish the down-regulation of Hox-A11 during myoblast differentiation. What might provide an additional control of Hox-A11 translation? It has been reported that a set of simultaneously expressed miRNAs can cooperatively regulate translation of a common mRNA target. This is, for example, the case for Myotrophin in pancreatic β-cells (Krek *et al.*, 2005). As predicted *in silico*, (PicTar), a total of 13 isoforms of nine miRNA species have potential binding sites in Hox-A11 mRNA. We have found that four of them, in addition to miR-181, are expressed and/or up-regulated in terminally differentiating myoblasts (Figure 30.5d and Table 30.1). The coordinate action of these miRNAs might lead to the complete disappearance of Hox-A11 upon differentiation.

Our results also suggest that Hox-A11 is not the only target of miR-181. Restoration of muscle differentiation in LNA treated cells by down-regulation of Hox-A11 with anti-Hox-A11 siRNA was only partial, probably as a result of the residual level of Hox-A11 protein observed in these cells (Figure 30.5c). Alternatively, other proteins regulated by miR-181 might be involved in terminal muscle differentiation.

Among the targets predicted *in silico*, we note some proteins relevant to muscle differentiation, such as bHLH-related inhibitor of differentiation and DNA binding Id2 or homolog of *Drosophila eyes absent* transcription activator Eya1 (see http://www.microrna.org), and many others, whose roles in this process are not as evident. On the other hand, in addition to its role in skeletal myogenesis, miR-181 was reported to be involved in B lymphocyte differentiation (Chen *et al.*, 2004). It will be important to determine whether mir-181 dependent pathways are common to the two cell types, and could be even more general for terminal differentiation processes. Interestingly, miR-181 has not been found in adult splenic B-lymphocytes. Up-regulation of miR-181 occurred very transiently, during the passage from the E4 to the E6 stage of pro-B lymphocyte differentiation (Monticelli *et al.*, 2005). In adult muscle, miR-181 is barely detectable, but it appears when active myoblast differentiation is launched upon regeneration.

Conclusions

Taken together, our data demonstrate that miR-181 is required for skeletal myoblast terminal differentiation. One of its important targets in this process is the homeobox protein Hox-A11. However, neither over-expression of miR-181 in proliferating myoblasts, nor down-regulation of Hox-A11 triggered myoblast terminal differentiation (Naguibneva *et al.*, 2006). We can conclude that miR-181 is necessary, but not sufficient, for differentiation and that it acts, like other miRNAs, in concert with a complex transcriptional network orchestrating muscle differentiation.

ACKNOWLEDGEMENTS

This work was supported by the Ligue Contre le Cancer and EU grant RIGHT IP, LSHB-CT-2004-005276.

REFERENCES

Ambros, V. (2003). MicroRNA pathways in flies and worms: growth, death, fat, stress, and timing. *Cell*, **113**, 673–676.

Bains, W., Ponte, P., Blau, H. and Kedes, L. (1984). Cardiac actin is the major actin gene product in skeletal muscle cell differentiation in vitro. *Molecular Cell Biology*, **4**, 1449–1453.

Beauchemin, M., Noiseux, N., Tremblay, M. and Savard, P. (1994). Expression of Hox A11 in the limb and the regeneration blastema of adult newt. *The International Journal of Developmental Biology*, **38**, 641–649.

Biemar, F., Zinzen, R., Ronshaugen, M. *et al.* (2005). Spatial regulation of microRNA gene expression in the *Drosophila* embryo. *Proceedings of the National Academy of Sciences USA*, **102**, 15 907–15 911.

Brennecke, J., Hipfner, D. R., Stark, A., Russell, R. B. and Cohen, S. M. (2003). bantam encodes a developmentally regulated microRNA that controls cell proliferation and regulates the proapoptotic gene hid in *Drosophila*. *Cell*, **113**, 25–36.

Brennecke, J., Stark, A., Russell, R. B. and Cohen, S. M. (2005). Principles of microRNA-target recognition. *Public Library of Science Biology*, **3**, 85.

Buckingham, M., Bajard, L., Chang, T. *et al.* (2003). The formation of skeletal muscle: from somite to limb. *Journal of Anatomy*, **202**, 59–68.

Charge, S. B. and Rudnicki, M. A. (2004). Cellular and molecular regulation of muscle regeneration. *Physiological Reviews*, **84**, 209–238.

Chen, C. Z., Li, L., Lodish, H. F. and Bartel, D. P. (2004). MicroRNAs modulate hematopoietic lineage differentiation. *Science*, **303**, 83–86.

Dinsmore, J., Ratliff, J., Jacoby, D., Wunderlich, M. and Lindberg, C. (1998). Embryonic stem cells as a model for studying regulation of cellular differentiation. *Theriogenology*, **49**, 145–151.

Hashimoto, K., Yokouchi, Y., Yamamoto, M. and Kuroiwa, A. (1999). Distinct signaling molecules control Hoxa-11 and Hoxa-13 expression in the muscle precursor and mesenchyme of the chick limb bud. *Development*, **126**, 2771–2783.

Houbaviy, H. B., Murray, M. F. and Sharp, P. A. (2003). Embryonic stem cell-specific MicroRNAs. *Developmental Cell*, **5**, 351–358.

Krek, A., Grun, D., Poy, M. N. *et al.* (2005). Combinatorial microRNA target predictions. *Nature Genetics*, **37**, 495–500.

Krichevsky, A. M., King, K. S., Donahue, C. P., Khrapko, K. and Kosik, K. S. (2003). A microRNA array reveals extensive regulation of microRNAs during brain development. *RNA*, **9**, 1274–1281.

Kurreck, J., Wyszko, E., Gillen, C. and Erdmann, V. A. (2002). Design of antisense oligonucleotides stabilized by locked nucleic acids. *Nucleic Acids Research*, **30**, 1911–1918.

Lagos-Quintana, M., Rauhut, R., Yalcin, A. *et al.* (2002). Identification of tissue-specific microRNAs from mouse. *Current Biology*, **12**, 735–739.

Lagos-Quintana, M., Rauhut, R., Meyer, J., Borkhardt, A. and Tuschl, T. (2003). New microRNAs from mouse and human. *RNA*, **9**, 175–179.

Lewis, B. P., Shih, I. H., Jones-Rhoades, M. W., Bartel, D. P. and Burge, C. B. (2003). Prediction of mammalian microRNA targets. *Cell*, **115**, 787–798.

Mansfield, J. H., Harfe, B. D., Nissen, R. *et al.* (2004). MicroRNA-responsive "sensor" transgenes uncover Hox-like and other developmentally regulated patterns of vertebrate microRNA expression. *Nature Genetics*, **36**, 1079–1083.

Monticelli, S., Ansel, K. M., Xiao, C. *et al.* (2005). MicroRNA profiling of the murine hematopoietic system. *Genome Biology*, **6**, R71.

Naguibneva, I., Ameyar-Zazoua M., Polesskaya A. *et al.* (2006a). The small non-coding RNA miR-181 targets the homeobox protein Hox-A11 during mammalian myoblast differentiation. *Nature Cell Biology*, **8**, 278–284.

Naguibneva, I., Ameyar-Zazoua, M., Nonne, N. *et al.* (2006b). An LNA-based loss-of-function assay for micro-RNAs. *Biomedicine & Pharmacotherapy*, **60**, 633–638.

Petersen, M. and Wengel, J. (2003). LNA: a versatile tool for therapeutics and genomics. *Trends in Biotechnology*, **21**, 74–81.

Polesskaya, A., Cuvellier, S., Naguibneva, I. *et al.* (2007). Lin-28 binds IGF-2 mRNA and participates in skeletal myogenesis by increasing translation efficiency. *Genes and Development*, **21**, 1125–1138.

Poy, M. N., Eliasson, L., Krutzfeldt, J. *et al.* (2004). A pancreatic islet-specific microRNA regulates insulin secretion. *Nature*, **432**, 226–230.

Ramkissoon, S. H., Mainwaring, L. A., Ogasawara, Y. *et al.* (2005). Hematopoietic-specific microRNA expression in human cells. *Leukemia Research*, **30**, 643–647.

Schulman, B. R., Esquela-Kerscher, A. and Slack, F. (2005). Reciprocal expression of lin-41 and the microRNAs let-7 and mir-125 during mouse embryogenesis. *Developmental Dynamics*, **234**, 1046–1054.

Sempere, L. F., Sokol, N. S., Dubrovsky, E. B., Berger, E. M. and Ambros, V. (2003). Temporal regulation of microRNA expression in Drosophila melanogaster mediated by hormonal signals and broad-Complex gene activity. *Developmental Biology*, **259**, 9–18.

Sempere, L. F., Freemantle, S., Pitha-Rowe, I. *et al.* (2004). Expression profiling of mammalian microRNAs uncovers a subset of brain-expressed microRNAs with possible roles in murine and human neuronal differentiation. *Genome Biology*, **5**, R13.

Tajbakhsh, S., Rocancourt, D., Cossu, G. and Buckingham, M. (1997). Redefining the genetic hierarchies controlling skeletal myogenesis: Pax-3 and Myf-5 act upstream of MyoD. *Cell*, **89**, 127–138.

Takahashi, Y., Hamada, J., Murakawa, K. *et al.* (2004). Expression profiles of 39 HOX genes in normal human adult organs and anaplastic thyroid cancer cell lines by quantitative real-time RT-PCR system. *Experimental Cell Research*, **293**, 144–153.

Wienholds, E., Kloosterman, W. P., Miska, E. *et al.* (2005). MicroRNA expression in zebrafish embryonic development. *Science*, **309**, 310–311.

Xie, X., Lu, J., Kulbokas, E. J. *et al.* (2005). Systematic discovery of regulatory motifs in human promoters and 3′ UTRs by comparison of several mammals. *Nature*, **434**, 338–345.

Xu, P., Vernooy, S. Y., Guo, M. and Hay, B. A. (2003). The *Drosophila* microRNA Mir-14 suppresses cell death and is required for normal fat metabolism. *Current Biology*, **13**, 790–795.

Yamamoto, M. and Kuroiwa, A. (2003). Hoxa-11 and Hoxa-13 are involved in repression of MyoD during limb muscle development. *Development, Growth & Differentiation*, **45**, 485–498.

Yang, D. H. and Moss, E. G. (2003). Temporally regulated expression of Lin-28 in diverse tissues of the developing mouse. *Gene Expression Patterns*, **3**, 719–726.

Zhao, Y., Samal, E. and Srivastava, D. (2005). Serum response factor regulates a muscle-specific microRNA that targets Hand2 during cardiogenesis. *Nature*, **436**, 214–220.

31 Identification and potential function of viral microRNAs

Finn Grey*, Alec J. Hirsch and Jay A. Nelson

The discovery of RNA interference (RNAi) and microRNAs (miRNAs) is undoubtedly one of the most significant recent advances in the field of biology. miRNAs were initially identified in *Caenorhabditis elegans* with the discovery of a small RNA, lin-4, that was shown to regulate the heterochronic gene *lin-14* (Lee *et al.*, 1993; Wightman *et al.*, 1993). Further investigations led to the identification of a second small RNA, let-7, that played a similar role in regulation of developmental genes (Reinhart *et al.*, 2000). Subsequent studies using extensive cloning strategies and bioinformatics methods have identified hundreds of miRNA genes in plants and animals suggesting that post-transcriptional regulation through expression of small RNAs is an evolutionarily conserved and common mechanism of gene regulation (Bartel, 2004; Pfeffer *et al.*, 2004; Pfeffer *et al.*, 2005). More recent studies have revealed that a surprisingly large percentage of genes in higher organisms may be regulated by miRNAs (Brennecke *et al.*, 2005; Grün *et al.*, 2005; Krek *et al.*, 2005; Lewis *et al.*, 2005; Xie *et al.*, 2005). Given the widespread prevalence and influential effects of miRNAs on gene expression it is unsurprising that viruses exploit RNAi pathways by expressing their own small RNAs. In this chapter we will review what is currently known about virally encoded miRNAs, including examination of their expression, genomic position and degree of evolutionary conservation between related viruses. We will also discuss the potential functions of viral miRNAs.

Identification of viral miRNAs

The initial discovery of miRNAs was achieved by using forward genetic studies that mapped specific genomic regions responsible for observed developmental phenotypes (Lee *et al.*, 1993). Subsequent strategies have utilized extensive cloning and sequencing of small RNAs as well as bioinformatics approaches. These studies have discovered numerous miRNA genes throughout both the animal and plant kingdoms. So far over 1000 miRNA genes have been identified, with

---

* Author to whom correspondence should be addressed.

MicroRNAs: From Basic Science to Disease Biology, ed. Krishnarao Appasani. Published by Cambridge University Press. © Cambridge University Press 2008.

estimates based on bioinformatics studies suggesting that this number may rise significantly in the future (Bartel, 2004; Bentwich *et al.*, 2005; Berezikov *et al.*, 2005). There are a number of advantages and disadvantages associated with both methods of identifying miRNA genes – some of which are particularly relevant to identifying miRNA genes within viruses. Direct cloning of miRNAs relies on size fractionation of total cellular RNA followed by ligation of "linker" oligonucleotides to the 5′ and 3′ ends of the small RNAs. Following reverse transcription and PCR amplification using primers specific to the ligated oligonucleotide linkers, the cDNAs are concatamerized and cloned. Multiple clones are sequenced and analyzed for their ability to form stem-loop structures within the context of the genomic sequence surrounding the identified cDNA. The advantage of this method is the low number of "false-positives," since any sequence identified is almost certainly expressed. Direct cloning also enables precise determination of the 5′ and 3′ ends of the miRNA, an important facet when determining miRNA target sites and function. However, direct cloning is biased toward identifying miRNAs of high abundance and may fail to clone miRNAs that are expressed at low levels, tightly regulated or refractory to the cloning process.

In contrast to direct cloning, bioinformatic-based techniques do not rely on an ability to isolate and clone miRNAs and are therefore not limited by miRNA levels within a specific sample. Analysis of known miRNA transcripts has led to the identification of structural and sequence motifs characteristic of miRNA precursors. These motifs include sequences that are able to form short stem-loop secondary structures, evolutionary conservation of the stem-loop and a specifically high level of sequence conservation of the miRNA sequence within this stem-loop. These characteristics have been exploited to design computational methods to predict miRNA genes based on sequence information (Lim *et al.*, 2003). The disadvantage of bioinformatics methods is a potentially higher rate of "false-negative" sequences and a lower confidence in the exact 5′ and 3′ ends of the miRNA within the stem-loop sequence. Subsequent biochemical techniques such as specific PCR amplification or RNase protection are required to map the exact ends of predicted miRNAs. Both techniques have been successfully utilized in identifying virally encoded miRNAs. In the following sections we will review the viruses that have been demonstrated to express miRNAs.

Herpesviruses

Herpesviruses belong to a large family of enveloped, double-stranded DNA viruses disseminated throughout nature (Roizman and Pellet, 2001). Membership of the herpesviridae family is determined by virion morphology; however, the complex pattern of transcriptional gene expression during acute infection and during the establishment, maintenance and reactivation from latency is more intriguing when considering these viruses in the context of miRNA function (Roizman and Pellet, 2001). Given the requirements for multiple levels of transcriptional regulation during the different stages of the viral lifecycle, it is not surprising that these viruses have evolved the ability to encode miRNAs. It is, in fact, likely that herpesviruses have been utilizing RNA interference to regulate gene expression for millions of years.

Epstein–Barr virus

The initial identification of viral miRNAs was achieved by cloning small RNAs from a B cell line latently infected with Epstein–Barr virus (EBV) (Pfeffer *et al.*, 2004). EBV was initially identified as the etiological agent responsible for the lymphoproliferative disease, Burkitt's lymphoma. EBV is the prototypic member of the gamma herpesvirus subfamily and in addition to Burkitt's lymphoma, is associated with a number of lymphoproliferative diseases in humans, including: Castleman's disease, Hodgkin's disease and primary effusion lymphoma (PEL). Following initial infection of the mucosal epithelia, EBV establishes a latent infection in B lymphocytes and has a powerful transforming ability in these cells. Following the initial investigation by Pfeffer *et al.*, EBV was found to encode five miRNAs clustered within two genomic regions (Figure 31.1a). miR-BHRF1-1, 2 and 3 are located within the untranslated region (UTR) of BHRF1, an anti-apoptosis Bcl-2 homolog. miR-BHRF1-1 is located within the 5′ UTR with miR-BHRF1-2 and 3 encoded within the 3′ UTR. miR-BART-1 and 2 are located within intronic regions of the BART family of transcripts, which are extensively spliced. The BART family of transcripts have been suggested to encode a number of proteins, although it remains to be convincingly demonstrated that these proteins are expressed during viral infection (Smith *et al.*, 2000).

Interestingly, miR-BART-2 is directly complementary to the viral DNA polymerase transcript BALF5. Given this perfect complementarity it would be predicted that co-expression would result in cleavage of BALF5, suggesting a clear functional role for this miRNA in the control of acute gene expression and inhibition of viral DNA replication (Pfeffer *et al.*, 2004). Furthermore, an earlier study identified a smaller BALF5 transcript exhibiting a 3′ end that mapped precisely to the predicted miR-BART2 cleavage site (Furnari *et al.*, 1992). However, the biological significance of these findings remains unclear as the expression of the BALF5 cleavage product was detected in only a single EBV strain, B95-8, whereas full length BALF5 transcript was readily detected in all other strains (Furnari *et al.*, 1992; Sullivan and Ganem, 2005). These findings indicate that either BALF5 is resistant to cleavage by miR-BART2 or levels of BALF5 expression during acute replication overcome negative regulation by miR-BART-2. As it is unclear whether miR-BART2 is expressed during acute replication it may also be possible that miR-BART2 and BALF5 transcripts are expressed in a mutually exclusive manner, with BALF5 being expressed following reactivation of latent virus, and EBV miR-BART-2 expressed only during latent infection. A large deletion exists upstream of the BART transcripts that is unique to B95-8, which could conceivably affect the regulation of miR-BART-2 and lead to co-expression with BALF5 following reactivation. This could result in accumulation of the BALF5 cleavage product. However, the original study by Furnari *et al.*, suggested that the aberrant BALF5 transcript of B95-8 is stable and poly-adenylated, a situation that would not be expected to occur following miRNA cleavage. Clearly further analysis of gene

HSV-1

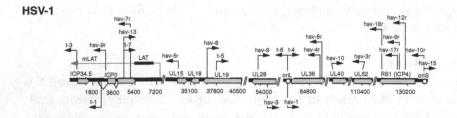

HCMV

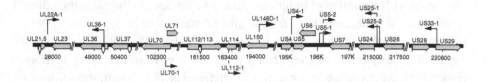

EBV   **KSHV**

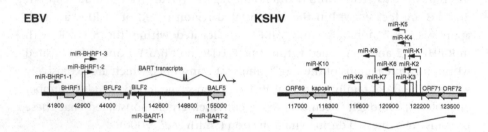

MHV-68

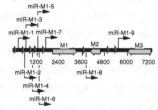

Figure 31.1. Schematic representations of herpesvirus miRNAs (black arrows) in relation to the surrounding ORFs (green arrows). Minor LAT of HSV-1 is shown in red. Position of MHV-68 "tRNAs" are shown as red arrows. Nucleotide coordinates refere to Genbank accession numbers; HSV-1 – NC_001806; HCMV – X17403; EBV – AJ507799; KSHV – AF14880. Coordinates for UL148D-1 of HCMV correspond to genbank accession number AC146906. (See color plate 22)

regulation during latency and reactivation of EBV may result in a better understanding of how these miRNAs and the associated target transcripts are regulated.

Kaposi's sarcoma-associated herpesvirus

Cloning studies by multiple groups identified viral miRNAs encoded by a second human gamma herpesvirus, Karposi's sarcoma-associated herpesvirus (KSHV) (Cai *et al.*, 2005; Pfeffer *et al.*, 2005; Samols *et al.*, 2005). Similar to EBV, KSHV infection is associated with a number of tumors, most notably Kaposi's sarcoma, a skin malignancy common amongst AIDS patients and older men of

Mediterranean descent (Moore and Chang, 2001). In common with all herpes-viruses, KSHV has a complex viral replication cycle involving initial acute replica-tion followed by life-long latent infection with the ability to reactivate (Moore and Chang, 2001). In three independent studies a total of 11 distinct miRNAs were cloned from two B cell lines derived from latently infected lymphomas. Ten of the 11 miRNAs are encoded within an approximately 4 kilobase (kb) region previously thought to be non-coding, between open reading frames (ORF) K12 (kaposin) and ORF71. A further miRNA resides within the coding region of the Kaposin gene. Like EBV, KSHV latent infection is characterized by limited gene expression. Many of the genes expressed during latent infection flank the geno-mic region encoding the viral miRNAs. Detailed mapping of the latent transcripts in two independent studies have subsequently identified three overlapping tran-scripts driven by two separate promoters that have the potential to encode 10 of the miRNAs within a single intron (Pearce *et al.*, 2005; Cai and Cullen, 2006). All 10 miRNAs were detected following transient transfection of the genomic region encoding the transcripts strongly suggesting that all 10 miRNAs are in fact encoded from within the identified intron (Cai and Cullen, 2006). The three spliced transcripts are also predicted to encode the kaposin ORF that contains the additional miRNA. As the transcript would be cleaved to release the pre-miRNA stem-loop, processing of this miRNA by Drosha would theoretically preclude expression of Kaposin protein leading to a novel form of gene regulation. However, as both miRNA and Kaposin protein are detected during latent infec-tion, sufficient Kaposin transcript apparently escapes Drosha processing to allow for translation of the Kaposin mRNA.

Regardless of the precise mechanism, the close association of the viral miRNAs with the expression of Kaposin does suggest the possibility of a functional associa-tion. The Kaposin gene encodes for a number of different proteins, including Kaposin B, which was recently shown to bind and activate MAPK-associated protein kinase 2 (MK2) (McCormick and Ganem, 2005). Activation of MK2 by Kaposin B leads to an increase in cytokine expression in KSHV infected cells by blocking the destabilizing effect of AU rich elements (AREs) within the 3' UTRs of cytokine mRNAs. Whether the miRNAs possess any functional relevance to this mechanism or to one of the Kaposin gene products, or whether the virus has evolved the mechanism as a simple means of utilizing the same latent promoters to express multiple gene products awaits further investigation.

Murine herpesvirus-68
A third member of the gamma herpesviruses, and a distant relative of EBV and KSHV, has been shown to express miRNAs during latent infection (Pfeffer *et al.*, 2005). Murine herpesvirus-68 (MHV-68) infects and establishes a latent infection within B cells of the germinal centers of the spleen. Following cDNA cloning of small RNAs from a stably infected B cell line, nine miRNAs were identified, again clustered within a genomic region of the virus known to encode latency-associated transcripts. Of particular interest is the existence of eight virally encoded tRNA-like transcripts (vtRNA) co-linear with the identified miRNAs.

Northern blot analysis and secondary RNA structure predictions suggest that these miRNAs are encoded within two terminal stem-loops down-stream of each vtRNA transcript. These sequences are also predicted to encode a further six potential miRNAs. However, these miRNAs have so far not been detected. Each of the vtRNAs are driven by an RNA polymerase III promoter, and would therefore be predicted to drive expression of the associated miRNAs. This situation is unique to viruses, with all other cellular miRNA transcripts identified being driven by RNA polymerase II transcription. *In vivo* analysis of mouse tissues following MHV-68 infection has identified expression of vtRNAs during both lytic and latent infection (Bowden *et al.*, 1997). Although no function has been identified for either the MHV-68 miRNAs or the vtRNAs, vtRNAs 1-4 (and their associated miRNAs) can be deleted from the virus with no obvious effect on the virus's ability to establish or reactivate from latency, suggesting either a subtle role for these miRNAs or a certain level of redundancy in their actions (Simas *et al.*, 1998).

Human cytomegalovirus

A number of studies, including our own, have identified viral miRNAs encoded by the beta herpesvirus, human cytomegalovirus (HCMV) (Dunn *et al.*, 2005; Grey *et al.*, 2005; Pfeffer *et al.*, 2005). In contrast to the other herpesviruses, HCMV miRNA expression was identified during the lytic phase of replication rather than the latent stage. HCMV infects a large percentage of the population and although benign in the majority of immunocompetent individuals, can cause serious disease in immunocompromised patients. Similar to other herpesviruses, HCMV is thought to have an initial acute replication phase followed by the establishment of a life long latent infection, although a low level of persistent infection has not been ruled out as a relevant mechanism of viral maintenance. Strategies using both direct cloning and bioinformatics have identified 12 miRNAs, four of which originate from both arms of a single pre-miRNA stem-loop. Unlike the gamma herpesviruses, miRNAs encoded by HCMV are located throughout the viral genome (Grey *et al.*, 2005; Pfeffer *et al.*, 2005). Whether this is a reflection of the differences between gamma and beta herpesviruses or between the expression patterns of viral miRNAs during lytic versus latent infection is unclear. However, bioinformatics analysis, which is based purely on sequence data and is therefore unaffected by viral replication timing, did not identify any clusters of predicted miRNAs, suggesting that the genomic positioning more likely reflects differences between gamma herpesviruses and beta herpesviruses rather than lytic versus latent (Grey *et al.*, 2005; Pfeffer *et al.*, 2005).

The bioinformatics approach taken by our laboratory utilized a computer algorithm called Stem-loop Finder (SLF) (Combimatrix) to predict potential RNA transcripts from the HCMV genome that could form stem-loop secondary structures. This algorithm uses free energy calculations to determine the theoretical stability of the base pairing within the stem region, including pairing between G:U bases, while maintaining a maximum and minimum length between the complementary base pairing to determine loop size. A scoring matrix then weights beneficial or detrimental folding structures and attributes a cumulative score to each potential stem-loop sequence. Analysis of the HCMV genome using SLF identified 406 potential stem-loop sequences. Previously identified miRNAs are often extensively conserved between many different species (Pasquinelli *et al.*, 2000). Consequently, we hypothesized that functionally important miRNAs expressed by HCMV would be conserved between HCMV and closely related viruses such as chimpanzee cytomegalovirus (CCMV). The sequence of each of the 406 candidate SLF-derived HCMV stem-loop transcripts was compared with the CCMV genome for potential homology. A minimum score of 60% homology with CCMV was used to select stem-loop sequences for further analysis. Our preliminary studies determined that this level of homology was sufficiently stringent to identify significant sequence conservation without exclusion of any potential miRNAs. Of the 406 sequences analyzed, 110 potential stem-loop sequences scored higher than 60% homology.

The 110 HCMV stem-loop sequences selected using the criteria (detailed above), and the corresponding homologous CCMV sequences were then analyzed using a bioinformatics program (MiRscan) (Lim *et al.*,2003) to predict which of the conserved stem-loop sequences had a high probability of encoding genuine miRNA transcripts. The MiRscan program compares two sequences and provides a score based on a number of aspects such as the ability to form a stem-loop, symmetry of bulge loops and conservation of the predicted miRNA sequence within the stem-loop. Specific sequence bias within the 5' region of the miRNA, such as a propensity for the first base of the miRNA to be a uracil, was also a criteria for the program. The program then predicts a candidate miRNA that would be generated from the pre-miRNA stem-loop structure. In a similar previous study, stem-loop sequences that scored higher than 10 using the MiRscan algorithm were experimentally tested (Lim *et al.*, 2003). The 110 HCMV stem-loop sequences selected using the criteria mentioned above and the corresponding homologous CCMV sequences were analyzed using the MiRscan program. Of the 110 sequences analyzed, 13 scored higher than 10 using the MiRscan algorithm. To determine whether the predicted miRNAs were expressed by HCMV, Northern blot analysis of total RNA from infected cells was performed using end-labeled oligonucleotide probes. Following infection of primary fibroblast cells with HCMV AD169 (a laboratory strain), total RNA was harvested at 2, 8, 24, 48 and 72 h post infection (pi). Four of the predicted miRNAs, UL36-1, US4-1, US5-1 and US5-2, were detected in infected RNA samples, with an additional miRNA, UL70-1, detected using a probe to the complementary strand of the stem-loop (Figure 31.2). The majority of the viral miRNAs were first detected 24 h pi and continued to accumulate over time with peak levels being detected 72 h pi.

Similar to all herpesviruses, expression of HCMV transcripts can be grouped into three kinetic classes, immediate early, early or late, based on their requirement for expression of viral protein and viral DNA replication (DeMarchi *et al.*, 1980). To determine the specific kinetic class of each of the viral miRNAs, viral infections were performed in the presence of either cycloheximide, which blocks protein translation suggesting early gene expression, or Foscarnet which blocks DNA replication indicating late gene expression. Total RNA samples were harvested 36 h pi followed by Northern blot analysis. Figure 31.3 demonstrates that cycloheximide but not

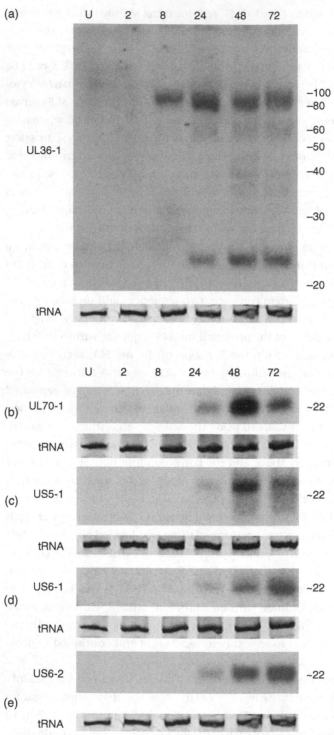

Figure 31.2. Northern blot analysis of HCMV expressed miRNAs. Human fibroblast cells were infected at a multiplicity of five plaque forming units per cell. Total RNA was harvested and subjected to Northern blot analysis using probes specific for predicted viral miRNA sequences. (a) Expression of UL36-1 miRNA as well as the ~80 base pre-miRNA species. (b)–(e) The expression of miRNAs UL70-1, US5-1, US6-1 and US6-2. Ethidium bromide staining of tRNA band shown as loading control for each blot. Lanes: U, uninfected; 2, 8, 24, 48, 72 h post infection. Position of radioactive size markers indicated on right of blot.

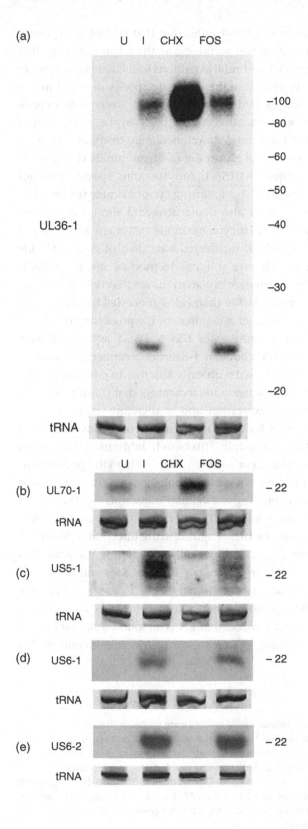

Foscarnet blocked the expression of UL36-1 indicating that UL36-1 is expressed with early kinetics. This observation was surprising as the UL36 transcript that contains the intron encoding the UL36-1 miRNA exhibits immediate early kinetics (Wathen and Stinski, 1982). Furthermore the ~80 base species detected in the initial Northern blot time course was detected at elevated levels in the cycloheximide sample. One potential explanation for the block in the production of mature miRNAs would be that the extended cycloheximide treatment depleted proteins, such as dicer, required for the processing of mature miRNAs. However, this expression pattern was unique to UL36-1; corresponding species were not detected for UL70-1, US4-1, US5-1 or US5-2 following cycloheximide treatment.

Investigation of earlier time points also demonstrated a similar expression pattern for UL36-1 with as little as 14 h cycloheximide treatment and levels of the cellular miRNA, miR-22, remained unaffected after 36 h of cycloheximide treatment, suggesting that the cells were still able to produce mature miRNAs (data not shown). Two possible alternative explanations are that the UL36-1 pre-miRNA is transcribed independently rather than being processed from the UL36 intron or that an early viral gene product is required for the processing of UL36-1 from the pre-miRNA transcript. Expression of US4-1, US5-1 and US5-2 were blocked by cycloheximide treatment but not Foscarnet treatment, indicating that these transcripts were also expressed with early kinetics. In contrast UL70-1 was expressed in the presence of cycloheximide indicating that this miRNA does not require *de novo* viral protein expression and is therefore expressed with immediate early kinetics. The probe for UL70-1 also consistently cross-hybridized to a species in the uninfected control sample. This cross hybridization is unique to the UL70-1 sequence and does not occur with any of the other viral probes. This observation may suggest that UL70-1 has some homology to a cellular miRNA or that the probe is non-specifically binding to a cellular transcript.

In contrast to the bioinformatics approach Pfeffer *et al.* used a cloning technique to identify miRNAs encoded by HCMV during acute infection of fibroblast cells. A total of 11 miRNAs were identified by cloning, including three miRNAs from the same stem-loops identified in our study (UL36-1, US5-1 and US5-2). To extend the initial identification of these additional miRNAs, Northern blot analysis was used to determine the expression kinetics during HCMV acute infection. Figure 31.4 indicates that the virally encoded miRNAs all follow a similar expression pattern with levels continuing to increase over time. The kinetic studies

---

Figure 31.3. (cont.)
Expression kinetics of HCMV miRNAs. Human fibroblast cells were treated with either cycloheximide, Foscarnet or with no drug and infected at a multiplicity of 5 pfu per cell. At 36 hours total RNA was harvested and subjected to Northern blot analysis. Total RNA from infected cells was harvested as a negative control. Lanes; U, uninfected; I, infected no drug; CHX, cycloheximide treated; FOS, foscarnet treated. (a) Expression of UL36-1 miRNA and potential pre-miRNA sequence. (b)–(e) Expression of UL70-1, US5-1, US6-1 and US6-2. Position of radioactive size markers indicated on the right of the blots. tRNA band loading control visualized by ethidium bromide staining of polyacrylamide gel.

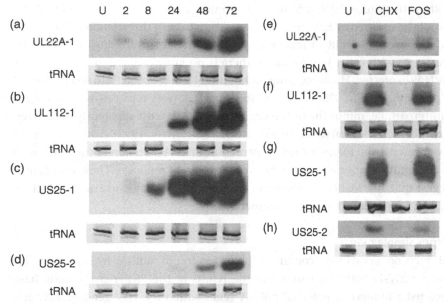

Figure 31.4. Northern blot analysis of HCMV expressed miRNAs. Human fibroblast cells were infected at a multiplicity of five plaque forming units per cell. Total RNA was harvested and subjected to Northern blot analysis using probes specific for predicted viral miRNA sequences. (a)–(d) The expression of miRNAs UL22A-1, UL112-1, US25-1 and US26-2. Ethidium bromide staining of tRNA band shown as loading control for each blot. Lanes: U, uninfected; 2, 8, 24, 48, 72 h post infection. (e)–(h) Expression of miRNAs in presence of cycloheximide, Foscarnet or no drug. Uninfected control shown. Lanes: U, uninfected; I, infected no drug; CHX, cycloheximide treated; FOS, Foscarnet treated.

also indicate that all the miRNAs with the exception of UL70-1 are expressed with early kinetics. We were unable to detect US33-1 by Northern blot analysis.

Four of the identified miRNAs are within coding regions of HCMV and as all four are directly antisense to the ORFs of the annotated genes they are predicted to cleave the expressed mRNAs. One of the open reading frames, UL70, is an essential gene, disruption of which leads to an inability of the virus to replicate (Dunn *et al.*, 2003; Yu *et al.*, 2003). UL114 encodes the uracil DNA glycosylase that is required for efficient DNA replication and lytic viral production (Prichard *et al.*, 1996; Courcelle *et al.*, 2001). The function of US29 is unknown, as is the function of UL150. We have recently demonstrated that transient expression of HCMV miRNAs are capable of down regulating reporter mRNAs containing exact complementary target sites within their 3' UTRs, suggesting that the viral transcripts antisense to the miRNAs would indeed be cleaved. Whether regulation of these transcripts by direct cleavage is the biologically relevant function of these miRNAs is not clear and will require further investigation. Also of interest is the miRNA encoded within the intronic region of UL36. UL36 belongs to a family of spliced transcripts that encode for proteins that block apoptosis signaling within infected cells (Goldmacher *et al.*, 1999; Skaletskaya *et al.*, 2001).

It has been suggested that co-regulation of miRNAs within introns may suggest a related function between the small RNA and the protein coding transcript it is derived from. The co-expression of UL36 and UL36-1 is analogous to the BHRF1 miRNAs that are generated from the intronic region of the anti-apoptotic Bcl-2 homolog expressed by EBV. It will be interesting to see if this relationship between coding region and miRNA may be a common strategy utilized by other herpesviruses. As with other herpesviruses, there is no known function for any of the miRNAs expressed. However, a number of miRNAs have been fortuitously deleted in previous studies. The coding region for UL36 was replaced with its cDNA counterpart, i.e. minus the intron encoding UL36-1, with apparently no deleterious effects on the viral replication (Patterson and Shenk, 1999). Also, US5-1 and 2 have been deleted following the construction of a BAC version of the virus, with no apparent negative effects on the virus's ability to replicate. These results may suggest that, like MHV-68, deletion of HCMV miRNAs may cause subtle effects or there may be redundancy in their function.

Herpes simplex virus

Although no reports have confirmed the expression of miRNAs by herpes simplex virus-1 (HSV-1) both our results and the results of the Pfeffer *et al.*, study have predicted a number of potential miRNA genes within the HSV genome. HSV is a member of the alpha herpesvirus subfamily and is neurotropic in nature, initially infecting the mucosal epithelia before establishing a latent infection in the sensory neurons of the trigeminal ganglia. Reactivation of HSV-1 is associated with the occurrence of the common cold sore around the mouth, but can cause more severe infections in immunocompromised patients. Unlike HCMV, a number of animal models of infection are available as HSV-1 infects a range of different species. A number of well-established *in vitro* models of latency also exist. Furthermore, HSV replicates rapidly *in vitro*, unlike EBV and KSHV, allowing for the analysis of the lytic stage of viral replication. These features make HSV an attractive model for the study of viral miRNAs. Of particular interest is the expression of a latency specific, large non-coding RNA known as the latency associated transcript (LAT). Expressed exclusively within infected neurons, LATs are the only transcript known to be expressed by HSV-1 during latency. Although no function has been definitively shown for this transcript it has been suggested to be involved in restriction of viral trans-activating proteins and in blocking cellular apoptosis in response to infection, both by unknown mechanisms (Branco and Fraser, 2005; Thompson and Sawtell, 2001). Both Pfeffer's study and our own predictions suggest the possibility of miRNAs being expressed within this transcript and it will be interesting to see if they prove to be correct.

Adenovirus

Adenoviruses have long been known to express small ($\sim$160 nt) non-coding transcripts designated virus associated (VA) RNAs (Shenk, 2001). The VA RNAs are transcribed by RNA polymerase III beginning in the early phase of viral infection, and synthesis continues at higher levels through late times post infection. Many adenovirus serotypes, including Ad2 and Ad5, encode two VA RNAs, designated VAI and VAII, while other serotypes encode a single VA RNA. VAI is expressed in abundance – reaching levels of 10^8 copies per cell, while VAII is present at roughly

10-fold lower concentrations. VAI has been shown to bind to and block the activation of protein kinase R (PKR), which is normally activated by double stranded RNA (presumably present in adenovirus infected cells owing to the bi-directional transcription of the viral genome). Activated PKR phosphorylates eIF-2a, resulting in a block in translation and inhibition of viral replication.

In 2004, Lu and Cullen asked if VAI could also inhibit miRNA function in mammalian cells (Lu and Cullen, 2004). Their initial approach used a luciferase reporter to assay miRNA function. In this system, sites complementary to a known miRNA are cloned into the 3' UTR of a luciferase gene in an expression plasmid. The miRNA of interest is co-expressed as a pri-miRNA or pre-miRNA resulting in a decrease in luciferase activity. Lu and Cullen found that co-expression of VAI restored luciferase activity in the presence of a pri-miRNA expressed from a RNA polymerase II driven expression plasmid, a pre-miRNA transcribed by RNA polymerase III, or a short hairpin RNA (also equivalent to a pre-miRNA) directly transfected into cells. Interestingly, a transfected siRNA duplex was still able to inhibit luciferase expression even in the presence of VAI. Consistent with these results, VAI was found to compete with miRNAs for Exportin 5 and Dicer. Both these proteins are required for miRNA biogenesis from either type of DNA plasmid, while Dicer is required for processing of the transfected short hairpin RNA. The fact that a transfected siRNA is still capable of inhibition of the luciferase reporter clearly shows that VAI does not interfere with the function of the RISC complex.

Soon after this report appeared, Andersson et al., reported similar results with VAI and VAII, and also found that both VA RNAs not only bind Dicer but are also processed into miRNAs (Andersson et al., 2005). Two additional groups confirmed these findings for VAI, and all three agree that the small RNAs are derived from the VA RNA terminal stem (Andersson et al., 2005; Aparicio et al., 2006; Sano et al., 2006). Interestingly, this region is distinct from the region associated with inhibition of PKR activity. The resulting miRNAs, when co-transfected into cells together with a luciferase reporter containing complementary sequence in the 3' UTR, are capable of suppressing luciferase expression, suggesting that the miRNAs are functional. Additionally, association of VA derived miRNAs with the RISC complex was shown by co-immunoprecipitation of the miRNAs with the RISC protein Argonaute as well as by detection of VA specific RISC activity in adenovirus infected cell extracts (Andersson et al., 2005; Aparicio et al., 2006). In order to show a role for the VA miRNAs in viral replication, Aparicio et al. transfected a 2'-O-methyl RNA antisense to the VAI miRNA into adenovirus infected cells. This resulted in a modest reduction in viral growth (Aparicio et al. 2006). Although the 2'-O-methyl RNA does not overlap the VAI region associated with PKR inhibition, it was not ruled out that the 2'-O-methyl RNA interferes with this activity. It is therefore unclear as to the conclusions that can be drawn regarding the functional effect of blocking the VAI miRNAs on viral infection via this method.

Sano et al., hypothesized that the role of the VA miRNAs may involve regulation of cellular genes, and used a computational approach to identify targets with perfect or imperfect complementarity to the VAI miRNA (Sano et al., 2006).

Although no perfect matches were found, several imperfect matches in the 3′ UTR were identified and these sites were cloned into the 3′ UTR of a luciferase reporter. Interestingly, the only site that mediated VA specific repression of luciferase was found in the 3′ UTR of the dicer gene. However, a 1.3 kb sequence derived from the native dicer 3′ UTR was unable to confer sensitivity to VAI miRNAs, leaving unanswered the question as to whether this represents a bona fide target.

Polyomavirus

Simian virus 40

A computational search for stem-loop structures with pre-miRNA characteristics identified only two potential candidates in the genome of the polyoma virus Simian virus 40 (SV40) (Sullivan *et al.*, 2005). One of the candidates was detected by Northern blot analysis of RNA from SV40 infected cells. The identified pre-miRNA overlaps a previously described small RNA called SV40 associated small RNA (SAS), most likely the pri-miRNA transcript. Additionally, the miRNA is fully complementary to the mRNA encoding the large T-antigen (T-Ag), which is transcribed from the opposite strand of the genome. Because of the exact complementarity, the miRNA would be expected to mediate cleavage of the T-Ag mRNA. Consistent with this hypothesis, the predicted 3′ cleavage fragment was detected in SV40 infected cells by Northern blot and RNase protection analysis.

In order to assess what role the miRNA plays during infection, a mutant virus was constructed in which the pre-miRNA hairpin has been disrupted by changes in the nucleotide sequence that leave the T-Ag amino acid sequence intact. In cells infected with this mutant, the 3′ cleavage product of the mRNA was no longer observed, and the mutant virus produced levels of T-Ag mRNA and protein higher than the wild-type virus throughout the course of infection. However, the mutant virus replicated with identical kinetics when compared to the wild type virus in cell culture, indicating that an excess of T-Ag is not deleterious to viral replication *in vitro*. The authors of this study reasoned that modulation of T-Ag expression may in fact be a mechanism of immune evasion, resulting in less T-Ag presentation by class I MHC to cytotoxic T lymphocytes. In order to test this, cells infected with wild-type SV40 or the mutant were incubated with a CTL clone specific for a region of T-Ag. As expected, cells expressing more T-Ag (i.e. those infected with the miRNA mutant) are more susceptible to CTL mediated cell lysis than cells infected with the wild type virus.

Retroviridae

Human immunodeficiency virus (HIV)

There are several reports which examine possible miRNA expression by HIV. Bennasser *et al.* identifed an miRNA encoded within the *env* gene and confirmed its expression by Northern blot analysis of RNA from HIV infected cells (Bennasser *et al.*, 2005). This hairpin is conserved in HIV-2 and SIV genomes, suggesting potential functional relevance. Transfection of constructs encoding short hairpin versions of the viral miRNAs and GFP reporters with cognate miRNA recognition sites results in the expected reduction of GFP expression. Additionally, transfection of the short hairpin decreases env mRNA during viral infection, an effect reversed by transfection of a 2′-O-methyl RNA oligonucleotide antisense to the viral miRNA. A second miRNA was identified by Omoto *et al.*, within the coding region of the viral nef gene (Omoto and Fujii, 2005; Omoto *et al.*, 2004). This miRNA was also detected in infected cells. Interestingly, the cloning strategy of Pfeffer *et al.* did not detect any HIV encoded miRNAs (Pfeffer *et al.*, 2005). As noted above, however, this approach can suffer from false negatives if the miRNAs are not highly expressed or are somehow refractory to cloning.

Additionally (and somewhat paradoxically), HIV-1 appears to encode a suppressor of RNAi (Bennasser *et al.*, 2005). The HIV Tat protein is a transactivator of the viral LTR that functions through binding of the Tat activator region (TAR) present at the 5′ end of all viral mRNAs. Bennaser *et al.* found that Tat was able to suppress RNAi in reporter assays and deletion mutagenesis of Tat mapped the region required for RNAi suppression to amino acids 48–72 of the 101 amino acid protein. One point mutation was described (TatK51A) that retains transcriptional transactivator ability but has lost the ability to suppress RNAi.

Tat may suppress RNAi by competing with Dicer for pre-miRNAs as Tat has a known affinity for RNA stem-loops; the TAR domain shares many characteristics with pre-miRNAs. Tat binding of the viral miRNAs could theoretically relieve the inhibition of viral transcription resulting in increased replication. However, further studies will be required to determine the actual function of these miRNAs and the role of Tat in RNAi inhibition.

Cellular miRNAs and viral replication

Primate foamy virus-1

Recent evidence has shown that cellular miRNAs also play a role in virus biology in mammalian cells. It has been hypothesized that since small RNA based defenses against viruses exist in plants and insects, such anti-viral mechanisms might also be found in mammals. Lecellier *et al.* examined the replication of the retrovirus primate foamy virus-1 (PFV-1) in the presence and absence of P19, a known suppresser of RNAi (Lecellier *et al.*, 2005). In P19 expressing cells, an increased level of PFV-1 RNA was detected, suggesting that RNAi induced inhibition of PFV-1 does occur. Suspecting that a virally encoded miRNA might be responsible for this phenomenon, the investigators cloned fragments of the viral genome into the 3′ UTR of a GFP reporter, predicting that the region encoding the miRNA would result in a decrease in GFP expression in the presence of virus. Surprisingly, no region of the viral genome was found to result in virus dependent inhibition of GFP. However, one region of the virus, close to the 3′ end of the genome, did result in a reduction in the level of expressed GFP protein (but not RNA) independently of PFV-1 infection. Use of a computational algorithm identified a possible (imperfect) match with the human miRNA miR-32. Mutation of the putative miR-32 site in the GFP reporter resulted in a restoration of GFP protein expression.

Additionally, targeting miR-32 with antisense RNA, or altering the target site within the PFV-1 genome resulted in increased levels of viral RNA within the infected cell. As the miR-32 target is located in the 3′ UTR of all viral mRNAs, translational inhibition of all the viral proteins would be expected to occur. The observed reduction in viral mRNAs, therefore, is likely the secondary consequence of translational inhibition of the viral transactivator, Tas. Interestingly, it was also found that the viral Tas protein is capable of suppressing miRNA activity in transfected mammalian and plant cells. Although it is possible that this protein functions to overcome miR-32 suppression of PFV-1 replication, the fact that PFV-1 RNA is substantially increased in the presence of P19 suggests that inhibition of miRNA activity is partial, at best. However, given the data from SV40, it is possible that PFV-1 has evolved not to escape cellular miRNA restriction, but in fact uses it to reduce its own replication in order to reduce the host immune response.

Hepatitis C virus

Hepatitis C virus (HCV) is a member of the family *Flaviviridae*, which have single stranded positive sense RNA genomes and are entirely cytoplasmic during their replication cycle. These viruses would not, therefore, be expected to produce miRNAs, since they do not interact with the nuclear components of the miRNA biogenesis pathway. It is quite possible, however, that cytoplasmic viruses interact with cellular miRNAs, and that these interactions may have profound effects on viral replication. Because HCV is a hepatotropic virus, Jopling, *et al.*, asked if the liver-specific miRNA miR-122 plays a role in its replication (Jopling *et al.*, 2005). MiR-122 is found in human and mouse liver, and in Huh7 cells, a human hepatocellular carcinoma derived cell line that supports the replication of HCV replicons. A search for sequences that match the 5′ "seed sequence" of miR-122 found regions in the 5′ and 3′ UTRs of the HCV genome. Furthermore, these regions were conserved in all six HCV genotypes. Interestingly, in Huh7 cells harboring a genotype 1a or 1b HCV replicon, inactivation of miR-122 with 2′-O-methyl antisense RNA oligonucleotides has the effect of *decreasing* HCV RNA levels. This effect is specific for HCV, as placement of a miR-122 specific sequence in the 3′ UTR of a GFP reporter had the expected effect of transcript down-regulation. Mutation of the putative miR-122 binding regions in the HCV genome revealed that the 5′ UTR site is needed for efficient replication, while the 3′ UTR site is not. Additionally, expression of mutant versions of miR-122 that are complementary to the 5′ UTR mutations rescue HCV RNA replication. Further experiments conducted with replicons that lack a functional RNA-dependent RNA polymerase (and therefore measure only the ability of the HCV 5′ UTR to direct translation) show that wild type or an miR-122 mutant produce equal amounts of protein, indicating that the HCV IRES does not require miR-122 for its translational function.

Conclusion

Despite numerous studies identifying virally expressed miRNAs, little is known about their functional relevance in regards to viral replication and infection.

Cleavage of SV40 T antigen transcripts is the only verified report of a viral miRNA target and in this case the transcript was directly complementary to the miRNA (Sullivan *et al.*, 2005). Most viral miRNAs are not predicted to cleave known transcripts making identification of potential targets much more challenging. Determining the functional relevance of identified targets presents a further challenge. Although it has been proposed that the function of the SV40 miRNAs is to reduce presentation of T antigen, *in vivo* verification will be necessary to determine whether this is truly their primary role in SV40 infection (Sullivan *et al.*, 2005).

A critical question regarding viral miRNA function is whether they modulate host or viral gene expression. Although both scenarios are possible a number of observations suggest that "self-regulation" of viral transcripts is the primary function. First, none of the viral miRNAs identified to date share complete homology with any host transcripts, all but ruling out regulation by direct cleavage (Pfeffer *et al.*, 2005). Second, as single nucleotide changes are known to disrupt miRNA target function, variable host sequences would be unreliable target sites (Brennecke *et al.*, 2005). Viral miRNAs would therefore be predicted to bind conserved sequences within host transcripts resulting in evolutionary pressure on the virus to conserve miRNA sequences so as to maintain functional interactions with host target sites. However, viral miRNA sequences do not show high levels of sequence conservation, with only the closest of related viruses exhibiting any level of homology (Grey *et al.*, 2005). This rapid divergence of miRNA sequence would support a model of self-regulation of viral transcripts by viral miRNAs. With the viral genome encoding both the miRNA and target site, co-evolution of the sequences could occur rapidly, in contrast to the stringent conservation resulting from targeting of host transcripts. Further evidence to support this theory comes from observations by Sullivan and Ganem. Following the identification of miRNAs expressed by SV40, two additional miRNAs were identified in mouse polyoma virus that have the same function – cleavage of the early T antigen transcripts of the virus. However, these miRNAs are expressed from different regions of the viral genome and, therefore, share no homology with the SV40 miRNAs. This would preclude the possibility that they target the same cellular transcripts, rather suggesting that their sole function is to target viral genes (Sullivan and Ganem, 2005).

Presuming viral miRNAs target viral transcripts, a number of potential functions can be proposed. Current opinion in the field would suggest that the level of post-transcriptional effects of miRNAs range from relatively potent inhibition of specific gene products to more subtle, global effects on gene expression. Viral miRNAs could, therefore target a few specific viral trans-activating genes, leading to effective temporal repression over time as the levels of miRNAs increase (Figure 31.5a). Such a model could be viewed as analogous to endogenous miRNAs that act as temporal switches during developmental stages in embryogenesis (Bartel, 2004). Such targeting could have influential effects on viral replication and gene expression. Alternatively miRNAs may augment transcriptional repression of broad groups of viral genes resulting in more subtle effects.

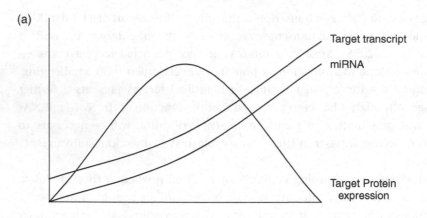

(a)

Target transcript

miRNA

Target Protein
expression

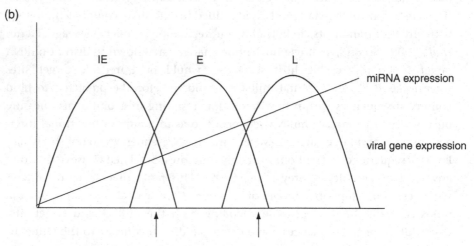

(b)

IE   E   L

miRNA expression

viral gene expression

miRNAs target viral genes of specific kinetic groups to
augmenting efficient transition of gene expression patterns

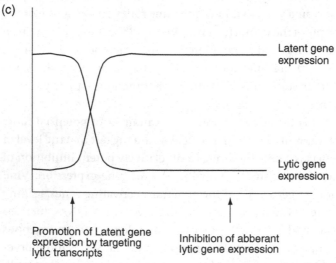

(c)

Latent gene
expression

Lytic gene
expression

Promotion of Latent gene
expression by targeting
lytic transcripts

Inhibition of abberant
lytic gene expression

Figure 31.5. Schematic representation of possible functional roles of viral miRNAs in the context
of herpesvirus infections. (a) Specific inhibition model – viral miRNA exerts efficient translational
inhibition of limited target genes in a temporal manner. (b) Global model – subtle global regulation
of groups of target transcripts enabling efficient switching of temporal transcription patterns.
(c) Potential functional roles of viral miRNAs in targeting viral transcripts during latency.

Herpesvirus gene expression occurs in a tightly regulated temporal manner with transcripts being expressed with immediate-early, early or late kinetics (Wathen and Stinski, 1982). miRNAs expressed by herpesviruses could therefore enhance efficient transition of temporal transcription phases by targeting mRNAs from previous temporal replication stages effectively blocking translation of redundant transcripts (Figure 31.5b).

Viral miRNA regulation may also promote establishment of latency in herpes-viruses by negatively regulating lytic viral gene expression. This model of cooperative gene regulation may also be more relevant when considering the function of viral miRNAs during latent infection. The majority of herpesvirus miRNAs were identified during latent infections, when viral transcription is already highly restricted (Cai et al., 2005; Pfeffer et al., 2004; Pfeffer et al., 2005; Samols et al., 2005). Expression of viral miRNAs during this stage of the viral replication cycle may protect against low levels of aberrant viral gene expression, safeguarding against presentation of antigens or inappropriate reactivation (Figure 31.5c).

Two recent studies have demonstrated the potential for cellular miRNAs to influence viral replication (Jopling et al., 2005; Lecellier et al., 2005). In one case a cellular miRNA was shown to inhibit efficient replication of the retrovirus PRV-1 by targeting a conserved sequence at the 3' end of the viral transcripts. This inhibitory effect of a host miRNA more likely represents an artifactual effect due to infection of a non-host, human cell line, rather than an innate immune response. Work by Fahr et al., and by Stark et al., indicates the significant effects of endogenous miRNAs on the evolution of transcript sequences (Farh et al., 2005; Stark et al., 2005). Both studies show that transcripts have evolved to selectively avoid containing potential "seed" target sites for miRNAs they are co-expressed with. This concept of "anti-targets" is of particular relevance to viruses as viral transcripts would be subject to the repressive action of an array of miRNAs already expressed by host cells. As suggested by Lecellier et al., such fortuitous interactions of host miRNAs with viral transcripts could be a factor in determining cellular tropism and species specificity of viruses. It will be interesting to determine whether viruses have evolved to avoid such sequences and what this may tell us about the actions of cellular miRNAs. Another possibility is that viruses have specifically evolved to utilize host miRNAs to negatively regulate their own transcripts. Viruses could evolve target sites for host miRNAs to achieve tissue specific gene expression profiles. For example, HSV could theoretically utilize a neuronal specific miRNA to restrict lytic gene expression, thereby promoting latent infection in the correct cell type.

Further studies will no doubt reveal the extent to which the expression of host and viral miRNAs impact on viral replication. The next crucial step will be the identification of targets of viral miRNAs. As the number of targets identified increases a clearer picture should be achieved as to the function of viral miRNAs and how they impact on replication, host response and evolution of viral sequences. Investigating miRNAs in the context of viral infections will allow us to gain insight into mechanisms of viral replication and gene regulation but, as has been the case in the past, may prove to be an effective model for the study of miRNA function in general. Studying miRNA function in the context of the relatively small genomes and limited coding capacity of viruses should prove a significant advantage over more complex organisms.

ACKNOWLEDGMENTS

This work was supported by a Public Service grant from the National Institutes of Health AI21640. We would like to thank Michael Jarvis for his critical review of this manuscript.

REFERENCES

Andersson, M. G., Haasnoot, P. C., Xu, N. *et al.* (2005). Suppression of RNA interference by adenovirus virus-associated RNA. *Journal of Virology*, **79**, 9556–9565.

Aparicio, O., Razquin, N., Zaratiegui, M., Narvaiza, I. and Fortes, P. (2006). Adenovirus virus-associated RNA is processed to functional interfering RNAs involved in virus production. *Journal of Virology*, **80**, 1376–1384.

Bartel, D. P. (2004). MicroRNAs: genomics, biogenesis, mechanism, and function. *Cell*, **116**, 281–297.

Bennasser, Y., Le, S. Y., Benkirane, M. and Jeang, K. T. (2005). Evidence that HIV-1 encodes an siRNA and a suppressor of RNA silencing. *Immunity*, **22**, 607–619.

Bentwich, I., Avniel, A., Karov, Y. *et al.* (2005). Identification of hundreds of conserved and nonconserved human microRNAs. *Nature Genetics*, **37**, 766–770.

Berezikov, E., Guryev, V., van de Belt, J. *et al.* (2005). Phylogenetic shadowing and computational identification of human microRNA genes. *Cell*, **120**, 21–24.

Bowden, R. J., Simas, J. P., Davis, A. J. and Efstathiou, S. (1997). Murine gammaherpesvirus 68 encodes tRNA-like sequences which are expressed during latency. *Journal of General Virology*, **78**, 1675–1687.

Branco, F. J. and Fraser, N. W. (2005). Herpes simplex virus type 1 latency-associated transcript expression protects trigeminal ganglion neurons from apoptosis. *Journal of Virology*, **79**, 9019–9025.

Brennecke, J., Stark, A., Russell, R. B. and Cohen, S. M. (2005). Principles of microRNA-target recognition. *Public Library of Science Biology*, **3**, e85.

Cai, X. and Cullen, B. R. (2006). Transcriptional origin of Kaposi's sarcoma-associated herpesvirus microRNAs. *Journal of Virology*, **80**, 2234–2242.

Cai, X., Lu, S., Zhang, Z. *et al.* (2005). Kaposi's sarcoma-associated herpesvirus expresses an array of viral microRNAs in latently infected cells. *Proceedings of the National Academy of Sciences USA*, **102**, 5570–5575.

Courcelle, C. T., Courcelle, J., Prichard, M. N. and Mocarski, E. S. (2001). Requirement for uracil-DNA glycosylase during the transition to late-phase cytomegalovirus DNA replication. *Journal of Virology*, **75**, 7592–7601.

DeMarchi, J. M., Schmidt, C. A. and Kaplan, A. S. (1980). Patterns of transcription of human cytomegalovirus in permissively infected cells. *Journal of Virology*, **35**, 277–286.

Dunn, W., Chou, C., Li, H. *et al.* (2003). Functional profiling of a human cytomegalovirus genome. *Proceedings of the National Academy of Sciences USA*, **100**, 14 223–14 228.

Dunn, W., Trang, P., Zhong, Q. *et al.* (2005). Human cytomegalovirus expresses novel microRNAs during productive viral infection. *Cellular Microbiology*, **7**, 1684–1695.

Farh, K. K., Grimson, A., Jan, C. *et al.* (2005). The widespread impact of mammalian microRNAs on mRNA repression and evolution. *Science*, **310**, 1817–1821.

Furnari, F. B., Adams, M. D. and Pagano, J. S. (1992). Regulation of the Epstein–Barr virus DNA polymerase gene. *Journal of Virology*, **66**, 2837–2845.

Goldmacher, V. S., Bartle, L. M., Skaletskaya, A. *et al.* (1999). A cytomegalovirus-encoded mitochondria-localized inhibitor of apoptosis structurally unrelated to Bcl-2. *Proceedings of the National Academy of Sciences USA*, **96**, 12 536–12 541.

Grey, F., Antoniewicz, A., Allen, E. *et al.* (2005). Identification and characterization of human cytomegalovirus-encoded microRNAs. *Journal of Virology*, **79**, 12 095–12 099.

Grun, D., Wang, Y. L., Langenberger, D., Gunsalus, K. C. and Rajewsky, N. (2005). MicroRNA target predictions across seven *Drosophila* species and comparison to mammalian targets. *Public Library of Science Computer Biology*, **1**, e13.

Jopling, C. L., Yi, M., Lancaster, A. M., Lemon, S. M. and Sarnow, P. (2005). Modulation of hepatitis C virus RNA abundance by a liver-specific microRNA. *Science*, **309**, 1577–1581.

Krek, A., Grun, D., Poy, M. N. *et al.* (2005). Combinatorial microRNA target predictions. *Nature Genetics*, **37**, 495–500.

Lecellier, C. H., Dunoyer, P., Arar, K. *et al.* (2005). A cellular microRNA mediates antiviral defense in human cells. *Science*, **308**, 557–560.

Lee, R. C., Feinbaum, R. L. and Ambros, V. (1993). The *C. elegans* heterochronic gene lin-4 encodes small RNAs with antisense complementarity to lin-14. *Cell*, **75**, 843–854.

Lewis, B. P., Burge, C. B. and Bartel, D. P. (2005). Conserved seed pairing, often flanked by adenosines, indicates that thousands of human genes are microRNA targets. *Cell*, **120**, 15–20.

Lim, L. P., Lau, N. C., Weinstein, E. G. *et al.* (2003). The microRNAs of *Caenorhabditis elegans*. *Genes & Development*, **17**, 991–1008.

Lu, S. and Cullen, B. R. (2004). Adenovirus VA1 noncoding RNA can inhibit small interfering RNA and microRNA biogenesis. *Journal of Virology*, **78**, 12 868–12 876.

McCormick, C. and Ganem, D. (2005). The kaposin B protein of KSHV activates the p38/MK2 pathway and stabilizes cytokine mRNAs. *Science*, **307**, 739–741.

Moore, P. and Chang, Y. (2001). Kaposi's sarcoma-associated Herpesvirus. *Fields Virology*, **2**, 2803–2831.

Omoto, S. and Fujii, Y. R. (2005). Regulation of human immunodeficiency virus 1 transcription by nef microRNA. *Journal of General Virology*, **86**, 751–755.

Omoto, S., Ito, M., Tsutsumi, Y. *et al.* (2004). HIV-1 nef suppression by virally encoded microRNA. *Retrovirology*, **1**, 44.

Pasquinelli, A. E., Reinhart, B. J., Slack, F. *et al.* (2000). Conservation of the sequence and temporal expression of let-7 heterochronic regulatory RNA. *Nature*, **408**, 86–89.

Patterson, C. E. and Shenk, T. (1999). Human cytomegalovirus UL36 protein is dispensable for viral replication in cultured cells. *Journal of Virology*, **73**, 7126–7131.

Pearce, M., Matsumura, S. and Wilson, A. C. (2005). Transcripts encoding K12, v-FLIP, v-cyclin, and the microRNA cluster of Kaposi's sarcoma-associated herpesvirus originate from a common promoter. *Journal of Virology*, **79**, 14 457–14 464.

Pfeffer, S., Zavolan, M., Grasser, F. A. *et al.* (2004). Identification of virus-encoded microRNAs. *Science*, **304**, 734–736.

Pfeffer, S., Sewer, A., Lagos-Quintana, M. *et al.* (2005). Identification of microRNAs of the herpesvirus family. *Nature Methods*, **2**, 269–276.

Prichard, M. N., Duke, G. M. and Mocarski, E. S. (1996). Human cytomegalovirus uracil DNA glycosylase is required for the normal temporal regulation of both DNA synthesis and viral replication. *Journal of Virology*, **70**, 3018–3025.

Reinhart, B. J., Slack, F. J., Basson, M. *et al.* (2000). The 21-nucleotide let-7 RNA regulates developmental timing in *Caenorhabditis elegans*. *Nature*, **403**, 901–906.

Roizman, B. and Pellet, E. (2001). The family Herpesviridae: a brief introduction. *Fields Virology*, **2**, 2381–2397.

Samols, M. A., Hu, J., Skalsky, R. L. and Renne, R. (2005). Cloning and identification of a microRNA cluster within the latency-associated region of Kaposi's sarcoma-associated herpesvirus. *Journal of Virology*, **79**, 9301–9305.

Sano, M., Kato, Y. and Taira, K. (2006). Sequence-specific interference by small RNAs derived from adenovirus VAI RNA. *Federation of European Biological Sciences Letters*, **580**, 1553–1564.

Shenk, T. (2001). Adenoviridae: the viruses and their replication. *Fields Virology*, **2**, 2265–2326.

Simas, J. P., Bowden, R. J., Paige, V. and Efstathiou, S. (1998). Four tRNA-like sequences and a serpin homologue encoded by murine gammaherpesvirus 68 are dispensable for lytic replication in vitro and latency in vivo. *Journal of General Virology*, **79**, 149–153.

Skaletskaya, A., Bartle, L. M., Chittenden, T. *et al.* (2001). A cytomegalovirus-encoded inhibitor of apoptosis that suppresses caspase-8 activation. *Proceedings of the National Academy of Sciences USA*, **98**, 7829–7834.

Smith, P. R., de Jesus, O., Turner, D. *et al.* (2000). Structure and coding content of CST (BART) family RNAs of Epstein–Barr virus. *Journal of Virology*, **74**, 3082–3092.

Stark, A., Brennecke, J., Bushati, N., Russell, R. B. and Cohen, S. M. (2005). Animal microRNAs confer robustness to gene expression and have a significant impact on 3′ UTR evolution. *Cell*, **123**, 1133–1146.

Sullivan, C. S. and Ganem, D. (2005). MicroRNAs and viral infection. *Molecular Cell*, **20**, 3–7.

Sullivan, C. S., Grundhoff, A. T., Tevethia, S., Pipas, J. M. and Ganem, D. (2005). SV40-encoded microRNAs regulate viral gene expression and reduce susceptibility to cytotoxic T cells. *Nature*, **435**, 682–686.

Thompson, R. L. and Sawtell, N. M. (2001). Herpes simplex virus type 1 latency-associated transcript gene promotes neuronal survival. *Journal of Virology*, **75**, 6660–6675.

Wathen, M. W. and Stinski, M. F. (1982). Temporal patterns of human cytomegalovirus transcription: mapping the viral RNAs synthesized at immediate early, early, and late times after infection. *Journal of Virology*, **41**, 462–477.

Wightman, B., Ha, I. and Ruvkun, G. (1993). Posttranscriptional regulation of the heterochronic gene lin-14 by lin-4 mediates temporal pattern formation in *C. elegans*. *Cell*, **75**, 855–862.

Xie, X., Lu, J., Kulbokas, E. J. *et al.* (2005). Systematic discovery of regulatory motifs in human promoters and 3′ UTRs by comparison of several mammals. *Nature*, **434**, 338–345.

Yu, D., Silva, M. C. and Shenk, T. (2003). Functional map of human cytomegalovirus AD169 defined by global mutational analysis. *Proceedings of the National Academy of Sciences USA*, **100**, 12 396–12 401.

32 Lost in translation: regulation of HIV-1 by microRNAs and a key enzyme of RNA-directed RNA polymerase

Yoichi Robertus Fujii

The annual labor of every nation is the fund which originally supplies it with all the necessaries and conveniences of life which it annually consumes, and which consist always either in the immediate produce of that labor, or in what is purchased with that produce from other nations.
      Adam Smith

Introduction

The human immunodeficiency virus type 1 (HIV-1) gene, *nef*, is located at the 3′ end of the viral genome and partially overlaps the 3′-long terminal repeat (LTR). The *nef* gene is expressed during HIV infection and often accounts for up to 80% of HIV-1-specific RNA transcripts during the early stages of viral replication (Robert-Guroff *et al.*, 1990). Because HIV-1 LTR activity is down-regulated just a few hours after induction by Tat, this may contribute to viral latency (Drysdale and Pavlakis, 1991). However, the mechanism of this down-regulation of Tat transactivation as well as accumulation of *nef* RNAs is not completely clear. MicroRNAs (miRNAs) are 21–25 nucleotides (nt) long and directed selectively to targeting sequences of mRNAs to trigger either translational repression or RNA cleavage through RNA interference (RNAi) (Zeng *et al.*, 2003; Yekta *et al.*, 2004). However, RNAi had not been found in mammals by functional assays, as observed naturally in plants, fungi, insects or nematodes (Fire *et al.*, 1998; Tuschl *et al.*, 1999; Ketting *et al.*, 1999; Tabara *et al.*, 1999; Aravin *et al.*, 2001; Baulcombe, 2001; Sijen and Plasterk, 2003). From our data with HIV-1-infected mammalian cells, we predicted that RNAi-based processes might have the power to overwhelm viruses by turning off harmful genes. The feasibility of this approach is now reflected in the announcement of RNAi-based phase I therapeutic trials for the treatment of age-related macular degeneration, which showed no toxic side-effects and resulted in disease stabilization in all examined patients.

A key enzyme, a cellular RNA-directed RNA polymerase (RdRP), well-known in the RNAi pathways of *C. elegans*, *A. thaliana* and *N. crassa*, replicates double-stranded RNA (dsRNA) from a single-stranded RNA (ssRNA) template to initiate

MicroRNAs: From Basic Science to Disease Biology, ed. Krishnarao Appasani. Published by Cambridge University Press. © Cambridge University Press 2008.

and amplify the gene silencing process (Sijen *et al.*, 2001; Matzke and Birchler, 2005). However, RdRP has not been identified in mammals. It may facilitate RNA polymerase chain reactions (R-PCR) not only as part of the anti-viral strategy of RNAi with miRNA but also for gene silencing in mammalian genomes (Figure 32.1).

I shall, in this chapter, endeavor to describe the cloning of HIV-1 viral miRNA and confirmation of its expression. I will try to point out its value in basic research and potential exploitation as a novel therapeutic approach. Here, we also show that reverse transcriptases (RTs) possess RNA polymerase activity *in vitro*. The identification of viral miRNA and RdRP activity of retroviral RTs strongly supports the hypothetical model that viral latency is caused by an RNA polymerase chain reaction (R-PCR) of viral miRNAs in mammalian cells.

Methods

Protocol of viral miRNA detection

Reagents
A 100 μM 3′-linker oligonucleotide A 5′-P-cuguguAGGCGTCGACATG-ddC-3′ (cugugu, RNA; P, phosphate; ddC, di-deoxy C; GTCGAC, *Sal* I site), 10 μM oligo-nucleotide B (5′-TCTGAGCATAGGCGGCCGAGGAGATGTTCATGTCGACGCC-3′) (GTCGAC, *Sal* I site), 10 μM 5′RACE Abridged Anchor Primer (AAP), C (5′-GGCC ACGCGTCGACTAGTACGGGIIGGGIIGGGIIG-3′) (I, deoxyinosine; GTCGAC, *Sal* I site), 10 μM oligonucleotide D (5′-TCTGAGCATAGGCGGCCGAG-3′), 10 μM Abridged Universal Amplification Primer (AUAP) E (5′-GGCCACGCGTCGAC-TAGTAC-3′) (GTCGAC, *Sal* I site) and 10 μM oligonucleotide F (5′-GAGATGTT-CATGTCGACGCC-3′) (GTCGAC, *Sal* I site) were purchased from Invitrogen.

Preparation of HIV-1-infected cells
HIV-1 (IIIB, SF2)-infected cells were cultured for at least 6 months with RPMI-1640 medium (GIBCO) containing heat-inactivated fetal bovine serum (FBS) (GIBCO) (Omoto and Fujii, 2005).

Total RNA extraction from HIV-1-infected cells
Total RNA was extracted with 1 ml of TRIZOL reagent (Invitrogen) from 5×10^6 cells.

Preparation of size-selected RNA from total RNA
A 10 μl sample of total RNA was subjected to gel electrophoresis using 15% acrylamide with 7 M urea. The part of the gel corresponding to 21–25 nt RNA was excised and the gel slice was transferred into 400 μl (this represents at least four times the volume of the gel slice) gel elution buffer (20 mM Tris-HCl, pH 7.5, 5 mM EDTA, 400 mM sodium acetate), and then agitated vigorously for 10 min. The gel slice homogenate was filtered through a SUPREC-01 filter (Takara). The sample was usually extracted with phenol/chloroform, then precipitated with ethanol in each final step.

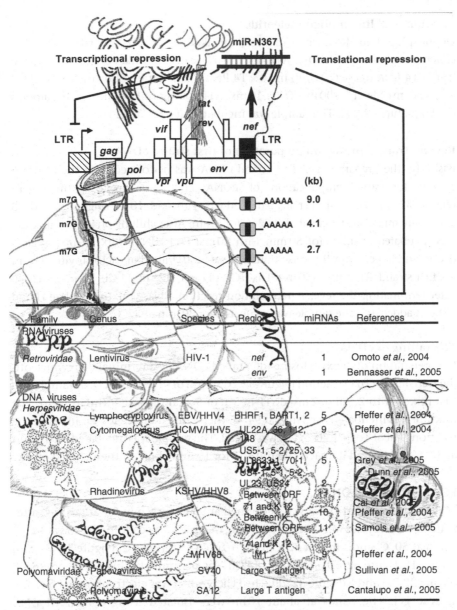

Family	Genus	Species	Region	miRNAs	References
RNA viruses					
Retroviridae	Lentivirus	HIV-1	nef	1	Omoto et al., 2004
			env	1	Bennasser et al., 2005
DNA viruses					
Herpesviridae	Lymphocryptovirus	EBV/HHV4	BHRF1, BART1, 2	5	Pfeffer et al., 2004
	Cytomegalovirus	HCMV/HHV5	UL22A, 36, 112, 148	9	Pfeffer et al., 2004
			US5-1, 5-2, 25, 33		
			UL36-3-1, 70-1	5	Grey et al., 2005
			US4-1, 5-1, 5-2		Dunn et al., 2005
			UL23, US24	2	
	Rhadinovirus	KSHV/HHV8	Between ORF 71 and K 12	11	Cai et al., 2005
			Between K	10	Pfeffer et al., 2004
			Between ORF 71 and K 12	11	Samols et al., 2005
		MHV68	M1	9	Pfeffer et al., 2004
Polyomaviridae	Papovavirus	SV40	Large T antigen	1	Sullivan et al., 2005
	Polyomavirus	SA12	Large T antigen	1	Cantalupo et al., 2005

Figure 32.1. Schematic representation of *nef* miRNA action and viral miRNAs. The hypothesis of the proviral DNA elimination model by HIV-1 *nef* miRNA (upper panel) is further represented in Figure 32.5. Reported viral miRNAs are shown in the Table (lower panel). These viral and cellular miRNAs might maintain viral persistence. The dsRNA for the systemic RNAi machinery is amplified by RdRP. The cellular RdRP synthesizes dsRNA from an ssRNA template to initiate or amplify the RNAi process. These cellular enzymes are found in plants, represented here as flowers, but not in *D. melanogaster* or mammals (symbolized by the Japanese "Kimono"). On the other hand, genomic DNA of some RNA viruses encodes viral RdRPs. In the case of infection, mammalian cells temporarily acquire RdRP. The structure of the RdRP of poliovirus has the motifs, the thumb, fingers and palm, looking like the broad belt of a Kimono, named "Ohbi." Structures of Klenow polymerase of *E. coli*, HIV-1 RT, T7 phage RNA polymerase and poliovirus RdRP all conserve similar motifs. It is hypothesized that RT of some retroviruses (including endogenous retrotransposons) may possess RdRP activity, so that the RT could act as an RdRP and the amplified viral miRNAs might disturb intracellular defense systems through the RNAi mechanisms of the host. Subsequently, the viral miRNAs may induce a latent state by proviral DNA elimination, i.e. generate viral mutants.

Ligation of 3′-linker oligonucleotide
Dephosphorylated RNA with calf intestinal alkaline phosphatase (CIAP) (20 U/μl) (Roche) was mixed with 60 μM ATP, 1 μl of 100 μM RNA 3′-linker oligonucleotide A, 2 μl of T4 RNA ligase (10 U/μl) in 10 × T4 RNA ligase buffer (500 mM Tris-HCl, pH 7.5, 100 mM MgCl$_2$, 100 mM DTT, 10 mM ATP, 600 μg/ml bovine serum albmin (BSA)) (Figure 32.2A). The sample was incubated at 37 °C for 1 h.

Reverse transcription and preparation of first strand cDNA
Basically, the preparation of first strand cDNA was performed with the 5′ RACE system for rapid amplification of cDNA ends, version 2.0 (Invitrogen) (Figure 32.2B). The 3′-linker ligated product (4 μl) and 10 μM oligonucleotide B (1 μl) were mixed and incubated at 72 °C for 2 min. After chilling on ice, the mixture was incubated at 42 °C for 50 min with 20 mM DTT, dNTP Mix (10 mM of each dNTP) and SuperScript II reverse transcriptase (200 U/μl) (1 μl each) (Invitrogen) in 5 × first-strand RT buffer (50 mM Tris-HCl, pH 8.4, 125 mM KCl, 7.5 mM MgCl$_2$). After terminating the reverse transcription reaction, RNase mixture (RNase H and T1) was added and the sample was incubated for 30 min at 37 °C.

TdT tailing of cDNA
For preparation of deoxycytosine (dC)-tailed cDNA by a terminal deoxynucleotidyl transferase (TdT) reaction (Figure 32.2C), 10 μl of first strand cDNA sample was first mixed with 2.5 μl of 2 mM dCTP in 5 μl of 5 × tailing buffer (Invitrogen). The mixture was incubated for 3 min at 94 °C, then chilled on ice for 1 min. After the reaction, 1 μl of TdT (Invitrogen) was added. The sample was incubated for 10 min at 37 °C and heated to inactivate the TdT for 10 min at 65 °C, then chilled on ice.

PCR of dC-tailed cDNA
PCR of the dC-tailed cDNA was performed at 94 °C for 5 min for annealing, after which 25 PCR cycles were carried out using the following conditions: 45 s at 94 °C, 1 min at 50 °C, and 1 min at 72 °C using 10 μM AAP (primer C) (Figure 32.2C) and primer D (Figure 32.2D). The nested-PCR was performed using AUAP primers E (Figure 32.2E) and F (Figure 32.2F). The PCR product was electrophoresed on a 2% agarose gel. The bands at about 70 bp were isolated using a QIAEXII DNA Extraction Kit (QIAGEN) and the cDNA from the gel was cut with *Sal* I restriction enzyme (20 U/μl) (Takara) (Figure 32.2). The digested cDNA was inserted into a conventional vector (Omoto and Fujii, 2005).

Protocol for RdRP assay

Reagents
Purified firefly luciferase and D-luciferin were purchased from Promega (Madison, WI). UTP, GTP or ATP stocks and RNaseOUT were obtained from Invitrogen (San Diego, CA). Coenzyme A, adenosine 5′-phosphosulfate (APS), adenosine 5′-triphosphate (ATP) sulfurylase (1 U produces 1 mmol of ATP from APS and free pyrophosphate (PPi) per min at pH 8 at 37 °C), and 3′-2′-deoxyuridine triphosphate

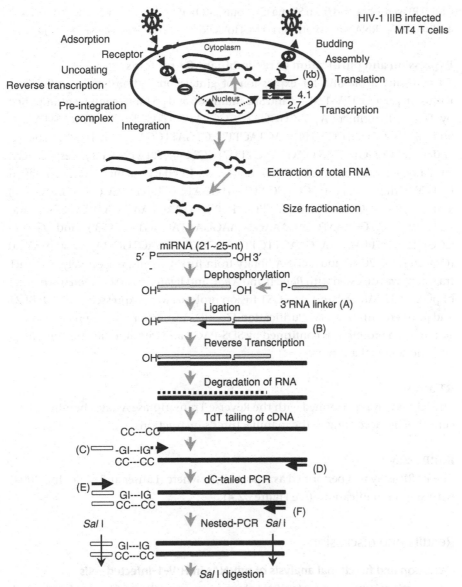

Figure 32.2. Replication pathway of HIV-1 and cloning strategy of HIV-1-encoded miRNA. Productive HIV-1 infections begin with the adsorption of virions to the cells and Env protein interaction with both CD4 and chemokine receptors. After invasion, the virions partially uncoat and then reverse transcription occurs in the cytoplasm. The proviral dsDNA is integrated into the host genome by integrase, a component of the pre-integration complex. The integrated proviral DNA leads to the production of viral mRNAs (c. 9.0, 4.1 and 2.7 kb). These mRNAs are translated into viral proteins. After packaging of the viral genome RNA (9 kb), progeny viruses bud from cells and are matured. For cloning of HIV-1-encoded miRNA, first, total RNAs were extracted from HIV-1-infected MT4 T cells. After DNase treatment and size-selection, 5'-phosphorylated mature miRNA was dephosphorylated and ligated with 3'-linker oligonucleotide (A). Next, the RNA was reverse transcribed to the first strand cDNA with oligonucleotide (B). Deoxycytosines were transferred to the cDNA by TdT reactions. The dC-tailed cDNA was amplified by PCR with primers (C) and (D). Finally, nested-PCR with primers (E) and (F) was performed. The cDNA was digested with restriction enzymes and the fragments were ligated into plasmids.

(ddUTP) were obtained from Sigma (St. Louis, MO). Poly A, C or U RNA homopolymer and oligo U_{12} RNA were synthesized by Hokkaido Systems Science (Sapporo, Japan).

Expression and purification of recombinant RT proteins

To construct a plasmid for expression of glutathione S-transferase (GST)-HIV-1 RT fusion protein, HIV-1 NL4-3 and SF2 strain p66 and p51 fragments were amplified by PCR with primers A (5′-CGGGATCCCCCATTAGTCCTATTGAGACTGTAC-3′) and B (5′-CCGCTCGAGTTATAGTACTTTCCTGATTCCAAGCACTG-3′), and A and C (5′-CCGCTCGAGTTAGAAAGTTTCTGCTCCTATTATGGG-3′), respectively. Similarly, feline FV and HERV-K (Berkhout *et al.*, 1999) RT fragments were amplified by PCR with primers D (5′-CGTGGATCCATGGATCTGCTGAAGCCGTTGACGG-3′) and E (5′-CCGCTCGAGCTAATATTGTTTTGGATAACCAACAGGTAATC-3′), and F (5′-CGGGATCCAAATCAAGAAAGAGAAGGAATAGGGTATCC-3′) and G (5′-CCGCTCGAGTTAAGCATGAAGTTCTTGTGCTTTTATGAGTGC-3′) using pSKY3.0 (Omoto *et al.*, 2004) and a cDNA library from Jurkat T cells, respectively. Each RT fragment was digested with *Bam*HI and *Xho*I, and ligated into the *Bam*HI-*Xho*I site of pGEX-4T2 (Amersham). The GST fusion proteins were expressed in *E. coli* BL21 and purified with Bulk GST Purification Modules (Amersham). The precise expression of each protein was confirmed by SDS-PAGE and immunoblotting with anti-GST monoclonal antibody.

RT assay

The RT assay was performed with the Reverse Transcriptase Assay, chemiluminescent (Roche) according to the manufacturer's instructions.

RdRP assay

The RdRP assay was performed as reported elsewhere (Lahser and Malcolm, 2004) with slight modifications (see Figure 32.4).

Results and discussion

Detection and functional analysis of miRNAs in HIV-1-infected cells

The functional importance of miRNAs in mammalian cells remained to be clarified. As a tool for elucidation of the function of miRNAs, we focused on cells persistently infected with HIV-1. We have shown that HIV-1 *nef* double-stranded (ds) RNA from AIDS patients who are long-term non-progressors (LTNPs) inhibits HIV-1 transcription (Yamamoto *et al.*, 2002). We hypothesized that these data could be reflecting the presence of RNAi machinery in mammals and some miRNAs diced from the *nef* dsRNA. On the other hand, it has been reported that mammalian post-transcriptional gene silencing (PTGS) mechanisms can be programmed with artificial 20–25 nt duplexes of small interfering RNAs (siRNAs) corresponding to cognate viral mRNA to induce an effective *in vitro* response against viral replication (Gitlin *et al.*, 2002; Hu *et al.*, 2002; Jacque *et al.*, 2002; Lee *et al.*, 2002; Novina *et al.*, 2002; Paddison *et al.*, 2002; Capodici *et al.*, 2002; Coburn and Cullen, 2002). However, at that time I had a question to the one posed: "Nobody has yet

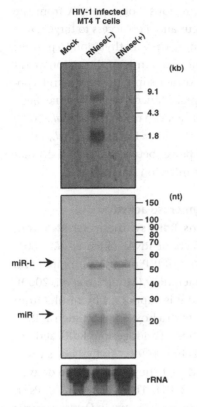

Figure 32.3. Detection of *nef*-derived miRNAs produced by HIV-1. Northern blot analysis of nef miRNAs in HIV-1-infected MT4 T cells. Total RNAs were treated or not treated with a mixture of RNase A and T1. The approximate sizes of the HIV-1 transcripts and mRNAs are indicated on the right as determined using Decade Markers (Ambion). The probe was anti-sense miR-N367. The positions of mature miRNAs (miR) and their predicted fold-back precursors (miR-L) are indicated on the left.

known whether RNAi mechanisms participate in natural ecological interactions." Thus, to investigate viral miRNA expression in T cells persistently infected with HIV-1 IIIB, total RNA was extracted and Northern blotting carried out using a *nef* miR-N367 probe. Small RNA molecules approximately 25 nt in size, as well as putative precursor bands, were detected (Figure 32.3). These bands were resistant to single-stranded RNA-specific RNase, indicating that they could be miRNAs. Next, we investigated whether the miR-N367 could reduce HIV-1 transcription. Use of the STYLE vector expressing miR-N367 showed that miR-N367 suppressed HIV-1 LTR-driven *luc* activity in MT-4 T cells infected with HIV-1 IIIB (Omoto and Fujii, 2005). This finding increased the likelihood that miRNAs may regulate viral gene expression. Although HIV-1 RNA elongation in transcription is supported by Tat and host factors, such as the Tat-responsive-region (TAR)-binding protein (TRBP), more recently TRBP has been reported to benefit Tat-dependent HIV-1 replication. The mechanism seems to be prevention of RNAi, acting as a Dicer-inhibiting factor against anti-HIV-1 miRNAs from host cells (Chendrimada *et al.*, 2005; Haase *et al.*, 2005; Ong *et al.*, 2005). On the other hand, promoter interference independent of Tat is commonly found in retroviral LTR, the P_L promoter of phage, prokaryotic operons and yeast (Cullen *et al.*, 1984). To examine further the effect of miR-N367 on post-transcriptional control by *nef* at the 3'-UTR, *cis*-action of miR-N367 was tested. It was found that miR-N367 suppressed HIV-1 LTR promoter activity via a negative response element (NRE) in the U3 of the 5'-LTR and the *nef* sequences located at the 3'-UTR of the targeted regions. Some miRNA from the HIV-1 genome might be transported to the nucleus and gain access to target DNA sequences, resulting in RNA–DNA pairing containing CpG methylation to prevent initiation and/or termination of transcription, independently of the action of Tat and TRBP (Omoto and Fujii, 2005). These data further supported my own hypothesis that RNAi machinery is present in mammals. Since then, RNAi has been reported to function as an antiviral immune defense mechanism (Ding *et al.*, 2004; Wilkins *et al.*, 2005; Lu *et al.*, 2005; Jopling *et al.*, 2005; Hariharan *et al.*, 2005; Lecellier *et al.*, 2005), suggesting that balance of power between viral and cellular miRNAs might be able to maintain the persistent infection state.

Reverse transcriptase exhibits *in vitro* RNA polymerase activity

It has been reported that, under certain conditions, RdRP synthesizes dsRNA from ssRNA templates to initiate or amplify the effects of RNAi (Sijen *et al.*, 2001; Matzke and Birchler, 2005). By amplifying miRNA, RdRP might cause rapid and effective gene silencing with conserved RNAi machinery (Mourrain *et al.*, 2000). The possibility of the existence of a cellular RdRP is supported by results from investigations of plant viroid replication systems with a host enzyme, such as RNA polymerase II (Pol II) (Diener, 1991) and by isolation of endogenous RdRP activity in plants (Schiebel *et al.*, 1993; Cogoni and Macino, 1999). Analysis of siRNAs produced during RNAi in *C. elegans* has revealed that they appear to be derived from the action of a cellular RdRP (Sijen *et al.*, 2001). There are four RdRPs in *C. elegans*, six in *A. thaliana* and three in *N. crassa* but none known in *D. melanogaster* or mammals. Intriguingly, cellular DNA-dependent RNA polymerase II has been postulated to carry out RNA-dependent RNA replication of hepatitis delta virus (HDV) (Modahl *et al.*, 2000; Filipovska and Konarska, 2000); however, the absence of homologous genomic sequences of RdRP-like genes implies that RNAi reactions can still proceed without RdRP in mammals. Further, embryonic extracts from *Drosophila* with no significant RdRP activity showed complete RNAi reactions (Zamore *et al.*, 2000). The sole responsibility of RdRP in RNAi machinery is therefore questionable.

As one possible pathway for RNA replication in mammalian cells in addition to pol II, we hypothesized that reverse transcriptase (RT) of human retrotransposons could be a candidate for a cellular RdRP. This was suggested because the protein structure of human immunodeficiency virus type-1 (HIV-1) RT resembles pol II as well as viral RdRP (Hansen *et al.*, 1997). To test this hypothesis, human retrotransposon (HIV-1 and human endogenous retrovirus type-K, HERV-K)-encoded RT proteins were expressed as GST fusion proteins in *E. coli*. First, RT assays with purified GST-RT proteins were performed, showing that HIV-1 NL4-3 p66 (24 U/µg) and p51 (7 U/µg) did mediate RT activity (Le Grice *et al.*, 1991). Similar results were obtained with HIV-1 SF2 p66 and p51. The HERV-K RT also showed RT activity (12 U/µg). These results suggest that the expressed RT proteins were all enzymatically active.

Next, RdRP assays were performed with purified RT proteins. Measurable enzyme activity was detected with NL4-3 p66 (1.6 U/µg) or p51 (1.5 U/µg)

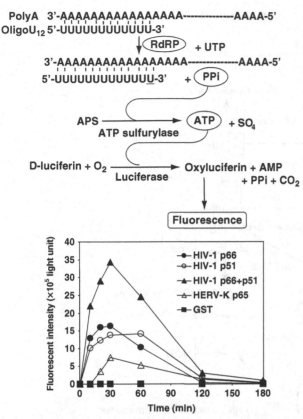

PolyA 3'-AAAAAAAAAAAAAAAAA--------------AAAA-5'
OligoU$_{12}$ 5'-UUUUUUUUUUUUU-3'

$\downarrow$ RdRP + UTP

3'-AAAAAAAAAAAAAAAAA--------------AAAA-5'
5'-UUUUUUUUUUUUU-3' + PPi

APS $\longrightarrow$ ATP + SO$_4$
ATP sulfurylase

D-luciferin + O$_2$ $\longrightarrow$ Oxyluciferin + AMP
Luciferase + PPi + CO$_2$

$\downarrow$

Fluorescence

Figure 32.4. The RdRP assay and RNA polymerase activity of RTs. The assay is based on the detection of PPi generated by the RdRP reaction (upper panel). Free PPi is converted to ATP in a reaction catalyzed by ATP sulfurylase using APS. This ATP pool provides the energy for a luciferase-catalyzed reaction producing photons. Final concentrations of RdRP reaction condition buffer were 50 mM Tris-HCl (pH 8.0), 10 mM MgCl$_2$, 10 mM MnCl$_2$, 20 mM KCl, 20 U/ml RNaseOUT, 10 mM DTT, 0.5 mM Coenzyme A. The oligo U$_{12}$ (0.5 μg/ml) and polyA (5 μg/ml) RNA components were preincubated in the buffer at room temperature for 15 min prior to the addition of recombinant proteins, 1 nM luciferase and 0.1 U ATP sulfurylase. The reactions were initiated by the addition of 1 μM UTP mixture, 5 μM APS and 100 μM D-luciferin. One hundred microliters of the RdRP reaction mixtures were incubated at 37 °C in 96-well plates. Fluorescence signals were detected at each time on a Multi-label Counter 1420 ARVO (Pharmacia). The background activity obtained with the GST control sample was subtracted from the RdRP activity for the GST-RT protein. One U of RdRP activity is calculated as the amount of protein required for the incorporation of 1 nmol of UTP in 10 min at 37 °C using poly(rA)-oligo(rU)$_{12}$ RNA template-primer duplex. RNA polymerase activity of retrotransposon RTs (lower panel). RNA polymerase activity of the HIV-1 NL4-3 RTs (p66, p51 and p66/p51). RdRP assays were undertaken with HIV-1 p66 (filled circles), p51 (open circles), a combination of p66 and p51 (filled triangles), HERV-K p65 RT (open triangles) and control GST (filled squares) (250 ng/μl). Triplicate samples were quantified and background values obtained with the control GST protein were subtracted. In these RdRP assays, less than 15% variation was measured among triplicate samples. Similar results were obtained in three independent experiments.

(Figure 32.4). The RNA polymerase activity of p66 and/or p51 increased with increasing protein concentration in a linear fashion, but could not be detected in reaction mixtures lacking oligo U$_{12}$RNA. Control GST protein did not show any activity. These results suggest that HIV-1 p66 and/or p51 may harbor RdRP

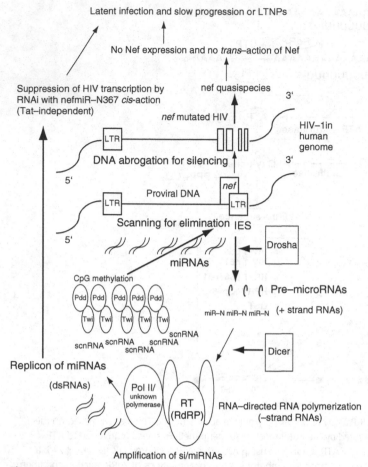

Figure 32.5. The RT in retroviruses may induce amplification of intervening repeat sequences as well as DNA elimination in heterochromatin. The panels represent an si/miRNA mutation model for internal eliminated segment (IES) in the LTR. The pre-miRNA from the *nef* region is processed by Dicer into miR-N microRNAs. The miRNA as a scanRNA (scnRNA) can target Twi and Pdd-like proteins to the cognate DNA sequence which, together with CpG methylation of DNA, then results in target DNA segment degeneration. Actually *nef* is a quasispecies, such as D36 PBMC (GenBank accession number of U37271), C18 PBMC (U37267), etc., derived from long-term nonprogressors (LTNPs) of AIDS. The deletion of genomic DNA may implicate viral mutation as well as ultimate gene silencing, which may induce latency or long-term nonprogression for infected individuals. This silencing requires RdRP to amplify miRNAs using a template of pre-miRNAs or miRNAs. Although RT can involve RdRP activity, cellular pol II at least in part, has been reported to be an enzyme producing dsRNAs.

activity. Intriguingly, a combination of p66 and p51 was approximately two-fold more active (3.5 U/µg: RT, 25 U/µg) than only p66 or p51 (Figure 32.4), indicating that p66/p51 heterodimers might mediate greater RNA polymerase activity than p66 or p51 proteins alone. Kinetic analysis revealed that enzyme activity of HIV-1 p66 and/or p51 peaked 30 min after initiation of the enzyme reaction (Figure 32.4). Similar results were obtained with HIV-1 SF2 p66 and p51. HERV-K RT also showed RNA polymerase activity (0.8 U/µg) (Figure 32.4); in addition,

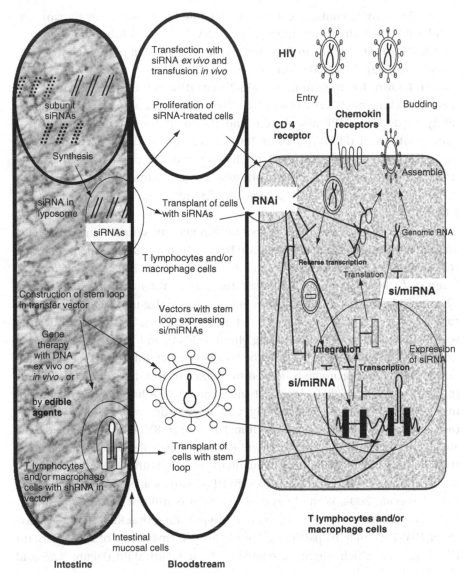

Figure 32.6. Challenges to applications for therapeutic si/miRNAs against AIDS. Given the significance of RNAi, it appears that HIV could be controlled predominantly by si/miRNAs. These biological approaches may be applicable in many different ways to therapy of AIDS. One proposal currently under consideration is to treat blood precursor cells, such as CD34$^+$ stem cells from the bone marrow, with liposomes (Sioud and Sorensen, 2003) or lentivectors (Wiznerowicz and Trono, 2005), to genetically equip the cells with anti-HIV si/miRNAs and then transfuse them back into the patient. Secondly, in nematodes, ingested dsRNA-expressing vectors have been reported to be delivered into internal organs (Timmons and Fire, 1998). Thus, oral siRNA agents could possibly be effective against HIV infection in the intestine *in situ* by protecting macrophages and dendritic cells against infection (Aquaro *et al.*, 2002). Thirdly, they could be applied via a mucosal route, such as through the nose; indeed, nasally administered si/miRNA to influenza H5N1 and respiratory syncytial viruses effectively inhibited infection in mice (Ge *et al.*, 2004; Bitko *et al.*, 2005). Finally, ingested phage DNA can be taken up via intestinal mucosa and transferred to spleen and liver in mice (Schubert *et al.*, 1997). Because HIV is transmitted both venereally and hematogenously and causes opportunistic infections (Gottlieb *et al.*, 1981), it is reasonable to speculate that prevention of HIV-1 invasion of the mucosal immune system by siRNA treatment may result in retention of native

measurable activity could be detected with feline FV RT (data not shown). The cellular RdRP activity of retrotransposon HERV-K RT was lower compared with viral RdRP activity of human hepatitis C virus (HCV) NS5B protein expressed in *E. coli* (200 U/μg) as a positive control.

Well known RT inhibitor, 3'-azido-2'-deoxythimidine triphosphate, completely blocked the RT activity exhibited by HIV-1 p66 and p51. To test susceptibility to ddUTP, RdRP assays were undertaken in the presence of different concentrations of ddUTP (1, 10 or 100 μM). ddUTP clearly blocked RNA polymerase activity in a dose-dependent manner, but did not block RT activity. And 3'-azido-2'-deoxythimidine triphosphate did not inhibit RNA polymerase activity, suggesting that RNA polymerase activity could be derived from the HIV-1 RT protein.

Quite recently, not only hundreds of human genome-encoded but also virally encoded miRNAs have been discovered in HIV-1 (Omoto and Fujii, 2006; Bennasser *et al.*, 2005) and other viruses (Pfeffer *et al.*, 2004; Cai *et al.*, 2005; Grey *et al.*, 2005; Dunn *et al.*, 2005; Samols, 2005; Sullivan *et al.*, 2005; Cantalupo *et al.*, 2005; see also Chapter 31 in this book by Nelson and his colleagues) (Figure 32.1). Although the roles of these viral miRNAs are not well understood, it is known that infection with RNA and DNA viruses including viroids leads to an inhibition of transcription of cellular protein-coding genes by host RNA pol II. Intriguingly, almost half of the mammalian genome is derived from ancient retrotransposable elements (Bannert and Kurth, 2004). Among them, some HERV-K has retained open reading frames coding for functionally and enzymatically active RT proteins (Berkhout *et al.*, 1999). If RTs derived from HIV-1 or HERVs in the human genome are able to replicate and amplify the miR-N367, accumulation of miRNAs from the 3'-LTR U3 might be predicted to maintain persistent infection long term, even for the 7–15 years seen in AIDS. Although HIV-1 can escape from RNAi by mutation (Das *et al.*, 2004; Westerhout *et al.*, 2005), it is still not known why siRNAs induce mutation in the HIV-1 genome. We hypothesize that RNA replicase activity of HIV-1 RT may be partly responsible for the emergence of RNAi-resistant HIV-1 strains, which might be related to their latent state (Figure 32.5 and Figure 32.6).

Figure 32.6. (cont.)
immunological competence against opportunistic pathogens. Transgenic plant products of genes encoding antigens of microbial pathogens can stimulate mucosal and systemic immune responses that fully or partly protect animals from subsequent exposure to the real pathogens very effectively when they were provided as food (Yusibov *et al.*, 1997; Hanlon *et al.*, 2002). Moreover, the use of plant-based systems also avoids many of the potential safety issues associated with contaminating mammalian viruses (Hooper, 1999). The focus of our research currently in progress lies in our investigation of expression of dsRNA segments corresponding to HIV-I *nef*, demonstrating that *nef* siRNA could be detected in embryonic cell aggregates of *A. thaliana* (Fujii, in press). Generation of transgenic fruits may become of benefit in most parts of the developing world known as "a resource poor setting" (Wiedle *et al.*, 2002) and may have advantages over conventional drug therapy.

REFERENCES

Aquaro, S., Calio, R., Balzarini, J. *et al.* (2002). Macrophages and HIV infection: therapeutical approaches toward this strategic virus reservoir. *Antiviral Research*, **55**, 209–225.

Aravin, A. A., Naumova, N. M., Tulin, A. V. *et al.* (2001). Double-stranded RNA-mediated silencing of genomic tandem repeats and transposable elements in the *D. melanogaster* germline. *Current Biology*, **11**, 1017–1027.

Bannert, N. and Kurth, R. (2004). Retroelements and the human genome: new perspectives on an old relation. *Proceedings of the National Academy of Sciences USA*, **101**, 14 572–14 579.

Baulcombe, D. C. (2001). RNA silencing. Diced defence. *Nature*, **409**, 295–296.

Bennasser, Y., Le, S. Y., Benkirane, M. and Jeang, K. T. (2005). Evidence that HIV-1 encodes an siRNA and a suppressor of RNA silencing. *Immunity*, **22**, 607–619.

Berkhout, B, Jebbink, M. and Zsiros, J. (1999). Identification of an active reverse transcriptase enzyme encoded by a human endogenous HERV-K retrovirus. *Journal of Virology*, **73**, 2365–2375.

Bitko, V., Musiyenko, A., Shulyayeva, O. and Barik, S. (2005). Inhibition of respiratory viruses by nasally administered siRNA. *Nature Medicine*, **11**, 50–55.

Cai, X., Lu, S., Zhang, Z. *et al.* (2005). Kaposi's sarcoma-associated herpesvirus expresses an array of viral microRNAs in latently infected cells. *Proceedings of the National Academy of Sciences USA*, **102**, 5570–5575.

Cantalupo, P., Doering, A., Sullivan, C. S. *et al.* (2005). Complete nucleotide sequence of polyomavirus SA12. *Journal of Virology*, **79**, 13094–13104.

Capodici, J., Kariko, K. and Weissman, D. (2002). Inhibition of HIV-1 infection by small interfering RNA-mediated RNA interference. *Journal of Immunology*, **169**, 5196–5201.

Chendrimada, T. P., Gregory, R. I., Kumaraswamy, E. *et al.* (2005). TRBP recruits the Dicer complex to Ago2 for microRNA processing and gene silencing. *Nature*, **436**, 740–744.

Coburn, G. A. and Cullen, B. R. (2002). Potent and specific inhibition of human immuno-deficiency virus type 1 replication by RNA interference. *Journal of Virology*, **76**, 9225–9231.

Cogoni, C. and Macino, G. (1999). Gene silencing in *Neurospora crassa* requires a protein homologous to RNA-dependent RNA polymerase. *Nature*, **399**, 166–169.

Cullen, B. R., Lomedico, P. T. and Ju, G. (1984). Transcriptional interference in avian retro-virus-implication for the promoter insertion model of leukaemogenesis. *Nature*, **307**, 241–245.

Das, A. T., Brummelkamp, T. R., Westerhout, E. M. *et al.* (2004). Human immunodeficiency virus type 1 escapes from RNA interference-mediated inhibition. *Journal of Virology*, **78**, 2601–2605.

Diener, T. O. (1991). Subviral pathogens of plants: viroids and viroidlike satellite RNAs. *Federation of the American Society for Experimental Biologists Journal*, **5**, 2808–2813.

Ding, S. W., Li, H., Lu, R., Li, F. and Li, W. X. (2004). RNA silencing: a conserved antiviral immunity of plants and animals. *Virus Research*, **102**, 109–115.

Drysdale, C. M. and Pavlakis, G. N. (1991). Rapid activation and subsequent down-regula-tion of the human immunodeficiency virus type-1 promoter in the presence of Tat – possible mechanisms contributing to latency. *Journal of Virology*, **65**, 3044–3051.

Dunn, W., Trang, P., Zhong, Q. *et al.* (2005). Human cytomegalovirus expresses novel microRNAs during productive viral infection. *Cellular Microbiology*, **7**, 1684–1695.

Filipovska, J. and Konarska, M. M. (2000). Specific HDV RNA-templated transcription by pol II in vitro. *RNA*, **6**, 41–54.

Fire, A., Xu, S., Montgomery, M. K. *et al.* (1998). Potent and specific genetic interference by double-stranded RNA in *Caenorhabditis elegans*. *Nature*, **391**, 806–811.

Ge, Q., Filip, L., Bai, A. *et al.* (2004). Inhibition of influenza virus production in virus-infected mice by RNA interference. *Proceedings of the National Academy of Sciences USA*, **101**, 8676–8681.

Gitlin, L., Kareisky, S. and Andino, R. (2002). Short interfering RNA confers intracellular antiviral immunity in human cells. *Nature*, **418**, 430–434.

Gottlieb, M. S., Schroff, R., Schanker, H. M. *et al.* (1981). Pneumocystis carinii pneumonia and mucosal candidiasis in previously healthy homosexual men. *New England Journal of Medicine*, **305**, 1425–1431.

Grey, F., Antoniewic, A., Allen, E. *et al.* (2005). Identification and characterization of human cytomegalovirus-encoded microRNAs. *Journal of Virology*, **79**, 12 095–12 099.

Haase, A. D., Jaskiewicz, L., Zhang, H. *et al.* (2005). TRBP, a regulator of cellular PKR and HIV-1 virus expression, interacts with Dicer and functions in RNA silencing. *European Molecular Biology Organization Reports*, **6**, 961–967.

Hanlon, C. A., Niezgoda, M., Morrill, P. and Rupprecht, C. E. (2002). Oral efficacy of an attenuated rabies virus vaccine in skunks and raccoons. *Journal of Wild Diseases*, **38**, 420–427.

Hansen, J. L., Long, A. M. and Schultz, S. C. (1997). Structure of the RNA-dependent RNA polymerase of poliovirus. *Structure*, **5**, 1109–1122.

Hariharan, M., Scaria, V., Pillai, B. and Brahmachari, S. K. (2005). Targets for human encoded microRNAs in HIV genes. *Biochemical Biophysical Research Communications*, **337**, 1214–1218.

Hooper, E. (1999). *The River; a Journey to the Source of HIV and AIDS*. Little, Brown and Company.

Hu, W. Y., Myers, C. P., Kilzer, J. M., Pfaff, S. L. and Bushman, E. D. (2002). Inhibition of retroviral pathogenesis by RNA interference. *Current Biology*, **12**, 1301–1311.

Jacque, J. M., Triques, K. and Stevenson, M. (2002). Modulation of HIV-1 replication by RNA interference. *Nature*, **418**, 435–438.

Jopling, C. L., Yi, M., Lancaster, A. M., Lemon, S. M. and Sarnow, P. (2005). Modulation of hepatitis C virus RNA abundance by a liver-specific microRNA. *Science*, **309**, 1577–1581.

Ketting, R. F., Haverkamp, T. H., van Luenen, H. G. and Plasterk, R. H. (1999). *mut-7* of *C. elegans*, required for transposon silencing and RNA interference, is a homolog of Werner syndrome helicase and RNaseD. *Cell*, **99**, 133–141.

Lahser, F. C. and Malcolm, B. A. (2004). A continuous nonradioactive assay for RNA-dependent RNA polymerase activity. *Analytical Biochemistry*, **325**, 247–254.

Lecellier, C.-H., Dunoyer, P., Arar, K. *et al.* (2005). A cellular microRNA mediates antiviral defense in human cells. *Science*, **308**, 557–560.

Lee, N. S., Dohjima, T., Bauer, G. *et al.* (2002). Expression of small interefering RNAs targeted against HIV-1 rev transcripts in human cells. *Nature Biotechnology*, **20**, 500–505.

Le Grice, S. F., Naas, T., Wohlgensinger, B. and Schatz, O. (1991). Subunit-selective mutagenesis indicates minimal polymerase activity in heterodimer-associated p51 HIV-1 reverse transcriptase. *European Molecular Biology Organization Journal*, **10**, 3905–3911.

Lu, R., Maduro, M., Li, F. *et al.* (2005). Animal virus replication and RNAi-mediated antiviral silencing in *Caenorhabditis elegans*. *Nature*, **436**, 1040–1043.

Matzke, M. A. and Birchler, J. A. (2005). RNAi-mediated pathways in the nucleus. *Nature Reviews in Genetics*, **6**, 24–35.

Modahl, L. E., Macnaughton, T. B., Zhu, N., Johnson, D. L. and Lai, M. M. (2000). RNA-dependent replication and transcription of hepatitis delta virus RNA involve distinct cellular RNA polymerases. *Molecular Cell Biology*, **20**, 6030–6039.

Mourrain, P., Beclin, C., Elmayan, T. *et al.* (2000). *Arabidopsis* SGS2 and SGS3 genes are required for posttranscriptional gene silencing and natural virus resistance. *Cell*, **101**, 533–542.

Novina, C. D., Murray, M. F., Dykxfoorn, D. M. *et al.* (2002). siRNA-directed inhibition of HIV-1 infection. *Nature Medicine*, **8**, 681–686.

Omoto, S. and Fujii, Y. R. (2005). Regulation of human immunodeficiency virus 1 transcription by nef microRNA. *Journal of General Virology*, **86**, 751–755.

Omoto, S. and Fujii, Y. (2006). Cloning and detection of HIV-1 encoded microRNA. *Method in Molecular Biology*, **342**, 255–265.

Omoto, S., Ito, M., Tsutsumi, Y. *et al.* (2004). HIV-1 nef suppression by virally encoded microRNA. *Retrovirology*, **1**, 44.

Ong, C. L., Thorpe, J. C., Gorry, P. R. *et al.* (2005). Low TRBP levels support an innate human immunodeficiency virus type 1 resistance in astrocytes by enhanceing the PKR anti-viral response. *Journal of Virology*, **79**, 12 763–12 772.

Paddison, P. J., Caudy, A. A. and Hannon, G. J. (2002). Stable suppression of gene expression by RNAi in mammalian cells. *Proceedings of the National Academy of Sciences USA*, **99**, 1443–1448.

Pfeffer, S., Zavolan, M., Grasser, F. A. *et al.* (2004). Identification of virus-encoded microRNAs. *Science*, **304**, 734–736.

Robert-Guroff, N., Popovic, N., Gartner, S. *et al.* (1990). Structure and expression of tat-, rev-, and nef-specific transcripts of human immunodeficiency virus type 1 in infected lymphocytes and macrophages. *Journal of Virology*, **64**, 3391–3398.

Samols, M. A., Hu, J., Skalsky, R. L. and Renne, R. (2005). Cloning and identification of a microRNA cluster within the latency-associated region of Kaposi's sarcoma-associated herpesvirus. *Journal of Virology*, **79**, 9301–9305.

Schiebel, W., Pelissier, T., Riedel, L. *et al.* (1993). Isolation of an RNA-directed RNA polymerase-specific cDNA clone from tomato. *Plant Cell*, **10**, 2087–2101.

Schubbert, R., Renz, D., Schmitz, B. and Doerfler, W. (1997). Foreign (M13) DNA ingested by mice reaches peripheral leukocytes, spleen, and liver via the intestinal wall mucosa and can be covalently linked to mouse DNA. *Proceedings of the National Academy of Sciences USA*, **94**, 961–966.

Sijen, T. and Plasterk, R. H. (2003). Transposon silencing in the *Caenorhabditis elegans* germ line by natural RNAi. *Nature*, **426**, 310–314.

Sijen, T., Fleenor, J., Simmer, F. *et al.* (2001). On the role of RNA amplification in dsRNA-triggered gene silencing. *Cell*, **107**, 465–476.

Sioud, M. and Sorensen, D. R. (2003). Cationic liposome-mediated delivery of siRNAs in adult mice. *Biochemical Biophysical Research Communications*, **312**, 1220–1225.

Sullivan, C. S., Grundhoff, A. T., Tevethia, S., Pipas, J. M. and Ganem, D. (2005). SV40-encoded microRNAs regulate viral gene expression and reduce susceptibility to cyto-toxic T cells. *Nature*, **435**, 682–686.

Tabara, H., Sarkissian, M., Kelly, W. G. *et al.* (1999). The *rde*-1 gene, RNA interference, and transposon silencing in *C. elegans*. *Cell*, **99**, 123–132.

Timmons, L. and Fire, A. (1998). Specific interference by ingested dsRNA. *Nature*, **395**, 854.

Tuschl, T., Zamore, P. D., Lehmann, R., Bartel, D. P. and Sharp, A. (1999). Targeted mRNA degradation by double-stranded RNA *in vitro*. *Genes & Development*, **13**, 3191–3197.

Westerhaout, E. M., Ooms, M., Vink, M., Das, A. T. and Berkhout, B. (2005). HIV-1 can escape from RNA interference by evolving an alternative structure in its RNA genome. *Nucleic Acids Research*, **33**, 796–804.

Wiedle, P. J., Mastro, T. D., Grant, A. D., Nkengasong, J. and Macharia, D. (2002). HIV/AIDS treatment and HIV vaccines for Africa. *Lancet*, **359**, 2261–2267.

Wilkins, C., Dishongh, R., Moore, S. C. *et al.* (2005). RNA interference is an antiviral defence mechanism in *Caenorhabditis elegans*. *Nature*, **436**, 1044–1047.

Wiznerowicz, M. and Trono, D. (2005). Harnessing HIV for therapy, basic research and biotechnology. *Trends in Biotechnology*, **23**, 42–47.

Yamamoto, T., Omoto, S., Mizuguchi, M. *et al.* (2002). Double-stranded nef RNA interferes with human immunodeficiency virus type 1 replication. *Microbiology and Immunology*, **46**, 809–817.

Yekta, S., Shih, I. and Bartel, D. P. (2004). MicroRNA-directed cleavage of HOXB8 mRNA. *Science*, **304**, 594–596.

Yusibov, V., Modelska, A., Steplewski, K. *et al.* (1997). Antigen produced in plants by infection with chimeric plant viruses immunize against rabies virus and HIV-1. *Proceedings of the National Academy of Sciences USA*, **94**, 5784–5788.

Zamore, P. D., Tuschl, T., Sharp, P. A. and Bartel, D. P. (2000). RNAi: double-stranded RNA directs the ATP-dependent cleavage of mRNA at 21- and 22-nucleotide RNAs. *Genes & Development*, **15**, 188–200.

Zeng, Y., Yi, R. and Cullen, B. R. (2003). MicroRNAs and small interfering RNAs can inhibit mRNA expression by similar mechanisms. *Proceedings of the National Academy of Sciences USA*, **100**, 9779–9784.

VI

MicroRNAs in stem cell development

33 MicroRNAs in the stem cells of the mouse blastocyst

Hristo B. Houbaviy

Introduction

The earliest differentiation event in mammalian embryogenesis is the formation of the inner cell mass (ICM) and the trophoblast compartments of the blastocyst (Theiler, 1989; Kaufman and Bard, 1999). While the ICM gives rise to the embryo proper and to the extraembryonic membranes found in mammals as well as reptiles and birds, the trophoblast contributes exclusively to the placenta, an organ that exists only in eutherian mammals.

Pluripotent stem cell lines can be derived from both the ICM and the trophoblast (Evans and Kaufman, 1981; Martin, 1981; Tanaka *et al.*, 1998). Embryonic stem (ES) cells, the derivatives of the ICM, differentiate into all cell types of the embryo proper when injected into recipient blastocysts. Their counterpart, the trophoblastic stem (TS) cells contribute only to the placental lineages of recipient blastocysts. At least some of this *in vivo* developmental potential can be recapitulated *in vitro* and the directed differentiation of ES and TS cells into defined cell types is a very active field of study with obvious therapeutic implications.

ES cells are almost identical to the so-called embryonic germ (EG) cell lines, which are derived from the primordial germ cells (PGC) of the embryo (Labosky *et al.*, 1994; Stewart *et al.*, 1994). Thus, ES cells are thought to share some characteristics with the germ line stem cells. Indeed, molecules required for the maintenance of the pluripotent ES cell state, such as the transcription factors Nanog and Oct-4, are also expressed in the germ line and its precursors (Niwa *et al.*, 2000; Pesce and Scholer, 2001; Chambers *et al.*, 2003; Mitsui *et al.*, 2003; Kehler *et al.*, 2004).

The rationale for the elucidation of the microRNA (miRNA) and short interfering RNA (siRNA) repertoire of ES and TS cells is two fold. First, it is likely that miRNAs specifically expressed in ES and/or TS cells might be important for early development as well as for the maintenance of the pluripotent stem cell phenotype as is the case for transcription factors such as Nanog and Oct-4 (Chambers *et al.*, 2003; Mitsui *et al.*, 2003; Niwa *et al.*, 2000). miRNAs are important developmental regulators in invertebrates (reviewed in Bartel (2004)). Classical examples

MicroRNAs: From Basic Science to Disease Biology, ed. Krishnarao Appasani. Published by Cambridge University Press. © Cambridge University Press 2008.

are lin-4 and let-7 whose loss of function causes heterochronic phenotypes in
C. elegans (Lee *et al.*, 1993; Reinhart *et al.*, 2000). In *D. melanogaster* the PAZ/PIWI
domain family of proteins, which are the effector components of the RNA
induced silencing complex (RISC), have been implicated in germ line maintenance (Cox *et al.*, 1998; Cox *et al.*, 2000). The corresponding mammalian
homologs are required for spermatogenesis (Deng and Lin, 2002; Kuramochi-
Miyagawa *et al.*, 2004). Recent data show that the nuclease Dicer-1, which is
responsible for miRNA production, is involved in the maintenance of the germ
stem cells in *Drosophila* (Hatfield *et al.*, 2005).

The second reason to look for short RNAs in early embryos and the corresponding stem cell lines, particularly in ES cells, concerns the connection between RNAi
and transcriptional silencing. RNAi dependent repressive chromatin modification has been very well documented in *Tetrahymena* and *S. pombe* (Mochizuki
et al., 2002; Volpe *et al.*, 2002; Mochizuki and Gorovsky, 2004; Verdel *et al.*, 2004)
and there are compelling data that RNAi is involved in transcriptional silencing
in *D. melanogaster* (Pal-Bhadra *et al.*, 2002; Pal-Bhadra *et al.*, 2004). Evidence for
RNAi-dependent transcriptional silencing in mammals has also been presented
(Kawasaki and Taira, 2004; Morris *et al.*, 2004). However, with the notable exception of *Tetrahymena* and *S. pombe*, the nature of the siRNAs that mediate silencing
is unknown. This is because siRNAs are thought to be involved in the initiation of
transcriptional silencing which is restricted spatially and temporarily (Hall *et al.*,
2002). Thus, such siRNAs are postulated to be scarce and, in most cases, relevant
biological material cannot be obtained in sufficient quantities to allow cloning.
ES cells, on the other hand, can be grown in sufficient quantities for cloning and
recapitulate various epigenetic silencing phenomena that take place during
embryogenesis such as X-chromosome inactivation and transcriptional silencing
of oncoretroviruses (e.g. Moloney Murine Leukemia Virus, MMLV) (Lee *et al.*,
1996; Cherry *et al.*, 2000; Wutz and Jaenisch, 2000). X-chromosome inactivation
depends on the interaction of a pair of complementary non-coding RNAs – *Xist*
and *Tsix* (Lee *et al.*, 1999). *Xist* is downregulated posttranscriptionally and because
of its complementarity to *Tsix* it is tempting to speculate that dsRNA is formed
and is subsequently degraded by the RNAi machinery (Panning *et al.*, 1997).
Studies in *C. elegans* and *D. melanogaster* have implicated the RNAi machinery in
the silencing of repetitive elements in the germ line (Ketting *et al.*, 1999; Aravin
et al., 2001; Aravin *et al.*, 2004). Given the relationships between ES cells and the
mammalian germ line and the presence of multiple MMLV proviral copies in the
mouse, it is tempting to speculate that the transcriptional silencing of oncoretroviruses in mouse ES cells reflects a general mechanism for repetitive element
silencing, perhaps involving the RNAi machinery.

Obstacles that preclude the identification of siRNAs involved in transcriptional silencing in ES cells

One approach for identifying siRNAs involved in epigenetic phenomena is to
perform Northern hybridizations with probes derived from hypothetical

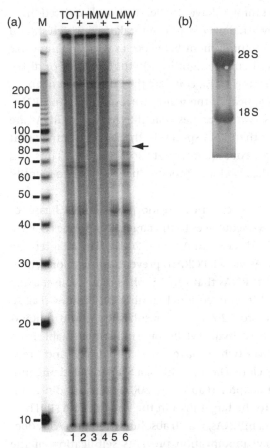

Figure 33.1. Short RNA profile of ES cells. (a) Total ES cell RNA (lanes 1 and 2) or the high molecular weight (HMW, lanes 3 and 4) and low molecular weight (LMW, lanes 5 and 6) fractions obtained after precipitation with 10% polyethylene glycol were either phosphorylated directly (odd lanes) or dephosphorylated by alkaline phosphatase treatment and then phosphorylated (even lanes). The molecular weight marker is shown on the left (M). The arrow points to a band unmasked by dephosphorylation. (b) Denaturing agarose gel electrophoretic analysis of the total RNA from the experiment shown in (a) shows intact 28 S and 18 S rRNA bands and, thus, lack of degradation.

candidate loci. We have been unable to detect siRNAs originating from the *Xist* locus or from repetitive elements in ES cells via short RNA Northerns under conditions that readily detect miRNAs (data not shown). This leaves short RNA cloning as the only remaining option, which does not depend on an underlying hypothesis as to the origin of the siRNAs, and should, in theory, identify even very rare RNA species provided that a sufficiently large number of clones is obtained.

When 5'- radioactively phosphorylated total ES cell RNA is analyzed by denaturing polyacrylamide electrophoresis, a smear of bands over a wide range of molecular weights is observed (Figure 33.1a, odd lanes). Additional discrete bands corresponding to abundant non-coding RNAs (e.g. tRNA, 5S rRNA, 5.8S rRNA and snRNAs) are also seen. Because labeling by 5' phosphorylation requires free 5'-hydroxyls, siRNAs and miRNAs that already have 5' phosphates would not be observed in the above experiment (Elbashir *et al.*, 2001a). To radioactively label siRNAs and miRNAs, the RNA samples have to be dephosphorylated first. Phosphatase treatment followed by 5' radioactive phosphorylation, which should label siRNAs and miRNAs, unmasks a prominent RNA species with apparent size of approximately 82 nt but does not change significantly the intensity or distribution of most remaining bands (Figure 33.1a, compare the odd and even lanes). Importantly, no additional bands appear in the region corresponding to siRNAs and miRNAs (21–23 nt). This observation argues that siRNAs and miRNAs are orders of magnitude less abundant than RNA species in the 21–23 nt range that have free 5'-hydroxyls and which are consequently not Dicer products. Thus, it is essential to use methods that specifically clone Dicer products in order to identify siRNAs and miRNAs.

The short RNA cloning procedure relies on the ligation, by T4 RNA ligase, of adapter oligonucleotides with known sequences to the target RNA species (Lagos-Quintana et al., 2001; Lau et al., 2001; Lee and Ambros, 2001). The adapters are then used to amplify the target RNAs via RT-PCR. To prevent self-ligation (circularization and concatamerization) of RNAs that carry 5'-phosphates, the starting RNA material is dephosphorylated in conventional cloning procedures (Lagos-Quintana et al., 2001; Lee and Ambros, 2001). As a result siRNAs and miRNAs (which have a 5'-phosphate and a free 3' hydroxyl) become indistinguishable from RNAs that are not processed by Dicer (which have a free 5'-hydroxyl and phosphorylated 3' ends). To specifically clone Dicer products, a 5' adenylated oligonucleotide is used as the 3' cloning adaptor (Lau et al., 2001). This modification permits ligation of the 3' adaptor to the target RNA in the absence of ATP. Thus, self-ligation of the target siRNAs and miRNAs, which absolutely requires ATP, does not occur. A 3'-deoxy or 3'-O-alkyl modification prevents self-ligation of the 3'-adaptor and results in a target RNA–adaptor conjugate that can ligate only at its 5' end to the 5' adaptor during the next step which is carried out in the presence of ATP.

Does the above procedure result in the cloning only of Dicer products? The observed size distribution of the short RNA clones argues that this is not the case. The size histograms of three short RNA libraries respectively obtained from ES cells grown on feeders (library L1), ES cells grown without feeders but in the presence of LIF (library L2) and ES cells differentiated by retinoic acid treatment in monolayer (library L3) are shown in Figure 33.2a. All three size distributions have prominent peaks at 22 nt, suggesting that Dicer products are indeed a significant fraction of the libraries. However, species with sizes outside the 21–23 nt range are a major part of all three libraries. In libraries L2 and L3 a peak at 17 nt is actually more prominent than the one centered at 22 nt. RNAs with sizes significantly outside the 21–23 nt range are unlikely to be produced by a mammalian Dicer enzyme and thus they represent "noise" in the RNA libraries (Zhang et al., 2002; Zhang et al., 2004).

Many short RNA species cloned from ES cells are fragments of rRNAs and tRNAs. rRNA genes are present as tandem arrays in the genome. In plants and S. pombe some rRNA copies appear to be silenced transcriptionally via an siRNA dependent mechanism (Xie et al., 2004; Cam et al., 2005). Thus, it is tempting to speculate

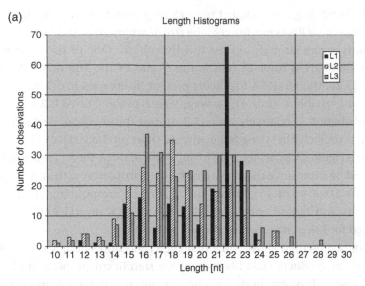

(b) 18S rRNA 5'-end 15 observations
    Average: 20.5
    RMSD: 2.3

```
ATACCTGGTTGATCCTGCCAGTAGC
 TACCTGGTTGATCCTGCCAGTAGC
 TACCTGGTTGATCCTGCCAGTAG
 TACCTGGTTGATCCTGCCAGTA
 TACCTGGTTGATCCTGCCAGT
 TACCTGGTTGATCCTGCCAGT
 TACCTGGTTGATCCTGCCAGT
 TACCTGGTTGATCCTGCCAG
 TACCTGGTTGATCCTGCCAG
 TACCTGGTTGATCCTGCCAG
 TACCTGGTTGATCCTGCCAG
 TACCTGGTTGATCCTGCCA
 TACCTGGTTGATCCTGCC
 TACCTGGTTGATCCTGCC
 TACCTGGTTGATCCTG
```

(c) tRNA-Ile-ATT 3'-end 7 observations
    Average: 20.4
    RMSD: 1.1

```
TCGCGGGTTCGATCCCCGTACGGGCCACCA
     TCGATCCCCGTACGGGCCACCA
     TCGATCCCCGTACGGGCCACCA
      ATCCCCGTACGGGCCACCA
       TCCCCGTACGGGCCACCA
        CCCCGTACGGGCCACCA
         CCGTACGGGCCACCA
```

(d) miR-21 21 observations
    Average: 22.6
    RMSD: 1.1

```
TAGCTTATCAGACTGATGTTGAC
TAGCTTATCAGACTGATGTTGAC
TAGCTTATCAGACTGATGTTGAC
TAGCTTATCAGACTGATGTTGAC
TAGCTTATCAGACTGATGTTGAC
TAGCTTATCAGACTGATGTTGAC
TAGCTTATCAGACTGATGTTGAC
TAGCTTATCAGACTGATGTTGAC
TAGCTTATCAGACTGATGTTGAC
TAGCTTATCAGACTGATGTTGAC
TAGCTTATCAGACTGATGTTGAC
TAGCTTATCAGACTGATGTTGAC
TAGCTTATCAGACTGATGTTGAC
TAGCTTATCAGACTGATGTTGAC
TAGCTTATCAGACTGATGTTGA
TAGCTTATCAGACTGATGTTGA
TAGCTTATCAGACTGATGTTGA
TAGCTTATCAGACTGATGTTGA
TAGCTTATCAGACTGATGTTG
 TTATCAGACTGATGTTGA
```

Figure 33.2. Size distributions of the short RNA clones. (a) Length histograms of the clones from short RNA libraries obtained from ES cells grown on feeders (L1), ES cells grown without feeders but in the presence of LIF (L2) and ES cells differentiated in monolayer with retinoic acid (L3). The vertical lines indicate the area of the sequencing gel (18–26 nt) from which the initial cloning material was extracted. Aligned individual clones corresponding to the 5'-end of the 18 S rRNA, the 3'-end of tRNA-Ile-ATT and miR-21 are shown in (b), (c) and (d), respectively.

that the rRNA fragments cloned from ES cells reflect rRNA gene transcriptional silencing via the RNAi machinery. The size distribution of the cloned short rRNA fragments argues strongly against this hypothesis. One of the most frequently cloned short RNAs from ES cells is the 5′ end of the 18S rRNA. As shown in Figure 33.2b, the 15 rRNA fragments present in libraries L1-L3 have a constant 5′ end and variable 3′ ends. The average length of these RNAs is 20.5 nt and their size distribution is relatively wide (2.3 nt root mean square deviation, RMSD). Thus, it is unlikely that these fragments are Dicer products. Likewise, the 7 clones shown in Figure 33.2c, which map to the 3′ end of tRNA-Ile-ATT, are unlikely to be produced by Dicer since they have a wide size distribution with an average of 20.4 nt and RMSD of 4.6 nt. In contrast, the 21 clones corresponding to miR-21 have a very tight distribution with an average size of 22.6 nt and RMSD of 1.1 nt as expected for RNAs processed by Dicer (Figure 33.2d).

The data presented above show that even procedures designed to clone specifically Dicer products yield libraries with a significant proportion of irrelevant RNA clones. Thus, methods for filtering out the latter are needed. miRNAs are processed from characteristic stem-loop precursors (pre-miRNAs) (Lee *et al.*, 1993; Hutvagner *et al.*, 2001; Lagos-Quintana *et al.*, 2001; Lau *et al.*, 2001; Lee and Ambros, 2001). This property makes it easy to find potential miRNAs within a library bioinformatically. To identify clones corresponding to potential miRNAs, genomic sequences extending approximately 50 nt from the ends of each short RNA clone are retrieved from the genome databases and folded *in silico* (Zuker *et al.*, 1999). Clones that yield stable stem-loop hairpin structures represent potential miRNAs.

Unlike miRNAs, siRNAs do not share structural or primary sequence identifying features. Dicer processes siRNAs from long (> 100 bp) double stranded RNAs without any sequence preference (Zamore *et al.*, 2000; Elbashir *et al.*, 2001b). The double stranded RNA precursor is made either owing to the transcription of both strands of a genomic locus or by an RNA dependent RNA polymerase that synthesizes the antisense strand of an RNA transcript (reviewed in Baulcombe (2004)). Apart from the requirement that siRNAs possess ends consistent with a type III RNAse cleavage, length is the only criterion that can distinguish potential siRNAs from RNAs not generated by Dicer (Zamore *et al.*, 2000; Elbashir *et al.*, 2001b). Like miRNA clones, clustered siRNAs would have tight size distributions with maxima at 22 nt (Figure 33.3a) whereas clustered mRNA degradation products would have broad size distributions (Figure 33.3b) similar to the rRNA and tRNA clones discussed above. A locus can potentially produce both *bona fide* siRNAs and degradation products. In this case, a broad distribution with a peak at 22 nt would be diagnostic.

So far no siRNAs potentially involved in transcriptional silencing have been identified in ES cells. It is possible that increasing the size of the short RNA libraries would result in the identification of such siRNAs. Unfortunately, as the size of the libraries increases so does the number of irrelevant clones. The problem is further compounded by the fact that miRNAs are the most abundant Dicer products in ES cells. Thus, potential siRNAs can be identified only via application

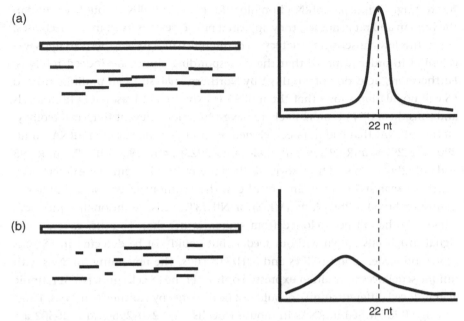

Figure 33.3. Distinguishing clustered siRNAs from clustered degradation products. Drawings of hypothetical loci yielding clustered siRNAs or clustered degradation products together with the corresponding predicted size distributions of the short RNA clones are shown in (a) and (b) respectively.

of the bioinformatic criteria discussed above as libraries with good statistics become available.

miRNAs in ES cells

As discussed above, unlike siRNAs, potential miRNAs are easy to identify bioinformatically in short RNA libraries. However, given the high proportion of non-Dicer products, folding into a hairpin precursor is not sufficient to prove that a clone is indeed a miRNA. This is in part because of the fact that precise rules for what constitutes a good hairpin are difficult to formulate. Thus, additional formal criteria for miRNA identification have been suggested (Ambros *et al.*, 2003). At a minimum, the miRNA clone and its corresponding hairpin precursor must be phylogenetically conserved. The evidence that a clone represents a miRNA is further strengthened by the detection of a discrete band in the 21–23 nt range corresponding to the clone by Northern analysis. Finally, it is desirable, but not absolutely required, to show that production of a putative miRNA is dependent on the key processing enzymes Dicer and Drosha.

Two studies have reported miRNAs in mouse and human ES cells (Houbaviy *et al.*, 2003; Suh *et al.*, 2004). Houbaviy *et al.* identified 53 distinct miRNAs in undifferentiated mouse ES cells and mouse ES cells that underwent retinoic acid induced differentiation (Houbaviy *et al.*, 2003). The majority of these (32) were identical or closely related to miRNAs previously cloned from sources other than ES cells. However, 15 miRNAs (miR-290–miR-302) had unique sequences.

Interestingly, while pre-miRNA homologs for these 15 miRNAs could be found in the human and rat genomes, they appeared not to be conserved in invertebrates. Given this low conservation, detection of discrete bands by Northern analysis was essential in order to prove that the corresponding clones are indeed miRNAs. Furthermore, detection of a miRNA by Northern analysis in an undifferentiated ES cell population shows that the miRNA is expressed at reasonably high levels and thus the miRNA is more likely to have a function relevant to ES cell biology. Of the miRNAs that had not been cloned previously, nine mature miRNAs (miR-290, miR-291-s, miR-291-as, miR-292-s, miR-292-as, miR-293, miR-294, miR-295 and miR-296) processed from seven distinct pre-miRNA hairpins were detected by Northern analysis in ES cells grown either with or without a feeder layer but not in mouse embryonic fibroblasts (MEFs) or NIH/3T3 cells. Additionally, miR-302, which also hadn't been cloned from sources other than ES cells, gave a weak signal in ES cells grown without feeders but could not be detected in ES cells grown on feeders and in MEFs and NIH/3T3 cells. The remaining miRNAs with unique sequences were either expressed only after the ES cells underwent retinoic acid induced differentiation or could not be detected by Northern analysis. Thus, among all identified miRNAs in mouse ES cells, miR-290–296 and miR-302 are most likely to be pertinent to ES cell biology. Further analysis revealed that pre-miR-290–295 are located within 2 kb of each other and are probably expressed from a common transcription unit. While miR-296 and miR-302 are well conserved between mouse and human the predicted human homolog of the pre-miR-290–295 cluster contains only three pre-miRNA stem-loops with sequences that are significantly divergent from their mouse counterparts.

The miRNA repertoire of human ES cells is qualitatively similar to that of mouse ES cells. Suh *et al.* identified 36 miRNAs in undifferentiated human ES cells, including the human homolog of miR-296 as well as four distinct homologs of mouse miR-302 (designated miR-302a-d) (Suh *et al.*, 2004). Nine miRNAs (miR-367–373, miR-373* and miR374) had sequences sufficiently different from previously identified miRNAs to be given new miRNA designations. More detailed analysis, however, reveals that most of these are in fact related to miRNAs cloned previously from other sources. Analysis of the genomic locus where miR-370 maps reveals that this miRNA is a member of a large miRNA cluster within the imprinted Dlk1-Gtl2 domain, that includes miR-300 which was cloned from mouse ES cells but could not be detected by Northern analysis, as well as numerous miRNAs with brain specific expression (Lagos-Quintana *et al.*, 2002; Houbaviy *et al.*, 2003; Seitz *et al.*, 2003; Kim *et al.*, 2004; Seitz *et al.*, 2004). Since a common primary transcript likely spans this entire cluster, expression of miR-300 and miR-370 is probably not restricted to ES cells. Bioinformatic analysis showed that pre-miR-302a–d together with pre-miR-367 form a cluster that is also likely to be expressed from a common primary transcript. While only a single miRNA that maps within this cluster (mouse miR-302 which is identical to human miR-302a) was cloned from undifferentiated mouse ES cells, the remaining pre-miRNA hairpins are present in the mouse genome suggesting that miR-302b–d and miR-367 are expressed in mouse ES cells. Even though they received new miRNA designations, miR-371–373 and miR-373* are clearly the predicted human homologs of the mouse miR-290–295 cluster (Houbaviy *et al.*, 2003). Thus, only 3 of the miRNAs cloned from human ES cells (miR-368, miR-369 and miR-374) are completely unrelated to miRNAs previously cloned from other sources, including undifferentiated mouse ES cells. Of these only miR-368 was detected in ES cells but not in somatic cell lines by Northern analysis, whereas miR-369 gave no signal and miR-374 was also expressed in HeLa cells. Neither the mouse nor the human ES cell miRNA libraries completely saturate the pool of available miRNAs. Thus, the absence of miR-374 from the mouse ES cell library probably does not reflect any fundamental differences between species.

In summary, cloning and Northern analysis suggest that the following miRNAs might have functions relevant to ES cell biology: miR-296, miR-368, miRNAs expressed from the cluster comprised of pre-miR-367 and miR-302a–302d and miRNAs expressed from the homologous clusters comprised of pre-miR-290–295 in the mouse and pre-miR-371–373 in human.

miRNA clusters pre-miR-290–295/pre-miR-371–373 and pre-miR-367-pre-miR-302a–d

While the data presented above suggest that miR-296 and miR-367 might have ES cell specific functions, these miRNAs are expressed at relatively low levels in ES cells as both were cloned only once in each library. In contrast, miRNAs from the cluster comprising pre-miR-290–295 account for 25% of the miRNAs cloned from mouse ES cells grown on feeders and 29% of the miRNAs cloned from mouse ES cells grown without feeders (Houbaviy *et al.*, 2003). No other miRNA locus has such a high cloning frequency in the mouse ES cell libraries. Likewise, miRNAs from the pre-miR-367-pre-miR-302a–d cluster are the most frequently cloned species from human ES cells and represent 86% of all clones (Suh *et al.*, 2004). Consistent with the cloning data, miRNAs from the pre-miR-290–295 and miR-367-pre-miR-302a–d clusters gave robust signals in Northern hybridizations of RNAs from mouse and human ES cells respectively. Interestingly, the human homolog of pre-miR-290–295, the pre-miR-371–373 cluster was represented by only 4% of the human ES cell clones whereas miRNAs belonging to the pre-miR-367-pre-miR-302a–d cluster were represented by only a single clone from ES cells differentiated with retinoic acid. Thus, while the miRNA repertoires of human and mouse ES cells are qualitatively the same, an important quantitative difference is the relative expression of clusters pre-miR-290–295/pre-miR-371–373 and pre-miR-367-pre-miR-302a–d.

As mentioned above, mouse miR-302 was detected by Northern analysis only in ES cells grown without feeders but not in ES cells grown on a feeder layer of MEFs. Thus, it is tempting to speculate that the pre-miR-367-pre-miR-302a–d cluster is involved in interactions between the ES cells and the feeder layer. For ethical reasons, human ES cells can not be grown on human embryonic fibroblasts. The ES cell lines from which the human miRNA libraries were derived were grown on STO feeders – an immortalized mouse fibroblast-like cell line (Bernstein *et al.*,

1976). Thus, the elevated miR-367-miR-302a–d levels in human ES cells may reflect the inability of the mouse feeders to interact with the human ES cell lines.

Multiple sequence alignment of the mouse and human pre-miRNA hairpins reveals that pre-miR-302a–d is related to pre-miR-290–295/pre-miR-371–373 (Figure 33.4). The most conserved area in the alignment, the motif AAGTGC, corresponds to positions 2–7 at the predominant 5′ ends of the mature miRNAs processed from the 3′ strand of the hairpin stems. This so-called seed region is the major determinant of miRNA target recognition (Lewis *et al.*, 2003; Doench and Sharp, 2004; Lewis *et al.*, 2005). Therefore, miRNAs processed from the 3′ strands of the stems of pre-miR-290–295/pre-miR-371–373 and pre-miR-302a–d might repress the same mRNAs. (However, there is some heterogeneity in the 5′ ends of the mature miR-290–296/miR-371–373 miRNAs which changes the seed, and thus the theoretical targets completely.) The area corresponding to the seeds of miRNAs processed from the 5′ strands of the hairpin stems is less conserved, the predominant seed of the pre-miR-290–295/pre-miR-371–373 cluster being CTCAAA and that of the pre-miR-367-pre-miR-302a–d cluster CTTTAA. Although conserved between mouse and human, pre-miR-367 is unrelated to pre-miR-302a–d and pre-miR-290–295/pre-miR-371–373 and has a completely different seed (Figure 33.4). Despite their similarities the two clusters cloned from ES cells are clearly different genes. They form different branches in the phylogenetic tree (Figure 33.4b) and have different organization (Houbaviy *et al.*, 2003; Suh *et al.*, 2004).

Homologs of pre-miR-367-pre-miR-302a–d are easily identifiable in tetrapods, but not in fishes, by standard BLAST searches of the genome databases. Multiple sequence alignment shows that the sequences of the pre-miR-302a–d and pre-miR367 hairpins from frog, chicken, mouse and human are very well conserved (Figure 33.5). The hairpins are more conserved at the 3′ stem than the 5′ stem with the mature miRNAs processed from the 3′ strand of the stem conserved almost to the nucleotide level. Given this high level of conservation, it is unlikely that BLAST searches would have missed any pre-miR-367-pre-miR-302a–d homologs present in the fish genomes. Bioinformatic analysis of the genomic sequences flanking the pre-miR-367-pre-miR-302a–d cluster shows that the pre-miRNAs are antisense to an intron of a protein-coding gene (ENSEMBL identifiers *Homo sapiens*: ENSG00000174720, *Mus musculus*: ENSMUSG00000027968, *Gallus gallus*: ENSGALG00000012048, *Xenopus*: ENSXETG00000024097). While reciprocal BLAST analysis predicts fish homologs for this gene (*Danio rerio*: ENSDARG00000017315, *Tetraodon fugu*: NEWSINFRUG00000127190) systematic folding of the corresponding fish genomic sequences does not reveal any potential stem-loop precursors. Thus, it is likely that the pre-miR-367-pre-miR-302a–d cluster does not exist in fishes and performs a function unique to tetrapod vertebrates.

The high conservation of pre-miR-367-pre-miR-302a–d allows the easy detection of the hairpins by dotplot analysis (Sonnhammer and Durbin, 1995). A dotplot is a graphical representation of the homology between two sequences. Each axis of the plot represents positions within the sequences to be compared. Dots are placed at coordinates where the local homology, defined as per cent identity in the forward or reverse orientation between two short (10–20 bp)

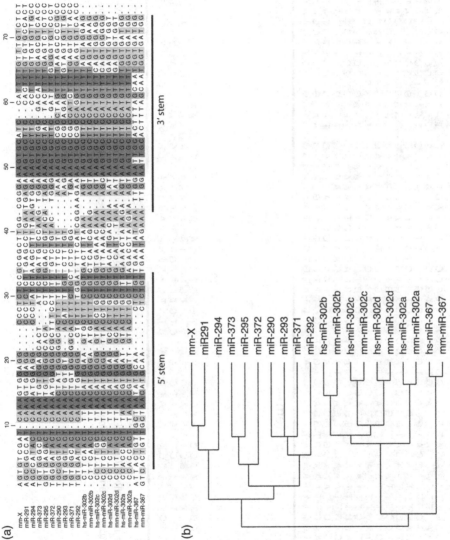

Figure 33.4. Conservation of pre-miR-290–295/pre-miR-371–373 and pre-miR-302a–d. (a) Multiple sequence alignment of the human and mouse pre-miR-367-pre-miR-302a–d hairpins, the human pre-miR-371–373 hairpins and the mouse pre-miR-290–295 hairpins. mm-X is the predicted seventh pre-miRNA within the pre-miR-290–295 cluster. The approximate positions of the two strands of the hairpin stems are underlined. (b) Phylogenetic tree corresponding to the alignment shown in (a).

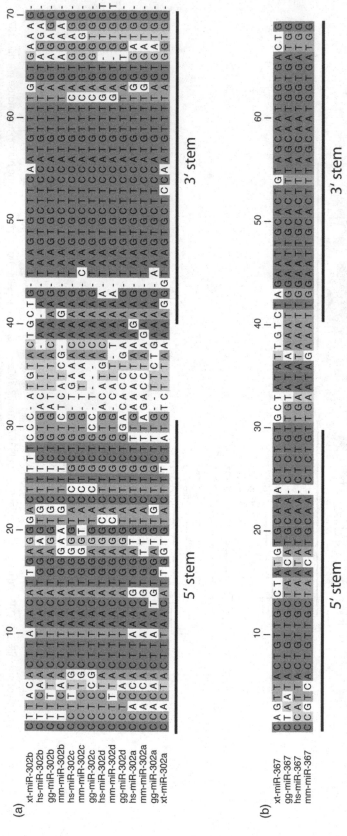

Figure 33.5. Conservation of the pre-miR-367-pre-miR-302a–d cluster. Multiple sequence alignments of the pre-miR-302a–d hairpins from frog (*Xenopus tropicalis*, xt), chick (*Gallus gallus*, gg), mouse (*Mus musculus*, mm) and human (*Homo sapiens*, hs) are shown in (a). The multiple sequence alignment of the corresponding pre-367 hairpins is shown in (b).

sequence intervals centered at each coordinate, exceeds a given threshold. Lines parallel to the main diagonal, thus, correspond to regions that are homologous in the sense orientation whereas lines perpendicular to the main diagonal represent regions that are homologous in the antisense orientation. Because the two strands of a pre-miRNA hairpin stem are partially complementary to each other, homologous pre-miRNAs appear as X-shaped features in dotplots. Dotplot comparison of the mouse and human loci corresponding to the pre-miR-367-pre-miR-302a–d cluster reveals lines parallel to the main diagonal showing that the two loci are well conserved beyond the miRNA cluster (Figure 33.6a). The 4 × 4 array of X-shaped features corresponds to the four homologous pre-miR-302a–d hairpins. The additional X at the main diagonal corresponds to the pre-miR-367 hairpin. Dotplot comparison of the chick and mouse pre-miR-367-pre-miR-302a–d loci shows less overall conservation but the pre-miRNA positions are still clearly visible (Figure 33.6b).

The pre-miR-290–295/pre-miR-371–373 cluster is much less conserved than the pre-miR-368-pre-miR-302a–d cluster. Dotplot comparison of the mouse pre-miR-290–295 and human pre-miR-371–373 loci reveals no obvious global homologies and very weak X-shaped features corresponding to the pre-miRNAs (Figure 33.6c). This high sequence variability of the pre-miR-290–295/pre-miR-371–373 hairpins presents special problems in identifying homologous clusters by genome searches. BLAST searches of the human genome with the sequences of the mature miR-290–295 miRNAs found only one hairpin within the human cluster (pre-miR-371) (Houbaviy *et al.*, 2003). The remaining two hairpins were predicted by systematic folding of the genomic sequence flanking pre-miR-371. Because the pre-miRNA sequences contain more information than the mature miRNAs, they give better results in BLAST searches. If the pre-miR-290–295 hairpin sequences are used for BLAST searches instead of the sequences of the mature miRNAs, two out of the three pre-miR-371–373 hairpins are detected in the human genome. This method identifies at least one homologous hairpin within the cow and dog genomes and additional homologous pre-miRNA hairpins can be identified by systematic folding of the flanking genomic sequences (Houbaviy *et al.*, 2005). In addition, systematic folding detected a seventh pre-miRNA hairpin within the mouse pre-miR-290–295 (mm-X) that was missed in the original analysis of the cluster. No miRNAs from this hairpin were cloned from mouse suggesting that it may not be processed or that the corresponding mature miRNAs are unstable. The genomic organization of the mouse pre-miR-290–295 cluster, the human pre-miR-371–373 cluster and their predicted homologs from cow and dog are shown in Figure 33.7a. The multiple sequence alignment of all predicted pre-miRNA hairpins within the four clusters is given in Figure 33.7b. No hairpins that conform to the above alignment can be identified by BLAST searches in non-mammalian genomes and in the genome of the marsupial *Monodelphis domestica*. Thus, either the pre-miR-290–295/pre-miR-371–373 cluster exists only in eutherian mammals or it is so divergent in marsupials and non-mammalian vertebrates that BLAST searches are not sensitive enough to detect it. An argument that suggests the former possibility as more likely is made below.

(a)

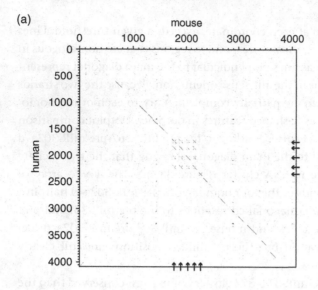

(b)

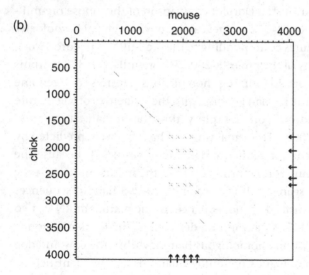

(c)

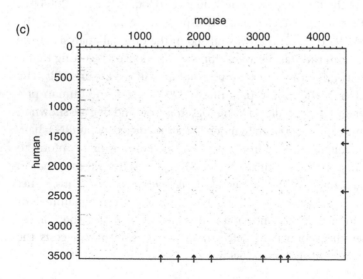

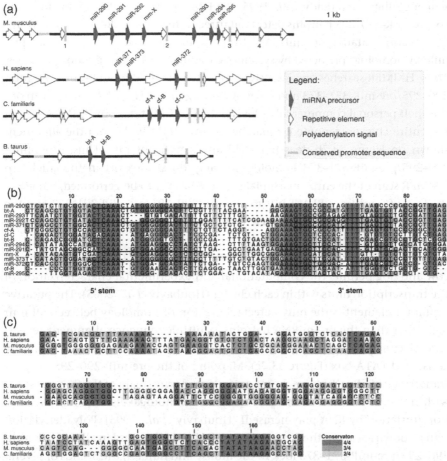

Figure 33.7. Conservation of the pre-miR-290–295/pre-miR-371–373 cluster. (a) Genomic organization of the pre-miR-290–295/pre-miR-371–373 homologs in mouse (*Mus musculus*, mm), human (*Homo sapiens*, hs), dog (*Canis familiaris*, cf) and cow (*Bos taurus*, bt). Sequence features are according to the legend. Alignments of the pre-miRNA hairpins and putative promoter regions are shown in (b) and (c) respectively. The positions of cloned mature miRNAs as well as the approximate locations of the hairpin stems are underlined in (b).

The most sensitive method for the detection of pre-miRNA homologs relies on the genome-wide identification of potential pre-miRNA hairpins via computational folding (Bentwich *et al.*, 2005). Once all putative pre-miRNAs are identified, a search can be made for those that fit the alignment in Figure 33.7b. The above, however, is a relatively computationally intensive procedure due to the time it takes to fold sequences. A faster method relies on the program HMMER which uses hidden Markov models (HMMs) to find sequences that fit a given multiple

Figure 33.6. (cont.)
Dotplots of the pre-miR-290–295/pre-miR-371–373 and pre-miR-367-pre-miR-302a–d loci. Dotplot comparisons of the mouse and human pre-miR-367-pre-miR-302a–d loci (a), the mouse and chick pre-miR-367-pre-miR-302a–d loci (b) and the mouse pre-miR-290–295 and human pre-miR-371–373 loci (c) are shown. Arrows indicate the positions of the individual hairpins on each axis.

sequence alignment (Eddy *et al.*, 1995). An HMM built from the alignment of the six pre-miR-290–295 hairpins detects all three human pre-miR-371–373 hairpins, the seventh putative pre-miRNA within the mouse cluster, as well as all pre-miRNA homologs predicted by systematic folding in the cow and dog genomes. Thus, HMMER searches are as sensitive as systematic folding in finding pre-miR-290–295/pre-miR-371–373 homologs. A whole genome HMMER search on an off-the-shelf personal computer takes several days of CPU time. An HMMER scan of the entire chicken genome for matches to the HMM built from the alignment shown on Figure 33.7b does not yield any significant hits. Thus, a pre-miR-290–295/pre-miR-371–373 homolog probably does not exist in the chick. An HMMER scan of the entire marsupial genome has yet to be performed, but given the lack of significant BLAST hits to pre-miR-290–295/pre-miR-371–373 and the result of the chick genome search it is likely that pre-miR-290–295/pre-miR-371–373 does not exist in marsupials.

The identification of pre-miR-290–295/pre-miR-371–373 homologs in the cow and dog permits the computational and experimental characterization of the transcription units within each cluster (Houbaviy *et al.*, 2005). The putative promoter element is the only detectable region of homology between all four loci other than the pre-miRNA hairpins. Multiple sequence alignment reveals several conserved sequence motifs within this element including a highly conserved TATA-box (Figure 33.7c). Mapping of the pre-miR-290–295 primary transcript by 5'-RACE places the start site 35 nt downstream of the first T-residue within the conserved TATA motif, which is the canonical distance for transcription initiation by RNA polymerase II (Houbaviy *et al.*, 2005). Polyadenylation signals downstream of the 3'-most pre-miRNAs can be found in all four pre-miR-290–295/pre-miR-371–373 loci. In the mouse and human, the 3' ends of several ESTs align immediately adjacent to the putative polyadenylation signals providing experimental evidence for transcription termination at these sites. A primary transcript consistent with initiation from the identified promoter and termination at the polyadenylation signal in mouse ES cells can be detected by Northern analysis and has been mapped by hybridization with oligonucleotide probes and RNAse protection assays (Houbaviy *et al.*, 2005).

Potential early embryo specific function of pre-miR-290–295/ pre-miR-371–373

The failure to identify pre-miR-290–295/pre-miR-371–373 homologs in birds and marsupials suggests that this miRNA cluster plays a role in developmental events that are unique to eutherian mammals such as the differentiation of the blastocystic lineages and the development of the placenta (Rossant and Cross, 2001). In contrast, the pre-miR-367-pre-miR-302a–d cluster probably controls events that are common to all tetrapods. Failure to clone miRNAs belonging to either cluster from adult somatic tissue implies early embryonic expression of both pre-miR-290–295/ pre-miR-371–373 and pre-miR-367-pre-miR-302a–d. The potential function of pre-miR-290–295/pre-miR-371–373 specifically in the peri-implantation embryo and in ES and TS cells is supported by additional evidence. Whereas pre-miR-367-pre-miR-302a–d is expressed in ES as well as embryonal carcinoma (EC) cells, pre-miR-290–295/pre-miR-371–373 is expressed only in the former (Suh *et al.*, 2004). Both ES and EC cells have considerable developmental potential, but only ES cells can contribute to all embryonic lineages (Andrews *et al.*, 2001; Smith, 2001). Thus, the more restricted expression of pre-miR-290–295/pre-miR-371–373 implies that it may have a more ES cell specific function.

Analysis of the ESTs derived from the primary transcripts spanning the pre-miR-290–295 and pre-miR-371–373 clusters provides further evidence for their early embryo specific expression. All ESTs that map within the mouse cluster are derived from ES cells, EG cells and pre-implantation embryos. ESTs in the human cluster are exclusively from testicular tumors and human ES cells. Significantly, testicular tumors contain cells resembling primordial germ cells, which in turn are related to ES cells (Andrews *et al.*, 2001). There are no ESTs from pre-implantation embryos within the human cluster, however, this is probably due to limited EST data.

Consistent with a role in the development of the trophoblastic lineage miR-290–295 have been detected in trophoblastic stem cells (Houbaviy *et al.*, 2005). Presently, it is not known if miR-367-miR-302a–d is expressed in TS cells.

While it is tempting to speculate that pre-miR-290–295/pre-miR-371–373 is involved in the maintenance of the stem cell state, the phenotype of Dicer null mutants argues against this hypothesis. Two specialized Dicer paralogs exist in flies (Lee *et al.*, 2004). Dicer-1 is responsible for the production of miRNAs, whereas Dicer-2 produces siRNAs. Loss of Dicer-1, but not Dicer-2, function in the *Drosophila* germ stem cells results in cell cycle delay (Hatfield *et al.*, 2005). However, the Dicer-1 null germ line stem cells remain in the stem cell compartment of the ovary and retain their stem cell identity. Consistent with the *Drosophila* data, Dicer null mouse ES cells are viable, have normal ES cell morphology and express the appropriate ES cell markers (Kanellopoulou *et al.*, 2005; Murchison *et al.*, 2005). However, their cell cycle is slowed down and the cells fail to differentiate. Growth defects due to Dicer loss of function are not unique to ES cells. Conditional depletion of Dicer in the mouse limb results in severe developmental delays that are consistent with growth defects but does not affect the basic patterning of the limb (Harfe *et al.*, 2005). Thus, if Dicer dependent cell cycle defects in ES cells are due to the loss of miRNA, as opposed to siRNA function, it is likely that they involve miRNAs the expression of which is not limited to ES cells and early embryos. Given the Dicer knock out phenotypes, the pre-miR-290–295/pre-miR-371–373 cluster likely regulates important events in early mammalian development rather than stem cell maintenance. Interestingly, no ES cells could be obtained *de novo* from Dicer null blastocysts, suggesting that miRNAs may be required for ES cell derivation (Bernstein *et al.*, 2003).

Concluding remarks

Cloning of short RNAs from ES cells failed to identify potential siRNAs but revealed several miRNAs with apparent ES cell specific expression. Among these, two miRNA clusters, pre-miR-367-pre-miR-302a–d, which seems to exist only in tetrapods and pre-miR-290–295/pre-miR-371–373, which can be found only in eutherian mammals, appear especially interesting. Both of these clusters are likely to play important roles in early development which makes them attractive candidates for knockout studies in the mouse. Their identification in ES cells implies that these clusters are expressed in the blastocyst ICM. The expression of the pre-miR-290–295/pre-miR-371–373 cluster in TS cells strongly suggests that it is expressed in the trophoblast. However, the expression pattern of these clusters in the early mouse embryo at single cell resolution is unknown. Recently, the technology to do miRNA *in situ* hybridization studies has become available and can be applied to whole mount pre-implantation embryos (i.e. blastocysts) and to sections of early post-implantation embryos (i.e. egg-cylinder stage) (Wienholds *et al.*, 2005). Of particular interest would be the determination of differentiation events that correlate with repression of these miRNA clusters. *In situ* hybridization should also show if there is a gradient of expression of the pre-miR-290–295/pre-miR-371–373 and the pre-miR-367-pre-miR-302a–d clusters in pre-implantation embryos and whether the corresponding miRNAs are present in the primitive endoderm of the blastocyst.

One key question that remains unanswered concerns the identities of the targets repressed by the pre-miR-367-pre-miR-302a–d and pre-miR-290–295/pre-miR-371–373 clusters. As discussed above, the predominant seed of the mature miRNAs processed from the 3′ pre-miRNA stems is the same for both clusters. It is shared with miR-93, the expression of which is not restricted to ES and TS cells or early embryos (Lewis *et al.*, 2003; Lewis *et al.*, 2005). Pre-miR-93 is unrelated to pre-miR-367-pre-miR-302a–d and pre-miR-290–295/pre-miR-371–373 as it encodes the mature miR-93 within the 5′ strand of the hairpin stem. Thus pre-miR-367-pre-miR-302a–d and pre-miR-290–295/pre-miR-371–373 have evolved independently from pre-miR-93 and likely perform specific functions. Efficient translational repression by miRNAs requires multiple sites complementary to the seed within the mRNA target (Doench *et al.*, 2003; Doench and Sharp, 2004). Thus, it is possible that miRNAs processed from the 3′ strands of the pre-miR-302a–d stems act co-operatively with miR-367 and with miRNAs processed from the 5′ strands of the pre-miR-302a–d stems on targets that miR-93 alone cannot repress efficiently. A similar argument can be made for pre-miR-290–295/pre-miR-371–373. This cluster produces mature miRNAs with unique seeds not only from the 5′ strands of the hairpin stems but also from the 3′ strands due to the variability of the 5′ end of these miRNAs. Remarkably, the alternative seeds of two miRNAs processed from the 3′ strands of the hairpins, miR-293 and miR-371, are conserved between mouse and human suggesting functional significance. No ES cell relevant pattern can be discerned among the predicted targets of miR-93 and no targets have been predicted for the alternative seeds discussed above. The computational and experimental identification of the targets of the two ES cell specific miRNA clusters would complement studies of the knockout phenotypes in elucidating their function.

REFERENCES

Ambros, V., Bartel, B., Bartel, D. P. *et al.* (2003). A uniform system for microRNA annotation. *RNA*, **9**, 277–279.

Andrews, P. W., Przyborski, S. A. and Thomson, J. A. (2001). Embryonal carcinoma cells as embryonic stem cells. In *Stem Cell Biology*, D. R. Marshak, R. L. Gardner and D. Gottlieb (eds.). Cold Spring Harbor, New York: Cold Spring Harbor Laboratory Press, pp. 231–265.

Aravin, A. A., Naumova, N. M., Tulin, A. V. *et al.* (2001). Double-stranded RNA-mediated silencing of genomic tandem repeats and transposable elements in the *D. melanogaster* germline. *Current Biology*, **11**, 1017–1027.

Aravin, A. A., Klenov, M. S., Vagin, V. V. *et al.* (2004). Dissection of a natural RNA silencing process in the *Drosophila melanogaster* germ line. *Molecular Cell Biology*, **24**, 6742–6750.

Bartel, D. P. (2004). MicroRNAs: genomics, biogenesis, mechanism, and function. *Cell*, **116**, 281–297.

Baulcombe, D. (2004). RNA silencing in plants. *Nature*, **431**, 356–363.

Bentwich, I., Avniel, A., Karov, Y. *et al.* (2005). Identification of hundreds of conserved and nonconserved human microRNAs. *Nature Genetics*, **37**, 766–770.

Bernstein, A., MacCormick, R. and Martin, G. S. (1976). Transformation-defective mutants of avian sarcoma viruses: the genetic relationship between conditional and nonconditional mutants. *Virology*, **70**, 206–209.

Bernstein, E., Kim, S. Y., Carmell, M. A. *et al.* (2003). Dicer is essential for mouse development. *Nature Genetics*, **35**, 215–217.

Cam, H. P., Sugiyama, T., Chen, E. S. *et al.* (2005). Comprehensive analysis of heterochromatin- and RNAi-mediated epigenetic control of the fission yeast genome. *Nature Genetics*, **37**, 809–819.

Chambers, I., Colby, D., Robertson, M. *et al.* (2003). Functional expression cloning of Nanog, a pluripotency sustaining factor in embryonic stem cells. *Cell*, **113**, 643–655.

Cherry, S. R., Biniszkiewicz, D., van Parijs, L., Baltimore, D. and Jaenisch, R. (2000). Retroviral expression in embryonic stem cells and hematopoietic stem cells. *Molecular Cell Biology*, **20**, 7419–7426.

Cox, D. N., Chao, A., Baker, J. *et al.* (1998). A novel class of evolutionarily conserved genes defined by piwi are essential for stem cell self-renewal. *Genes & Development*, **12**, 3715–3727.

Cox, D. N., Chao, A. and Lin, H. (2000). piwi encodes a nucleoplasmic factor whose activity modulates the number and division rate of germline stem cells. *Development*, **127**, 503–514.

Deng, W. and Lin, H. (2002). miwi, a murine homolog of piwi, encodes a cytoplasmic protein essential for spermatogenesis. *Developmental Cell*, **2**, 819–830.

Doench, J. G. and Sharp, P. A. (2004). Specificity of microRNA target selection in translational repression. *Genes & Development*, **18**, 504–511.

Doench, J. G., Petersen, C. P. and Sharp, P. A. (2003). siRNAs can function as miRNAs. *Genes & Development*, **17**, 438–442.

Eddy, S. R., Mitchison, G. and Durbin, R. (1995). Maximum discrimination hidden Markov models of sequence consensus. *Journal of Computational Biology*, **2**, 9–23.

Elbashir, S. M., Lendeckel, W. and Tuschl, T. (2001a). RNA interference is mediated by 21- and 22-nucleotide RNAs. *Genes & Development*, **15**, 188–200.

Elbashir, S. M., Martinez, J., Patkaniowska, A., Lendeckel, W. and Tuschl, T. (2001b). Functional anatomy of siRNAs for mediating efficient RNAi in *Drosophila melanogaster* embryo lysate. *European Molecular Biology Organization Journal*, **20**, 6877–6888.

Evans, M. J. and Kaufman, M. H. (1981). Establishment in culture of pluripotential cells from mouse embryos. *Nature*, **292**, 154–156.

Hall, I. M., Shankaranarayana, G. D., Noma, K. *et al.* (2002). Establishment and maintenance of a heterochromatin domain. *Science*, **297**, 2232–2237.

Harfe, B. D., McManus, M. T., Mansfield, J. H., Hornstein, E. and Tabin, C. J. (2005). The RNaseIII enzyme Dicer is required for morphogenesis but not patterning of the vertebrate limb. *Proceedings of the National Academy of Sciences USA*, **102**, 10898–10903.

Hatfield, S. D., Shcherbata, H. R., Fischer, K. A. *et al.* (2005). Stem cell division is regulated by the microRNA pathway. *Nature*, **435**, 974–978.

Houbaviy, H. B., Murray, M. F. and Sharp, P. A. (2003). Embryonic stem cell-specific microRNAs. *Developmental Cell*, **5**, 351–358.

Houbaviy, H. B., Dennis, L., Jaenisch, R. and Sharp, P. A. (2005). Characterization of a highly variable eutherian microRNA gene. *RNA*, **11**, 1245–1257.

Hutvagner, G., McLachlan, J., Pasquinelli, A. E. *et al.* (2001). A cellular function for the RNA-interference enzyme Dicer in the maturation of the let-7 small temporal RNA. *Science*, **293**, 834–838.

Kanellopoulou, C., Muljo, S. A., Kung, A. L. *et al.* (2005). Dicer-deficient mouse embryonic stem cells are defective in differentiation and centromeric silencing. *Genes & Development*, **19**, 489–501.

Kaufman, M. H. and Bard, J. B. L. (1999). *The Anatomical Basis of Mouse Development*. San Diego: Academic Press.

Kawasaki, H. and Taira, K. (2004). Induction of DNA methylation and gene silencing by short interfering RNAs in human cells. *Nature*, **431**, 211–217.

Kehler, J., Tolkunova, E., Koschorz, B. *et al.* (2004). Oct4 is required for primordial germ cell survival. *European Molecular Biology Organization Reports*, **5**, 1078–1083.

Ketting, R. F., Haverkamp, T. H., van Luenen, H. G. and Plasterk, R. H. (1999). Mut-7 of *C. elegans*, required for transposon silencing and RNA interference, is a homolog of Werner syndrome helicase and RNaseD. *Cell*, **99**, 133–141.

Kim, J., Krichevsky, A., Grad, Y. *et al.* (2004). Identification of many microRNAs that copurify with polyribosomes in mammalian neurons. *Proceedings of the National Academy of Sciences USA*, **101**, 360–365.

Kuramochi-Miyagawa, S., Kimura, T., Ijiri, T. W. *et al.* (2004). Mili, a mammalian member of piwi family gene, is essential for spermatogenesis. *Development*, **131**, 839–849.

Labosky, P. A., Barlow, D. P. and Hogan, B. L. (1994). Mouse embryonic germ (EG) cell lines: transmission through the germline and differences in the methylation imprint of insulin-like growth factor 2 receptor (Igf2r) gene compared with embryonic stem (ES) cell lines. *Development*, **120**, 3197–3204.

Lagos-Quintana, M., Rauhut, R., Lendeckel, W. and Tuschl, T. (2001). Identification of novel genes coding for small expressed RNAs. *Science*, **294**, 853–858.

Lagos-Quintana, M., Rauhut, R., Yalcin, A. *et al.* (2002). Identification of tissue-specific microRNAs from mouse. *Current Biology*, **12**, 735–739.

Lau, N. C., Lim, L. P., Weinstein, E. G. and Bartel, D. P. (2001). An abundant class of tiny RNAs with probable regulatory roles in *Caenorhabditis elegans*. *Science*, **294**, 858–862.

Lee, J. T., Strauss, W. M., Dausman, J. A. and Jaenisch, R. (1996). A 450 kb transgene displays properties of the mammalian X-inactivation center. *Cell*, **86**, 83–94.

Lee, J. T., Davidow, L. S. and Warshawsky, D. (1999). Tsix, a gene antisense to Xist at the X-inactivation centre. *Nature Genetics*, **21**, 400–404.

Lee, R. C. and Ambros, V. (2001). An extensive class of small RNAs in *Caenorhabditis elegans*. *Science*, **294**, 862–864.

Lee, R. C., Feinbaum, R. L. and Ambros, V. (1993). The *C. elegans* heterochronic gene lin-4 encodes small RNAs with antisense complementarity to lin-14. *Cell*, **75**, 843–854.

Lee, Y. S., Nakahara, K., Pham, J. W. *et al.* (2004). Distinct roles for *Drosophila* Dicer-1 and Dicer-2 in the siRNA/miRNA silencing pathways. *Cell*, **117**, 69–81.

Lewis, B. P., Shih, I. H., Jones-Rhoades, M. W., Bartel, D. P. and Burge, C. B. (2003). Prediction of mammalian microRNA targets. *Cell*, **115**, 787–798.

Lewis, B. P., Burge, C. B. and Bartel, D. P. (2005). Conserved seed pairing, often flanked by adenosines, indicates that thousands of human genes are microRNA targets. *Cell*, **120**, 15–20.

Martin, G. R. (1981). Isolation of a pluripotent cell line from early mouse embryos cultured in medium conditioned by teratocarcinoma stem cells. *Proceedings of the National Academy of Sciences USA*, **78**, 7634–7638.

Mitsui, K., Tokuzawa, Y., Itoh, H. *et al.* (2003). The homeoprotein Nanog is required for maintenance of pluripotency in mouse epiblast and ES cells. *Cell*, **113**, 631–642.

Mochizuki, K., Fine, N. A., Fujisawa, T. and Gorovsky, M. A. (2002). Analysis of a piwi-related gene implicates small RNAs in genome rearrangement in tetrahymena. *Cell*, **110**, 689–699.

Mochizuki, K. and Gorovsky, M. A. (2004). Small RNAs in genome rearrangement in Tetrahymena. *Current Opinions in Genetics and Development*, **14**, 181–187.

Morris, K. V., Chan, S. W., Jacobsen, S. E. and Looney, D. J. (2004). Small interfering RNA-induced transcriptional gene silencing in human cells. *Science*, **305**, 1289–1292.

Murchison, E. P., Partridge, J. F., Tam, O. H., Cheloufi, S. and Hannon, G. J. (2005). Characterization of Dicer-deficient murine embryonic stem cells. *Proceedings of the National Academy of Sciences USA*, **102**, 12135–12140.

Niwa, H., Miyazaki, J. and Smith, A. G. (2000). Quantitative expression of 10/-3/4 defines differentiation, dedifferentiation or self-renewal of ES cells. *Nature Genetics*, **24**, 372–376.

Pal-Bhadra, M., Bhadra, U. and Birchler, J. A. (2002). RNAi related mechanisms affect both transcriptional and posttranscriptional transgene silencing in *Drosophila*. *Molecular Cell*, **9**, 315–327.

Pal-Bhadra, M., Leibovitch, B. A., Gandhi, S. G. *et al.* (2004). Heterochromatic silencing and HP1 localization in *Drosophila* are dependent on the RNAi machinery. *Science*, **303**, 669–672.

Panning, B., Dausman, J. and Jaenisch, R. (1997). X chromosome inactivation is mediated by Xist RNA stabilization. *Cell*, **90**, 907–916.

Pesce, M. and Scholer, H. R. (2001). Oct-4: gatekeeper in the beginnings of mammalian development. *Stem Cells*, **19**, 271–278.

Reinhart, B. J., Slack, F. J., Basson, M. *et al.* (2000). The 21-nucleotide let-7 RNA regulates developmental timing in *Caenorhabditis elegans*. *Nature*, **403**, 901–906.

Rossant, J. and Cross, J. C. (2001). Placental development: lessons from mouse mutants. *Nature Reviews Genetics*, **2**, 538–548.

Seitz, H., Youngson, N., Lin, S. P. *et al.* (2003). Imprinted microRNA genes transcribed antisense to a reciprocally imprinted retrotransposon-like gene. *Nature Genetics*, **34**, 261–262.

Seitz, H., Royo, H., Bortolin, M. L. *et al.* (2004). A large imprinted microRNA gene cluster at the mouse Dlk1-Gtl2 domain. *Genome Research*, **14**, 1741–1748.

Smith, A. (2001). Embryonic stem cells. In *Stem Cell Biology*, D. R. Marshak, R. L. Gardner and D. Gottlieb (eds.). Cold Spring Harbor, New York: Cold Spring Harbor Laboratory Press, pp. 205–230.

Sonnhammer, E. L. and Durbin, R. (1995). A dot-matrix program with dynamic threshold control suited for genomic DNA and protein sequence analysis. *Gene*, **167**, GC1–10.

Stewart, C. L., Gadi, I. and Bhatt, H. (1994). Stem cells from primordial germ cells can reenter the germ line. *Developmental Biology*, **161**, 626–628.

Suh, M. R., Lee, Y., Kim, J. Y. *et al.* (2004). Human embryonic stem cells express a unique set of microRNAs. *Developmental Biology*, **270**, 488–498.

Tanaka, S., Kunath, T., Hadjantonakis, A. K., Nagy, A. and Rossant, J. (1998). Promotion of trophoblast stem cell proliferation by FGF4. *Science*, **282**, 2072–2075.

Theiler, K. (1989). *The House Mouse: Atlas of Embryonic Development*. New York: Springer-Verlag.

Verdel, A., Jia, S., Gerber, S. *et al.* (2004). RNAi-mediated targeting of heterochromatin by the RITS complex. *Science*, **303**, 672–676.

Volpe, T. A., Kidner, C., Hall, I. M. *et al.* (2002). Regulation of heterochromatic silencing and histone H3 lysine-9 methylation by RNAi. *Science*, **297**, 1833–1837.

Wienholds, E., Kloosterman, W. P., Miska, E. *et al.* (2005). MicroRNA expression in zebrafish embryonic development. *Science*, **309**, 310–311.

Wutz, A. and Jaenisch, R. (2000). A shift from reversible to irreversible X inactivation is triggered during ES cell differentiation. *Molecular Cell*, **5**, 695–705.

Xie, Z., Johansen, L. K., Gustafson, A. M. *et al.* (2004). Genetic and functional diversification of small RNA pathways in plants. *Public Library of Science Biology*, **2**, E104.

Zamore, P. D., Tuschl, T., Sharp, P. A. and Bartel, D. P. (2000). RNAi: double-stranded RNA directs the ATP-dependent cleavage of mRNA at 21 to 23 nucleotide intervals. *Cell*, **101**, 25–33.

Zhang, H., Kolb, F. A., Brondani, V., Billy, E. and Filipowicz, W. (2002). Human Dicer preferentially cleaves dsRNAs at their termini without a requirement for ATP. *European Molecular Biology Organization Journal*, **21**, 5875–5885.

Zhang, H., Kolb, F. A., Jaskiewicz, L., Westhof, E. and Filipowicz, W. (2004). Single processing center models for human Dicer and bacterial RNase III. *Cell*, **118**, 57–68.

Zuker, M., Mathews, D. H. and Turner, D. H. (1999). Algorithms and thermodynamics for RNA secondary structure prediction: a practical guide. In *RNA Biochemistry and Biotechnology*, J. Barciszewski and B. F. C. Clark (eds.). Kluwer Academic Publishers.

34 The role of miRNA in hematopoiesis

Michaela Scherr and Matthias Eder

Introduction

RNA interference (RNAi) represents a highly conserved cellular mechanism to specifically regulate eukaryotic gene expression either by inducing sequence-specific degradation of complementary mRNA or by inhibiting its translation (reviewed in Hannon, 2002; Hutvagner & Zamore, 2002). RNAi is triggered by two classes of small RNAs: one class, called siRNA (small interfering RNA), can be derived from longer double-stranded RNAs that are transcribed from different kinds of vector or introduced directly into cells by transfection, whereas the second class, miRNA (microRNA), is processed from stem-loop precursors that are encoded within the host genome (Elbashir et al., 2001; Ambros et al., 2003).

Non-coding miRNAs negatively regulate the expression of genes at the post-transcriptional level through the RNAi pathway (Bartel, 2004). The first miRNA, lin-4, was discovered in 1993 by Ambros and colleagues (Lee et al., 1993) in a study of developmental timing in the nematode worm C. elegans and soon led to the identification of the first miRNA target lin-14 (Wightman et al., 1993). The second miRNA discovered, let-7, is involved in regulation of intracellular signal transduction and has recently been shown to inhibit expression of let-60, the nematode RAS homolog (Johnson et al., 2005). Hundreds of miRNAs have been identified in flies, worms, plants, fish, and mammals by cloning of size-fractionated RNAs or bioinformatic prediction strategies (Lagos-Quintana et al., 2001; Llave et al., 2002; Lim et al., 2003a, 2003b; Watanabe et al., 2005). More recently, miRNAs have also been identified in the Epstein Barr virus and are likely to be found in other viruses as well (Pfeffer et al., 2004; Cai et al., 2005). It is becoming more and more evident that miRNAs play an important role in a wide range of biological processes regulating cell cycle progression and proliferation, differentiation, cell survival, and development, among others (Brennecke et al., 2003; Xu et al., 2003; Chen et al., 2004; Yekta et al., 2004).

MiRNA biogenesis

MiRNA genes are usually transcribed by RNA polymerase II. As shown in Figure 34.1, the initial transcript of a miRNA gene, the precursor miRNA (pri-miRNA), passes

MicroRNAs: From Basic Science to Disease Biology, ed. Krishnarao Appasani. Published by Cambridge University Press. © Cambridge University Press 2008.

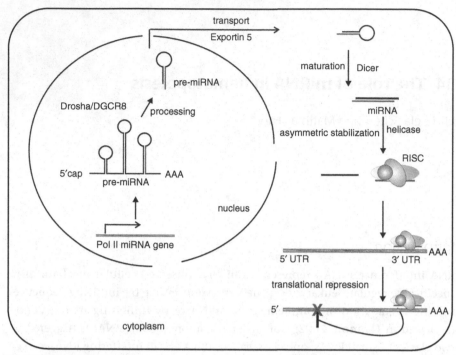

Figure 34.1. The miRNA biogenesis pathway. Pri-miRNAs are transcribed by Pol II and processed in the nucleus into stem-loop precursors (pre-miRNA) by Drosha (ribonuclease III) and DGCR8 (double-stranded RNA binding protein). The pre-miRNA is transported into the cytoplasm by Exportin 5 and further processed by a second RNAse III called Dicer into mature miRNA. One strand is loaded into a multiprotein complex called RISC (RNA induced silencing complex) and finally guides RISC to target mRNAs resulting in translational attenuation (or mRNA degradation).

through various processing steps before it is exported to the cytoplasm (Lee *et al.*, 2004; Cai *et al.*, 2004). The pri-miRNA transcripts contain a 5′-7methyl guanosine cap and a 3′-poly A tail, range from hundreds to thousands of nucleotides in length, and consist of monocistronic or polycistronic miRNAs. The majority of miRNAs are transcribed from genomic regions distinct from protein-coding sequences whereas others are located in introns of protein encoding genes (Rodriguez *et al.*, 2004). Pri-miRNAs contain about 70–80 nucleotide stem-loop structures (with the mature miRNA included in their arms), which are recognized by the nuclear RNAse III Drosha for processing to the hairpin precursor miRNA (pre-miRNA) (Lee *et al.*, 2003). Efficient pri-miRNA processing by Drosha requires DCGR8 /Pasha (Denli *et al.*, 2004; Han *et al.*, 2004) and results in pre-miRNAs with 5′-phosphate and a 2-nucleotide 3′-hydroxyl overhang. The pre-miRNAs are then transported from the nucleus into the cytoplasm by Exportin-5, in a Ran-GTP-dependent manner (Yi *et al.*, 2003; Bohnsack *et al.*, 2004). In the cytoplasm, a second RNAse III – Dicer – produces the mature miRNA, an about 21 nucleotide RNA duplex with identical structure as siRNA-duplexes with the exception that the miRNA is usually imperfectly paired. Current models suggest that the PAZ domain of Dicer recognizes the 2-nucleotide 3′-hydroxyl overhang of the double-stranded pre-miRNA (Zhang *et al.*, 2004). A putative helicase activity

Table 34.1. miRNAs expressed in hematopoietic cells

miRNA	Target gene	Function, lineage expression, and/or disease	Reference
miR-181	?	B-cell differentiation	Chen *et al.* (2004)
miR-142	?	myeloid, B-lymphoid cells	Chen *et al.* (2004)
miR-223	NFI-A	granulocytic differentiation	Chen *et al.* (2004), Fazi *et al.* (2005)
miR-155	PU.1 (?)	(B-cell differentiation)	Eis *et al.* (2005), van den Berg *et al.*
	C/EBPβ (?)	B-NHL	(2003), Metzler *et al.* (2004)
miR-15a, miR-16-1	BCL-2	CLL	Calin *et al.* (2002, 2005), Cimmino *et al.* (2005)
miR-17-92	?	B-lymphoid cells, B-NHL	He *et al.* (2005)

NHL: non Hodgkin's lymphoma, CLL: chronic lymphocytic leukemia.

produces the mature single-stranded miRNA, which is loaded in an effector complex termed RNA-induced silencing complex (RISC). The other RNA strand is, in contrast, degraded. RISC is able to regulate the fate of target mRNAs either by translational repression or mRNA destruction. In general, incomplete hybridization due to base pair mismatches between RISC and target mRNA inhibits mRNA translation rather than inducing mRNA destruction. Apparently, Argonaute proteins present in RISC modulate the fate of mRNAs bound to miRNA loaded RISC. For example, miR-196 can cleave HOXB8 target mRNA in murine embryos in the same way as siRNAs (Yekta *et al.*, 2004). The details of miRNA biogenesis have recently been reviewed and are described in other chapters of this book as well (Du and Zamore, 2005).

Currently, there are over 250 confirmed genes that encode miRNAs in the human genome and several others are predicted based on sequence annotation. Using bioinformatic approaches a single miRNA is predicted to regulate on average about 200 target genes which implies that miRNAs modulate the expression of at least 30% of protein-encoding genes (Xie *et al.*, 2005; Lewis *et al.*, 2005, Krek *et al.*, 2005). The absence or presence of specific miRNAs in cells has been shown to be linked with human diseases including cancer and viral infection (McManus, 2003; Lu *et al.*, 2005; He *et al.*, 2005; Jopling *et al.*, 2005). In the following sections we will summarize the current knowledge about expression and function of miRNAs in mammalian hematopoiesis (see also Table 34.1).

miRNAs and differentiation of hematopoietic cells

The role of miRNAs in hematopoiesis was systematically analyzed in a recent study (Chen *et al.*, 2004). To determine miRNA expression in hematopoietic cells the authors first isolated and cloned about 100 miRNAs from murine bone marrow. In the subsequent expression analysis of various organs and tissues three miRNAs were largely restricted to hematopoietic cells, namely miR-181 to the thymus (as well as lung and brain), miR-223 almost exclusively to bone marrow cells, and miR-142 to all hematopoietic tissues analyzed (bone marrow, spleen, fetal liver, thymus). Furthermore, sorting of bone marrow cells based on lineage restricted surface markers revealed lineage-specific miRNA expression: mature miR-181 expression is found in immature lineage-negative (lin-) cells but specifically increases during B-lymphoid differentiation. In contrast, miR-142 is expressed in erythroid and T-lymphoid cells at very low level, whereas high expression is found in the myelo-monocytic and B-lymphoid lineages in accordance with its tissue expression pattern. Finally, miR-223 expression is almost restricted to granulocytic and monocytic cells consistent with the lack of or very low expression in spleen and thymus. Interestingly, the expression of all three miRNAs is low in lin-cells as compared to more mature lineage restricted cells the individual miRNA is expressed in.

To study miRNA function in hematopoietic cells, lineage-restricted miRNAs were overexpressed in lin- bone marrow cells by retroviral gene transfer of an optimized Pol III dependent miRNA expression cassette. Transduced cells were tracked *in vitro* in suspension cultures and/or *in vivo* upon transplantation into lethally irradiated recipients by monitoring retrovirally encoded GFP-fluorescence. Interestingly, ectopic miRNA expression altered lineage differentiation with a doubling of CD19+ B-lymphoid cells and a concomitant reduction of Thy-1.2+ mainly CD8+ T-lymphocytes upon expression of miR-181. Similarly, engraftment of miR-181 expressing lin- bone marrow cells in transplanted recipients resulted in about 80% GFP+/ CD19+ double positive cells as compared to 30% in control animals. Again, Thy-1.2+ cells were dramatically reduced whereas no or only minor effects were observed in the myeloid lineages. This study demonstrates that certain miRNAs are differentially expressed in hematopoietic cells and can modulate lineage differentiation.

A paper published by Fazi *et al.* in 2005 revealed some molecular mechanisms involved in miR-223 mediated granulocytic differentiation (Fazi *et al.*, 2005). The authors analyzed miR-223 expression and function in human cells during all trans retinoic-acid (ATRA)-induced granulocytic differentiation. Specific chromosomal translocations involving the retinoic-acid receptor α (RARα) are the hallmark of acute promyelocytic leukemia (APL). In APL, chimeric fusion proteins, mainly PML-RARα, disrupt the function of PML and inhibit transcription of differentiation associated genes by aberrant recruitment of histone deacytelase- and DNA-methyltransferase- activity. Pharmacological doses of ATRA can release the differentiation block resulting in terminal differentiation of PML/RARα positive APL blasts *in vitro* and *in vivo*. Using primary normal and APL CD34+ cells as well as the human APL cell line NB4 and some NB4-derived ATRA-resistant clones Fazi *et al.* first showed that miR-223 is mainly expressed in CD34- bone marrow cells similar to what had been described in the mouse by Chen *et al.* Next they demonstrated that miR-223 expression is induced by RA in APL cells both *in vitro* and *in vivo*. This close association of miR-223 expression with granulocytic differentiation is specific since (i) no changes in miR-223 levels have been found in RA-resistant cells, (ii) no miR-223 induction has been observed during monocytic differentiation induced by vitamin D3, and (iii) the expression of miR-15 and miR-16a did not change during RA-induced differentiation.

Subsequent promoter analysis revealed two putative C/EBPα binding sites about 700 bp upstream of the miR-223 coding sequence. C/EBPα is a transcription factor involved in regulation of terminal granulocytic differentiation (Tenen, 2003) whose activity is reduced by PML/RARα. Using functional reporter assays and chromatin immunoprecipitation the authors went on to show that these binding sites are necessary and sufficient for C/EBPα binding and that RA treatment induces recruitment of C/EBPα to the miR-223 promoter region thereby activating transcription of the miR-223 gene. Furthermore, the 3' C/EBPα binding site can alternatively be recognized by NFI-A, a transcription factor known to have binding sequences similar to that of C/EBP-family members. Indeed, NFI-A and C/EBPα compete for promoter binding with antagonistic effects on gene transcription, and promoter activation by RA treatment correlates with replacement of NFI-A by C/EBPα. In addition, NFI-A itself was identified as a target for miR-223 based on sequence analysis and reporter assays using expression constructs harboring the NFI-A 3' UTR. Finally, overexpression and sequestration of miR-223 revealed that miR-223 is crucial for controlling granulocytic differentiation of NB4 cells in terms of CD11b and G-CSF receptor expression, as well as morphology. The data suggest a model where NFI-A produces low levels of miR-223, whereas C/EBPα is the differentiation-specific activator leading to enhanced miR-223 expression which inhibits NFI-A expression in an autoregulatory loop. This study adds miR-223 to the well known components regulating the molecular events during myeloid differentiation. Finally, a study by Eis *et al.* (2005) suggests based on sequence comparison that expression of PU.1 and C/EBPβ, two transcription factors involved in regulation of myeloid and/or late B-cell differentiation (Tenen, 2003; Xie *et al.*, 2004) may also be regulated by a miRNA, namely miR155 (see below).

The fundamental work by Chen *et al.* (2004) and Fazi *et al.* (2005) shares some interesting quantitative features. For example, a 3- to 5-fold induction of miR-223 results in an increase of differentiated granulocytic cells very similar to that observed upon miR-181 expression in B-lymphoid cells. Both studies provide some evidence that differentiation of hematopoietic cells requires fine tuning of gene expression more than an on-off switch – at least on the level of progenitor cells. Differentiation specific patterns of miRNA expression have been described in non-hematopoietic cells such as embryonic stem cells (Houbaviy *et al.*, 2003; Suh *et al.*, 2004). With some precaution due to technical reasons, miRNAs may be grouped into three different classes based on their expression profile during ES cell differentiation: miRNAs mainly (i) expressed in undifferentiated ES cells and not found in adult tissues, (ii) expressed in both undifferentiated ES cells as well as in adult tissues, and (iii) those the expression of which increases dramatically during differentiation. In addition, miRNAs are involved in the regulation of germline stem cell division in *Drosophila melanogaster* by reducing expression of Dacapo (Dap) – a homologue of the p21/p27 family of cyclin-dependent kinase inhibitors (Hatfield *et al.*, 2005). It will be interesting to learn to what extent lineage specific and differentiation related subsets of miRNA exist and which role miRNAs may play in the regulation of hematopoietic stem cell maintenance and during commitment and maturation of hematopietic progenitor cells.

MiRNAs in B-cell lymphoma and leukemia

MiRNAs were initially recognized as factors linked to hematologic malignancies by molecular genetic analysis of B-cell chronic lymphocytic leukemia cells (CLL) (Calin *et al.*, 2002). In this and subsequent studies the authors identified genomic deletions at 13q14 encompassing the miR-16-1-miR-15a genes and also a germ-line mutation in the miR-16-1-miR-15a primary precursor causing low level expression both *in vivo* and *in vitro* with some hint to a family history of CLL and breast cancer (Calin *et al.*, 2005). The details of this work have been published (Calin *et al.*, 2004a; Calin et.al., 2005; Cimmino *et al.* 2005) and are described in Chapter 23 of this book by Croce and his colleagues. Conclusions from this work indicate that BCL2 is a target for miR-15 and miR-16 and that expression of thirteen miRNAs, including miR-15a-miR16-1 and miR-155, may define prognostic risk groups in CLL. Taken together with the fact that about 50% of the known miRNA genes are located in cancer associated regions (Calin *et al.*, 2004b) the work on CLL demonstrates that alterations in miRNA sequence and/or expression may play a functional role in transformation of hematopoietic cells.

Beside miR-15a and miR-16-1 altered expression of miR-155 and the miR 17-92 polycistron has been reported in Hodgkin's lymphoma (HD), pediatric Burkitt's lymphoma (BL), diffuse large B-cell lymphoma (DLBCL) as well as other non-Hodgkin's B-cell lymphoma (NHL) (van den Berg *et al.*, 2003; Metzler *et al.*, 2004; Eis *et al.*, 2005; He *et al.*, 2005). MiR-155 is generated from the non-coding polyadenylated and spliced transcript of the *BIC* gene initially identified as a common retroviral insertion site in avian lymphoma and known to cooperate with the c-myc gene in cellular transformation (Tam *et al.*, 2002). BIC transcripts are over-expressed in HD-derived cell lines and, more than 100-fold, in pediatric BL. Along this line, the study by Eis *et al.* demonstrates that both BIC transcripts and miR-155 are over-expressed in DLBCL and the small number of follicular and marginal zone B-cell lymphoma analyzed. In addition, the expression level of both RNAs may discriminate between prognostically different subgroups of DLBCL.

Furthermore, He *et al.* reported the over-expression of a miRNA polycistron (miR-17-92) located at 13q31-32, a region often amplified in human B-cell lymphoma, in 46 NHL including DLBCL and follicular lymphoma (He *et al.*, 2005). The authors further demonstrate functional cooperation between overexpression of the miR-17-92 polycistron and the c-myc gene in a bone marrow transplantation model using transgenic mice expressing c-myc under control of the Ig heavy-chain enhancer (*Eµ-myc*). Whereas transplantation with control *Eµ-myc* bone marrow cells induces B-cell lymphoma with long latency and low penetrance, *Eµ-myc* cells retrovirally transduced to express the miR-17-92 polycistron cause fatal B-cell lymphoma in all mice within 100 days after transplantation. Interestingly, the transcription of the miR-17-92 polycistron is regulated by the c-myc gene product, and miR-17-5p and miR-20a, two members of this cluster, negatively regulate expression of E2F1, an additional transcriptional target of c-myc that promotes cell cycle progression (O'Donnell *et al.*, 2005; see also Chapter 22 in this book by O'Donnell and Mendell). Finally, it should be mentioned that miRNA profiling can also identify subgroups in acute lymphoblastic leukemia (ALL), potentially reflecting different genetic mechanisms of transformation (Lu *et al.*, 2005). In this study miRNA profiling was superior in classifying cancer as compared with genome wide mRNA expression analysis establishing miRNA profiling as a potential novel tool in cancer diagnosis.

In summary, data from genetic and functional analyses indicate that: (i) genetic alterations in miRNAs may include mutations and genomic deletions leading to aberrant miRNA expression, (ii) miRNAs may be causally involved in transformation of B-cells, and (iii) altered miRNA expression may have prognostic significance at least in human B-cell lymphoma and leukemia. Without doubt we will see additional studies on miRNA function and expression in other hematologic malignancies such as acute myeloid leukemia, myeloproliferative disorders, and myelodysplastic syndromes in the near future.

Concluding remarks

With increasing numbers of miRNA targets being identified, the regulation of miRNA expression as well as the signaling circuitry of miRNAs are being clarified, and some functional effects of miRNAs in hematopoiesis different from lineage differentiation are being described. Our understanding of miRNA-regulated processes in normal and malignant hematopoietic cells will markedly increase in the near future and should eventually allow us to modulate miRNA function for molecularly defined therapeutic approaches.

REFERENCES

Ambros, V., Bartel, B., Bartel, D. P. *et al.* (2003). A uniform system for microRNA annotation. *RNA*, **9**, 277–279.

Bartel, D. P. (2004). MicroRNAs: genomics, biogenesis, mechanism, and function. *Cell*, **116**, 281–297.

Bohnsack, M. T., Czaplinski, K. and Gorlich, D. (2004). Exportin 5 is a RanGTP-dependent dsRNA-binding protein that mediates nuclear export of pre-miRNAs. *RNA*, **10**,185–191.

Brennecke, J., Hipfner, D. R., Stark, A., Russell, R. B. and Cohen, S. M. (2003). Bantam encodes a developmentally regulated microRNA that controls cell proliferation and regulates the proapoptotic gene hid in *Drosophila*. *Cell*, **113**, 25–36.

Cai, X., Hagedorn, C. H. and Cullen, B. R. (2004). Human microRNAs are processed from capped, polyadenylated transcripts that can also function as mRNAs. *RNA*, **10**, 1957–1966.

Cai, X., Lu, S., Zhang, Z. *et al.* (2005). Kaposi's sarcoma-associated herpesvirus expresses an array of viral microRNAs in latently infected cells. *Proceedings of the National Academy of Sciences USA*, **102**, 5570–5575.

Calin, G. A., Dumitru, C. D., Shumizu, M. *et al.* (2002). Frequent deletions and downregulation of micro-RNA genes miR15 and miR16 at 13q14 in chronic lymphocytic leukemia. *Proceedings of the National Academy of Sciences USA*, **99**, 15 524–15 529.

Calin, G. A., Liu, C. G., Sevignani, C. *et al.* (2004a). A microRNA profiling reveals distinct signatures in B cell chronic lymphocytic leukemias. *Proceedings of the National Academy of Sciences USA*, **101**, 11 755–11 760.

Calin, G. A., Sevignani, C., Dunitru, C. D. *et al.* (2004b). Human microRNA genes are frequently located at fragile sites and genomic regions involved in cancers. *Proceedings of the National Academy of Sciences USA*, **101**, 2999–3004.

Calin, G. A., Ferracin, M., Cimmino, A. *et al.* (2005). A microRNA signature associated with prognosis and progression in chronic lymphocytic leukemia. *New England Journal of Medicine*, **353**, 1793–1801.

Chen, C. Z., Li, L., Lodish, H. F. and Bartel, D. P. (2004). MicroRNAs modulate hematopoietic lineage differentiation. *Science*, **303**, 83–86.

Cimmino, A., Calin, G. A., Fabbri, M. *et al.* (2005). miR-15 and miR-16 induce apoptosis by targeting BCL2. *Proceedings of the National Academy of Sciences USA*, **102**, 13 944–13 949.

Denli, A. M., Tops, B. B., Plasterk, R. H., Ketting, R. F. and Hannon, G. J. (2004). Processing of primary microRNAs by the Microprocessor complex. *Nature*, **432**, 231–235.

Du, T. and Zamore, P. D. (2005). microPrimer: the biogenesis and function of microRNA. *Development*, **132**, 4645–4652.

Eis, P. S., Tam, W., Sun, L. *et al.* (2005). Accumulation of miR-155 and BIC RNA in human B cell lymphomas. *Proceedings of the National Academy of Sciences USA*, **102**, 3627–3632.

Elbashir, S. M., Lendeckel, W. and Tuschl, T. (2001). RNA interference is mediated by 21- and 22-nucleotide RNAs. *Genes & Development*, **15**, 188–200.

Fazi, F., Rosa, A., Fatica, A. *et al.* (2005). A minicircuitry comprised of microRNA-223 and transcription factors NFI-A and C/EBPalpha regulates human granulopoiesis. *Cell*, **123**, 819–831.

Han, J., Lee, Y., Yeom, K. H. *et al.* (2004). The Drosha-DGCR8 complex in primary microRNA processing. *Genes & Development*, **18**, 3016–3027.

Hannon, G. J. (2002). RNA interference. *Nature*, **418**, 244–251.

Hatfield, S., Shcherbata, H. R., Fischer, K. A. *et al.* (2005). Stem cell division is regulated by the microRNA pathway. *Nature*, **435**, 974–978.

He, L., Thomson, J. M., Hemann, M. T. *et al.* (2005). A microRNA polycistron as a potential human oncogene. *Nature*, **435**, 828–833.

Houbaviy, H. B., Murray, M. F. and Sharp, P. A. (2003). Embryonic stem cell-specific microRNAs. *Developmental Cell*, **5**, 351–358.

Hutvagner, G. and Zamore, P. D. (2002). A microRNA in a multiple-turnover RNAi enzyme complex. *Science*, **297**, 2056–2060.

Johnson, S. M., Grosshans, H., Shingara, J. *et al.* (2005). RAS is regulated by the let-7 microRNA family. *Cell*, **120**, 635–647.

Jopling, C. L., Yi, M., Lancaster, A. M., Lemon, S. M. and Sarnow, P. (2005). Modulation of hepatitis C virus RNA abundance by a liver-specific microRNA. *Science*, **309**, 1577–1581.

Krek, A., Grün, D., Poly, M. N. *et al.* (2005). Combinatorial microRNA target predictions. *Nature Genetics*, **37**, 495–500.

Lagos-Quintana, M., Rauhut, R., Lendeckel, W. and Tuschl, T. (2001). Identification of novel genes coding for small expressed RNAs. *Science*, **294**, 853–858.

Lee, R. C., Feinbaum, R. L. and Ambros, V. (1993). The *C. elegans* heterochronic gene lin-4 encodes small RNAs with antisense complementarity to lin-14. *Cell*, **75**, 843–854.

Lee, Y., Ahn, C., Han, J. *et al.* (2003). The nuclear RNase III Drosha initiates microRNA processing. *Nature*, **425**, 415–419.

Lee, Y., Kim, M., Han, J. *et al.* (2004). MicroRNA genes are transcribed by RNA polymerase II. *European Molecular Biology Organization Journal*, **23**, 4051–4060.

Lewis, B. P., Burge, C. B. and Bartel, D. P. (2005). Conserved seed pairing, often flanked by adenosines, indicates that thousands of human genes are microRNA targets. *Cell*, **120**, 15–20.

Lim, L. P., Lau, N. C., Weinstein, E. G. *et al.* (2003a). The microRNAs of *Caenorhabditis elegans*. *Genes & Development*, **17**, 991–1008.

Lim, L. P., Glasner, M. E., Yekta, S., Burge, C. B. and Bartel, D. P. (2003b). Vertebrate microRNA genes. *Science*, **299**, 1540.

Llave, C., Xie, Z., Kasschau, K. D. and Carrington, J. C. (2002). Cleavage of Scarecrow-like mRNA targets directed by a class of *Arabidopsis* miRNA. *Science*, **297**, 2053–2056.

Lu, J., Getz, G., Miska, E. A. *et al*. (2005). MicroRNA expression profiles classify human cancers. *Nature*, **435**, 834–838.

McManus, M. T. (2003). MicroRNAs and cancer. *Seminars in Cancer Biology*, **13**, 253–258.

Metzler, M., Wilda, M., Busch, K., Viehmann, S. and Borkhardt, A. (2004). High expression of precursor microRNA-155/BIC RNA in children with Burkitt lymphoma. *Genes, Chromosomes & Cancer*, **39**, 167–169.

O'Donnell, K. A., Wentzel, E. A., Zeller, K. I., Dang, C. V. and Mendell, J. T. (2005). c-Myc-regulated microRNAs modulate E2F1 expression. *Nature*, **435**, 839–843.

Pfeffer, S., Zavolan, M., Grasser, F. A. *et al*. (2004). Identification of virus-encoded microRNAs. *Science*, **304**, 734–736.

Rodriguez, A., Griffiths-Jones, S., Ashurst, J. L. and Bradley, A. (2004). Identification of mammalian microRNA host genes and transcription units. *Genome Research*, **14**, 1902–1910.

Suh, M. R., Lee, Y., Kim, J. Y. *et al*. (2004). Human embryonic stem cells express a unique set of microRNAs. *Developmental Biology*, **270**, 488–498.

Tam, W., Hughes, S. H., Hayward, W. S. and Besmer, P. (2002). Avian bic, a gene isolated from a common retroviral site in avian leukosis virus-induced lymphomas that encodes a noncoding RNA, cooperates with c-myc in lymphomagenesis and erythro-leukemogenesis. *Journal of Virology*, **76**, 4275–4286.

Tenen, D. G. (2003). Disruption of differentiation in human cancer: AML shows the way. *Nature Reviews Cancer*, **3**, 89–101.

van den Berg, A., Kroesen, B. J., Kooista, K. *et al*. (2003). High expression of B-cell receptor inducible gene BIC in all subtypes of Hodgkin lymphoma. *Genes, Chromosomes & Cancer*, **37**, 20–28.

Wightman, B., Ha, I. and Ruvkun, G. (1993). Posttranscriptional regulation of the heterochronic gene lin-14 by lin-4 mediates temporal pattern formation in *C. elegans*. *Cell*, **75**, 855–862.

Watanabe, Y., Yachie, N., Numata, K. *et al*. (2005). Computational analysis of microRNA targets in *Caenorhabditis elegans*. *Gene* (Epub ahead of print).

Xie, H., Ye, M., Feng, R. and Graf, T. (2004). Stepwise reprogramming of B cells into macrophages. *Cell*, **117**, 663–676.

Xie, X., Lu, J., Kulbokas, E. J. *et al*. (2005). Systematic discovery of regulatory motifs in human promoters and 3' UTRs by comparison of several mammals. *Nature*, **243**, 338–345.

Xu, P., Vernooy, S. Y., Guo, M. and Hay, B. A. (2003). The *Drosophila* microRNA Mir-14 suppresses cell death and is required for normal fat metabolism. *Current Biology*, **13**, 790–795.

Yekta, S., Shih, I. H. and Bartel, D. P. (2004). MicroRNA-directed cleavage of HOXB8 mRNA. *Science*, **304**, 594–596.

Yi, R., Qin, Y., Macara, I. G. and Cullen, B. R. (2003). Exportin-5 mediates the nuclear export of pre-microRNAs and short hairpin RNAs. *Genes & Development*, **17**, 3011–3016.

Zhang, H., Kolb, F. A., Jaskiewicz, L., Westhof, E. and Filipowicz, W. (2004). Single processing center models for human Dicer and bacterial RNase III. *Cell*, **118**, 57–68.

35 MicroRNAs in embryonic stem cell differentiation and prediction of their targets

Yvonne Tay, Andrew M. Thomson and Bing Lim*

Introduction

Embryonic stem cells (ESCs) exhibit the capacity for unlimited self-renewal and the ability to differentiate into multiple cell lineages, and are termed pluripotent (Smith, 2001; Czyz *et al.*, 2003). The study of mammalian ESCs may facilitate understanding of early developmental events and contribute to the advance of cell-based regenerative medicine (Loebel at al., 2003). ESCs are isolated from the inner cell mass (ICM) of the pre-implantation embryo (Evans and Kaufman, 1981) and share many properties with pluripotent cells from the ICM (Figure 35.1). These include the expression of key regulators of pluripotency and self-renewal, *Oct4*, *Sox2* and *Nanog*, and the capacity to differentiate to all cell types of the embryo, including mesoderm, ectoderm and endoderm (Orkin, 2005). *Oct4* is activated at the four cell stage of mouse pre-implantation development and is essential for the formation of the ICM (Chew *et al.*, 2005). Increased Oct4 levels in mouse ESCs (mESCs) result in different-iation to primitive endoderm and mesoderm, whereas a reduction leads to differentiation to trophectoderm (Niwa *et al.*, 2000). Co-regulation of genes by *Oct4* and *Sox2* appears to be essential in maintaining self-renewal of ESCs (Chew *et al.*, 2005). Nanog acts in concert with Oct4 and Sox2, where Nanog overexpression negates the requirement for leukemia-inhibitory factor (LIF)-activated signaling by mESCs for self-renewal (Mitsui *et al.*, 2003). *Nanog* knock-out results in differentiation of mESCs into parietal/visceral endoderm (Mitsui *et al.*, 2003).

Despite the major role and intricate interplay of these transcription factors in regulating self-renewal and differentiation of ESCs, the precise molecular mechan-isms governing ESC fate remain poorly understood. Posttranscriptional regula-tory mechanisms are pivotal in the modulation of expression of many genes, emphasized by the many points of control in eukaryotes (Hollams *et al.*, 2002; Proudfoot *et al.*, 2002). It is now apparent that non-coding RNAs, including

---

* Author to whom correspondence should be addressed.

MicroRNAs: From Basic Science to Disease Biology, ed. Krishnarao Appasani. Published by Cambridge University Press. © Cambridge University Press 2008.

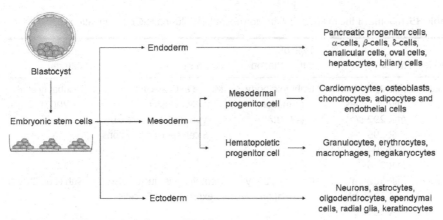

Figure 35.1. Differentiation of mouse ES cells.

microRNAs (miRNAs), have an important role in modulating mRNA decay and translation rates in eukaryotes (Seitz *et al.*, 2004). Nascent primary miRNA transcripts (pri-miRNAs) are processed sequentially by nuclear Drosha and cytoplasmic Dicer RNase III endonuclease-containing complexes to yield mature miRNAs, which are incorporated into the RNA interference silencing complex (RISC) (Gregory *et al.*, 2005; He *et al.*, 2005). The activated RISC-miRNA complexes bind to cognate mRNA *cis*-elements, and these complexes are directed to P-bodies where mRNA degradation or translational repression occurs (Cheng *et al.*, 2005; Gregory *et al.*, 2005; Liu *et al.*, 2005a,b; Marx, 2005; Rand *et al.*, 2005). The mode of silencing is dependent on which of the four Argonaute (AGO) RNA binding proteins is present in the RISC, where AGO2 alone mediates miRNA-directed mRNA degradation whilst AGO1-4 are involved in translational repression (Meister *et al.*, 2004; Rand *et al.*, 2005).

Recent studies suggest a role for miRNAs in the regulation of vertebrate development (Miska, 2005; Schulman *et al.*, 2005). The restoration of a single miRNA (miR-430) in zebrafish modified to prevent production of endogenous miRNAs ameliorated deficits in neuroectodermal development and neuronal differentiation (Giraldez *et al.*, 2005). In mammals, specific miRNAs have been shown to regulate B cell differentiation (Chen *et al.*, 2004), adipocyte differentiation (Esau *et al.*, 2004), and insulin secretion (Poy *et al.*, 2004). MiRNA regulation of *Hox* expression modulates developmental patterning processes to allow the generation of asymmetric morphology (Mansfield *et al.*, 2004; Yekta *et al.*, 2004). Alterations in miRNA expression are associated consistently with several forms of human cancers and cancer cell lines (McManus, 2003; Metzler *et al.*, 2004; Takamizawa *et al.*, 2004; Lu *et al.*, 2005; Miska, 2005). In the context of mESCs, the loss of mature miRNAs in Dicer1 null mESCs results in a failure of mESCs to differentiate (Kanellopoulou *et al.*, 2005). These data alone highlight the importance of regulated miRNA expression in controlling ESC growth and differentiation. ESC-specific miRNAs have been identified in murine and human ESCs (Houbaviy *et al.*, 2003; Suh *et al.*, 2004), however, their functional significance has not been evaluated (Tables 35.1 and 35.2).

Table 35.1. Some of the miRNAs that are down-regulated during ESC differentiation

Species	MiRNA	Type of differentiation	Function	Reference
Mouse	miR-290 miR-291-s/as miR-292-s/as miR-293 miR-294 miR-295	Embryoid body formation (+/− RA)	Mouse ESC-specific miRNA cluster, hypothesized to have ES cell-specific functions	Houbaviy *et al.* (2003)
Human	miR-302a/a* miR-302b/b* miR-302c/c* miR-302d miR-367	Embryoid body formation	Homologous cluster, also expressed in ES cells	Suh *et al.* (2004)
Human	miR-371 miR-372 miR-373/3*	Embryoid body formation	Human ESC-specific miRNA cluster, homologous to the mouse ESC-specific cluster above	Suh *et al.* (2004)

Table 35.2. Some of the miRNAs that are up-regulated during ESC differentiation

Species	MiRNA	Type of differentiation	Function	Reference
Mouse	let-7a-3	Neural		Lee *et al.* (2005)
Mouse	miR-9/9*	Neural	Affects neural lineage differentiation, may act in the STAT3 signaling pathway	Krichevsky *et al.* (2005)
Mouse	miR-21	Embryoid body formation (+/− RA)		Houbaviy *et al.* (2003)
Mouse	miR-22	Embryoid body formation (+/− RA)		Houbaviy *et al.* (2003)
		Neural		Krichevsky *et al.* (2005)
Mouse	miR-124a	Neural	Affects neural lineage differentiation, may act in the STAT3 signaling pathway	Krichevsky *et al.* (2005)
Mouse	miR-125b	Neural		Lee *et al.* (2005)
		Neural		Krichevsky *et al.* (2005)
Mouse	miR-134	RA induction		Our lab (unpublished data)

MiRNAs expressed in ESCs

The expression patterns of miRNAs in ESCs can be classified into four groups: (1) miRNAs that are expressed in ES cells as well as in embryonic carcinoma (EC) cells, which may have conserved roles in mammalian pluripotent stem cells; (2) miRNAs expressed specifically in ES cells but not in other cells including EC cells. These may have functions specific to ES cells; (3) miRNAs that are rare in ES cells and increase upon differentiation, which may be involved in the differentiation process; and (4) miRNAs that are present in ES cells and remain at a constant level during differentiation. These may be involved in general aspects of cell physiology.

MiRNAs expressed in mouse ESCs

MiRNAs were first identified in ESCs using cDNA cloning (Tables 35.1 and 35.2). Houbaviy *et al.* describe 53 miRNAs in mES cells, of which 15 are novel (Houbaviy *et al.*, 2003; see also Chapter 33 in this book by Houbaviy). Although the levels of many previously described miRNAs remain constant or increase upon differentiation, eight of the novel miRNAs (miR-290, miR-291s, miR-291as, miR-292, miR-292as, miR-293, miR-294, miR-295) appear to be ES cell or early embryo specific by four criteria: (1) their sequences are distinct from those of previously described miRNAs, including miRNAs cloned from adult mouse organs; (2) they cannot be detected in adult mouse organs by Northern analyses; (3) they are repressed during ESC differentiation *in vitro*; and (4) all ESTs that map within the cluster are derived from ESCs or preimplantation embryos (Houbaviy *et al.*, 2005).

MiR-290 to miR-295 is a cluster of partially homologous pre-miRNA hairpins encoded by genomic loci clustered within 2.2 kb of each other (Houbaviy *et al.*, 2003). This entire Early Embryonic microRNA Cluster (EEmiRC) is spanned by a spliced, capped and polyadenylated primary transcript and transcription is directed by a conserved promoter element containing a TATA box. Sequence analysis shows that the EEmiRC transcription unit is remarkably variable and can only be identified bioinformatically in placental (eutherian) mammals (Houbaviy *et al.*, 2005). The only conserved regions within the locus are the pre-miRNA hairpins and the putative minimal promoter. The number and precise sequences of the pre-miRNAs, their distance from the promoter and the polyadenylation sites, the regions flanking the hairpins, and the types, positions and numbers of repetitive element insertions vary in species belonging to different mammalian orders (Houbaviy *et al.*, 2005).

MiRNAs expressed in human ESCs (hESCs)

Experimental support for the *in silico* prediction of an EEmiRC counterpart in the human genome was provided by the cDNA cloning of the corresponding miRNA homologs (miR-371, miR-372, miR-373 and miR-373*) from hESCs (Suh *et al.*, 2004). Although mouse and human EEmiRC miRNAs are sufficiently different from each other to warrant different numerical designations, multiple sequence alignment reveals related sequences both within each cluster and across species.

Furthermore, sequence conservation extends beyond the mature miRNAs to the entire pre-miRNA hairpin sequences (Houbaviy *et al.*, 2003).

The miR-371 to miR-373* cluster is located within a 1050 bp region on chromosome 19 (Suh *et al.*, 2004). Intriguingly, these are also highly expressed in cancer cell lines, and miR-372 and miR-373 have the ability to protect cells from oncogenic stress and transform primary human cells (Kim, 2005; Voorhoeve *et al.*, 2006). This suggests that they may operate key regulatory networks conserved in pluripotent stem cells and cancer cells, and serve as molecular markers for undifferentiated hESCs and specific cancers (Tables 35.1 and 35.2).

Suh *et al.* report another cluster of eight highly related miRNAs (miR-302b, miR-302b*, miR-302c, miR-302c*, miR-302a, miR-302a*, miR-302d and miR-367) located within a ∼700 bp region on chromosome 4. MiR-302b, miR-302c and miR-302d appear to be the close homologs of miR-302 that was cloned from mESCs. Although these are the most abundant miRNAs in hESCs, their murine homolog miR-302 appears to be less abundant (Suh *et al.*, 2004).

Conservation of miRNA clusters in mouse and human ESCs

It is intriguing that two miRNA clusters are conserved and specifically expressed in both mouse and human ESCs. Although the variation in the numbers and sequences of the homologs may implicate divergence of the conserved regulatory pathways, these conserved miRNAs are likely to play central roles in the regulation of mammalian ESCs.

There are considerable differences between the cloned miRNAs from human and mouse ESCs. This may imply fundamental differences between the regulatory networks in hESCs and mESCs, or may simply be because of the cloning techniques employed, which may not be sensitive enough to identify the complete set of miRNAs in these cell lines. This limitation has been overcome by recently developed miRNA microarray techniques which have enabled global miRNA expression profiling. Microarray analysis of 124 mammalian microRNAs has identified six additional miRNAs, miR-214, miR-134, miR-25, miR-182, miR-204, miR-132, which demonstrate mouse ESC-specific expression (Thomson *et al.*, 2004).

MiRNAs upregulated during ESC differentiation

Induction of miR-125b and let-7 during neural differentiation of mESCs

MiR-125b and let-7 miRNAs are induced, whereas their putative targets, *lin-28* and *lin-41*, are decreased during the *in vitro* neuronal differentiation of mESCs. Both miRNAs were weakly expressed in ESCs, slightly decreased in embryoid bodies (EBs), and then increased with differentiation through the neural progenitor (NP) stage into mature neurons. However, they are neither sufficient nor necessary for the downregulation of *lin-28* and *lin-41* during differentiation (Lee *et al.*, 2005).

MiR-124a and miR-9 affect neural differentiation in mESCs

MiRNA expression profiling during *in vitro* mESC-derived neurogenesis shows that a number of miRNAs are simultaneously co-induced during the differentiation of neural progenitor cells to neurons and astrocytes (Krichevsky *et al.*, 2005). Seventy of the 134 vertebrate miRNAs tested showed a greater than two-fold change between any two stages of the five-step neural differentiation protocol. MiR-124a and miR-9, which are almost exclusively expressed in the brain, affect neural lineage differentiation of the mESC cultures. Krichevsky *et al.* hypothesize that early overexpression of miR-124a in NPs prevents gliogenesis whereas miR-9 expression contributes to neurogenesis (Krichevsky *et al.*, 2005). Additionally, STAT3, a member of the signal transducer and activator of transcription (STAT) pathway, is involved in the function of these miRNAs. These data demonstrate a functional role for distinct miRNAs in the determination of neural fates during ESC differentiation.

MiRNAs upregulated during retinoic acid-induced ESC differentiation

In our laboratory, we observed that retinoic acid (RA)-induced differentiation in mESCs was accompanied by the upregulation of a number of microRNAs, including miR-134 (Table 35.2). MiR-134 was shown to induce differentiation of mESCs to a primitive ectodermal phenotype, a precursor to neuronal differentiation. Interestingly, miR-134 has been shown recently to direct neural stem cell differentiation (Schratt *et al.*, 2006).

MiRNA expression profiling in single ESCs

An important caveat to note is that seemingly uniform cells, such as ESCs, may differ from each other. The recent development of a real-time PCR-based 220-plex method to profile miRNA expression in single ESCs will be an invaluable tool for the unambiguous miRNA profiling of specific cell types (Tang *et al.*, 2006).

Transcriptional regulation of microRNAs in ESCs

Not much is known about the transcriptional regulation of microRNAs in ESCs. However, our data and those of others have demonstrated that microRNA levels are modulated during differentiation. The levels of the transcription factors Oct4, Sox2 and Nanog are invariably reduced during ESC-differentiation. Interestingly, in hESCs, it has been found that at least 14 microRNA loci are bound closely by Oct4, Sox2 or Nanog, where two of the putative promoters of two of these microRNAs, miR-137 and miR-301, are co-occupied by all three transcription factors (Boyer *et al.*, 2005) (Table 35.3). However, no one has as yet demonstrated that the transcription factors do indeed regulate the transcription rate of these microRNA loci. Again, it was observed in mESCs that five microRNA loci were bound by Nanog and Oct4, where the putative promoters of two of these microRNAs, miR-296 and miR-302, are co-occupied by Nanog and Oct4 (Loh *et al.*, 2006) (Table 35.3). However no regulatory data is available.

RA-induced miR-134 is produced from a primary transcript that contains 20 microRNAs (Seitz *et al.*, 2004). To date nothing is known about the promoter of this transcript, although a rudimentary search suggests that there may be a weak

Table 35.3. MicroRNAs that have loci near regions bound by OCT4, SOX2 and NANOG transcription factors, although it is still unknown if these transcription factors do regulate transcription of these microRNAs

	Transcription factors		
MicroRNAs	OCT4	SOX2	NANOG
Hs-miRNA			
miR-7-1		+	
miR-10a	+		
miR-22		+	+
miR-32		+	+
miR-128a			+
miR-135b		+	+
miR-137	+	+	+
miR-196a-1			+
miR-196b	+		
miR-204		+	+
miR-205		+	+
miR-301	+	+	+
miR-361			+
miR-448	+		
Mm-miRNA			
miR-9-2			+
miR-124a			+
miR-135			+
miR-296	+		+
miR-302	+		+

Source: From Boyer *et al.*, 2005 (Hs); Loh *et al.*, 2006 (Mm).

RA-receptor (RAR) binding site upstream of the transcript. Work is underway to determine if RA acts through its cognate receptor, RAR, to regulate transcription of the microRNA cluster transcript containing miR-134.

MicroRNAs and their target mRNAs – prediction and analysis in ESCs

Increasingly, more microRNAs are being discovered and, as such, tools designed to predict their identity, abundance, and/or potential mRNA *cis*-element targets are the main focus *in silico*. To this end, many predictive models are being developed and tested, and it is a highly contested area of research (see Table 35.4). Tools that predict the identity and abundance of microRNAs are less prevalent than those that attempt to predict mRNA *cis*-element targets.

Predicting microRNAs from the genome

Owing to the variables in many of the models, estimates of the number of microRNAs that may exist vary quite considerably from a few hundred to scores of thousands. One of the main problems facing researchers in the area is that many of these microRNAs may be expressed at very low levels in cells, in only certain cell

Table 35.4. Prediction tools – microRNAs and their mRNA *cis*-element targets: success of prediction tools has been somewhat limited, although the rna22 algorithm has shown greatest potential and number of targets tested

Reference/*Species*	No. of miRs tested	No. of predicted targets	No. of targets validated	Method basis
Lewis *et al.* (2003) *Homo sapiens, Mus musculus, Rattus norvegicus, Fugu rubripes*	8	15	11	TargetScan
Kiriakidou *et al.* (2004) *Homo sapiens*	10	23	12	Diana-MicroT
Enright *et al.* (2003) *Drosophila melanogaster* John *et al.* (2004) *Homo sapiens*	Showed computationally an ability to recover a few previously reported targets with associated false positive rate of 24% and 39%		None	MiRanda
Krek *et al.* (2005) *Mus musculus*	2	13	7	PicTar
Rehmsmeier *et al.* (2004) *Drosophila melanogaster*	Able to recover few of previously reported targets		None	RNAhybrid
Miranda *et al.* (2006)	3	253	180	rna22 – "Teirisias"
(http://cbcsrv.watson.ibm. com/rna22.html) *Mus musculus*	miR-134	160	120	

types, or only at certain time points in the cell cycle. Therefore, obtaining biological data to test the models is somewhat limiting. However, knowing that a 10 base loop is required for correct recognition and processing of pre-microRNAs into mature forms (Zeng *et al.*, 2005), using RNA secondary structure modeling one can start allowing prediction across the genome of which transcripts would be expected to be processed into microRNAs. As stated previously, because of the lack of coverage in annotated gene expression databases, there is a paucity of data with which to test working models. However, deep sequencing projects are now discovering many small non-coding RNAs, which will help confirm the computational models based on predicting microRNAs (Rigoutsos, 2006).

Predicting mRNA *cis*-elements to which microRNAs bind
A unique feature of microRNAs is that they do not require complete complementarity with their cognate mRNA *cis*-element, suggesting a sliding scale is required for predicting the mRNA target sequences of microRNAs (Stark *et al.*, 2005). Therefore, a microRNA can have the potential to be active against many mRNAs with little sequence homology. This is an ingenious mechanism allowing coordinate control of many genes, however, the algorithms required to predict microRNAs' target mRNAs will be complicated. To date, most predictive tools have identified only a few genes successfully (see Table 35.4). It appears that most of the first and second generation tools designed over the past 2.5 years are

now moving towards using the "Teiresias" based set of algorithms, originally designed for studying protein–protein interactions, for developing predictive models (Rigoutsos and Floratos, 1998).

Testing predictions of microRNA target mRNAs in ESCs

In our laboratory we are investigating the role of non-coding RNAs in embryonic stem cell differentiation. We wish to know which microRNAs are being expressed in ESCs, and how their mode of expression influences cell growth or differentiation. This knowledge should increase our capacity to manipulate and develop ESCs as therapeutics for regenerative medicine. Therefore, we are interested in testing tools that can predict novel microRNAs and mRNA *cis*-element targets. We have tested several predictive tools, and to date one that has shown greatest potential is the novel algorithm, rna22, which happens to be based on the "Teiresias" model.

In our research, we have used classic luciferase reporter based assays with the predicted mRNA *cis*-element incorporated into the 3'-untranslated region (-UTR) of the luciferase mRNA (Balmer *et al.*, 2001). Transfection of cells with precursors to our microRNA of interest, an unrelated microRNA, and a non-microRNA along with the luciferase constructs were used to investigate the degree of suppression of luciferase reporter, indicative of translation repression. Some researchers have applied reporter constructs containing the whole 3'-UTR of the mRNA to be analyzed, which may allow a mRNA to fold to its correct form and present the *cis*-element in a more relevant biological context (Brennecke *et al.*, 2003). However, this still may not be an efficient reporter system if numerous microRNAs are required to target this 3'-UTR to elicit translation suppression. Using rna22, we tested three different murine microRNAs on predicted targets, of which miR-134 was predicted to be active against over 5000 mRNA species. Out of 253 *cis*-elements tested, we had successful suppression of luciferase reporter activity with 183 *cis*-elements, where miR-134 showed suppression of 114 out of 160 target *cis*-elements tested. This is the first time that the promiscuous nature of a microRNA has been demonstrated in any cell type or species. One of the miR-134 targets predicted was Lmk1, and this has since been proven to be a major target of miR-134 in murine neural stem cells (Schratt *et al.*, 2006). Interestingly, when we investigated the endogenous protein and mRNA levels of four predicted miR-134 target genes, we found that pre-miR-134 induced a translation down regulation in these genes, some of which are important in directing ESC growth or differentiation. We surmise that the actual positive mRNA target number is higher than that demonstrated in our luciferase reporter assays, and that it is the variation of the transfection assay that is limiting our ability to discern all true interactions. As such, some of the true positive targets may be missed currently, suggesting the need to design novel assay methods. Nevertheless, the value of some of the tools, especially rna22, is proving to be a very powerful and accurate predictor of miRNA/mRNA interactions in ESCs.

Other microRNA target regions in mRNAs

Currently, most predictive models have concentrated their efforts on the 3'-UTR of mRNAs as they tend to be longer than 5'-UTRs, and are known to direct mRNA stability, translation efficiency and localization. However, the 5'-UTR also directs mRNA translation (Muckenthaler *et al.*, 1998; Rouault, 2005; Thomson *et al.*, 2005). The 5'-UTR and coding regions could potentially be targets for miRNAs, but at the time of writing no work has been reported on the targeting of 5'-UTR or coding regions by miRNAs. We are investigating this in ESCs utilizing rna22 predictions that many microRNAs target the 5'-UTR and coding sequence of multiple mRNA species.

MicroRNAs and RNA-binding proteins and their RNA cargoes in ESCs

It has been established that *trans*-acting RNA binding proteins that bind to *cis*-acting mRNA elements are major determinants of mRNA translation efficiency, stability, and localization to P-bodies (regions of translational silencing) or polyribosomes (translation active mRNAs) or other cellular compartments (Yeap *et al.*, 2002; Giles *et al.*, 2003; Gore *et al.*, 2005; Thomson *et al.*, 2005). Additionally, correct transport and localization of microRNAs have been shown to be important in dendritic spine function and development, and in directing mRNAs to P-bodies for translation silencing (Schratt *et al.*, 2006). Therefore, investigators are now beginning to study the association of microRNAs and RNA-binding protein complexes that bind to mRNAs. To this end, microRNAs have been shown to be associated with Argonaute complexes that bind to a number of RNA-binding proteins, including fragile X mental retardation protein, an essential RNA-binding protein for neural stem cell (NSC) differentiation and function of those NSC-derived cells (Jin *et al.*, 2004; John *et al.*, 2004). A recent exciting publication has shown that localization of mRNAs is vitally important in early vertebrate development for determining cell fate (Gore *et al.*, 2005). Thus from an embryonic stem cell and developmental research point of view, it is essential to define the microRNA–RNA-binding protein–mRNA associations that dictate mRNA localization, which may have a critical role during differentiation of ESCs to specific cell phenotypes.

Conclusions and future work

All living organisms have devised elaborate and exquisitely controlled systems to regulate cell response and homeostasis. The microRNAs, which are highly conserved across vertebrates and invertebrates, play a central role in modulating the translation efficiency, localization, and stability of their target mRNAs produced in a cell. The discovery that localization of mRNAs, which is directed by microRNA and RNA-binding protein association, is of vital importance in determining cell fate is exciting for stem cell biologists. It is vital that we define these posttranscriptional mechanisms in ESCs such that we can develop methods that allow us to manipulate ESCs for directed differentiation and use as therapeutics.

Since the mature microRNA sequence is short and exact complementarity is not required for action, it is proposed that a large number of mRNAs can be engaged by a single species of miRNA. Indeed, computational predictions suggest that some of these microRNAs may each regulate thousands of genes, and our data in ESCs are beginning to show that this is true. This is particularly exciting as one microRNA, miR-134, targets a number of genes associated with inducing differentiation in ESCs. Differentiation of ESCs is also associated with changes in numerous microRNAs, and microRNAs acting in concert are an excellent example of how cells have crafted a control system to regulate co-ordinately multiple genes.

We are beginning to appreciate the vital role that posttranscriptional mechanisms play in directing cell fate. The past 10 years have seen an explosion on the information of the biology of microRNAs, their mRNA *cis*-element targets, and their role in cell cycle regulation and differentiation. We look forward to an exciting future work which will develop better predictive tools for identifying microRNAs and clarify their mRNA and RNA-binding protein partners, and unravel the mysteries of their function and regulation in directing ESC growth and differentiation.

REFERENCES

Balmer, L., Beveridge, D. J., Jazayeri, J. A. *et al.* (2001). Identification of a novel AU-rich element in the 3′ UTR of epidermal growth factor receptor mRNA that is the target for regulated RNA-binding proteins. *Molecular and Cellular Biology*, **21**, 2070–2084.

Boyer, L. A., Lee, T. I., Cole, M. F., Johnstone, S. E. and Levine, S. S. (2005). Core transcriptional regulatory circuitry in human embryonic stem cells. *Cell*, **122**, 947–956.

Brennecke, J., Hipfner, D. R., Stark, A., Russell, R. B. and Cohen, S. M. (2003). bantam encodes a developmentally regulated microRNA that controls cell proliferation and regulates the proapoptotic gene hid in *Drosophila*. *Cell*, **113**, 25–36.

Chen, C. Z., Li, L., Lodish, H. F. and Bartel, D. P. (2004). MicroRNAs modulate hematopoietic lineage differentiation. *Science*, **303**, 83–86.

Cheng, L., Tavazole, M. and Doetsch, F. (2005). Stem cells: from epigenetics to microRNAs. *Neuron*, **46**, 363–367.

Chew, J. L., Loh, Y. H., Zhang, W. *et al.* (2005). Reciprocal transcriptional regulation of Pou5f1 and Sox2 via the Oct4/Sox2 complex in embryonic stem cells. *Molecular and Cellular Biology*, **25**, 6031–6046.

Czyz, J., Wiese, C., Rolletschek, A. *et al.* (2003). Potential of embryonic and adult stem cells *in vitro*. *Biological Chemistry*, **384**, 1391–1409.

Enright, A. J., John, B., Gaul, U. *et al.* (2003). MicroRNA targets in *Drosophila*. *Genome Biology*, **5**, R1.

Esau, C., Kang, X., Peralta, E. *et al.* (2004). MicroRNA-143 regulates adipocyte differentiation. *Journal of Biological Chemistry*, **279**, 52361–52365.

Evans, M. J. and Kaufman, M. H. (1981). Establishment in culture of pluripotential cells from mouse embryos. *Nature*, **292**, 154–156.

Giles, K. M., Daly, J. M., Beveridge, D. J. *et al.* (2003). The 3′ untranslated region of p21WAF1 mRNA is a composite cis-acting sequence bound by RNA-binding proteins from breast cancer cells, including HuR and poly(C)-binding protein. *Journal of Biological Chemistry*, **278**, 2937–2946.

Giraldez, A. J., Cinalli, R. M., Glasner, M. E. *et al.* (2005). MicroRNAs regulate brain morphogenesis in zebrafish. *Science*, **308**, 833–838.

Gore, A. V., Maegawa, S., Cheong, A. *et al.* (2005). The zebrafish dorsal axis is apparent at the four-cell stage. *Nature*, **438**, 1030–1035.

Gregory, R. I., Chendrimuka, T. P., Cooch, N. and Shiekhattar, R. (2005). Human RISC couples microRNA biogenesis and posttranscriptional gene silencing. *Cell*, **123**, 631–640.

He, L., Thomson, J. M., Hemann, M. T. *et al.* (2005). A microRNA polycistron as a potential human oncogene. *Nature*, **435**, 828–33.

Hollams, E. M., Giles, K. M., Thomson, A. M. and Leedman, P. J. (2002). mRNA stability and the control of gene expression: implications for human disease. *Neurochemical Research*, **27**, 957–980.

Houbaviy, H. B., Murray, M. F. and Sharp, P. A. (2003). Embryonic stem cell-specific microRNAs. *Developmental Cell*, **5**, 351–358.

Houbaviy, H. B., Dennis, L., Jaenisch, R. and Sharp, P. A. (2005). Characterization of a highly variable eutherian microRNA gene. *RNA*, **11**, 1–13.

Jin, P., Zarnescu, D. C., Ceman, S. *et al.* (2004). Biochemical and genetic interaction between the fragile X mental retardation protein and the microRNA pathway. *Nature Neuroscience*, **7**, 113–117.

John, B., Enright, A. J., Aravin, A. *et al.* (2004). Human microRNA targets. *Public Library of Science Biology*, **2**, e363.

Kanellopoulou, C., Muljo, S. A., Kung, A. L. *et al.* (2005). Dicer-deficient mouse embryonic stem cells are defective in differentiation and centromeric silencing. *Genes & Development*, **19**, 489–501.

Kim, K. (2005). Personal communication from Keystone Symposia on Stem Cells, Senescence and Cancer, Singapore, 2005.

Kiriakidou, M., Nelson, P. T., Kouranov, A. *et al.* (2004). A combined computational-experimental approach predicts human microRNA targets. *Genes & Development*, **18**, 1165–1178.

Krek, A., Grun, D., Poy, M. N. *et al.* (2005). Combinatorial microRNA target predictions. *Nature Genetics*, **37**, 495–500.

Krichevsky, A., Sonntag, K., Isacson, O. and Kosik, K. S. (2005). Specific microRNAs modulate ES cell-derived neurogenesis. *Stem Cells*, **Dec 15** (Epub ahead of print).

Lee, Y. S., Kim, H. K., Chung, S., Kim, K. and Dutta, A. (2005). Depletion of human microRNA miR-125b reveals that it is critical for the proliferation of differentiated cells but not for the downregulation of putative targets during differentiation. *Journal of Biological Chemistry*, **280**, 16 635–16 641.

Lewis, B. P., Shih, I. H., Jones-Rhoades, M. W. *et al.* (2003). Prediction of mammalian microRNA targets. *Cell*, **115**, 787–798.

Liu, J., Rivas, F. V., Wohlschlegel, J. *et al.* (2005a). A role for the P-body component GW182 in microRNA function. *Nature Cell Biology*, **Nov 13** (Epub ahead of print).

Liu, J., Valencia-Sanchez, M. A., Hannon, G. J. and Parker, R. (2005b). MicroRNA-dependent localization of targeted mRNAs to mammalian P-bodies. *Nature Cell Biology*, **7**, 719–723.

Loebel, D. A., Watson, C. M., De Young, R. A. and Tam, P. P. (2003). Lineage choice and differentiation in mouse embryos and embryonic stem cells. *Developmental Biology*, **264**, 1–14.

Loh, Y. H., Wu, Q., Chew, J. L. *et al.* (2006). The Oct4 and Nanog transcription network regulates pluripotency in mouse embryonic stem cells. *Nature Genetics*, **38**, 431–440.

Lu, J., Getz, G., Miska, E. A. *et al.* (2005). MicroRNA expression profiles classify human cancers. *Nature*, **435**, 834–838.

Mansfield, J. H., Harfe, B. D., Nissen, R. *et al.* (2004). MicroRNA-responsive "sensor" transgenes uncover Hox-like and other developmentally regulated patterns of vertebrate microRNA expression. *Nature Genetics*, **36**, 1079–1083.

Marx, J. (2005). Molecular biology. P-bodies mark the spot for controlling protein production. *Science*, **310**, 764–765.

McManus, M. T. (2003). MicroRNAs and cancer. *Seminars in Cancer Biology*, **13**, 253–258.

Meister, G., Landthaler, M., Patkaniowska, A. *et al.* (2004). Human Argonaute2 mediates RNA cleavage targeted by miRNAs and siRNAs. *Molecular Cell*, **15**, 185–197.

Metzler, M., Wilda, M., Busch, K., Viehmann, S. and Borkhardt, A. (2004). High expression of precursor microRNA-155/BIC RNA in children with Burkitt lymphoma. *Genes, Chromosomes & Cancer*, **39**, 167–169.

Miranda, K. C., Huynh, T., Tay, Y. *et al.* (2006). A pattern-based method for the identification of microRNA binding sites and their corresponding heteroduplexes. *Cell*, **126**, 1203–1217.

Miska, E. A. (2005). How microRNAs control cell division, differentiation and death. *Current Opinion in Genetics & Development*, **15**, 563–568.

Mitsui, K., Tokuzawa, Y., Itoh, H. *et al.* (2003). The homeoprotein Nanog is required for maintenance of pluripotency in mouse epiblast and ES cells. *Cell*, **113**, 631–642.

Muckenthaler, M., Gray, N. K. and Hentze, M. W. (1998). IRP-1 binding to ferritin mRNA prevents the recruitment of the small ribosomal subunit by the cap-binding complex eIF4F. *Molecular Cell*, **2**, 383–388.

Niwa, H., Miyazaki, J. and Smith, A. G. (2000). Quantitative expression of Oct-3/4 defines differentiation, dedifferentiation or self-renewal of ES cells. *Nature Genetics*, **24**, 372–376.

Orkin, S. H. (2005). Chipping away at the embryonic stem cell network. *Cell*, **122**, 828–830.

Poy, M. N., Eliasson, L., Krutzfeldt, J. *et al.* (2004). A pancreatic islet-specific microRNA regulates insulin secretion. *Nature*, **432**, 226–230.

Proudfoot, N. J., Furger, A. and Dye, M. J. (2002). Integrating mRNA processing with transcription. *Cell*, **108**, 501–512.

Rand, T. A., Petersen, S., Du, F. and Wang, X. (2005). Argonaute2 cleaves the anti-guide strand of siRNA during RISC activation. *Cell*, **123**, 621–629.

Rehmsmeier, M., Steffen, P., Hochsmann, M. and Giegerich, R. (2004). Fast and effective prediction of microRNA/target duplexes. *RNA*, **10**, 1507–1517.

Rigoutsos, I. and Floratos, A. (1998). Combinatorial pattern discovery in biological sequences: The TEIRESIAS algorithm. *Bioinformatics*, **14**, 55–67.

Rouault, T. A. (2005). The intestinal heme transporter revealed. *Cell*, **122**, 649–651.

Schratt, G. M., Tuebing, F., Nigh, E. A. *et al.* (2006). A brain-specific microRNA regulates dendritic spine development. *Nature*, **439**, 283–289.

Schulman, B. R., Esquela-Kerscher, A. and Slack, F. J. (2005). Reciprocal expression of lin-41 and the microRNAs let-7 and mir-125 during mouse embryogenesis. *Developmental Dynamics*, **234**, 1046–1054.

Seitz, H., Royo, H., Bortolin, M. L. *et al.* (2004). A large imprinted microRNA gene cluster at the mouse Dlk1-Gtl2 domain. *Genome Research*, **14**, 1741–1748.

Smith, A. G. (2001). Embryo-derived stem cells: of mice and men. *Annual Review of Cell and Developmental Biology*, **17**, 435–462.

Stark, A., Brennecke, J., Bushati, N., Russell, R. B. and Cohen, S. M. (2005). Animal microRNAs confer robustness to gene expression and have a significant impact on 3′ UTR evolution. *Cell*, **123**, 1133–1146.

Suh, M. R., Lee, Y., Kim, J. Y. *et al.* (2004). Human embryonic stem cells express a unique set of microRNAs. *Developmental Biology*, **270**, 488–498.

Takamizawa, J., Konishi, H., Yanagisawa, K. *et al.* (2004). Reduced expression of the *let-6* microRNAs in human lung cancers in association with shortened postoperative survival. *Cancer Research*, **64**, 3753–3756.

Tang, F., Hajkova, P., Barton, S. C., Lao, K. and Surani, M. A. (2006). MicroRNA expression profiling of single whole embryonic stem cells. *Nucleic Acids Research*, **34**, e9.

Thomson, A. M., Cahill, C. M., Cho, H. *et al.* (2005). The acute box cis-element in human heavy ferritin mRNA 5′-untranslated region is a unique translation enhancer that binds poly(c)-binding proteins. *Journal of Biological Chemistry*, **280**, 30 032–30 045.

Thomson, J. M., Parker, J., Perou, C. M. and Hammond, S. M. (2004). A custom microarray platform for analysis of microRNA gene expression. *Nature Methods*, **1**, 1–7.

Voorhoeve, P. M., Sage, C. L., Schrier, M. *et al.* (2006). A genetic screen implicates miRNA-372 and miRNA-373 as oncogenes in testicular germ cell tumors. *Cell*, **124**, 1169–1181.

Yeap, B. B., Voon, D. C., Vivian, J. P. *et al.* (2002). Novel binding of HuR and poly(C)-binding protein to a conserved UC-rich motif within the 3′-untranslated region of the androgen receptor messenger RNA. *Journal of Biological Chemistry*, **277**, 27 183–27 192.

Yekta, S., Shih, I. H. and Bartel, D. P. (2004). MicroRNA-directed cleavage of HOXB8 mRNA. *Science*, **304**, 594–596.

Zeng, Y., Yi, R. and Cullen, B. R. (2005). Recognition and cleavage of primary microRNA precursors by the nuclear processing enzyme Drosha. *European Molecular Biology Organization Journal*, **24**, 138–148.

36 Generation of single cell microRNA expression profile

Fuchou Tang, Kaiqin Lao* and M. Azim Surani*

Introduction

MicroRNAs (miRNAs) are a class of 17–25 nt non-coding RNAs that have been shown to have critical functions during development in a wide variety of biological processes, including cell cycle regulation, cell differentiation, apoptosis, maintenance of stemness and imprinting (Ambros, 2004; Alvarez-Garcia and Miska, 2005; Harfe, 2005; Plasterk, 2006). Profiling of miRNAs is a prerequisite for dissecting their biological function. A number of methods have been developed for this purpose, such as the miRNA microarray techniques during the past three years (Krichevsky et al., 2003; Babak et al., 2004; Barad et al., 2004; Calin et al., 2004; Liu et al., 2004; Miska et al., 2004; Nelson et al., 2004; Sempere et al., 2004; Sioud and Rosok, 2004; Sun et al., 2004; Thomson et al., 2004; Baskerville and Bartel, 2005; Liang et al., 2005; Lu et al., 2005; Castoldi et al., 2006; Neely et al., 2006). Most of these methods need microgram (µg) amounts of total RNA for the assay, although a few of them are more sensitive and only need nanogram (ng) amounts of RNA. However, all of the methods require purification of total RNA for the assays, which is not feasible for single cell miRNA expression profiling. In some cases, such as the analysis of very early embryos, adult stem cells, and cancer stem cells, it is often crucial to analyze miRNA profile in individual cells. In these cases, very limited amounts of material may be available for miRNA profiling assay. Quite often there is inherent variability among the cells making essential that analysis is carried out at the level of single cells. For these reasons, it is greatly desirable to develop a miRNA expression profiling assay for single cells.

Recently, we developed a real-time PCR-based miRNA profiling assay, which, as we demonstrated, works robustly for single embryonic stem cells (ES cells) (Chen et al., 2005; Lao et al., 2006; Tang et al., 2006a, b). We used a set of stem-loop-structured primers to specifically reverse transcribe mature miRNAs into corresponding cDNAs. We also designed a universal reverse primer to permit us to

---

* Authors to whom correspondence should be addressed.

MicroRNAs: From Basic Science to Disease Biology, ed. Krishnarao Appasani. Published by Cambridge University Press. © Cambridge University Press 2008.

amplify miRNA cDNAs evenly before quantifying their expression by TaqMan probe-based real-time PCR.

Methodology

Design of primers and probes

To specifically discriminate between mature miRNAs from precursor RNAs, we used stem-loop-structured primers for reverse transcription (Figure 36.1). The primers have eight nucleotides complementary to the 3′ end of corresponding miRNAs. The left part of the primer forms a stem-loop structure. Because in each individual cell the copy number of each miRNA is between dozens to several thousands, it is not feasible to split them into 220 parts for the analysis of 220 individual miRNAs. To circumvent this problem, we add a pre-PCR step to amplify these miRNA cDNAs evenly before splitting them to check the expression of individual miRNAs. During the pre-PCR step, a forward primer with 16 nucleotides complementary to 3′ end nucleotides of corresponding miRNA cDNAs, and a universal reverse primer with 18 nucleotides complementary to 5′ end nucleotides of the corresponding miRNA cDNAs were used to amplify these cDNAs. Here we introduced the universal reverse primer in the pre-PCR reaction to reduce the potential primer–primer interactions and make it feasible to simultaneously amplify cDNAs of 220 miRNAs evenly. We showed that 18 cycles of pre-PCR would amplify the miRNA cDNAs sufficiently for expression analysis of all 220 miRNAs without introducing any significant bias. Finally, during the real-time PCR step, a TaqMan probe complementary to the sequence of miRNA cDNA between the forward and reverse primers was used to specifically detect their expression (Figure 36.2).

Isolation of single cells

Here, as an example, we describe isolation of single embryonic stem cells (ES cells) for miRNA analysis, but the procedure can be adapted for the analysis of any other cell. ES colonies were trypsinized to obtain single cell suspension. These cells were re-suspended in 0.1% BSA in PBS.

Cell lysis and reverse transcription

Because each miRNA is present between dozens to several thousand copies in a single cell, any method requiring purification of RNAs from individual cells will result in a significant loss of miRNAs, and potentially introduce errors. We found that thermal lysis of cells at 95 °C for 5 min is sufficient to release miRNAs reliably and efficiently. Meanwhile, our stem-loop-structured primers are specific to mature miRNAs and resistant to miRNA precursors and primary transcripts, so it is not necessary to remove genomic DNAs and RNAs by purification and fractionation. With the use of glass capillaries to pick individual single cells, we transferred them into a RT reaction tube. The individual cells were lysed by heat treatment at 95 °C for 5 min. We then added the reverse transcriptase and other reagents and performed the reverse transcription reaction to convert mature miRNAs into corresponding cDNAs.

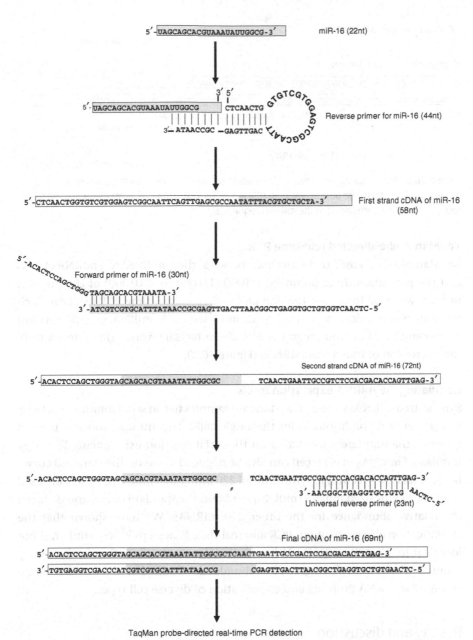

Figure 36.1. Schematic figure of the miRNA expression profiling assay, using miR-16 as an example. The letters with gray shading correspond to the miR-16 sequence.

Pre-PCR

For this step we used a mixture of 220 forward primers for each individual miRNA, and one universal reverse primer against the stem-loop part of the primers for reverse transcription, to amplify all miRNA cDNAs evenly by 18 PCR cycles (Figure 36.1). This protocol was optimized to ensure that the amplification did not introduce any significant bias, and it provides a good measure of the relative abundance for all miRNAs expressed at different levels.

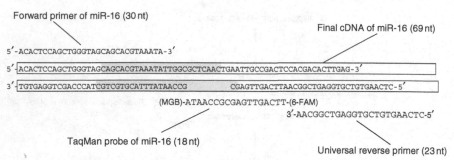

Forward primer of miR-16 (30 nt)

Final cDNA of miR-16 (69 nt)

5'-ACACTCCAGCTGGGTAGCAGCACGTAAATA-3'

5'-ACACTCCAGCTGGGTAGCAGCACGTAAATATTGGCGCTCAACTGAATTGCCGACTCCACGACACTTGAG-3'

3'-TGTGAGGTCGACCCATCGTCGTGCATTTATAACCG   CGAGTTGACTTAACGGCTGAGGTGCTGTGAACTC-5'

(MGB)-ATAACCGCGAGTTGACTT-(6-FAM)

3'-AACGGCTGAGGTGCTGTGAACTC-5'

TaqMan probe of miR-16 (18 nt)

Universal reverse primer (23 nt)

Figure 36.2. Schematic figure showing the relation between forward primer, universal reverse primer and TaqMan probe during real-time PCR reactions, using miR-16 as an example. The letters with gray shading correspond to the miR-16 sequence.

TaqMan probe-directed real-time PCR

TaqMan probe was used to discriminate between the amplified real miRNA cDNAs and the potential primer dimmers; 1/1000 (1/1000 to 1/10 000) of the pre-PCR product was used for a 10 μl TaqMan probe-directed real-time PCR reaction. Each reaction was duplicated or triplicated and used each miRNA specific forward primer and TaqMan probe, together with the universal reverse primer, to quantify the expression of individual miRNAs (Figure 36.2).

Calculating the miRNA expression levels

Synthesized miRNAs were used as standard samples to run serial dilution reactions under identical conditions as for the single cells. This information was used to generate the standard curve based on the serial dilution experiment. The copy number of miRNAs in each cell can also be deduced based on the standard curve. Because it is not practical to synthesize all 220 individual miRNAs, we usually used only one synthesized miRNA (miR-16) to obtain the standard curve, and deduced the relative abundance for the other 220 miRNAs. We have shown that the amplification efficiency of pre-PCR and real-time PCR steps is very similar, if not identical for all 220 miRNAs. Although this does not reflect the absolute copy number of each miRNA, their relative abundance obtained in this way is accurate enough for miRNA profiling and classification of diverse cell types.

Results and discussion

To prove the sensitivity of the assay, we ran miRNA expression profiling on individual mouse ES cells and individual mouse embryonic fibroblasts (MEFs) (Figure 36.3; Tang *et al.*, 2006a,b). The result showed that the assay clearly discriminated between miRNA profiles of ES cells compared with MEFs at the single cell level. To rigorously test the specificity of the assay for mature miRNAs, we ran the assay on Dicer knockout mouse embryonic fibroblasts (MEFs), which is known to result in the loss of all mature miRNAs and to promote the accumulation of miRNA primary transcripts and miRNA precursors (Cobb *et al.*, 2005; Kanellopoulou *et al.*, 2005; Muljo *et al.*, 2005; Murchison *et al.*, 2005; Yi *et al.*, 2006). We found that the assay for 202 (94%) miRNAs we tested showed negligible

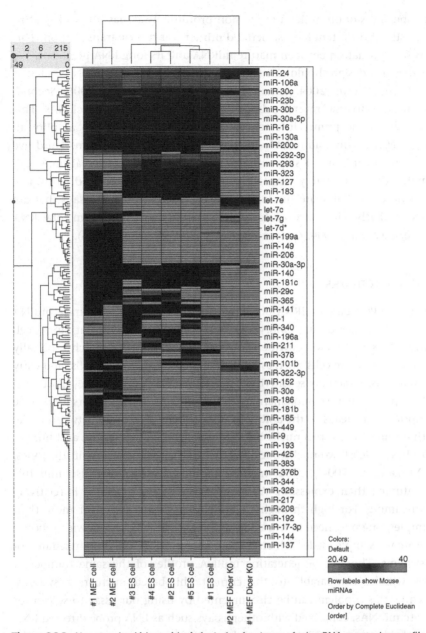

Figure 36.3. Unsupervised hierarchical clustering heatmap of microRNA expression profile for single embryonic stem cells (ES cells), single wildtype mouse embryonic fibroblasts (MEFs), and single Dicer knockout MEFs. The cluster heatmap was produced using expression levels (Ct value) of 214 miRNAs. Correlation coefficient was used as similarity measure and complete linkage method was used as clustering method. Note that higher Ct value means lower expression level. (See color plate 23)

signals compared with wildtype control MEFs (Figure 36.3; Tang *et al.*, 2006b). The remaining 13 (6%) miRNAs showed a background signal of more than 1% compared with the signal from wildtype control MEFs. This observation confirmed that our assay is highly specific for mature miRNAs and it works robustly without RNA purification and fractionation.

The robustness of our miRNA expression profiling assay has several key attributes. Firstly, it uses stem-loop-structured primers for reverse transcription. This permits discrimination between mature miRNA and its long RNA precursor. We expect that this design should also work for other classes of small RNAs, such as piRNAs (Vazquez *et al.*, 2004; Zamore and Haley, 2005; Carthew, 2006). Second, we can use a universal reverse primer for amplifying all miRNA cDNAs. This greatly reduces the primer interactions and makes it feasible to amplify all miRNA cDNAs evenly and reliably in the same initial PCR reaction. Third, we can combine the initial PCR step with the TaqMan probe-directed real-time PCR. This makes the assay highly sensitive, which has the potential to discriminate between miRNAs that share similar sequences. We expect that assays that are designed similarly will also work for other small RNAs at nanogram total RNA input (Vazquez *et al.*, 2004; Zamore and Haley, 2005; Carthew, 2006).

Concluding remarks

Our real-time PCR-based miRNA profiling assay has allowed us to trace miRNA expression dynamics during embryonic stem cell differentiation at single cell resolution. This method will provide a powerful tool for miRNA research, especially for the analysis of rare cells, such as adult stem cells, cancer stem cells, and early embryonic cells. To achieve sensitivity of detection at the single cell level, we used eight base pairs that are complementary to the miRNAs in the primers for reverse transcription. This leads to the ability to discriminate rigorously between molecules that differ by just two nucleotides. Most of the known mouse and human miRNAs have at least two-nucleotide difference in their sequences (Griffiths-Jones, 2004; Miska *et al.*, 2004), so our method should be sufficiently discriminative for quantifying their expression. Although this assay is robust, it is relatively time-consuming. For high throughput miRNAs profiling assays for more than 100 samples, one may need to use a 384-well real-time PCR system with robotics and automatic sample loader, which has similar throughput to chip-based microarray systems. The cost of generating miRNA profiles of the same number of samples will be comparable to the current chip-based microarray systems. Meanwhile, this problem can be circumvented by using our assay to screen for candidate miRNAs, combined with other assays such as LNA probe-directed RNA *in situ* hybridization. In this way, candidate miRNA expression dynamics can be analyzed in a wider range of samples. We expect and conclude that the single cell miRNA profiling assay will contribute significantly to decipher the functions of miRNAs for adult stem cells, cancer stem cells, and during early embryonic development.

REFERENCES

Alvarez-Garcia, I. and Miska, E. A. (2005). MicroRNA functions in animal development and human disease. *Development*, **132**, 4653–4662.

Ambros, V. (2004). The functions of animal microRNAs. *Nature*, **431**, 350–355.

Babak, T., Zhang, W., Morris, Q., Blencowe, B. J. and Hughes, T. R. (2004). Probing microRNAs with microarrays: tissue specificity and functional inference. *RNA*, **10**, 1813–1819.

Barad, O., Meiri, E., Avniel, A. *et al.* (2004). MicroRNA expression detected by oligonucleotide microarrays: system establishment and expression profiling in human tissues. *Genome Research*, **14**, 2486–2494.

Baskerville, S. and Bartel, D. P. (2005). Microarray profiling of microRNAs reveals frequent coexpression with neighboring miRNAs and host genes. *RNA*, **11**, 241–247.

Calin, G. A., Liu, C. G., Sevignani, C. *et al.* (2004). MicroRNA profiling reveals distinct signatures in B cell chronic lymphocytic leukemias. *Proceedings of the National Academy of Sciences USA*, **101**, 11 755–11 760.

Carthew, R. W. (2006). A new RNA dimension to genome control. *Science*, **313**, 305–306.

Castoldi, M., Schmidt, S., Benes, V. *et al.* (2006). A sensitive array for microRNA expression profiling (miChip) based on locked nucleic acids (LNA). *RNA*, **12**, 913–920.

Chen, C., Ridzon, D. A., Broomer, A. J. *et al.* (2005). Real-time quantification of microRNAs by stem-loop RT-PCR. *Nucleic Acids Research*, **33**, e179.

Cobb, B. S., Nesterova, T. B., Thompson, E. *et al.* (2005). T cell lineage choice and differentiation in the absence of the RNase III enzyme Dicer. *Journal of Experimental Medicine*, **201**, 1367–1373.

Griffiths-Jones, S. (2004). The microRNA Registry. *Nucleic Acids Research*, **32**, D109–111.

Harfe, B. D. (2005). MicroRNAs in vertebrate development. *Current Opinion in Genetics & Development*, **15**, 410–415.

Kanellopoulou, C., Muljo, S. A., Kung, A. L. *et al.* (2005). Dicer-deficient mouse embryonic stem cells are defective in differentiation and centromeric silencing. *Genes & Development*, **19**, 489–501.

Krichevsky, A. M., King, K. S., Donahue, C. P., Khrapko, K. and Kosik, K. S. (2003). A microRNA array reveals extensive regulation of microRNAs during brain development. *RNA*, **9**, 1274–1281.

Lao, K., Xu, N. L., Yeung, V. *et al.* (2006). Multiplexing RT-PCR for the detection of multiple miRNA species in small samples. *Biochemical Biophysical Research Communications*, **343**, 85–89.

Liang, R. Q., Li W., Li, Y. *et al.* (2005). An oligonucleotide microarray for microRNA expression analysis based on labeling RNA with quantum dot and nanogold probe. *Nucleic Acids Research*, **33**, e17.

Liu, C. G., Calin, G. A., Meloon, B. *et al.* (2004). An oligonucleotide microchip for genome-wide microRNA profiling in human and mouse tissues. *Proceedings of the National Academy of Sciences USA*, **101**, 9740–9744.

Lu, J., Getz, G., Miska, E. A. *et al.* (2005). MicroRNA expression profiles classify human cancers. *Nature* **435**, 834–838.

Miska, E. A., Alvarez-Saavedra, E., Townsend, M. *et al.* (2004). Microarray analysis of microRNA expression in the developing mammalian brain. *Genome Biology*, **5**, R68.

Muljo, S. A., Ansel, K. M., Kanellopoulou, C. *et al.* (2005). Aberrant T cell differentiation in the absence of Dicer. *Journal of Experimental Medicine*, **202**, 261–269.

Murchison, E. P., Partridge, J. F., Tam, O. H., Cheloufi, S. and Hannon, G. J. (2005). Characterization of Dicer-deficient murine embryonic stem cells. *Proceedings of the National Academy of Sciences USA*, **102**, 12 135–12 140.

Neely, L. A., Patel, S., Garver, J. *et al.* (2006). A single-molecule method for the quantitation of microRNA gene expression. *Nature Methods*, **3**, 41–46.

Nelson, P. T., Baldwin, D. A., Scearce, L. M. *et al.* (2004). Microarray-based, high-throughput gene expression profiling of microRNAs. *Nature Methods*, **1**, 155–161.

Plasterk, R. H. (2006). Micro RNAs in animal development. *Cell*, **124**, 877–881.

Sempere, L. F., Freemantle, S., Pitha-Rowe, I. *et al.* (2004). Expression profiling of mammalian microRNAs uncovers a subset of brain-expressed microRNAs with possible roles in murine and human neuronal differentiation. *Genome Biology*, **5**, R13.

Sioud, M. and Rosok, O. (2004). Profiling microRNA expression using sensitive cDNA probes and filter arrays. *Biotechniques*, **37**, 574–576, 578–580.

Sun, Y., Koo, S., White, N. *et al.* (2004). Development of a micro-array to detect human and mouse microRNAs and characterization of expression in human organs. *Nucleic Acids Research*, **32**, e188.

Tang, F., Hajkova, P., Barton, S. C., Lao, K. and Surani M. A. (2006a). MicroRNA expression profiling of single whole embryonic stem cells. *Nucleic Acids Research*, **34**, e9.

Tang, F., Hajkova, P., Barton, S. C. *et al.* (2006b). 220-plex microRNA expression profile of a single cell. *Nature Protocols*, **1**, 1154–1159.

Thomson, J. M., Parker, J., Perou, C. M. and Hammond, S. M. (2004). A custom microarray platform for analysis of microRNA gene expression. *Nature Methods*, **1**, 47–53.

Vazquez, F., Vaucheret, H., Rajagopalan, R. *et al.* (2004). Endogenous *trans*-acting siRNAs regulate the accumulation of *Arabidopsis* mRNAs. *Molecular Cell*, **16**, 69–79.

Yi, R., O'Carroll, D., Pasolli, H. A. *et al.* (2006). Morphogenesis in skin is governed by discrete sets of differentially expressed microRNAs. *Nature Genetics*, **38**, 356–362.

Zamore, P. D. and Haley, B. (2005). Ribo-gnome: the big world of small RNAs. *Science*, **309**, 1519–1524.

37 Piwi-interacting RNAs (piRNAs)

Ergin Beyret and Haifan Lin

Introduction

Small, non-coding RNAs of 18–32 nucleotides have emerged as evolutionarily conserved potent regulators of gene expression in the past decade. Studies of the function of small RNAs demonstrated their pivotal roles in various aspects of cell and developmental biology, as detailed in the previous chapters of this book. As the newest citizens of the small RNA world, Piwi-interacting RNAs (piRNAs) of mostly 26–32 nucleotides in length were discovered in 2006 in mammalian testes (Aravin *et al.*, 2006; Girard *et al.*, 2006; Grivna *et al.*, 2006a; Lau *et al.*, 2006; Watanabe *et al.*, 2006). They are so named because they interact with the Piwi sub-family proteins of the evolutionary conserved Argonaute/Piwi protein family. piRNAs also exist in large numbers in fly (Brennecke *et al.*, 2007; Gunawardane *et al.*, 2007) and fish gonads (Houwing *et al.*, 2007), implying the evolutionary conservation of their function. They differ from miRNAs and siRNAs in size, biogenesis, expression pattern, and possibly function. Although still remaining to be fully elucidated, clues about their biogenesis and function have started emerging. There are over 60 000 different species of piRNA identified so far, much exceeding the several hundreds of miRNAs that have been discovered. This fascinating complexity of piRNAs provides unprecedented opportunities for unraveling novel and diverse mechanisms of small RNA-mediated gene regulation. This chapter will summarize the latest progress on piRNAs.

Ago/Piwi protein family comprises two sub-families

Any description of small RNAs would be incomplete without an account of their protein partners: the Argonaute/Piwi (Ago/Piwi) family proteins. This highly conserved protein family was first discovered by the cloning of the *Piwi* gene in *Drosophila* and its homologs in *C. elegans* and humans, as well as by the identification of *Ago* and *Zwille* as *Piwi*-like genes in *Arabidopsis* (Cox *et al.*, 1998). Its members comprise the two signature domains: PAZ (standing for *Piwi Argonaute Zwille*) and PIWI domains, and are thus also known as PAZ–PIWI

MicroRNAs: From Basic Science to Disease Biology, ed. Krishnarao Appasani. Published by Cambridge University Press. © Cambridge University Press 2008.

Domain-containing proteins, or PPD proteins. Structural studies demonstrated that the PAZ domain comprises an oligonucleotide/oligosaccharide-binding (OB) fold whereas the PIWI domain resembles RNase H enzymes (Lingel *et al.*, 2003; Song *et al.*, 2003, 2004; Yan *et al.*, 2003; Ma *et al.*, 2004; Parker *et al.*, 2004). As implied by these structural analyses, PAZ domain interacts with small RNAs while PIWI domain possesses a RNA endonuclease activity (Song *et al.*, 2004; Miyoshi *et al.*, 2005; Saito *et al.*, 2006). Phylogenetic analysis of PPD proteins reveals a divergence of two sub-families based on their similarities to the *Drosophila* Piwi and *Arabidopsis* Ago proteins (Carmell *et al.*, 2002). Interestingly, although Ago proteins occur in diverse eukaryotes including ciliates, plants, fungi and animals, Piwi proteins seem to have been eliminated in plants and fungi (Mochizuki *et al.*, 2002).

Like Ago proteins, Piwi proteins are also involved in gene regulation. For instance, *Drosophila* Piwi has been shown to be involved in transcriptional gene silencing (Pal-Bhadra *et al.*, 2004). Normally, genes that are near heterochromatin are inactivated owing to heterochromatin spreading. This so-called 'positional effect variegation' is suppressed in the absence of Piwi. This deficiency in silencing is correlated with the lack of heterochromatinazion of the genes as assessed with decreased levels of heterochromatic modifications of histones and epigenetic repressor proteins such as heterochromatin protein 1 (HP1). In addition to transcriptional silencing, Piwi proteins might be involved in post-transcriptional silencing since Piwi (Saito *et al.*, 2006) and its rat homolog Riwi (Lau *et al.*, 2006) have the "slicer" activity. Moreover, mouse Piwi homologs, Miwi (Grivna *et al.*, 2006b) and Mili (Y. Unhavaithaya and H. Lin, unpublished), are cytoplasmic and interact with the translational machinery and mRNAs. As they have been shown to be necessary for the progression of spermatogenesis, this interaction may be necessary for the translational regulation of spermatogenesis-related genes.

Functions of Piwi proteins in *Drosophila*

Piwi proteins are predominantly detected in gonads and largely associated with stem cell maintenance, unlike the Ago members, which appear to be more ubiquitously expressed. As anticipated, Piwi proteins are necessary for germline development and the progression of gametogenesis; loss of their function leads to infertility. In addition, Piwi also has crucial functions in the soma. For instance, during *Drosophila* development, Piwi is required first as a maternal component for embryogenesis and the establishment of the germline lineage, and then as a zygotic factor for epigenetic regulation, genome stability, the maintenance of germline stem cells (GSC) and subsequent gametogenic events in both males and females (Lin and Spradling, 1997; Cox *et al.*, 2000; Pal-Bhadra *et al.*, 2004; Kalmykova *et al.*, 2005; Megosh *et al.*, 2006; Pelisson *et al.*, 2007). In early *Drosophila* embryos, maternal Piwi is a cytoplasmic protein present throughout the entire embryo and enriched in the germplasm as a component of a germline-specific organelle called the polar body (Megosh *et al.*, 2006). Depleting maternal Piwi leads to failure in establishing primordial germ cells (PGCs), yet doubling and tripling the dosage of Piwi proportionally increases the number of PGCs. After PGC formation, Piwi appears to relocate from cytoplasm to the nucleus, where it is required for the normal development of PGCs during embryogenesis (D. Cox, Ph.D. thesis). Subsequent to embryogenesis, Piwi is retained as a nuclear protein in both germline and somatic cells to exert germline and somatic functions (Lin and Spradling, 1997; Cox et al., 1998, 2000; Pal-Bhadra et al., 2004). Specifically, maternal Piwi is required for embryogenesis; depleting maternal Piwi leads to embryonic lethality, a fully penetrant phenotype due to various mitotic and morphogenetic defects (Cox et al., 1998).

During post-embryonic development, Piwi represses transposition in the genome and functions as an epigenetic regulator (Pal-Bhadra et al., 2004; Kalmykova et al., 2005; Pelisson et al., 2007). In the ovary of *piwi* mutants, GSCs enter oogenesis without self-renewal (Cox et al., 1998). In addition, the subsequent step of oogenesis in the mutants is often aberrant (Lin and Spradling, 1997), implicating a post-GSC function of Piwi during oogenesis. Interestingly, the function of Piwi for GSC maintenance resides in somatic niche cells: removing Piwi from the germline does not affect GSC self-renewal (Cox et al., 1998), yet specific expression of Piwi in somatic niche cells in *piwi* mutant rescues the GSC phenotype (Szakmary et al., 2005). Furthermore, overexpressing Piwi in somatic cells expends the GSC niche and leads to significant increase in the number of GSCs (Cox et al., 2000). In addition to somatic function, Piwi acts in the germline to promote GSC division (Cox et al., 2000), to ensure normal progression of subsequent steps of oogenesis (Lin and Spradling, 1997), and to supply maternal Piwi for future embryogenesis (Cox et al., 1998).

Functions of Piwi proteins in other organisms

The function of Piwi proteins in other organisms have mostly been characterized in the germline. In *C. elegans*, RNAi depletion of two of the Piwi homologs, *prg-1* and *prg-2*, results in a reduced germline mitotic proliferation zone, which is the *C. elegans* equivalent of GSCs (Cox et al., 1998). Three close homologs of Piwi in mouse are necessary for the progression of spermatogenesis (Deng and Lin, 2002; Kuramochi-Miyagawa et al., 2004; Carmell et al., 2007). Among them, Mili and possibly Miwi2 are expressed in the early germ cells from spermatogonial stem cells to the mid-pachytene stage of spermatogenesis (Kuramochi-Miyagawa et al., 2004; Carmell et al., 2007). Mili has been shown to be expressed also in PGCs. Miwi, the third of the four Piwi homologs in mouse, is expressed from mid-pachytene stage spermatocytes to round spermatids (Deng and Lin, 2002), complementary to the expression profiles of Mili and Miwi2 during spermatogenesis. Consistent with its expression pattern, in the absence of Miwi, germ cells complete meiosis but are uniformly arrested at the beginning of round spermatid development, signifying its role as a key regulator of spermiogenesis (Deng and Lin, 2002). Both Miwi and Mili are cytoplasmic proteins detected only in spermatogenic cells, indicating the germline-autonomy of their function. In contrast, Miwi2 is expressed in both the germline and Sertoli cells, which are somatic supporting cells within seminiferous tubules. However, the somatic expression of Miwi2 is not necessary for germline function (Carmell *et al.*, 2007).

Unlike *Drosophila* Piwi proteins, the mutant phenotypes of mouse Piwi proteins are male-specific, with spermatogenesis blocked at specific stage, leading to complete male-only infertility. Studies on the Piwi homologs in other organisms such as zebrafish, jellyfish and planaria, likewise reflect their involvement in an evolutionarily conserved germline function, some potentially at the stem cell level (Tan *et al.*, 2002; Seipel *et al.*, 2004; Reddien *et al.*, 2005; Rossi *et al.*, 2006). In human, there is no report of loss-of-function phenotypes for the Piwi proteins. However, with regard to the dosage effect, the human homologs Hiwi (Qiao *et al.*, 2002) and Hili (Lee *et al.*, 2006) have been shown to be correlated with testicular seminomas, although it is not clear whether the elevated levels of these proteins are the cause or the result of oncogenesis.

The somatic functions of Piwi proteins in higher organisms have been inferred from their expression in multiple somatic tissues (Sharma *et al.*, 2001; Lee *et al.*, 2006) and from the correlation between Hiwi overexpression and gastric malignancies (Liu *et al.*, 2006). Little is known beyond these studies. In contrast, a few studies have definitively demonstrated the somatic function of Ago proteins. For instance, mouse *ago2* shows embryonic lethality, reflecting the essential role of RNAi pathway during embryogenesis (Liu *et al.*, 2004). In addition to these phylogenetic, cellular and phenotypic variations, a clear distinction between the Piwi and Ago proteins is perhaps their differential ability to bind to piRNAs and si/miRNAs, respectively.

Discovery of piRNAs

PiRNAs arrived on the scene in the small RNA world with a grand entrance when five groups of researchers independently discovered them in the summer of 2006 (Aravin *et al.*, 2006; Girard *et al.*, 2006; Grivna *et al.*, 2006a; Lau *et al.*, 2006; Watanabe *et al.*, 2006). Three groups independently searched for small RNAs associated with Piwi proteins by co-immunoprecitating small RNAs associated with Miwi and Mili in mouse testis, where these Piwi proteins are abundantly expressed (Aravin *et al.*, 2006; Girard *et al.*, 2006; Grivna *et al.*, 2006a). They co-immunoprecipitated an abundant small RNA species of approximately 26–31 nt, but not miRNAs. These Miwi- and Mili-associated small RNAs display a median length of 30 nt and 26 nt, respectively. As expected, two equivalent populations of piRNAs are also present in human testis (Aravin *et al.*, 2006; Girard *et al.*, 2006), implicating their likely interaction with the counterparts of Miwi and Mili in humans, Hiwi and Hili, respectively. Two other groups, Watanabe *et al.* (2006) and Lau *et al.* (2006) reached the same conclusion in pursuit of the small RNA profiles of mouse and rat germlines, respectively. Lau *et al.* (2006) observed that these small RNAs, unlike miRNAs, co-fractionated with the Rat Piwi, Riwi. Because of their association with Piwi proteins, these small RNAs are named *Piwi*-interacting RNAs (piRNAs).

PiRNAs differ from miRNAs not only in size and abundance, but also in tissue preference of expression and complexity. PiRNAs are only detectable in the testis, and become especially highly abundant after the pachytene stage of spermatogenesis (Aravin *et al.*, 2006; Girard *et al.*, 2006; Grivna *et al.*, 2006a; Watanabe *et al.*, 2006). At present, at least 55 000 species of piRNA have been identifed to be associated with Miwi and/or Mili. The total number of piRNAs is expected to significantly exceed the known number, because many piRNAs are only represented by a single sequence hit. The number of piRNAs, at least two orders of magnitude greater than that of miRNAs, implies the potential complexity and diversity of their function.

PiRNAs have also been identified in *Drosophila* (Brennecke *et al.*, 2007; Gunawardane *et al.*, 2007), zebrafish (Houwing *et al.*, 2007), planaria (personal communication) and a ciliate protozoan, *Tetrahymena thermophila* (Mochizuki *et al.*, 2002; Mochizuki and Gorovsky, 2004a), indicating their conservation during evolution and their biological significance. In *Drosophila*, each of the three Piwi proteins, Piwi, Ago3 and Aub, interacts with a distinct population of piRNAs, just like Miwi and Mili. Approximately 75% of the *Drosophila* piRNAs are derived from the highly repetitive sequences in the genome; thus a small number of them were initially identified by some as *repeat*-associated siRNAs (rasiRNAs) (Vagin *et al.*, 2006). This is in contrast to the fact that only 15% of mammalian piRNAs are derived from the repetitive sequences despite approximately 40% of the mammalian genome being annotated as repeats.

Genomic mapping of the piRNA sequences showed that they are largely derived in clusters from the genome. Although individual piRNA sequences do not seem to be evolutionary conserved, unlike miRNAs, the piRNA clusters themselves show synteny among mouse, rat and human genomes. Remarkably, most piRNAs correspond to intergenic regions previously thought to be untranscribed and thus referred as "junk" DNA. These peculiar features of piRNAs raised several questions. How are they produced and regulated? Are these piRNAs merely "junk" RNAs or do they have any functions? If so, what are their functions? The following sections summarize current progress towards addressing these questions.

Biogenesis of piRNAs

The biogenesis of piRNAs appears to be distinct from that of siRNAs and miRNAs. Similar to miRNAs and siRNAs, piRNAs show a high bias for Uracil at their 5′ ends. This echoes the products of the RNase III enzymes, such as *Drosha* and *Dicer*. However, unlike miRNAs and siRNAs, mammalian piRNAs appear to be derived from single-stranded precursors, based on the fact that most, if not all, piRNAs within a cluster are derived from only one of the strands in an overlapping fashion, with a head-to-tail homology (Figure 37.1). Frequently, piRNAs in one half of the cluster correspond to one of the DNA strands and those in the other half correspond to the other strand with an opposite transcriptional direction, as if the two halves represent two divergently transcribed precursors from a central promoter region (Figure 37.1). In support of the existence of large single-stranded piRNA precursors, a large number of long ESTs without any known function have been identified in mouse testis. One such EST corresponds to one of the biggest clusters, which resides on chromosome 17 (Yin, Beyret, and Lin, unpublished).

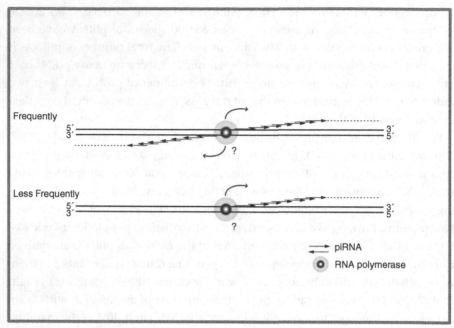

Frequently

Less Frequently

→← piRNA
⊙ RNA polymerase

Figure 37.1. Mammalian piRNA clusters. The majority of mammalian piRNA clusters seem to be under the control of a bidirectional promoter region. Unidirectional clusters are observed at a lower frequency. Individual piRNAs within a cluster are derived from only one of the DNA strands in the same direction, with a head-to-tail homology. These observations indicate the possible existence of long, single-stranded primary transcripts that are processed to form individual piRNAs.

Computational prediction for piRNA precursors so far has not revealed any configuration similar to miRNA precursors. No common secondary structure motif among piRNA clusters has been found, either. These bioinformatic results have been augmented with the experimental observation that *Drosophila Dicer1* or *Dicer2* loss-of-function mutants do not show any reduction in piRNA/rasiRNA levels (Vagin *et al.*, 2006). Although this does not exclude the possibility of redundancy between the two *Dicer* paralogs, a more convincing line of evidence came from zebrafish. Houwing *et al.* (2007) observed that zygotic *Dicer* is not necessary for piRNA biogenesis in zebrafish. However, in this case, the maternally loaded *Dicer* may be involved in piRNA production. Intriguingly, Grivna *et al.* (2006), and Aravin *et al.* (2007) showed that Miwi and Mili piRNAs are highly down-regulated in the Miwi and Mili-deficient mice, respectively, whereas miRNAs are unaffected. The same has been observed for at least one piRNA partner of Ziwi (Houwing *et al.*, 2007). Given that Piwi proteins possess RNA cleavage activities (Saito *et al.*, 2006; Gunawardane *et al.*, 2007) it is likely that Piwi proteins are the enzymes involved in cleaving the precursors into piRNAs. In addition, they may bind to mature piRNAs to form an effector complex. This interaction may be necessary for the stabilization of piRNAs, which might otherwise be rapidly degraded by cellular nucleases in the absence of their partner proteins.

In *Drosophila*, the biogenesis of certain transposon-derived piRNAs may involve an 'amplification mechanism' that accelerates the production of piRNAs from their precursors, as recently proposed (independently by Brennecke *et al.*, 2007; Gunawardane *et al.*, 2007) based on their bioinformatic analyses. Among three

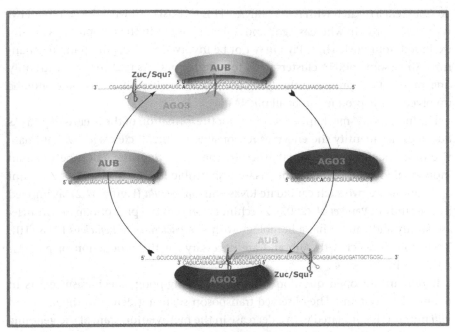

Figure 37.2. The "ping-pong" model for piRNA biogenesis and amplification. An Aub piRNA guides Aub to target Ago3 piRNA precursor with a complementary sequence. The precursor is then cleaved by Aub to produce a matching Ago3 piRNA, which can act on the Aub piRNA precursor to give rise to an Aub piRNA, initiating a positive feedback loop. Aub and Ago3 proteins of the Piwi sub-family in *Drosophila* exhibit such a complementarity on their 5′ ends for their first ten nucleotides. Shown are hypothetical sequences for precursors. The uracil bias for the first nucleotide of Aub piRNAs and the adenine bias for the tenth nucleotide of Ago3 piRNAs are highlighted.

Drosophila Piwi proteins, Aub and Piwi bind to piRNAs derived from the anti-sense strand of transposons or their derivatives. Ago3, on the other hand, associates with sense piRNAs. Although Aub and Piwi piRNAs show the 5′ uracil bias, Ago3 piRNAs do not display such a propensity. Instead, they frequently have adenine on their tenth nucleotide position. In particular, Aub piRNAs and Ago3 piRNAs that are derived from the same clusters are fully complementary to each other for the first ten nucleotides at their 5′ ends. This observation raised the possibility of their production in a 'slicer'-based mechanism, since the 'slicer' activity corresponds to the tenth nucleotide position of the small RNA in cleaving RISCs (Figure 37.2). In this model, Ago3 and Aub reciprocally act on the production of each other's piRNA partners. For instance, an Aub piRNA guides its partner protein Aub to the matching sequence in an Ago3 piRNA precursor. The 'slicer' activity of Aub incises the precursor, forming the 5′ end of the corresponding Ago3 piRNA. Upon the formation of the 3′ end by a yet to be identified enzyme, the resulting Ago3 piRNA is loaded onto Ago3, which in turn targets the Aub piRNA precursor and produces more Aub piRNAs. This positive feed-back loop, the so-called 'ping-pong mechanism', amplifies the piRNA pool from a minute amount of pre-existing mature piRNAs, as long as the matching precursors exist.

Consistent with this amplification model, matching piRNAs are indeed present at a higher abundance in the *Drosophila* piRNA library. Interestingly, Piwi piRNAs do not seem to match with Ago3 piRNAs. This is consistent with the fact that Piwi is a nuclear protein whereas Ago3 and Aub are present in the cytoplasm, spatially excluded from Piwi. Thus, Piwi may not be involved in the 'ping-pong mechanism'. Since some piRNA clusters seem to produce piRNAs that interact with only one of the Piwi proteins, the 'ping-pong' mechanism, if exists, may not be involved in the production of all piRNAs.

The 'ping-pong' mechanism accounts for the formation of the 5' ends of piRNAs but does not identify the enzymes responsible for the 3' cleavage. The ultimate size of a piRNA might be established by means of the interaction between an individual Piwi protein and the RNase responsible for the 3' cleavage. Zucchini and Squash are two such candidate RNases in *Drosophila* (Figure 37.2), as implied by the study of Pane *et al.* (2007). Zucchini belongs to the phospholipase-D/nuclease family and Squash has a homology with *Agrobacterium tumefaciens* RNase HII. They both interact with Aub and are necessary for the production of piRNAs/rasiRNAs in *Drosophila*.

It remains an open question whether the 'ping-pong' mechanism exists in mammalian systems. The observed transposon silencing defect in the *miwi* and *mili* mutants is correlated with a decrease in the methylation state of the genomic sequences, implying a transcriptional silencing mechanism. However, the formation of the transcript is necessary for the positive feedback loop in the 'ping-pong' model, which states a post-transcriptional silencing mode for transposons. Inhibition of the transcription by Piwi proteins should paralyze the loop.

A peculiar feature of piRNAs is that they are 2'-O-methylated at their 3' ends, similar to miRNAs in plants but different from animal miRNAs (Kirino and Mourelatos, 2007; Ohara *et al.*, 2007). The mechanism and functional significance of this methylation is not known, yet. Because the 2'-O-methylation at the 3' ends of plant miRNAs, processed by a methyl transferase called Hen1, is thought to confer stability against nucleases (Li *et al.*, 2005; Yu *et al.*, 2005; Yang *et al.*, 2006) such modification for piRNAs in animals may serve the same function. Alternatively, it might be a protection mechanism against RNA editing or simply a tag for the recognition by Piwi proteins but not Ago in animals. At present, a Hen1-like enzyme in charge of piRNA methylation remains unidentified.

Biological function of piRNAs

It is tempting to speculate that piRNAs might have diverse types of functions since they correspond to exons, introns, intergenic unique sequences, repetitive sequences, and transposable elements in the genome. PiRNAs are largely detected in gonads, implicating their function in gametogenesis. In mouse testes, piRNAs are mostly concentrated in the post-pachytene stage spermatocytes, and sharply downregulated in late spermatids. This may be indicative of their potential function during meiosis and early spermiogenesis. Such predicted function is supported by the roles of Mili and Miwi during spermatogenesis. Since piRNAs are not detected in epididymide, where mature sperm are stored, they are unlikely to be paternal factors for embryogenesis. Interestingly, no piRNA has been detected in mammalian ovary, or prophase II – metaphase II mature oocytes (Watanabe *et al.*, 2006). Watanabe *et al.* were able to clone some small RNAs matching to retro-transposonic sequences in oocytes. However, these small RNAs, to which they refer as germline-specific RNA (*gs*RNA), are 21.9 ± 1.3 nucleotides in length and smaller than their testicular gsRNAs, which are 25.5 ± 4.53 nucleotides in length and thus likely to be piRNAs. The apparent absence of piRNAs in mouse ovaries echoes the fact that mammalian Piwi proteins are expressed in testis but not ovary, and their mutations affect only spermatogenesis but not oogenesis. In contrast, the zebrafish Piwi (Ziwi) and the three *Drosophila* Piwi proteins are expressed in the female germline as well, with the mutants displaying defects in both male and female germlines.

Function of piRNAs in transposon silencing

A likely molecular activity of piRNAs in gametogenesis is transposon silencing, as indicated by studies in *Drosophila* (Vagin *et al.*, 2006; Brennecke *et al.*, 2007). The fact that most piRNA sequences in *Drosophila* match transposons suggests transposon silencing in the germline. Indeed, one of the piRNA clusters identified by Brennecke *et al.* (2007) corresponded to the well-known *flamenco* locus. This locus has been shown to be necessary for the silencing of the retrotransposons *gypsy*, *idefix* and *ZAM* in follicle cells in the fly ovary (Pelisson *et al.*, 1994; Prud'homme *et al.*, 1995; Desset *et al.*, 2003; Mevel-Ninio *et al.*, 2007). It consists of truncated and thus inactive versions of the transposons. PiRNAs derived from this cluster interact specifically with Piwi, and Piwi has been shown to be necessary for the silencing of some transposons including *gypsy* elements (Sarot *et al.*, 2004). Just like Ago proteins utilize miRNAs and siRNAs to target transcripts, Piwi proteins may use piRNAs to target the RNA intermediates of retrotransposons in the germline, cleaving them and ultimately leading to their degradation. Indeed, the level of *gypsy* RNA is elevated in the absence of Piwi but not Aub. Moreover, P-element induced mutagenesis in the *flamenco* locus resulted in *gypsy* RNA upregulation and a decrease in piRNA levels specifically derived from this locus. This provides correlative evidence for a role of gypsy piRNAs as partners of Piwi in silencing this transposon. Although the mode of this Piwi/piRNA mediated silencing still remains elusive, the "ping-pong" model proposed for the piRNA biogenesis in the fly is suitable for post-transcriptional silencing of trans-posons as well (Figure 37.2 and 37.3). PiRNAs that are antisense to transposons may guide Piwi or Aub to the retrotransposon transcripts. The "slicer" activities of these Piwi proteins may then cleave the transposon transcripts, forming the 5' end of another piRNA, which can potentially interact with Ago3. Hence, Piwi proteins may have a dual function in the "ping-pong" model: they are necessary for the production of piRNAs and for the post-transcriptional silencing of retrotransposons.

Although most of the fly piRNAs and approximately one third of zebrafish piRNAs correspond to repeat annotated sequences, only a minor fraction of mammalian piRNAs seem to be derived from repetitive regions. Although the models proposed for the fly piRNA biogenesis and function may be applicable for this fraction, they do not respond to the majority of mammalian piRNAs. It is

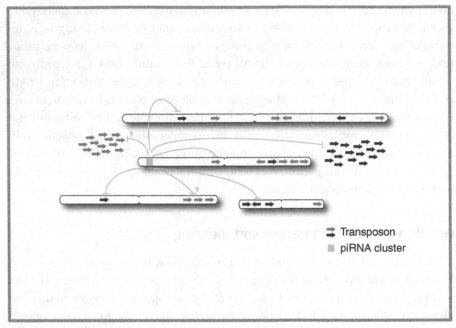

Figure 37.3. PiRNA-mediated transposon silencing in the germline. PiRNAs with complementary sequences to retrotransposons can guide Piwi proteins to target the corresponding matching transposons and lead to their silencing in the germline. This surveillance can be achieved at the transcriptional level by the epigenetic silencing function of Piwi as a chromatin binding protein or at the post-transcriptional level via the 'slicer' activities of Aub, Ago3, and possibly, Piwi.

possible that mammals may have undergone a decline in the piRNA-mediated transposon silencing while developing in some other aspects of piRNA-mediated functions. This adaptation may be an outcome of the elimination of all but a few major lines of mobile elements in the mammalian genome during evolution.

Function of piRNAs in transcriptional repression

In addition to transposon silencing via a post-trancriptional mechanism, piRNAs may be involved in transcriptional repression (Figure 37.3). Because Piwi promotes heterochromatinitization (Pal-Bhadra *et al.*, 2004), it is possible that it may be guided to the genomic sequences and/or nascent transcripts by piRNAs, leading to the transcriptional silencing of the targeted loci. In support of this, in the absence of Mili (Aravin *et al.*, 2007) or Miwi2 (Carmell *et al.*, 2007), mouse testis shows de-methylation of Line 1 elements. Correspondingly, Line1 and IAP type retro-transposons are up-regulated in the *Mili* or *Miwi2* knockout mutants . Zebrafish stands between the fly and mammals in terms of the representation of repeat-derived piRNAs. In this organism, approximately 34% of ovarian and 21% testicular piRNAs are derived from transposonic sequences. Among these piRNAs, approximately 60% of them correspond to antisense LTR elements including GypsyDR1 and GypsyDR2 elements both in ovary and testis, reinforcing the observations in mouse and fly. However, the effect of Ziwi on these transposons can not be assessed, as the germ cells are lost in the absence of Ziwi.

Function of piRNAs in DNA recombination/rearrangement

This potential role is largely suggested by the study of Lau *et al.* (2006), who observed that Riwi co-fractionated with RecQ DNA helicase. RecQ DNA helicases, capable of unwinding double-stranded DNAs, participate in a number of cellular processes including DNA replication, recombination and repair pathways. This observation supports the view of piRNA-guided recombination during meiosis. In this regard, it is important to identify the sub-cellular localization of Riwi. In addition to the Riwi study, earlier work in *Tetrahymena thermophila* have also suggested the role of *Tetrahymena* Piwi (Twi1P) and *scan* RNA (scnRNA) in eliminating particular DNA sequences in the newly formed macronucleus after conjugation (Mochizuki *et al.*, 2002; Mochizuki and Gorovsky, 2004a, b; Lee and Collins, 2006). It is possible that the scnRNAs are actually *Tetrahymena* piRNAs, which associate with Twi1P.

Function of piRNAs in translational regulation and mRNA stability

A subset of piRNAs may be involved in translational regulation and/or the control of mRNA stability, as largely implied by studies on Miwi and Mili. Both of these proteins are cytoplasmic and enriched in chromatoid bodies (Deng and Lin, 2002; Kuramochi-Miyagawa *et al.*, 2004; Grivna *et al.*, 2006b; Kotaja *et al.*, 2006a, b). Moreover, they interact with polyA-tailed RNAs. In *miwi* mutants, Miwi target mRNAs are drastically down-regulated (Deng and Lin, 2002), indicating that Miwi positively regulates the stability of its target mRNAs. This function of Miwi and Mili may involve their piRNAs. Indeed, Grivna *et al.* (2006b) demonstrated that a sub-population of piRNAs is co-fractionated with the translational machinery and that Miwi interacts with both these piRNAs and the translational machinery. Similar association has been observed for Mili as well (Unhavaithaya and Lin, unpublished data). Furthermore, Piwi (Megosh *et al.*, 2006), Miwi (Grivna *et al.*, 2006b), Mili (Kuramochi-Miyagawa *et al.*, 2004) and Ziwi (Tan *et al.*, 2002) all interact with Vasa or its homologs. Vasa is an ATP-dependent RNA helicase of the DEAD-box family specifically expressed in germ cells and implicated in translational regulation via interaction with dIF2, the *Drosophila* homolog of the translational initiation factor eIF5B (Carrera *et al.*, 2000). Because of these properties, it has been proposed to promote the translation of germline-specific RNAs. Lastly, the fly Piwi homolog Aub has been shown to regulate translation (Wilson *et al.*, 1996). These observations suggest that a sub-population of piRNAs and their protein partners may regulate translation and mRNA stability.

The mRNA stability control, either by means of piRNAs or not, may occur in the chromatoid body, a perinuclear structure known to contain GW182, *Mouse Vasa Homolog* (MVH), Dicer, miRNAs, and mRNAs as well as Miwi and Mili. Miwi and Mili both interact with Mael, a novel protein with unknown function, which is also localized to the chromatoid body in mouse and the chromatoid body-like structure, nuage, in the *Drosophila* germline (Findley *et al.*, 2003; Costa *et al.*, 2006). In addition to the Piwi proteins, Mael associates with the chromatin remodeler SNF5 and chromatin-associated protein SIN3B. The function of the chromatoid body is a mystery; however, it is tempting to speculate its role in mRNA storage or processing, given the known role of its analogous structure, the P body, in these processes in somatic cells. The chromatoid body forms in early spermatocytes as granules, which then fuse into one conspicuous sphere in round spermatids. It is a highly dynamic structure, which moves around the nuclear envelope and transiently associates with nuclear pore and Golgi complexes. Moreover, it has been shown to transit through the cytoplasmic bridges between early spermatids in rat, suggestive of a role as a cargo carrier between germ cells (Ventela *et al.*, 2003).

In addition to their role in gametogenesis, the maternal expression of Piwi proteins in zebrafish and *Drosophila* indicate that piRNAs might have maternal effect functions. Ziwi and at least one of its piRNA partners are maternally loaded (Houwing *et al.*, 2007), just like the *Drosophila* Piwi proteins (Brennecke *et al.*, 2007; Gunawardane *et al.*, 2007). It is certainly possible that the maternal role of Ziwi may be piRNA-independent, since Ziwi piRNAs are not observed in oogonia and PGCs even though Ziwi is present in these cells. Currently, it is not clear whether the discrepancy between the expression pattern of Piwi proteins and that of their piRNAs is due to functional divergence or simply a matter of sensitivity of the detection methods.

Concluding remarks

The field of piRNA research is progressing at an unprecedented rate. More insight about piRNA biogenesis and function will come from additional sequencing studies, cytological analysis and characterization of their partner Piwi proteins. For instance, what are the subcellular locations of specific piRNAs? What are the initiating piRNAs and other RNases involved in the 'ping-pong' model? Can this model be applicable to the mammalian systems? How are those piRNAs, whose biogenesis does not fit into this model, produced? Are there piRNAs present in somatic tissues? What are the consequences of perturbing the expression of specific piRNAs? An ultimate question is how a piRNA interacts with a Piwi protein and other factors to achieve a specific regulatory function. Answers to these questions should significantly advance our understanding of piRNA biogenesis and function as well as their potential implications in disease biology.

REFERENCES

Aravin, A., Gaidatzis, D., Pfeffer, S. *et al.* (2006). A novel class of small RNAs bind to MILI protein in mouse testes. *Nature*, **442**, 203–207.

Aravin, A. A., Sachidanandam, R., Girard, A., Fejes-Toth, K. and Hannon, G. J. (2007). Developmentally regulated piRNA clusters implicate MILI in transposon control. *Science*, **316**, 744–747.

Brennecke, J., Aravin, A. A., Stark, A. *et al.* (2007). Discrete small RNA-generating loci as master regulators of transposon activity in *Drosophila*. *Cell*, **128**, 1089–1103.

Carmell, M. A., Girard, A., van de Kant, H. J. *et al.* (2007). MIWI2 is essential for spermatogenesis and repression of transposons in the mouse male germline. *Developmental Cell*, **12**, 503–514.

Carmell, M. A., Xuan, Z., Zhang, M. Q. and Hannon, G. J. (2002). The Argonaute family: tentacles that reach into RNAi, developmental control, stem cell maintenance, and tumorigenesis. *Genes & Development*, **16**, 2733–2742.

Carrera, P., Johnstone, O., Nakamura, A. *et al.* (2000). VASA mediates translation through interaction with a Drosophila yIF2 homolog. *Molecular Cell*, **5**, 181–187.

Costa, Y., Speed, R. M., Gautier, P. *et al.* (2006). Mouse MAELSTROM: the link between meiotic silencing of unsynapsed chromatin and microRNA pathway? *Human Molecular Genetics*, **15**, 2324–2334.

Cox, D. N., Chao, A., Baker, J. *et al.* (1998). A novel class of evolutionarily conserved genes defined by piwi are essential for stem cell self-renewal. *Genes & Development*, **12**, 3715–3727.

Cox, D. N., Chao, A. and Lin, H. (2000). piwi encodes a nucleoplasmic factor whose activity modulates the number and division rate of germline stem cells. *Development*, **127**, 503–514.

Deng, W. and Lin, H. (2002). miwi, a murine homolog of piwi, encodes a cytoplasmic protein essential for spermatogenesis. *Developmental Cell*, **2**, 819–830.

Desset, S., Meignin, C., Dastugue, B. and Vaury, C. (2003). COM, a heterochromatic locus governing the control of independent endogenous retroviruses from Drosophila melanogaster. *Genetics*, **164**, 501–509.

Findley, S. D., Tamanaha, M., Clegg, N. J. and Ruohola-Baker, H. (2003). Maelstrom, a Drosophila spindle-class gene, encodes a protein that colocalizes with Vasa and RDE1/AGO1 homolog, Aubergine, in nuage. *Development*, **130**, 859–871.

Girard, A., Sachidanandam, R., Hannon, G. J. and Carmell, M. A. (2006). A germline-specific class of small RNAs binds mammalian Piwi proteins. *Nature*, **442**, 199–202.

Grivna, S. T., Beyret, E., Wang, Z. and Lin, H. (2006a). A novel class of small RNAs in mouse spermatogenic cells. *Genes & Development*, **20**, 1709–1714.

Grivna, S. T., Pyhtila, B. and Lin, H. (2006b). MIWI associates with translational machinery and PIWI-interacting RNAs (piRNAs) in regulating spermatogenesis. *Proceedings of the National Academy of Sciences USA*, **103**, 13415–13420.

Gunawardane, L. S., Saito, K., Nishida, K. M. *et al.* (2007). A slicer-mediated mechanism for repeat-associated siRNA 5' end formation in Drosophila. *Science*, **315**, 1587–1590.

Houwing, S., Kamminga, L. M., Berezikov, E. *et al.* (2007). A role for Piwi and piRNAs in germ cell maintenance and transposon silencing in Zebrafish. *Cell*, **129**, 69–82.

Kalmykova, A. I., Klenov, M. S. and Gvozdev, V. A. (2005). Argonaute protein PIWI controls mobilization of retrotransposons in the *Drosophila* male germline. *Nucleic Acids Research*, **33**, 2052–2059.

Kirino, Y. and Mourelatos, Z. (2007). Mouse Piwi-interacting RNAs are 2'-O-methylated at their 3' termini. *Nature Structural & Molecular Biology*, **14**, 347–348.

Kotaja, N., Bhattacharyya, S. N., Jaskiewicz, L. *et al.* (2006a). The chromatoid body of male germ cells: similarity with processing bodies and presence of Dicer and microRNA pathway components. *Proceedings of the National Academy of Sciences USA*, **103**, 2647–2652.

Kotaja, N., Lin, H., Parvinen, M. and Sassone-Corsi, P. (2006b). Interplay of PIWI/Argonaute protein MIWI and kinesin KIF17b in chromatoid bodies of male germ cells. *Journal of Cell Science*, **119**, 2819–2825.

Kuramochi-Miyagawa, S., Kimura, T., Ijiri, T. W. *et al.* (2004). Mili, a mammalian member of piwi family gene, is essential for spermatogenesis. *Development*, **131**, 839–849.

Lau, N. C., Seto, A. G., Kim, J. *et al.* (2006). Characterization of the piRNA complex from rat testes. *Science*, **313**, 363–367.

Lee, J. H., Schutte, D., Wulf, G. *et al.* (2006). Stem-cell protein Piwil2 is widely expressed in tumors and inhibits apoptosis through activation of Stat3/Bcl-XL pathway. *Human Molecular Genetics*, **15**, 201–211.

Lee, S. R. and Collins, K. (2006). Two classes of endogenous small RNAs in *Tetrahymena thermophila*. *Genes & Development*, **20**, 28–33.

Li, J., Yang, Z., Yu, B., Liu, J. and Chen, X. (2005). Methylation protects miRNAs and siRNAs from a 3'-end uridylation activity in *Arabidopsis*. *Current Biology*, **15**, 1501–1507.

Lin, H. and Spradling, A. C. (1997). A novel group of pumilio mutations affects the asymmetric division of germline stem cells in the *Drosophila* ovary. *Development*, **124**, 2463–2476.

Lingel, A., Simon, B., Izaurralde, E. and Sattler, M. (2003). Structure and nucleic-acid binding of the *Drosophila* Argonaute 2 PAZ domain. *Nature*, **426**, 465–469.

Liu, J., Carmell, M. A., Rivas, F. V. *et al.* (2004). Argonaute2 is the catalytic engine of mammalian RNAi. *Science*, **305**, 1437–1441.

Liu, X., Sun, Y., Guo, J. *et al.* (2006). Expression of hiwi gene in human gastric cancer was associated with proliferation of cancer cells. *International Journal of Cancer*, **118**, 1922–1929.

Ma, J. B., Ye, K. and Patel, D. J. (2004). Structural basis for overhang-specific small interfering RNA recognition by the PAZ domain. *Nature*, **429**, 318–322.

Megosh, H. B., Cox, D. N., Campbell, C. and Lin, H. (2006). The role of PIWI and the miRNA machinery in Drosophila germline determination. *Current Biology*, **16**, 1884–1894.

Mevel-Ninio, M., Pelisson, A., Kinder, J., Campos, A. R. and Bucheton, A. (2007). The flamenco locus controls the gypsy and ZAM retroviruses and is required for *Drosophila* oogenesis. *Genetics*, **175**, 1615–1624.

Miyoshi, K., Tsukumo, H., Nagami, T., Siomi, H. and Siomi, M. C. (2005). Slicer function of Drosophila Argonautes and its involvement in RISC formation. *Genes & Development*, **19**, 2837–2848.

Mochizuki, K., Fine, N. A., Fujisawa, T. and Gorovsky, M. A. (2002). Analysis of a piwi-related gene implicates small RNAs in genome rearrangement in *Tetrahymena*. *Cell*, **110**, 689–699.

Mochizuki, K. and Gorovsky, M. A. (2004a). Conjugation-specific small RNAs in Tetrahymena have predicted properties of scan (scn) RNAs involved in genome rearrangement. *Genes & Development*, **18**, 2068–2073.

Mochizuki, K. and Gorovsky, M. A. (2004b). Small RNAs in genome rearrangement in *Tetrahymena*. *Current Opinions in Genetics & Development*, **14**, 181–187.

Ohara, T., Sakaguchi, Y., Suzuki, T. *et al.* (2007). The 3′ termini of mouse Piwi-interacting RNAs are 2′-O-methylated. *Nature Structural & Molecular Biology*, **14**, 349–350.

Pal-Bhadra, M., Leibovitch, B. A., Gandhi, S. G. *et al.* (2004). Heterochromatic silencing and HP1 localization in Drosophila are dependent on the RNAi machinery. *Science*, **303**, 669–672.

Pane, A., Wehr, K. and Schupbach, T. (2007). Zucchini and squash encode two putative nucleases required for rasiRNA production in the Drosophila germline. *Developmental Cell*, **12**, 851–862.

Parker, J. S., Roe, S. M. and Barford, D. (2004). Crystal structure of a PIWI protein suggests mechanisms for siRNA recognition and slicer activity. *European Molecular Biology Organization Journal*, **23**, 4727–4737.

Pelisson, A., Sarot, E., Payen-Groschene, G. and Bucheton, A. (2007). A novel repeat-associated small interfering RNA-mediated silencing pathway downregulates complementary sense gypsy transcripts in somatic cells of the *Drosophila* ovary. *Journal of Virology*, **81**, 1951–1960.

Pelisson, A., Song, S. U., Prud'homme, N. *et al.* (1994). Gypsy transposition correlates with the production of a retroviral envelope-like protein under the tissue-specific control of the *Drosophila* flamenco gene. *European Molecular Biology Organization Journal*, **13**, 4401–4411.

Prud'homme, N., Gans, M., Masson, M., Terzian, C. and Bucheton, A. (1995). Flamenco, a gene controlling the gypsy retrovirus of *Drosophila melanogaster*. *Genetics*, **139**, 697–711.

Qiao, D., Zeeman, A. M., Deng, W., Looijenga, L. H. and Lin, H. (2002). Molecular characterization of hiwi, a human member of the piwi gene family whose overexpression is correlated to seminomas. *Oncogene*, **21**, 3988–3999.

Reddien, P. W., Oviedo, N. J., Jennings, J. R., Jenkin, J. C. and Sanchez Alvarado, A. (2005). SMEDWI-2 is a PIWI-like protein that regulates planarian stem cells. *Science*, **310**, 1327–1330.

Rossi, L., Salvetti, A., Lena, A. *et al.* (2006). DjPiwi-1, a member of the PAZ-Piwi gene family, defines a subpopulation of planarian stem cells. *Developmental Genes & Evolution*, **216**, 335–346.

Saito, K., Nishida, K. M., Mori, T. *et al.* (2006). Specific association of Piwi with rasiRNAs derived from retrotransposon and heterochromatic regions in the *Drosophila* genome. *Genes & Development*, **20**, 2214–2222.

Sarot, E., Payen-Groschene, G., Bucheton, A. and Pelisson, A. (2004). Evidence for a piwi-dependent RNA silencing of the gypsy endogenous retrovirus by the *Drosophila melanogaster* flamenco gene. *Genetics*, **166**, 1313–1321.

Seipel, K., Yanze, N. and Schmid, V. (2004). The germ line and somatic stem cell gene Cniwi in the jellyfish Podocoryne carnea. *International Journal of Developmental Biology*, **48**, 1–7.

Sharma, A. K., Nelson, M. C., Brandt, J. E. *et al.* (2001). Human CD34(+) stem cells express the hiwi gene, a human homologue of the *Drosophila* gene piwi. *Blood*, **97**, 426–434.

Song, J. J., Liu, J., Tolia, N. H. *et al.* (2003). The crystal structure of the Argonaute2 PAZ domain reveals an RNA binding motif in RNAi effector complexes. *Nature Structural Biology*, **10**, 1026–1032.

Song, J. J., Smith, S. K., Hannon, G. J. and Joshua-Tor, L. (2004). Crystal structure of Argonaute and its implications for RISC slicer activity. *Science*, **305**, 1434–1437.

Szakmary, A., Cox, D. N., Wang, Z. and Lin, H. (2005). Regulatory relationship among piwi, pumilio, and bag-of-marbles in *Drosophila* germline stem cell self-renewal and differentiation. *Current Biology*, **15**, 171–178.

Tan, C. H., Lee, T. C., Weeraratne, S. D. *et al.* (2002). Ziwi, the zebrafish homologue of the *Drosophila* piwi: co-localization with vasa at the embryonic genital ridge and gonad-specific expression in the adults. *Gene Expresion Patterns*, **2**, 257–260.

Vagin, V. V., Sigova, A., Li, C. *et al.* (2006). A distinct small RNA pathway silences selfish genetic elements in the germline. *Science*, **313**, 320–324.

Ventela, S., Toppari, J. and Parvinen, M. (2003). Intercellular organelle traffic through cytoplasmic bridges in early spermatids of the rat: mechanisms of haploid gene product sharing. *Molecular Biology of the Cell*, **14**, 2768–2780.

Watanabe, T., Takeda, A., Tsukiyama, T. *et al.* (2006). Identification and characterization of two novel classes of small RNAs in the mouse germline: retrotransposon-derived siRNAs in oocytes and germline small RNAs in testes. *Genes & Development*, **20**, 1732–1743.

Wilson, J. E., Connell, J. E. and Macdonald, P. M. (1996). Aubergine enhances oskar translation in the Drosophila ovary. *Development*, **122**, 1631–1639.

Yan, K. S., Yan, S., Farooq, A. *et al.* (2003). Structure and conserved RNA binding of the PAZ domain. *Nature*, **426**, 468–474.

Yang, Z., Ebright, Y. W., Yu, B. and Chen, X. (2006). HEN1 recognizes 21–24 nt small RNA duplexes and deposits a methyl group onto the 2′ OH of the 3′ terminal nucleotide. *Nucleic Acids Research*, **34**, 667–675.

Yu, B., Yang, Z., Li, J. *et al.* (2005). Methylation as a crucial step in plant microRNA biogenesis. *Science*, **307**, 932–935.

38 MicroRNAs in immunology, cardiology, diabetes, and unicellular organisms

Krishnarao Appasani

As detailed in this volume, the tiny RNAs known as microRNA molecules regulate several biological processes, such as development, differentiation and disease biology. However, how they do so has remained unclear. This book does not cover the recent reports on their involvement in the immune system and even stress responses in the heart. It is becoming clear that certain microRNAs are emerging as key players in stage-specific expression in the immune system. Almost two decades ago, biologists began to identify the roles of genes by knocking them out and studying these "knock-out" animals, which lacked the proteins encoded by the targeted genes. Now, by using the same strategy it is possible to remove portions of genes that make scraps of RNA. Overall, studies of knock-out mice so far have helped us to understand how genes govern health and disease. As the field of microRNomics is rapidly growing at such a rapid pace, the material contributed in this volume might already be slightly outdated. In fact by the time we finalized the book chapters, the roles of microRNAs in immunology, cardiology, diabetes and unicellular organisms had been reported. I will therefore review these topics here.

MicroRNAs in immunology

Over 30% of our genes are under the control of small molecules called microRNAs. They prevent specific genes from being turned into protein and regulate many crucial processes such as cell division and development. Although more than 500 microRNAs have been identified in mammals to date, how they function and how important individual microRNAs are has remained unclear until now.

Dicer is a key enzyme responsible for regulatory RNA biogenesis, and is required in the immune system. The function of dicer in the T cell instruction indicates that miRNAs do indeed play significant regulatory roles in lymphocytes (Muljo *et al.*, 2005; Cobb *et al.*, 2006). *Bic* was originally discovered as a recurrent integration site of avian leucosis virus in chicken lymphoma cells (Clurman and Hayward, 1989), and was recently observed *as* a primary miRNA precursor of *miR-155* (Tam *et al.*, 1997; Lagos-Quintana *et al.*, 2002). Increased expression

MicroRNAs: From Basic Science to Disease Biology, ed. Krishnarao Appasani. Published by Cambridge University Press. © Cambridge University Press 2008.

levels of *bic/miR-155* were observed in activated B cells, T cells, activated macrophages and dendritic cells (Stetson and Medzhitov, 2006; Taganov *et al.*, 2006). Over-expression of *miR-155* has been reported in B cell lymphomas, solid tumors and Hodgkin's disease suggesting that the locus may also be linked to cancer (Calin and Croce, 2006; Costinean *et al.*, 2006).

Until recently, there had been no reports of the consequences of deletion of miRNA genes in vertebrate organisms. Therefore, to define the *in vivo* role of microRNAs, Allan Bradley's group from the Sanger Institute, and Klaus Rajewsky's group from Harvard, independently generated mammalian knockouts to obtain *bic*-deficient mice (Rodriguez *et al.*, 2007; Thai *et al.*, 2007). Bradley's laboratory generated *bic/miR-155* allele deficient mice and observed that they are viable and fertile, and in the long run developed lung airway remodeling (as a result of bronchiolar sub-epithelial collagen deposition and an increase in the size of the ventricle wall). In addition, the number of leukocytes was significantly increased in bronchoalveolar lavage fluids. These phenotypes suggest that *bic/miR-155* may participate or play a role in regulating the homeostasis of the immune system. To understand the nature of defective immune responses *in vivo*, Bradley's group also explored the possibility of an intrinsic requirement for *bic/miR-155* in B cells and T cells. In this study a wide spectrum of *miR-155* target genes were identified with diverse molecular roles, such as T cell costimulation, chemotaxis, and signaling (Rodriguez *et al.*, 2007). Interestingly, the role of the transcription factor *c-Maf* as a potent transactivator in the interleukin 4 pathway was also described. The discovery of several new potential targets of *miR-155* in this study also supports the idea that *bic/miR-155* acts as a core regulator of gene expression in multiple cell types. It is interesting that the human *BIC/miR-155* gene maps to an asthma, pollen sensitivity, and atopic dermatitis susceptibility region on chromosome 21. This severe loss-of-function phenotype study not only shed light on understanding the role of miRNAs *in vivo* but was also directly relevant to human disease. Hopefully these transgenic mice studies will help to understand the potential immune disease locus in humans.

Germinal centers represent sites of antibody affinity maturation and memory B cell generation in T cell-dependent antibody responses (Rajewsky, 1996). In humans *bic/miR-155* was detected in germinal center B cells. To understand the physiological function of *bic/miR-155* Rajewsky's group created transgenic mice, and observed that B and T cells present in the gut-associated lymphoid tissue were activated (Thai *et al.*, 2007). Although normal expression levels of *bic/miR-155* were detected in B cells isolated from wild mice, the expression was up-regulated when these cells were stimulated with CD40 or with mitogens that bind Toll-like receptors. Normal *bic/miR-155* expression was observed in T cells; however, upon activation with cytokines (IL-4) the levels increased, suggesting that these animals are more prone to T helper 2 (Th2) differentiation. Both these transgenic experiments established through a combined genetic loss and gain-of-function approach, that *miR-155* is critically involved in the *in vivo* control of specific differentiation processes in the immune response. Importantly, these studies suggest that a single gene may be the target for more than one microRNA.

The relationship between inflammation, innate immunity, and miRNA expression is just beginning to be explored. Activation of mammalian innate and acquired immunity is regulated by several mechanisms (Taganov *et al.*, 2007). Recently, David Baltimore and his colleagues from the California Institute of Technology have shown that microRNAs are involved in the immune-response pathway (Taganov *et al.*, 2006). Expression profiling of 200 miRNAs in human monocytes revealed that several of them (*miR-146a/b, miR-132, and miR-155*) are endotoxin-responsive genes. Systematic analysis of *miR-146a* and *miR-146b* gene expression unveiled a pattern of induction in response to a variety of microbial components and pro-inflammatory cytokines (Taganov *et al.*, 2006). *MiR-146* was also identified as an NF-*kappaB*-dependent gene (Taganov *et al.*, 2006). Recently, it was observed that *miR-155* is the only microRNAs that were regulated by Toll-like receptor ligands (O'Connell *et al.*, 2007). From previous studies it is known that *miR-155* functions as an oncogene (Calin and Croce, 2006; Costinean *et al.*, 2006). However, the recent involvement of *miR-155* in innate immunity (Taganov *et al.*, 2007) suggests that there may be a potential link between inflammation and cancer, which should be explored further.

MicroRNAs in cardiology

The numerous miRNA genes present in animals and plants may turn out to mediate much of the gene regulation that generates cell diversity and developmental patterning. Although the functional roles of microRNAs in various forms of cancer were described by several researchers as well as documented in this volume (Part V), their role in cardiac hypertrophy was not established until recently.

MicroRNA research is now playing an important role in the field of cardiology; how these tiny RNA molecules influence heart development and disease is a topic of growing interest. Recently, four groups have independently demonstrated the importance of microRNAs in cardiology, particularly in heart development and function, the regulation of the expression of genes involved in electrical conductance, hypertrophy, and contractility. In general, external stimulus by cytokines or pressure overloads that are activated by diverse signal transduction pathways to the cardiac myocytes convey cardiac stress or hypertrophy (McKinsey and Olson, 2005). As a result these induce reprogramming of cardiac gene expression and the activation of 'fetal' cardiac genes (McKinsey and Olson, 2005). MicroRNAs *miR-133* and *miR-1* are specifically expressed in skeletal muscle and cardiac myocytes (Chen *et al.*, 2006; van Rooij *et al.*, 2006). The function of *miR-133* in skeletal myoblast proliferation and differentiation has previously been established but has never been investigated in cardiac myocyte hypertrophy.

Therefore, Peschle's group from Instituto Superiore Sanita of Italy and Condorelli's group at the University of California at San Diego jointly explored the expression profile of *miR-133* and *miR-1* in different tissues (Care *et al.*, 2007). These studies by in situ hybridization provided initial clues about the presence of *miR-133* messages in embryonic heart and skeletal muscle. In three mouse models of cardiac hypertrophy (an essential adaptive physiological response to mechanical and hormonal stress-whereby cardiac cells grow in size), reduced expression of both *miR-133* and *miR-1* (especially low in the right ventricle) was observed. When these studies were extended to human heart disease patients 50% of lower expression levels in disease ventricle and atria were noticed (Care *et al.*, 2007). Taken together, the murine and human data indicate an inverse correlation between *miR-133* and *miR-1* expression and myocardial hypertrophy.

Additionally, functional studies *in vitro* and *in vivo* were carried out by using an antagomir oligonucleotide approach. Recent studies have also suggested that miRNAs may function according to a "combinatorial circuitry model", whereby a single miRNA targets multiple mRNAs and several co-expressed miRNAs may target a single mRNA (Lin *et al.*, 2005; Stark *et al.*, 2005). Based on this concept, these studies also revealed three targets of *miR-133-RhoA*, *Cdc42* and *NELFA/Whsc2*-relevant to cardiac hypertrophy development. Since *miR-133* and *miR-1* have been revealed to be key regulators of cardiac hypertrophy they can therefore be potentially used for therapeutic applications in heart disease. In particular, antagomir-133 oligonucleotide administration offers a new possibility for future therapeutic application (Care *et al.*, 2007).

Cardiac contractility depends on the expression of α and β chains of myosin heavy chain (MHC) proteins. Therefore, it is rational to understand and decipher these gene expression mechanisms, and to develop therapeutic strategies (McKinsey and Olson, 2005). Previously, Olson's group from Southwestern Medical School have identified a signature pattern of miRNAs associated with pathological cardiac growth and remodeling (van Rooij *et al.*, 2006). In a recent study they have generated *miR-208* mutant mice by using *Cre*-mediated recombination, and observed that disruption of α MHC gene causes early embryonic lethality. *MiR-208* deleted mice were viable and did not exhibit any abnormalities (such as size, shape, or structure) in the heart up to 20 weeks of age. However, cardiac function, as monitored by echocardiography, showed a slight reduction in contractility, which was attributed to an increase in left ventricular diameter during systole (van Rooij *et al.*, 2007). From the age of 6 months onwards cardiac function declined further in mutant animals. It was also observed that the *miR-208* deleted mice were resistant to fibrosis and cardiomyocyte hypertrophy (van Rooij *et al.*, 2007) (Figure 38.1).

Using a transgenic approach, the up-regulation of β MHC expression was demonstrated by knocking out *miR-208*. These experiments also showed that *miR-208* potentiates β MHC expression through a mechanism involving thyroid hormone receptor signaling (van Rooij *et al.*, 2007). Bioinformatic approaches allowed thyroid hormone receptor associated protein 1 to be identified, as a putative target for *miR-208*. Therapeutic manipulation of *miR-208* expression or interaction with its mRNA targets could potentially enhance cardiac function by suppressing β MHC expression. Taken together these results demonstrate that *miR-208* is an essential cardiac-specific regulator of β MHC expression and a mediator of stress and T3 signaling in the heart (van Rooij *et al.*, 2007).

Organogenesis is a complex biological process that requires precise spatial and temporal control of gene expression (Mishima *et al.*, 2007). Although significant

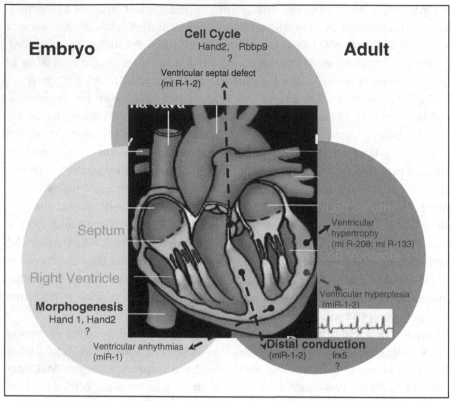

Figure 38.1 MicroRNAs in cardiogenesis and heart function. The importance of microRNAs in myogenesis of heart and their role in disease is stated in detail. Absence of *miR-1* leads to ventricular arrhythmia, where the electrical impulse of the heartbeat becomes irregular. Lack of *miR-1–2* can lead to defects in ventricular septum and distal conduction. Deficiency of *miR-208* and *miR-133* can lead to ventricular myocyte hypertrophy in which heart muscle cells enlarge significantly. Heart morphogenesis is also affected by *miR-1–2* defects; their identified potential targets are shown during embryonic development. Presumed miR-1 targets involved in cell cycle include *Hlf* and *Rbbp9* (hepatocyte leukemia factor and retinoblastoma-binding protein 9) and are also shown here. (This diagram is redrawn with modifications from the original drawings of Chien, K. R. (2007). *Nature*, 447, 389, and Mishima, Y., Stalhut, C. and Giraldez, A. J. (2007). *Cell*, 129, 248.)

dysregulation of miRNA expression has been reported in cardiac diseases, it remains unknown whether the heart requires miRNA function for normal development or maintenance (van Rooij *et al.*, 2006; Sayed *et al.*, 2007). A network of transcription factors regulate heart development and maintenance in a dose dependent manner, but the effects of translational regulation on the titration of these pathways are largely unknown (Y. Zhao *et al.*, 2007). Therefore to understand the role of miRNAs in cardiac development, two groups engineered mice that lacked the microRNA processing enzyme Dicer in heart tissue alone, and focused on *miR1-2*. MicroRNA *miR-1* has been implicated in muscle cell differentiation during myogenesis (Zhao *et al.*, 2005). Recently it has been reported that the levels of *miR-1* are significantly higher in adult rat hearts than in neonatal hearts, suggesting that it might have pathophysiological functions other than myogenesis (Chen *et al.*, 2006; Rao *et al.*, 2006; Yang *et al.*, 2007). The phenomenon of up-regulation of *miR-1* has also been observed in human individuals with coronary artery diseases in particular in the ischemic zone.

In a therapeutic approach, *miR-1*-specific anti-sense modified oligonucleotides were delivered *in vivo* into the infracted myocardium, and arrhythmias were found to be significantly suppressed. These results indicate that *miR-1* is an arrhythmogenic or proarrhythmic factor that is detrimental to the ischemic heart (Yang *et al.*, 2007). Deepak Srivastava and his colleagues from the University of California at San Francisco demonstrated the effect of a global loss of miRNAs during cardiac development by the deletion of *miR1–2* and observed that these mutants had normal cardiac function despite the fact that their heart walls were thickened (Y. Zhao *et al.*, 2007). In addition, the mice lacking *miR1-2* had a spectrum of abnormalities, including lethality, cardiac rhythm disturbances and a striking monocyte cell-cycle abnormality (Figure 38.1). By creating mice that were deficient in a muscle-specific miRNA, *miR-1–2*, Y. Zhao *et al.* (2007) uncovered a role for miRNAs in one of the commonest forms of congenital heart disease: ventricular septal defects, characterized by a hole in the wall between the left and right ventricles of the heart (Figure 38.1).

Both these studies identify *miR-1* targets that may, at least in part, account for the manifestation of the associated diseases. These targets are a gap-junction protein that functions in the electrical communication between cardiac cells (Yang *et al.*, 2007), and a cardiac transcription factor, *Irx5*, which regulates the formation and maturation of the electrical conduction system in the heart (Y. Zhao *et al.*, 2007) (Figure 38.1). These results have several important implications such as: (i) m*iR-1–2* is a key microRNA during embryonic and adult cardiac function; (ii) provided a set of potential *miR-1* target genes those play crucial roles during cardiac development and physiology; (iii) *miR-1* dosage is important during heart development. Overall, it suggests that a tight regulation of *miR-1* levels is crucial for the maintenance of normal cardiac conduction. Fourfold more expression of *miR-21* or *miR-18b* was observed in hypertrophic mouse hearts compared with a normal group, by using expression profiling (Cheng *et al.*, 2007; Tatsuguchi *et al.*, 2007).

In vivo approaches are one of the cornerstones of research in modern translational (bench-to-bedside) medicine in order to understand how complex physiological processes are coordinately controlled at the molecular level (Chien, 2007). Although the four teams previously described knocked out different microRNAs, we do still not know all of the gene targets of each molecule. Importantly, these studies revealed a lot more questions than answers. However, they provide clear evidence that a subset of miRNAs can modulate a diverse spectrum of cardiovascular functions *in vivo* (Figure 38.1). It has been previously shown that miRNAs indeed participate in the control of specific cell-fate decisions in embryonic heart progenitor cells (Y. Zhao *et al.*, 2005, 2007). We anticipate that assignment of these miRNAs to specific locations in the fate map of cardiovascular cell lineages will be forthcoming. Overall, these results have implications for translational medicine and will have an impact on our understanding of the pathways involved in heart disease, and potential opportunities to develop miRNA as therapeutic targets for cardiovascular diseases such as hypertension, ischemic heart disease, valvular disease, and endocrine disorders (Chien, 2007). MicroRNAs may also be potentially involved in Alzheimer's disease, for which we await for results. One day microRNAs may explain inherited defects in diseases for which genes have been elusive (Couzin, 2007).

Role of microRNAs in diabetes

Diabetes mellitus is a very common metabolic disorder that can lead to blindness, renal failure and lower limb amputations and is a major risk factor for cardiovascular disease and stroke. Type 2 diabetes is the most common form of the disease, resulting from the release of insufficient amounts of insulin from pancreatic β-cells and/or from a decreased sensitivity of insulin target tissues (Abderrahmani *et al.*, 2006). To study the role of miRNAs in pancreatic endocrine cells, Stoffel' and his colleagues have cloned and identified a novel, evolutionarily conserved and islet-specific miRNA, *miR-375* (Poy *et al.*, 2004). These results suggest that over-expression of *miR-375* suppressed glucose-induced insulin secretion, and conversely, inhibition of endogenous *miR-375* function enhanced insulin secretion. In this study myotrophin (Mtpn), a novel target for *miR-375*, was predicted and experimentally validated (Poy *et al.*, 2004). Inhibition of Mtpn by small interfering RNAs mimicked the effects of *miR-375* on glucose-stimulated insulin secretion and exocytosis. These results are exciting, particularly as they open new doors for understanding how to regulate insulin secretion in the body, which may offer avenues for treating diabetes (Poy *et al.*, 2004). Recently, Regazzi's group demonstrated by utilizing the global microRNAs expression profiling that *miR-9* controls the expression of key components required for insulin-secreting cells and highlighted their contribution to β-cell specificity and to the regulation of β-cell death (Plaisance *et al.*, 2006).

MicroRNAs in unicellular organisms

Planarians are simple metazoans that have the unique ability to regenerate lost tissues. Planarians are therefore an excellent model system for studying many aspects of regeneration, cell lineage decisions, and basic stem cell biology. Cloning and initial characterization of 71 miRNAs from the planarian *Schmidtea mediterranea* have been reported (Palakodeti *et al.*, 2006). Although several of these are members of miRNA families identified in other species, planarian-specific miRNAs were also identified. In fission yeast, a single class of siRNAs exists in the RNA-induced transcriptional silencing complex, and these have been demonstrated to have functions in heterochromatin assembly and epigenetic silencing (Lippman and Martiensen, 2004). Small RNAs called scan RNAs, which are *c.*26–31 nt in length, were identified in *Tetrahymena thermophila* (Mochizuki and Gorovky, 2004; Lee and Collins, 2006). The tiny RNA molecular pathways found in unicellular organisms appear to be simpler than those in multicellular organisms.

MicroRNAs were not detected in lower organisms until recently. The development and application of novel high-throughput parallel pyrosequencing methods helped to sift through thousands of RNA molecules in search of microRNAs (Margulies *et al.*, 2005). Using these deep sequencing methods, two groups have independently sequenced about four thousand small RNA sequences from *Chlamydomonas reinhardtii*, a green alga (Molnár *et al.*, 2007; T. Zhao *et al.*, 2007). Quite surprisingly, *c.*200 (5%) of them turned out to be microRNAs. This is the first study describing the presence of microRNAs in a lower unicellular organism. Besides miRNAs, many endogenous small RNAs derived from various genomic locations were also discovered (Zhao *et al.*, 2007). Interestingly, none of the *Chlamydomonas* miRNAs had an ortholog in other green algae, plants, or animals, suggesting that *C. reinhardtii* miRNA genes may have arisen independently after splitting from a common lineage (Molnar *et al.*, 2007; Siomi and Siomi, 2007). However, *Chlamydomonas* miRNAs share common features with plant miRNAs, such as: (i) being *c.*21 nt in length and having a preference for uracil at the 5' end; and (ii) being generated from a stem-loop by *Dicer*; and (iii) being competent in directing target RNA cleavage (T. Zhao *et al.*, 2007).

The biological functions of these miRNAs have yet to be understood. It has been observed that the expression of some of the miRNAs in *Chlamydomonas* altered during gametic differentiation and flagella assembly, suggesting an involvement in asexual reproduction and flagella development (T. Zhao *et al.*, 2007). These miRNAs may help *C. reinhardtii* to respond to extremely diverse environments, or metabolic pathways (Molnar *et al.*, 2007). The presence of microRNAs in algae indicates that they are much more ancient than previously thought; they may have persisted for more than a billion years (Odling-Smee, 2007). It seems that the RNA in simple unicellular organisms could be as complex as in higher creatures. Validation of miRNA targets and elucidation of their functions in *Chlamydomonas* biology is a major and exciting task ahead of us all. Ultimately, we want to know how these miRNAs or siRNAs present in *C. reinhardtii* connection are involved in the genotype and phenotype. Hopefully, within a few years the composition of these tiny RNAs present in unicellular organisms will be fully investigated, and will help us to understand more about the physiology of microRNAs in general.

REFERENCES

Abderrahmani, A., Cheviet, S., Ferdaoussi, M. *et al.* (2006). ICER induced by hyperglycemia represses the expression of genes essential for insulin exocytosis. *European Molecular Biology Organization Journal*, **25**, 977–986.

Calin, G. A. and Croce, C. M. (2006). MicroRNA signatures in human cancers. *Nature Reviews in Cancer*, **11**, 857–866.

Care, A., Catalucci, D., Felicetti, F. *et al.* (2007). MicroRNA-133 controls cardiac hypertrophy. *Nature Medicine*, **13**, 613–618.

Chen, J. F., Mandel, E. M., Thomson, J. M. *et al.* (2006). The role of microRNAs-1 and *microRNAs-133* in skeletal muscle proliferation and differentiation. *Nature Genetics*, **38**, 228–233.

Cheng, Y., Ji, R., Yue, J. *et al.* (2007). MicroRNAs are aberrantly expressed in hypertrophic heart: do they play a role in cardiac hypertrophy? *American Journal of Pathology*, **170**, 1831–1840.

Chien, K. R. (2007). MicroRNAs and the tell-tale heart. *Nature*, **447**, 389–390.

Clurman, B. E. and Hayward, W. S. (1989). Multiple proto-oncogene activations in avian leukosis virus-induced lymphomas: evidence for stage-specific events. *Molecular Cell Biology*, **9**, 2657–2664.

Cobb, B. S., Hertweck, A., Smith, J. *et al.* (2006). A role for Dicer in immune regulation. *Journal of Experimental Medicine*, **203**, 2519–2527.

Costinean. S., Zanesi, N., Pekarsky, Y. *et al.* (2006). Pre-B cell proliferation and lymphoblastic leukemia/high-grade lymphoma in E(mu)-*miR155* transgenic mice. *Proceedings of the National Academy of Sciences USA*, **103**, 7024–7029.

Couzin, J. (2007). Erasing microRNAs reveals their powerful punch. *Science*, **316**, 530.

Kwon, C., Han, Z., Olson, E. N. and Srivastava, D. (2005). MicroRNA1 influences cardiac differentiation in *Drosophila* and regulates Notch signaling. *Proceedings of the National Academy of Sciences USA*, **102**, 18986–18991.

Lagos-Quintana, M., Rauhut, R., Yalcin, A. *et al.* (2002). Identification of tissue-specific microRNAs from mouse. *Current Biology*, **12**, 735–739.

Lee, S. R. and Collins, K. (2006). Two classes of endogenous small RNAs in *Tetrahymena thermophila*. *Genes & Development*, **20**, 28–33.

Lin, H., Thomson, J. M., Hemann, M. T. *et al.* (2005). A microRNAs polycistron as a potential human oncogene. *Nature*, **435**, 828–833.

Lippman, Z. and Martiensen, R. (2004). The role of RNA interference in heterchromatic silencing. *Nature*, **431**, 364–370.

Margulies, M., Egholm, M., Altman, W. E. *et al.* (2005). Genome sequencing in microfabricated high-density picolitre reactors. *Nature*, **437**, 376–380.

McKinsey, T. A. and Olson, E. N. (2005). Toward transcriptional therapies for the failing heart: chemical screens to modulate genes. *Journal of Clinical Investigation*, **115**, 538–546.

Mishima, Y., Stahlhut, C. and Giraldez, A. J. (2007). miR-1–2 gets to the heart of the matter. *Cell*, **129**, 247–249.

Mochizuki, K. and Gorovky, M. A. (2004). Conjugation-specific small RNAs in *Tetrahymena* have predicted properties of scan (scn) RNAs involved in genome rearrangement. *Genes & Development*, **18**, 2068–2073.

Molnár, A., Schwach, F., Studholme, D. J., Thuenemann, E. C. and Baulcombe, D. C. (2007). miRNAs control gene expression in the single-cell alga *Chlamydomonas reinhardtii*. *Nature*, May 30 [Epub ahead of print]

Muljo, S. A., Ansel, K. M., Kanellopoulou, C. *et al.* (2005). Aberrant T cell differentiation in the absence of Dicer. *Journal of Experimental Medicine*, **202**, 261–269.

O'Connell, R. M., Taganov, K. D., Boldin, M. P., Cheng, G. and Baltimore, D. (2007). *MicroRNA-155* is induced during the macrophage inflammatory response. *Proceedings of the National Academy of Sciences USA*, **104**, 1604–1609.

Odling-Smee, L. (2007). Complex set of RNAs found in simple algae. *Nature*, **447**, 518.

Palakodeti, D., Smielewska, M. and Graveley, B. R. (2006). MicroRNAs from the planarian *Schmidtea mediterranea*: a model system for stem cell biology. *RNA*, **12**, 1640–1649.

Plaisance, V., Abderrahmani, A., Perret-Menoud, V. *et al.* (2006). MicroRNA-9 controls the expression of Granuphilin/Slp4 and the secretory response of insulin-producing cells. *Journal of Biological Chemistry*, **281**, 26 932–26 942.

Poy, M. N., Eliasson, L., Krutzfeldt, J. *et al.* (2004). A pancreatic islet-specific microRNA regulates insulin secretion. *Nature*, **432**, 226–230.

Rajewsky, K. (1996). Clonal selection and learning in the antibody system. *Nature*, **381**, 751–758.

Rao, P. K., Kumar, R. M., Farkhondeh, M., Baskerville, S. and Lodish, H. (2006). Myogenic factors that regulate expression of muscle-specific microRNAs. *Proceedings of the National Academy of Sciences USA*, **103**, 8721–8726.

Rodriguez, A., Vigorito, E., Clare, S. *et al.* (2007). Requirement of *bic/microRNAs-155* for normal immune function. *Science*, **316**, 608–611.

Sayed, D., Hong, C., Chen, I.-Y., Lipowy, J. and Abdellatif, M. (2007). MicroRNAs play an essential role in the development of cardiac hypertrophy. *Circulation Research*, **100**, 416–424.

Siomi, H. and Siomi, M. C. (2007). Expanding RNA physiology: microRNAs in a unicellular organism. *Genes & Development*, **21**, 1153–1156.

Stark, A., Brennecke, J., Bushati, N., Russell, R. B. and Cohen, S. M. (2005). Animal microRNAs confer robustness to gene expression and have a significant impact on 3′UTR evolution. *Cell*, **123**, 1133–1146.

Stetson, D. B. and Medzhitov, R. (2006). Recognition of cytosolic DNA activates an IRF3-dependent innate immune response. *Immunity*, **24**, 93–103.

Taganov, K. D., Boldin, M. P., Chang, K. J. and Baltimore, D. (2006). NF-κB-dependent induction of microRNA *miR-146*, an inhibitor targeted to signaling proteins of innate immune responses. *Proceedings of the National Academy of Sciences USA*, **103**, 12 481–12 486.

Taganov, K. D., Boldin, M. P. and Baltimore, D. (2007). MicroRNAs and immunity: tiny players in a big field. *Immunity*, **26**, 133–137.

Tam, W., Ben-Yehuda, D. and Hayward, W. S. (1997). *bic*, a novel gene activated by proviral insertions in avian leukosis virus-induced lymphomas, is likely to function through its noncoding RNA. *Molecular Cell Biology*, **17**, 1490–1502.

Tatsuguchi, T., Seok, H. Y., Callis, T. E. *et al.* (2007). Expression of microRNAs is dynamically regulated during cardiomyocyte hypertrophy. *Journal of Molecular and Cellular Cardiology*, **42**, 1137–1141.

Thai, T., Calado, D. P., Casola, S. *et al.* (2007). Regulation of germinal center response by microRNAs-155. *Science*, **316**, 604–608.

van Rooij, E., Sutherland, L. B., Liu, N. *et al.* (2006). A signature pattern of stress-responsive microRNAs that can evoke cardiac hypertrophy and heart failure. *Proceedings of the National Academy of Sciences USA*, **103**, 18 255–18 260.

van Rooij. E., Sutherland, L. B., Qi, X. *et al.* (2007). Control of stress-dependent cardiac growth and gene expression by a microRNAs. *Science*, **316**, 575–579.

Yang, B., Lin, H., Xiao, J. *et al.* (2007). The muscle specific microRNAs *miR-1* regulates cardiac arrhythmogenic potential by targeting GJA1 and KCNJ2. *Nature Medicine*, **13**, 486–491.

Zhao, T., Li, G., Mi, S. *et al.* (2007). A complex system of small RNAs in the unicellular green alga *Chlamydomonas reinhardtii*. *Genes & Development*, **21**, 1190–1203.

Zhao, Y., Samal, E. and Srivastava, D. (2005). Serum response factor regulates a muscle-specific mircroRNA that targets Hand2 during cardiogenesis. *Nature*, **436**, 214–220.

Zhao, Y., Ransom, J. F., Li, A. *et al.* (2007). Dysregulation of cardiogenesis, cardiac conduction, and cell cycle in mice lacking *miRNA-1-2*. *Cell*, **129**, 303–317.

Index

TAT motif 460
  TAT-box 460
Tat-responsive-region (TAR)-binding
      protein(TRBP) 433
T-cell receptor (TCR) 373
  T-cell development 325
  TCP transcription factors 142
telomerase 23
temporal expression 10
termed miRNp or RISC 85
Tet-On or Tet-Off system 39
Tetrahymena thermophila 446, 518
Tetrahymena Piwi(Twi1P) 507
Tetraodon fugu 454
therapeutic intervention 54
thermodynamic binding stability 166, 211216
thyroid hormone receptor associated
      protein 1 515
tibialis anterior (TA) 396
timing of development 13
tiny non-coding RNA (tncRNA) 23
tissue-specific expression patterns 296
  tissue-specific pattern 310
  tissue-specificity 45
Toll-like receptors 513
TPA-induced differentiation 381, 383
training set 214
transamination reactions 129
transcript cleavage 148
transcriptional regulation 7
  transcriptional repression 506
  transcriptional silencing 446, 450
    transcriptome 85
transcription factor binding sites 50, 445
  transcription factor DAF-16 16
  transcription factor die-1 15
  transcription factor MAFB 315
  transcription factors NFI-A and
      C/EBPα 315
  transcription factors Oct4 481
transfer RNA (tRNA) 23
transgene quelling 25
transgenic animal models 38
  transgenic mouse 302
  transgenic plants 147
transient assay 139
translation 23
  translational efficiency 485
  translation control 225
  translation initiation 94
  translation repression 86, 95, 269
  translation silencing 485
  translational machinery 507
  translational regulation 498
  translational suppression 365
translational medicine 517
translocations 314, 343
transposable elements 30, 222
  transposon activity 30
  transposon silencing 505

trilogy 20, 257
tRNA-like transcripts 409
trophoblasitc stem (TS) cells 445
  trophoblast 445, 446
tumorigenesis 47, 295
  tumor profiling 47
  tumor growth arrest 313
  tumor progression 351
tumor suppressors 295, 358
  tumor suppressor genes (TSG) 327
  tumor suppressor genes map 311
tumor-specific miRNA signature 317
tumor-suppressor function 318
two-component strategy 49
two-dimensional gel electrophoresis 383
type Z diabetes 518

U6 RNA kits 247
UL36-1 414
ungapped alignment 212
universal RT-primer 283
uracil DNA glycosylase 415
uRT specific primers 289
3′untranslated region (3′-UTR) 8, 27, 42, 70,
      102, 137, 139, 164, 166, 173, 175,
      176, 182, 184, 204, 207, 214, 222,
      225, 295, 317, 351, 364, 387, 484

V (D)J recombination 373, 374, 376, 378
valvular disease 518
vascular development 80
vertebrate development 477
vertebrate limb 58
viral DNA polymerase 407
  viral DNA replication 411, 416
viral infection 38, 164
  viral lifecycle 406
viral miRNAs 405
visuospatial cognition 125
vitro transcription 89
vulva 15

Watson–Crick base pairing 173, 199, 204,
      230, 247
weighted sequence motifs 215
Williams Syndrome (WS) 125

X chromosome 29
  X-chromosome inactivation 31, 446
X mental retardation syndrome 38
Xenopus oocytes 24, 95
X-linked condition 31

zebrafish 38
  brain development 296
  Piwi (Ziwi) 505
  Zwill as Piwi-like genes 497
zeta-probe GT membranes 72
zinc finger transcription factor 15
zone of polarizing activity (ZPA) 62, 66